Air Pollution and Environmental Analysis

Editor: Bernie Goldman

R CALLISTO REFERENCE

www.callistoreference.com

Callisto Reference,
118-35 Queens Blvd., Suite 400,
Forest Hills, NY 11375, USA

Visit us on the World Wide Web at:
www.callistoreference.com

ISBN: 978-1-63239-839-0 (Hardback)

The publisher's policy is to use permanent paper from mills that operate a sustainable forestry policy. Furthermore, the publisher ensures that the text paper and cover boards used have met acceptable environmental accreditation standards.

Printed in the United States of America.

Cataloging-in-publication Data

Air pollution and environmental analysis / edited by Bernie Goldman.
    p. cm.
Includes bibliographical references and index.
ISBN 978-1-63239-839-0
1. Air--Pollution. 2. Air quality management. 3. Environmental protection. 4. Environmental health.
5. Environmental engineering. I. Goldman, Bernie.
TD883 .E54 2017
628.53--dc23

Table of Contents

Preface

This text will discuss in detail the various problems faced by environmentalists while controlling air pollution. It will provide thorough information about the control strategies used to restrict air pollution. Air is an essential part for life sustenance on earth and therefore, controlling its degradation is the need of the hour. To achieve the goal of pollution free air, it is important to device new laws, regulations, techniques and methods. This book explores all the important aspects of this subject in the present day scenario. It strives to provide a fair idea about air pollution control and to help develop a better understanding of the latest advances within this field. As this area is emerging at a rapid pace, the contents of this text will help the readers understand the modern concepts and applications of the subject. Students, researchers, experts, professionals and anyone else associated with this field will benefit alike from this book.

All of the data presented henceforth, was collaborated in the wake of recent advancements in the field. The aim of this book is to present the diversified developments from across the globe in a comprehensible manner. The opinions expressed in each chapter belong solely to the contributing authors. Their interpretations of the topics are the integral part of this book, which I have carefully compiled for a better understanding of the readers.

At the end, I would like to thank all those who dedicated their time and efforts for the successful completion of this book. I also wish to convey my gratitude towards my friends and family who supported me at every step.

Editor

Pulmonary Function and Incident Bronchitis and Asthma in Children: A Community-Based Prospective Cohort Study

Yungling Leo Lee[1,2]*, Bing-Fang Hwang[3], Yu-An Chen[1], Jer-Min Chen[1,4], Yi-Fan Wu[1,4]

1 Institute of Epidemiology and Preventive Medicine, College of Public Health, National Taiwan University, Taipei, Taiwan, **2** Department of Public Health, College of Public Health, National Taiwan University, Taipei, Taiwan, **3** Department of Occupational Safety and Health, College of Public Health, China Medical University, Taichung, Taiwan, **4** Department of Family Medicine, Taipei City Hospital, Renai Branch, Taipei, Taiwan

Abstract

Background: Previous studies revealed that reduction of airway caliber in infancy might increase the risks for wheezing and asthma. However, the evidence for the predictive effects of pulmonary function on respiratory health in children was still inconsistent.

Methods: We conducted a population-based prospective cohort study among children in 14 Taiwanese communities. There were 3,160 children completed pulmonary function tests in 2007 and follow-up questionnaire in 2009. Poisson regression models were performed to estimate the effect of pulmonary function on the development of bronchitis and asthma.

Results: After adjustment for potential confounders, pulmonary function indices consistently showed protective effects on respiratory diseases in children. The incidence rate ratios of bronchitis and asthma were 0.86 (95% CI 0.79–0.95) and 0.91 (95% CI 0.82–0.99) for forced expiratory volume in 1 second (FEV_1). Similar adverse effects of maximal mid-expiratory flow (MMEF) were also observed on bronchitis (RR = 0.73, 95% CI 0.67–0.81) and asthma (RR = 0.85, 95% CI 0.77–0.93). We found significant decreasing trends in categorized FEV_1 (p for trend = 0.02) and categories of MMEF (p for trend = 0.01) for incident bronchitis. Significant modification effects of traffic-related air pollution were noted for FEV_1 and MMEF on bronchitis and also for MMEF on asthma.

Conclusions: Children with high pulmonary function would have lower risks on the development of bronchitis and asthma. The protective effect of high pulmonary function would be modified by traffic-related air pollution exposure.

Editor: James L. Kreindler, Abramson Research Center, United States of America

Funding: This work was supported by grants #98-2314-B-002-138-MY3 and #96-2314-B-006-053 from the Taiwan National Science Council. The funders had no role in study design, data collection and analysis, decision to publish, or preparation of the manuscript.

Competing Interests: The authors have declared that no competing interests exist.

* E-mail: leolee@ntu.edu.tw

Introduction

Previous researches showed that bronchitis and asthma were substantial respiratory health issues in children, which may cause major childhood morbidity and high socioeconomic costs [1,2]. Some studies have suggested a complex etiological pathway for certain respiratory diseases but until now, definite pathogenesis was still unclear [3]. There is growing evidence that air pollution is an important environmental factor that may be associated with childhood asthma [4,5]. Children with higher prevalence of bronchitis resided in areas with higher ambient air pollutants, such as nitrogen dioxide (NO_2), has also been reported [6,7]. In the previous finding from Taiwan Children Health Study (TCHS), we found that risks of the microsomal epoxide hydroxylase (*EPHX1*) gene for childhood respiratory health could be modified by ambient NO_2 levels [8], which suggested that air pollution may overwhelm the genetic effects and increase the risk of respiratory diseases in a subset of children.

Pulmonary function is a sensitive marker of respiratory health, which can detect and grade airway obstruction [9]. Bronchial hyper-reactivity was an obvious characteristic of asthma and could lead to the bronchial symptoms in both children and adults [10,11]. Previous studies showed that high pulmonary function indices such as the ratio of maximal mid-expiratory flow divided by forced vital capacity (MMEF/FVC) and MMEF were inversely associated with bronchial hyper-reactivity [12,13]. Although the association between pulmonary function during the first few weeks of life and onset of asthma in later life had been reported before [14–16], evidence for the effect of pulmonary function on incident respiratory diseases in school-aged children was inconclusive.

TCHS was a community-based longitudinal study that provided information about genetics, respiratory health and environmental factors, such as ambient air pollution. We had the opportunity to assess the hypothesis that high pulmonary function indices are associated with reduced risk of incident respiratory diseases. The modification effect for traffic-related air pollution could also be examined in the present study.

Materials and Methods

Ethics approval

The study protocol was approved by the institutional review board of our university hospital and complied with the principles outlined in the Helsinki Declaration [17].

Study design

TCHS is a prospective cohort study examining the determinants of children's respiratory health, which focused on outdoor air pollutants as primary interest. The study design and methods for TCHS have been described in detail previously [7,18]. A total of 4,134 seventh-grade children were recruited from public schools from 14 communities. At cohort entry in 2007, all participants were collected with baseline respiratory conditions and arranged to complete pulmonary function tests. We followed these children in 2009 with the identical questions concerning respiratory diseases and pulmonary function tests. The mean follow-up period for the current study was two years.

Questions regarding respiratory symptoms and diseases were modified after those used in the Children's Health Study in southern California [19]. The outcomes of interest were physician-diagnosed bronchitis and asthma. Children who were disease-free at baseline and reported a yes answer to the question "Has a physician ever diagnosed your child as having bronchitis?" on follow-up questionnaire were classified as having incident bronchitis. The incident asthma was defined by the question "Has a physician ever diagnosed your child as having asthma?" on baseline and follow-up questionnaire. Those with wheezing history but no diagnosis were considered to be at risk for a new diagnosis of bronchitis or asthma and were included in the study.

Pulmonary function measurement

Children's pulmonary function tests (PFT) were performed during the morning hours in indoor buildings to avoid daily and annual peak air pollution levels. For the baseline survey in 2007,

children returned parental questionnaire were included. After excluding subjects with incomplete questionnaire, recent symptomatic upper respiratory infections, or other acute pulmonary function or cardiac diseases, there were 3,261 children eligible for pulmonary function tests at the baseline survey.

Each subject was requested to perform three satisfactory blows, defined as both of the two largest FVC and forced expiratory volume in 1 second (FEV$_1$) agreeing within 150 ml, extrapolation volume less than 150 ml or 5% of FVC, and forced expiratory time exceeding 6 seconds. These criteria were based on American Thoracic Society and European Respiratory Society recommendations, updated in 2005, modified for children [20]. No more than 5 blows were attempted per time, no more than two times were asked per child. Resting for more than 10 minutes was required for every subject to prevent exercise bias. To predict subjects' pulmonary function, height and weight were measured at the time of testing. Two fully-trained technicians performed PFT, using two spirometers (Chestgraph HI-101, CHEST M.I., INC). Spirometers' calibrations were checked before, during, and after every morning's testing using 1L flow-volume syringes.

Three pulmonary function indices were analyzed: FVC, FEV$_1$ and MMEF. We obtained sex-specific percentage predicted pulmonary function indices by using linear regression models. The selection of the best prediction models from age, height, nature log values for height, the square of height, weight, nature log values for weight, the square of weight, body-mass index, nature log values for body-mass index and the square of body-mass index was based on the attained adjusted R^2 using sex-specific equations (Table S1). We calculated each percentage predicted pulmonary function index from dividing the observed values by the predicted values and was expressed in percentage. Then, each percentage predicted pulmonary function index was categorized into three groups based on the cut-off points 90% and 110% that provided an adequate distribution for analyses.

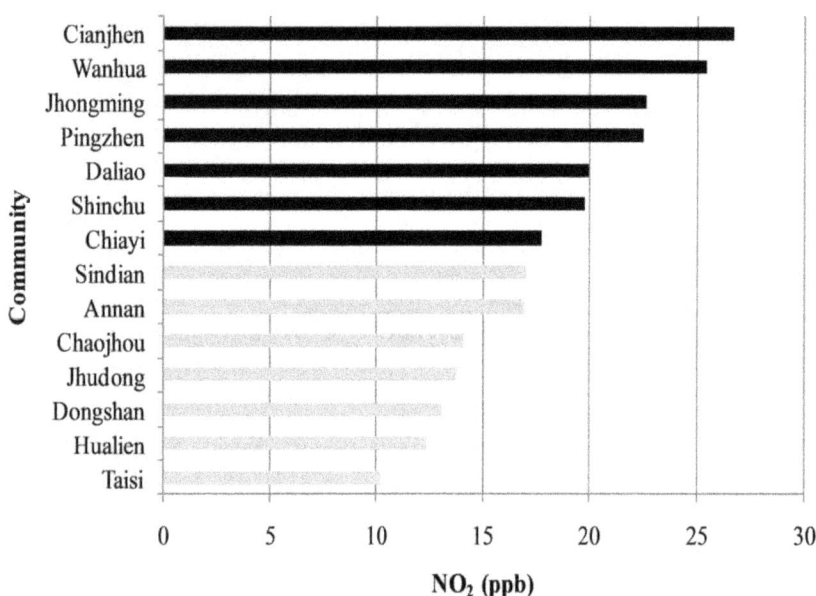

Figure 1. Annual average NO$_2$ levels (2007–2009) in 14 Taiwan Children Health Study communities. Dark bars: higher NO$_2$; mean, 22.1 ppb. Light bars: lower NO$_2$; mean, 14.0 ppb.

Table 1. Demographic characteristics of the study participants.

Selected characteristics at study entry[a]	Bronchitis-free cohort (N = 2,893)		Asthma-free cohort (N = 2,818)	
	n	%	n	%
Gender				
Boys	1466	51.5	1452	51.3
Girls	1380	48.5	1352	48.7
Age, yr				
12	2254	79.5	2200	79.5
13	576	20.3	561	20.3
14	5	0.2	5	0.2
Parental income[b]				
≤400,000	943	35.9	930	36.3
410,000~800,000	1059	40.4	1031	40.3
≥810,000	622	23.7	598	23.4
Parental education, yr				
≤12	1860	65.4	1813	65.3
13~15	554	19.5	547	19.7
≥16	432	15.2	417	15.0
Family history of asthma				
Yes	69	2.4	55	2.0
Family history of atopy[c]				
Yes	663	23.3	655	23.6
In utero exposure to maternal smoking				
Yes	110	3.9	103	3.7
Active smoking				
Yes	149	5.2	145	5.2
Current SHS				
Yes	1222	42.9	1209	43.5
Carpet use				
Yes	254	9.0	252	9.1
Dogs in home				
Yes	807	28.4	784	28.2
Cats in home				
Yes	149	5.2	146	5.3
Cockroaches				
Yes	2413	85.5	2367	85.9
Visible mould				
Yes	963	34.1	957	34.7
Mildewy odor				
Yes	416	14.7	413	14.9
Water stamp on the wall				
Yes	718	25.4	711	25.7

[a]Number of subjects do not add up to total N because of missing data.
[b]New Taiwan dollars per year ($1 US = $ 33 New Taiwan).
[c]Atopy is defined as allergic rhinitis or atopic eczema.

Air pollution monitoring data

Complete monitoring data for the criteria air pollutants are available from Taiwan Environmental Protection Agency (EPA) monitoring stations beginning in 1995. The average hourly levels of NO_2 were measured in 14 monitoring stations. We computed the annual average of ambient NO_2 levels between 2007 and 2009. The 14 communities were grouped by median NO_2 level into seven higher and seven lower ones (median level, 17.5 ppb). The mean annual ambient NO_2 level was 22.1 ppb in higher NO_2 communities and 14.0 ppb in lower NO_2 communities (Figure 1).

Statistical analysis

The association between pulmonary function indices and the incidence of bronchitis and asthma were examined by fitting Poisson regression models (PROC GENMOD) with a logarithmic link function. To account for over-dispersion of incidence rates in Poisson models, we performed all analyses by using deviance adjusted variance estimates [21]. On the basis of study design and *a priori* consideration of potential confounders, we included *in utero* exposure to maternal smoking, family history of asthma, family history of atopy, and community in all models. If estimates of pulmonary function indices changed by more than 10% when a covariate was added in base models, the covariate would be included in final models [22]. Subjects with missing covariate information were included in the model using missing indicators [23]. The incidence rate ratios were scaled by inter-quartile range (IQR) increase of the corresponding values in certain pulmonary function indices. We also conducted sensitivity analyses by restricting the cohort to children without a history of bronchitis, asthma or wheezing at study entry.

The influences of community-level air pollutants were estimated by fitting two-stage hierarchical models. The models assumed two sources of variation: the variation among subjects in the first stage, part of which could be explained by the individual confounders, and the variation of air pollution between communities in the second stage, part of which could be explained by variables measured at the community level. To investigate the effects of traffic-related air pollution on the relationship between pulmonary function indices and respiratory diseases, we conducted stratified analysis comparing the children in communities with lower and higher levels of NO_2. We used likelihood ratio tests to examine the interactive effects between pulmonary function indices and ambient NO_2 levels on respiratory diseases.

We also presented graphically the potential modification effects of traffic-related air pollution by plotting ambient NO_2 levels on X-axis and community-specific rate ratios on Y-axis. The regression curves were drawn through the predicted values derived from exponential regression models. All analyses were performed by SAS software version 9.1 (SAS Institute, Gary, NC, USA) and assumed a 0.05 significant level based on a two-sided estimate.

Results

Participant characteristics

In 2009, a total of 3,160 children completed the follow-up questionnaire and pulmonary function tests, with follow-up rate of 96.9%. There were 2,893 bronchitis-free children and 2,818 asthma-free children at cohort entry. The incidence rates were 8.1/1000 person-years for bronchitis and 7.2/1000 person-years for asthma. The characteristics of study participants were shown in Table 1. The proportion of gender was almost equal (51% male and 49% female). Approximately 24% of the children had family history of atopy which included any history of hay fever, allergies to food or medicine, inhaled dusts, pollen, molds, animal fur or dander, or skin allergies. The prevalence of current second-hand smoke (SHS) and active smoke exposure were around 43% and more than 5% among children. The distribution of all covariates was consistent in the two disease-free cohorts.

Pulmonary function and respiratory diseases

The relationships between pulmonary function indices and respiratory diseases were presented in Table 2. We found the consistently protective effects for all pulmonary function indices. Scaled across inter-quartile range elevation, the incidence rate ratio of bronchitis was 0.86 (95% CI 0.79−0.95) for FEV_1 and 0.73 (95% CI 0.67−0.81) for MMEF. Similar adverse effects were observed for asthma with incidence rate ratio 0.90 (95% CI 0.82−0.99) for FVC, 0.91 (95% CI 0.82−0.99) for FEV_1, and 0.85 (95% CI 0.77−0.93) for MMEF. Restricting the analyses to children without a history of bronchitis, asthma or wheezing did not substantially alter the above findings (Table S2).

Table 3 presents the associations between pulmonary function indices categories and incident bronchitis and asthma. The incidence rate ratios for bronchitis and asthma were decreased in the upper categories. In bronchitis-free cohort, significant decreasing trends were found on categories of FEV_1 (*p* for trend = 0.02) and on categories of MMEF (*p* for trend = 0.01).

Table 2. Association between incident bronchitis and asthma and pulmonary function indices at study entry.

Pulmonary function index[a]	Bronchitis		Asthma	
	RR[b]	95% CI	RR[b]	95% CI
FVC (% predicted)	1.08	0.99 – 1.18	0.90	0.82 – 0.99
FEV_1 (% predicted)	0.86	0.79 – 0.95	0.91	0.82 – 0.99
MMEF (% predicted)	0.73	0.67 – 0.81	0.85	0.77 – 0.93

[a]FVC, forced vital capacity; FEV_1, forced expiratory volume in 1s; MMEF, forced expiratory flow over the mid-range of expiration.
[b]Relative risks (RR) and 95% confidence intervals (CI) of outcomes were scaled across the inter-quartile range elevation by each pulmonary function index. All models were adjusted for community, *in utero* exposure to maternal smoking, family history of asthma, family history of atopy, active smoking and current SHS.

Table 3. Association between incident bronchitis and asthma and pulmonary function indices categories at study entry.

Pulmonary function index*	Bronchitis		Asthma	
	RR	95% CI	RR	95% CI
FVC (% predicted)				
≤90	1.00		1.00	
90–110	1.55	1.16 – 2.06	0.93	0.71 – 1.22
≧110	0.78	0.52 – 1.16	0.72	0.50 – 1.03
p for trend	0.38		0.08	
FEV_1 (% predicted)				
≤90	1.00		1.00	
90–110	0.80	0.63 – 1.02	0.57	0.45 – 0.76
≧110	0.68	0.49 – 0.95	0.70	0.51 – 0.95
p for trend	0.02		0.01	
MMEF (% predicted)				
≤90	1.00		1.00	
90–110	0.60	0.48 – 0.75	0.78	0.61 – 1.00
≧110	0.44	0.33 – 0.57	0.92	0.71 – 1.18
p for trend	0.01		0.47	

*FVC, forced vital capacity;
FEV_1, forced expiratory volume in 1s;
MMEF, forced expiratory flow over the mid-range of expiration.
All models were adjusted for community, *in utero* exposure to maternal smoking, family history of asthma, family history of atopy, active smoking and current SHS.

Table 4. Relative risks of incident bronchitis and asthma by pulmonary function indices, stratified by community-specific annual average NO_2 levels (2007–2009).

	Bronchitis						Asthma					
	Low NO_2[b]		High NO_2[b]		interaction p value		Low NO_2[b]		High NO_2[b]		interaction p value	
Pulmonary function index[a]	RR[c]	95% CI	RR[c]	95% CI			RR[c]	95% CI	RR[c]	95% CI		
FVC (% predicted)	0.83	0.72 – 0.95	1.25	1.10 – 1.42	0.12		0.82	0.72 – 0.93	1.00	0.88 – 1.14	0.40	
FEV$_1$ (% predicted)	0.56	0.49 – 0.65	1.10	0.96 – 1.24	0.005		0.85	0.75 – 0.97	0.94	0.82 – 1.08	0.47	
MMEF (% predicted)	0.43	0.37 – 0.50	0.96	0.85 – 1.10	<0.001		0.73	0.64 – 0.83	0.97	0.85 – 1.11	0.04	

[a]FVC, forced vital capacity; FEV$_1$, forced expiratory volume in 1s; MMEF, forced expiratory flow over the mid-range of expiration.
[b]Two NO_2 strata were defined as less than and greater than the median level (17.5 ppb).
[c]Relative risks (RR) and 95% confidence intervals (CI) of outcomes were scaled across the inter-quartile range elevation by each pulmonary function index.
All models were adjusted for community, *in utero* exposure to maternal smoking, family history of asthma, family history of atopy, active smoking and current SHS.

Interrelationship between traffic-related air pollution, pulmonary function and respiratory diseases

We found loss of protective effects by each pulmonary function index against bronchitis, chronic cough and asthma in the higher NO_2 communities (Table 4). Scaled across the inter-quartile range elevation of FEV$_1$, the incidence rate ratio of bronchitis was 0.56 (95% CI 0.49–0.65) in the lower NO_2 communities, whereas the effect was 1.10 (95% CI 0.96–1.24) in the higher NO_2 communities (*p* for interaction = 0.005). For each inter-quartile range changes in MMEF, the effects on incident bronchitis were 0.43 (95% CI 0.37–0.50) in the lower NO_2 communities and 0.96 (95% CI 0.85–1.10) in the higher NO_2 communities (*p* for interaction <0.001). Similarly, the incidence rate ratio of asthma was significantly lower in the lower NO_2 communities (RR = 0.73, 95% CI 0.64–0.83), compared with the effect in the higher NO_2 communities (RR = 0.97, 95% CI 0.85–1.11) (*p* for interaction = 0.04) (Table 4). In addition, we found no statistical significant differences on the effects of pulmonary function indices for incident bronchitis in relation to exposure to the other air pollutants in TCHS (PM$_{2.5}$, PM$_{10}$ and 8-hour O$_3$). We calculated the communities-specific rate ratios of each pulmonary function index on incident bronchitis. The loss of protective effects from high pulmonary function indices would be interpreted through Figure 2, with significant interactions between pulmonary function indices and ambient NO_2 levels. We found the protective effects

would disappear in communities over ambient NO_2 level of 25 ppb.

Discussion

We examined the relationship between pulmonary function and respiratory diseases among children under different traffic-related air pollutant levels. Pulmonary function indices, such as FEV$_1$ and MMEF, were found to predict the new-onset bronchitis and asthma. Exposure to ambient NO_2 would reduce the protective effect of high pulmonary function on the occurrence of bronchitis. To our best knowledge, this is the first report to describe the association between pulmonary function indices and incident bronchitis.

Our work is a prospective cohort study with a very high follow-up rate (96.9%). The participants were representative of the native population of Taiwan. Some individual characteristics, such as *in utero* exposure to maternal smoking [24,25], family history of asthma or atopy [26,27], and residential community may influence the occurrence of asthma in childhood. Active smoking and current SHS were also found to interfere in the association with respiratory health [28–32]. We adjusted the above covariates in Poisson regression models to minimize the confounding effects.

Some studies assessed the role of pulmonary function indices in respiratory health. The airway caliber in infancy has been reported as a decreasing risk for wheezing [16]. The association

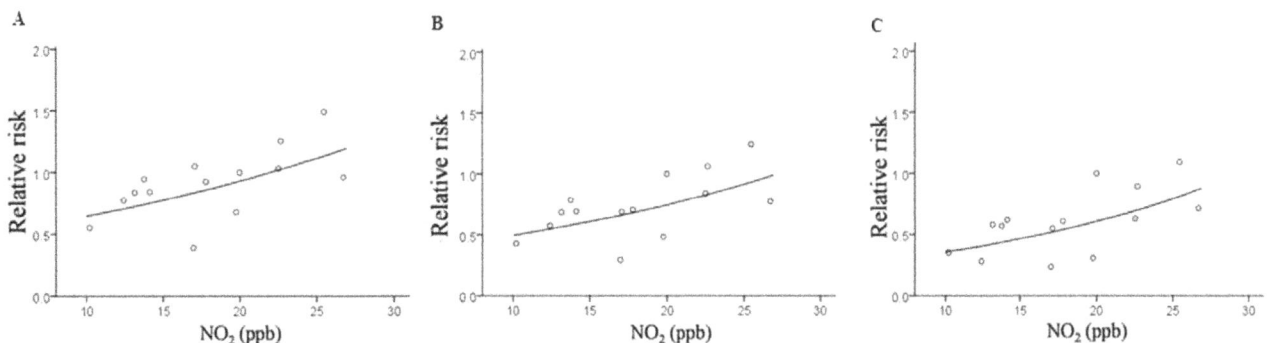

Figure 2. Community-specific relative risks (RR) of incident bronchitis across inter-quartile range elevation for pulmonary function indices by ambient NO_2 levels. Each circle represents one community. (A) FVC, forced vital capacity; (B) FEV$_1$, forced expiratory volume in 1s; (C) MMEF, forced expiratory flow over the mid-range of expiration.

between the reduced airway function in the first few weeks of life and asthma in later life had also been described [14,15]. In previous epidemiological studies, high flow rates were noted to be significantly correlated with a low prevalence of airway hyper-reactivity [13,33], and age-related decline on prevalence of airway hyper-reactivity was larger among those with higher FEV1% [34]. Furthermore, Islam *et al.* reported that high pulmonary function provided protective effects in asthma development during childhood [35]. The relationship of pulmonary function indices on development of bronchitis in children has not been previously investigated.

Pulmonary function indices have been clinically proved as important phenotypes of respiratory health. Bronchial hyper-reactivity could lead to the respiratory symptoms in childhood [10,11]. In a previous cross-sectional study from the United States, methacholine airway responsiveness was noted to be inversely associated with high pulmonary function indices such as MMEF/FVC [12]. Another hospital-based study from Italy also revealed that MMEF was related to bronchial hyper-reactivity consistently among patients with asthma and rhinitis [13]. All the above evidences were consistent with our finding that pulmonary function indices could well predict the occurrence of respiratory diseases in childhood.

The respiratory system is particularly sensitive to air pollutants. Traffic-related air pollution has been reported as a key factor for the incidence of respiratory diseases [36]. NO_2, the representative traffic-related air pollutant, is through interaction with impairment of respiratory response or the immune system to infection, and damages the epithelial cells by oxidative injury [7,37,38]. In Taiwan, we have previously reported that an increase of 8.79 ppb of ambient NO_2 exposure would result in 80% increase in the prevalence of bronchitic symptoms [7]. We have interpreted our data as a protective effect of high pulmonary function that was attenuated by the effects of traffic-related air pollution (Table 4). This finding was also consistent with previous analyses in our cohort that the risk of respiratory morbidity associated with the *EPHX1* 139Arg allele could be modified by ambient NO_2 levels [8]. The pathogenesis of bronchitis and asthma included airway remodeling characterized by extraordinary mucous membrane secretion and mild bronchial wall thickening. Long-term exposure to higher level of air pollution has been noted to be associated with small airway remodeling, followed by chronic bronchial obstruction [39]. It is likely that evolutionary selection has resulted in pulmonary characteristics that lead to better respiratory health. Higher pulmonary function may be a marker for lower susceptibility to respiratory diseases.

Another concern for present study was the insufficient power to detect relatively small modification effects of air pollution on incident asthma. The primary pathophysiological characteristics of bronchitis may include chronic pulmonary inflammation and cellular oxidative stress [36], which might result from chemical irritants such as NO_2. The relatively higher predictive power for bronchitis rather than asthma may be due to the essentially different pathophysiology between these two kinds of diseases.

Our study has some limitations. The assessment of respiratory diseases was based on questionnaire reports. Nevertheless, questionnaire was widely used to define respiratory outcomes in epidemiologic studies in children. Bronchitis and asthma were through physician-diagnoses, which would reduce the possibility of differential symptom identifications and recall bias. Another limitation was the measurements of pulmonary function depending on many known or unknown factors, such as the quality of the spirometry equipment, technician's skill, and children's cooperation [9]. In order to avoid the possible errors of PFT measurements, the children's PFT were performed by the same well-trained technicians and spirometers in indoor building within the same season. We minimized these potential information biases in our study design. It is possible that the cases of new-onset bronchitis or asthma in our study were undiagnosed cases and had low pulmonary function at study entry. This seems unlikely because effect estimates did not changed substantially when we restricted the analyses to children without a history of bronchitis, asthma or wheezing. Because incident cases were defined without knowledge of pulmonary function measurement, differential misclassification is probably not a major source of bias that accounts for our results.

In conclusion, our study provides new evidence for the protective effects of high pulmonary function indices on incidence of bronchitis and asthma. What is important to investigate is that traffic-related air pollution is a substantial modifiable risk for childhood respiratory diseases. Children living in communities with higher traffic-related air pollution would suffer from increased risks for new-onset bronchitis and asthma than those in lower polluted communities. Additional research is necessary to clarify the mechanism by NO_2 modifies the protective effect of high pulmonary function on the development of bronchitis and asthma.

Supporting Information

Table S1 Predictive equations proposed in this study using age, height, weight, and body mass index variables.

Table S2 Association between incident bronchitis and asthma and pulmonary function indices, after excluding those with bronchitis, asthma or wheezing at study entry.

Acknowledgments

The authors thank the field workers, teachers, and other school staff who supported the data collection, and all the parents and children who participated in this study.

Author Contributions

Conceived and designed the experiments: YLL BFH. Performed the experiments: YLL JMC YFW. Analyzed the data: YLL BFH YAC. Contributed reagents/materials/analysis tools: YLL. Wrote the paper: YLL.

References

1. Akinbami L (2006) The state of childhood asthma, United States, 1980–2005. Adv Data. pp 1–24.
2. Carroll KN, Gebretsadik T, Griffin MR, Wu P, Dupont WD, et al. (2008) Increasing burden and risk factors for bronchiolitis-related medical visits in infants enrolled in a state health care insurance plan. Pediatrics 122: 58–64.
3. Reed CE (2006) The natural history of asthma. J Allergy Clin Immunol 118: 543–548; quiz 549–550.
4. Gilmour MI, Jaakkola MS, London SJ, Nel AE, Rogers CA (2006) How exposure to environmental tobacco smoke, outdoor air pollutants, and increased pollen burdens influences the incidence of asthma. Environ Health Perspect 114: 627–633.
5. McConnell R, Berhane K, Yao L, Jerrett M, Lurmann F, et al. (2006) Traffic, susceptibility, and childhood asthma. Environ Health Perspect 114: 766–772.
6. Sunyer J, Jarvis D, Gotschi T, Garcia-Esteban R, Jacquemin B, et al. (2006) Chronic bronchitis and urban air pollution in an international study. Occup Environ Med 63: 836–843.
7. Hwang BF, Lee YL (2010) Air pollution and prevalence of bronchitic symptoms among children in taiwan. Chest 138: 956–964.

8. Tung KY, Tsai CH, Lee YL (2011) Microsomal epoxide hydroxylase genotypes/diplotypes, traffic air pollution, and childhood asthma. Chest 139: 839–848.

9. MacIntyre NR, Selecky PA (2010) Is there a role for screening spirometry? Respir Care 55: 35–42.

10. Henderson J, Granell R, Heron J, Sherriff A, Simpson A, et al. (2008) Associations of wheezing phenotypes in the first 6 years of life with atopy, lung function and airway responsiveness in mid-childhood. Thorax 63: 974–980.

11. Stern DA, Morgan WJ, Halonen M, Wright AL, Martinez FD (2008) Wheezing and bronchial hyper-responsiveness in early childhood as predictors of newly diagnosed asthma in early adulthood: a longitudinal birth-cohort study. Lancet 372: 1058–1064.

12. Litonjua AA, Sparrow D, Weiss ST (1999) The FEF25-75/FVC ratio is associated with methacholine airway responsiveness. The normative aging study. Am J Respir Crit Care Med 159: 1574–1579.

13. Cirillo I, Klersy C, Marseglia GL, Vizzaccaro A, Pallestrini E, et al. (2006) Role of FEF 25%–75% as a predictor of bronchial hyperreactivity in allergic patients. Ann Allergy Asthma Immunol 96: 692–700.

14. Young S, Arnott J, O'Keeffe PT, Le Souef PN, Landau LI (2000) The association between early life lung function and wheezing during the first 2 yrs of life. Eur Respir J 15: 151–157.

15. Haland G, Carlsen KC, Sandvik L, Devulapalli CS, Munthe-Kaas MC, et al. (2006) Reduced lung function at birth and the risk of asthma at 10 years of age. N Engl J Med 355: 1682–1689.

16. Martinez FD, Wright AL, Taussig LM, Holberg CJ, Halonen M, et al. (1995) Asthma and wheezing in the first six years of life. The Group Health Medical Associates. N Engl J Med 332: 133–138.

17. Anonymous (1997) World Medical Association declaration of Helsinki. Recommendations guiding physicians in biomedical research involving human subjects. JAMA 277: 925–926.

18. Tsai CH, Huang JH, Hwang BF, Lee YL (2010) Household environmental tobacco smoke and risks of asthma, wheeze and bronchitic symptoms among children in Taiwan. Respir Res 11: 11.

19. McConnell R, Berhane K, Gilliland F, London SJ, Vora H, et al. (1999) Air pollution and bronchitic symptoms in Southern California children with asthma. Environ Health Perspect 107: 757–760.

20. Miller MR, Hankinson J, Brusasco V, Burgos F, Casaburi R, et al. (2005) Standardisation of spirometry. Eur Respir J 26: 319–338.

21. Cheng CC, Lin NN, Lee YF, Wu LY, Hsu HP, et al. (2010) Effects of Shugan-Huayu powder, a traditional Chinese medicine, on hepatic fibrosis in rat model. Chin J Physiol 53: 223–233.

22. Tong IS, Lu Y (2001) Identification of confounders in the assessment of the relationship between lead exposure and child development. Ann Epidemiol 11: 38–45.

23. Huberman M, Langholz B (1999) Application of the missing-indicator method in matched case-control studies with incomplete data. Am J Epidemiol 150: 1340–1345.

24. Dezateux C, Stocks J, Dundas I, Fletcher ME (1999) Impaired airway function and wheezing in infancy: the influence of maternal smoking and a genetic predisposition to asthma. Am J Respir Crit Care Med 159: 403–410.

25. Lodrup Carlsen KC (2002) The environment and childhood asthma (ECA) study in Oslo: ECA-1 and ECA-2. Pediatr Allergy Immunol 13 Suppl 15: 29–31.

26. London SJ, James Gauderman W, Avol E, Rappaport EB, Peters JM (2001) Family history and the risk of early-onset persistent, early-onset transient, and late-onset asthma. Epidemiology 12: 577–583.

27. Lee YL, Lin YC, Hsiue TR, Hwang BF, Guo YL (2003) Indoor and outdoor environmental exposures, parental atopy, and physician-diagnosed asthma in Taiwanese schoolchildren. Pediatrics 112: e389.

28. Cerveri I, Accordini S, Corsico A, Zoia MC, Carrozzi L, et al. (2003) Chronic cough and phlegm in young adults. Eur Respir J 22: 413–417.

29. Coultas DB, Mapel D, Gagnon R, Lydick E (2001) The health impact of undiagnosed airflow obstruction in a national sample of United States adults. Am J Respir Crit Care Med 164: 372–377.

30. David GL, Koh WP, Lee HP, Yu MC, London SJ (2005) Childhood exposure to environmental tobacco smoke and chronic respiratory symptoms in non-smoking adults: the Singapore Chinese Health Study. Thorax 60: 1052–1058.

31. Segala C, Poizeau D, Neukirch F, Aubier M, Samson J, et al. (2004) Air pollution, passive smoking, and respiratory symptoms in adults. Arch Environ Health 59: 669–676.

32. Ho SY, Lam TH, Chung SF, Lam TP (2007) Cross-sectional and prospective associations between passive smoking and respiratory symptoms at the workplace. Ann Epidemiol 17: 126–131.

33. Ciprandi G, Cirillo I, Vizzaccaro A, Monardo M, Tosca MA (2005) Early bronchial airflow impairment in patients with persistent allergic rhinitis and bronchial hyperreactivity. Respir Med 99: 1606–1612.

34. Ulrik CS, Backer V (1998) Longitudinal determinants of bronchial responsiveness to inhaled histamine. Chest 113: 973–979.

35. Islam T, Gauderman WJ, Berhane K, McConnell R, Avol E, et al. (2007) Relationship between air pollution, lung function and asthma in adolescents. Thorax 62: 957–963.

36. Schwartz J (2004) Air pollution and children's health. Pediatrics 113: 1037–1043.

37. Jerrett M, Shankardass K, Berhane K, Gauderman WJ, Kunzli N, et al. (2008) Traffic-related air pollution and asthma onset in children: a prospective cohort study with individual exposure measurement. Environ Health Perspect 116: 1433–1438.

38. McConnell R, Islam T, Shankardass K, Jerrett M, Lurmann F, et al. (2010) Childhood incident asthma and traffic-related air pollution at home and school. Environ Health Perspect 110: 1021–1026.

39. Churg A, Brauer M, del Carmen Avila-Casado M, Fortoul TI, Wright JL (2003) Chronic exposure to high levels of particulate air pollution and small airway remodeling. Environ Health Perspect 111: 714–718.

2

Access Rate to the Emergency Department for Venous Thromboembolism in Relationship with Coarse and Fine Particulate Matter Air Pollution

Nicola Martinelli[1], Domenico Girelli[1], Davide Cigolini[1], Marco Sandri[1], Giorgio Ricci[2], Giampaolo Rocca[2], Oliviero Olivieri[1]*

1 Section of Internal Medicine, Department of Medicine, University of Verona, Verona, Italy, 2 Emergency Department, Hospital of Verona, Verona, Italy

Abstract

Particulate matter (PM) air pollution has been associated with cardiovascular and respiratory disease. Recent studies have proposed also a link with venous thromboembolism (VTE) risk. This study was aimed to evaluate the possible influence of air pollution-related changes on the daily flux of patients referring to the Emergency Department (ED) for VTE, dissecting the different effects of coarse and fine PM. From July 1st, 2007, to June 30th, 2009, data about ED accesses for VTE and about daily concentrations of PM air pollution in Verona district (Italy) were collected. Coarse PM ($PM_{10-2.5}$) was calculated by subtracting the finest $PM_{2.5}$ from the whole PM_{10}. During the index period a total of 302 accesses for VTE were observed (135 males and 167 females; mean age 68.3 ± 16.7 years). In multiple regression models adjusted for other atmospheric parameters $PM_{10-2.5}$, but not $PM_{2.5}$, concentrations were positively correlated with VTE (beta-coefficient = 0.237; P = 0.020). During the days with high levels of $PM_{10-2.5}$ ($\geq 75^{th}$ percentile) there was an increased risk of ED accesses for VTE (OR 1.69 with 95%CI 1.13–2.53). By analysing days of exposure using distributed lag non-linear models, the increase of VTE risk was limited to $PM_{10-2.5}$ peaks in the short-term period. Consistently with these results, in another cohort of subjects without active thrombosis (n = 102) an inverse correlation between $PM_{10-2.5}$ and prothrombin time was found (R = −0.247; P = 0.012). Our results suggest that short-time exposure to high concentrations of $PM_{10-2.5}$ may favour an increased rate of ED accesses for VTE through the induction of a prothrombotic state.

Editor: Pieter H. Reitsma, Leiden University Medical Center, Netherlands

Funding: These authors have no support or funding to report.

Competing Interests: The authors have declared that no competing interests exist.

* E-mail: oliviero.olivieri@univr.it

Introduction

In the last decade, air pollution has been shown to affect not only respiratory but also cardiovascular morbidity and mortality [1]. Since 2004 the American Heart Association (AHA) recognized that exposure to fine particulate matter (PM) air pollution, i.e. air-dispersed particle with aerodynamic diameter less than 10 μm (PM_{10}), is associated with an increased risk for cardiovascular events, particularly myocardial infarction (MI), stroke, arrhythmia, and heart failure [2]. Importantly, exacerbation within hours to days of exposure in susceptible individuals was claimed as important as higher long-term average PM levels, both these factors being potentially relevant in reducing life expectancy [1–4].

The pathogenic properties of PM_{10} are thought to be influenced by their size, leading to subdivision in 2 groups: a coarse component with aerodynamic diameter between 2.5 and 10 μm ($PM_{10-2.5}$) and a finest component with aerodynamic diameter less than 2.5 μm ($PM_{2.5}$). The two components have different sources and composition. Indeed, while $PM_{2.5}$ particles result mainly from combustion of fossil fuels from a variety of activities (e.g. traffic and industry), $PM_{10-2.5}$ particles are associated with non-combustion surface or fugitive releases by a variety of human (e.g agriculture) and natural (e.g. erosion) activities [5]. Moreover, $PM_{10-2.5}$ particles deposit preferentially in the upper and larger airways, while the $PM_{2.5}$ particles may reach the smallest airways and alveoli, where the finest component (ultrafine particles <0.1 μm) can spread even into the systemic circulation throughout the alveolar-capillary wall [6]. Originally, in two landmark cohort-based mortality studies, the Harvard Six Cities [7] and the ACS studies [8], $PM_{10-2.5}$ was not related to mortality, while $PM_{2.5}$ was associated with mortality for both cardiovascular and pulmonary causes. On the basis of similar observations and results, $PM_{2.5}$ has been considered as the true or, at least, the main culprit of the adverse effects of PM air pollution on the human health [1].

In very recent years, PM air pollution has been also associated with changes in the global coagulation function, suggesting an activation of the hemostatic system and an unbalanced hyperco-agulability even after short-term exposure [3,4,9]. In addition, exposure to the air pollution seems to significantly affect the risk of venous thromboembolism (TVE) with an approximately linear exposure-response relationship over the PM range [10–12]. If this is the case, the amount of individuals referring to the hospital for VTE may vary accordingly to ambient PM concentrations even on relatively short periods of time.

The aims of the present study were therefore to evaluate possible influence of air pollution-related changes on the daily flux of patients referring to Emergency Department (ED) for VTE, dissecting the potentially different effects of coarse and fine PM.

Results

During the index period (n = 640 days), 302 patients were recognized to be affected by VTE (135 males and 167 females; mean age 68.3 ± 16.7 years). For each day, data of both PM_{10} and $PM_{2.5}$ (and therefore also of $PM_{10-2.5}$) were available. Figure 1 shows the seasonal trend of PM concentration during the study period. As expected, PM_{10} and $PM_{2.5}$ concentrations were higher during the cool seasons (Figure 1A), while no clear-cut trend was found for $PM_{10-2.5}$ concentration (Figure 1B).

At univariate analysis, $PM_{10-2.5}$ concentration showed a significant correlation with VTE (beta coefficient = 0.163; $P = 0.039$ – Table 1), while no significant association was found for either PM_{10} (beta coefficient = 0.167; $P = 0.064$) or $PM_{2.5}$ (beta coefficient = 0.091; $P = 0.221$).

Considering the strong correlation among PM data, including $PM_{10-2.5}$, and atmospheric parameters, a specific adjustment was performed. $PM_{10-2.5}$ concentration remained significantly associated with ED daily admissions for VTE after adjustment for atmospheric parameters (beta coefficient = 0.151, $P = 0.002$ – Table 1), as well as after the inclusion in the regression model of $PM_{2.5}$ data (beta coefficient = 0.237; $P = 0.020$, respectively – Table 1).

Subsequent analyses showed that mean levels of $PM_{10-2.5}$, as well as the proportion of days with high concentration of $PM_{10-2.5}$ (defined as greater than the 75th percentile, i.e. 19 $\mu g/m^3$), progressively raised by increasing the number of ED daily admissions because of VTE (Figure 2). Interestingly, there was no sex or age difference between VTE patients during days with high or low $PM_{10-2.5}$ concentration (data not shown). Considering the daily hospital referral for VTE as a dichotomic variable (yes/ no), the days characterized by higher concentration of $PM_{2.5-10}$, i.e. ≥19 $\mu g/m^3$, showed an increased probability (by more than 50%) to observe at least one VTE access than the days with

$PM_{2.5-10} < 19$ $\mu g/m^3$ (OR 1.69 with 95%CI 1.13–2.53 after adjustment for atmospheric parameters and $PM_{2.5}$ data).

The association between daily hospital referral for VTE and $PM_{2.5-10}$ concentrations was investigated also at different time-lags by using distributed lag non-linear models (DLNM). In such analysis the increase of VTE risk was limited to short-time exposure. Indeed, only the current-day (lag 0) $PM_{2.5-10}$ levels presented a significant association with VTE, while considering the previous days no significant association was found (Figure 3). Moreover, in order to evaluate the stability of the estimated effects, we fitted a series of DLNM models with different configurations of parameters which showed that the reported effects remained substantially unchanged under model perturbations (data not shown).

Finally, on the basis of such results, we further investigated the associations of PM concentrations and coagulation times, both PT and aPTT. In a subgroup of study population – admitted for mild respiratory symptoms without active thrombosis and without warfarin therapy (n = 102)– an inverse correlation between $PM_{10-2.5}$ and PT was detected, while no significant associations were found for $PM_{2.5}$ (Table 2). Noteworthy, the correlation between $PM_{10-2.5}$ and PT remained significant also in a linear regression model adjusted for sex, age, and the other atmospheric variables (standardized beta-coefficient −0.251; $P = 0.035$).

Discussion

The main result of the present study was the association between 24-h levels of coarse $PM_{10-2.5}$ and referral rate for VTE independently from age, gender, and the other atmospheric parameters. Such association appeared to be limited to short-time exposure. Moreover, consistently with this result, there was an inverse correlation between $PM_{10-2.5}$ levels and PT values, with

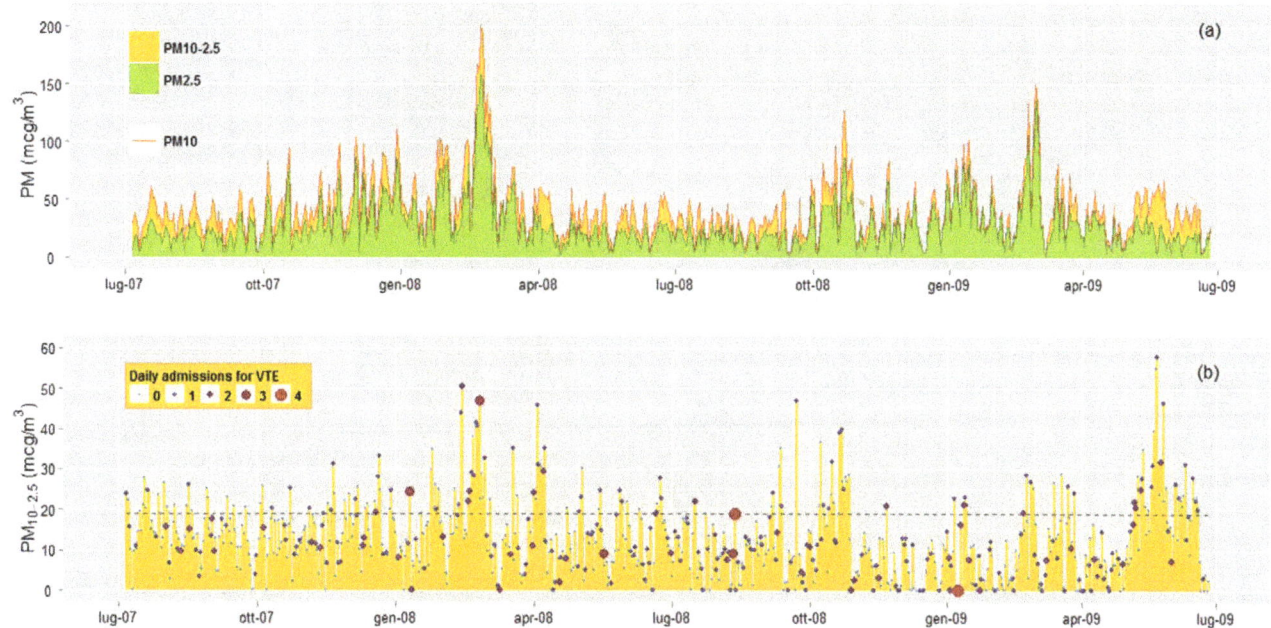

Figure 1. Seasonal trend of total, fine, and coarse particulate matter (PM$_{10}$, PM$_{2.5}$, and PM$_{10-2.5}$) concentrations during the study period. In Figure 1A, PM$_{10}$ levels are represented by the orange line and the area under the curve is divided in the 2 components, PM$_{2.5}$ represented by the green area and PM$_{10-2.5}$ represented by the ochre yellow area. In Figure 1B, the seasonal trend of PM$_{10-2.5}$ concentrations is separately represented and related with data of daily admissions for venous thromboembolism (VTE). The dashed line represents the PM$_{10-2.5}$ 75th percentile, at 19 $\mu g/m^3$.

Table 1. Correlation of coarse particulate matter (PM$_{10-2.5}$) pollution with Emergency Department daily admissions for venous thromboembolism by unadjusted and multiple adjusted Poisson regression.

PM$_{10-2.5}$	Unadjusted		Adjusted			
			Model 1†		Model 2‡	
	Beta-coefficient	P	Beta-coefficient	P	Beta-coefficient	P
Venous thromboembolism	0.163	0.039	0.246	0.003	0.237	0.020

†: adjusted for atmospheric variables (i.e. mean temperature, humidity, and air pressure).
‡: adjusted for atmospheric variables and PM$_{2.5}$.

shorter PT during the periods characterized by higher coarse PM air pollution.

Exposure to air pollution is well known to be associated with adverse effects on health. As regards to cardiovascular health, particulate air pollution increases morbidity and mortality due to atherothrombotic events [13]. Moreover, recent preliminary findings suggest an association of particulate air pollution also with VTE [9–12]. Particulate air pollution has been linked with several pathological mechanisms (reviewed in [1] and [13]) from endothelial dysfunction to prothrombotic diathesis [14–21]. The majority of these associations has been ascribed to the fine PM$_{2.5}$, while the coarse particles have received less attention and emphasis.

However, things may be more complex than they appear. Indeed, also coarse PM generates oxidative stress [22], has cytotoxic properties [23] and stimulates cytokine production [24]. Many experimental findings have demonstrated that coarse PM may exert specific and characteristic toxic effects as finest PM, so that more

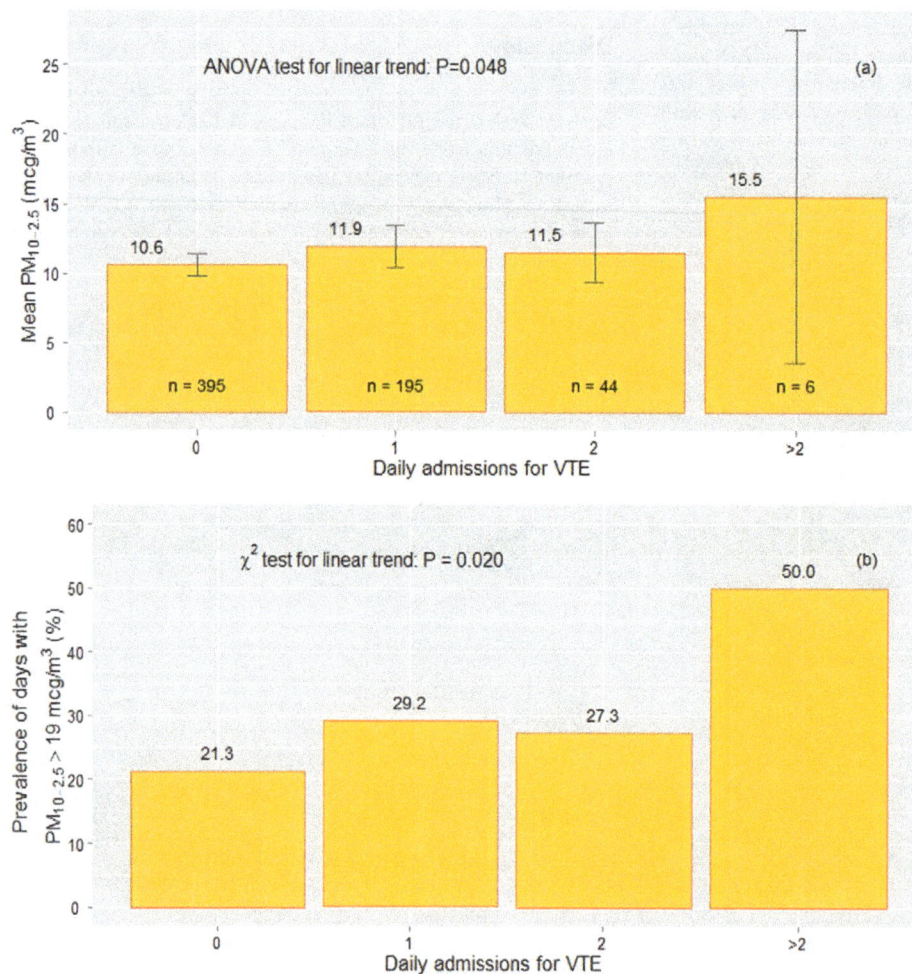

Figure 2. Coarse particulate matter (PM$_{10-2.5}$) and daily admissions to the Emergency Department for venous thromboembolism (VTE). Data are presented as mean level of PM$_{10-2.5}$ concentration (Figure 2A) and prevalence of days with high PM$_{10-2.5}$ concentration, defined as higher than the 75th percentile – 19 mcg/m^3 (Figure 2B), according to the number of daily admissions to the Emergency Department for VTE.

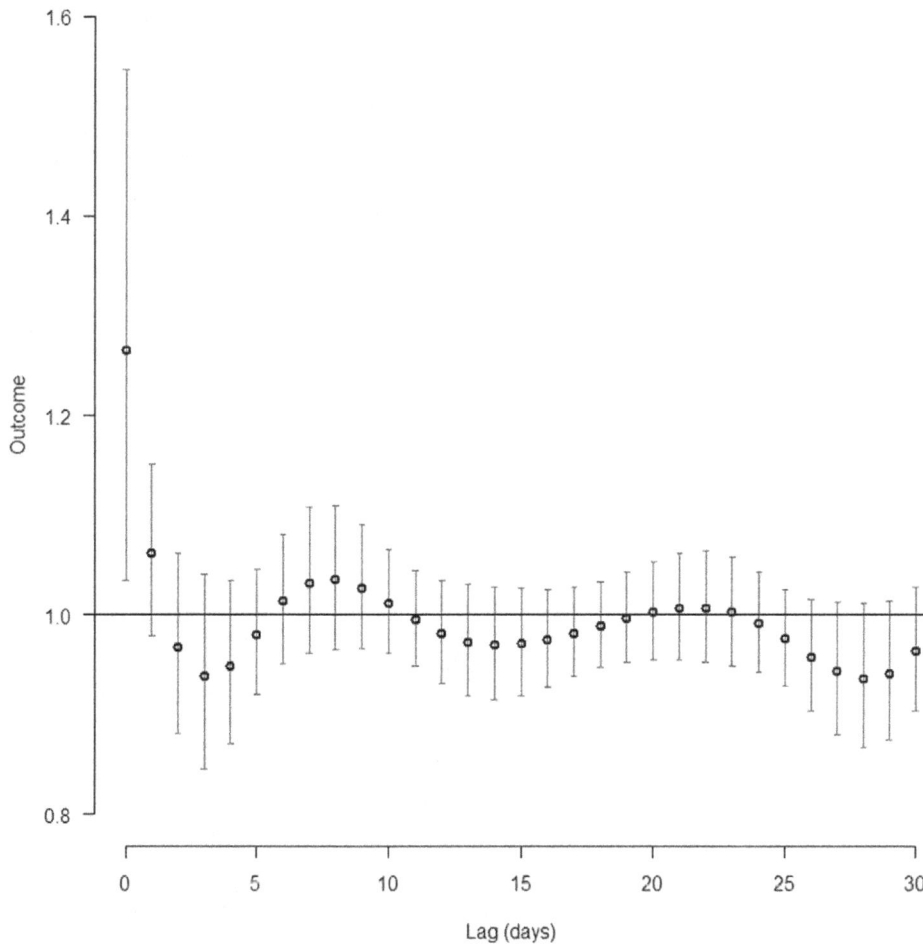

Figure 3. Association between daily hospital referral for venous thromboembolism (VTE) and coarse particulate matter (PM$_{10-2.5}$) concentration at different time-lags. The association was estimated by using distributed lag non-linear models with VTE risk as outcome and time-lags expressed as the number of previous days. Only the current-day (lag 0) PM$_{2.5-10}$ levels presented a significant association with VTE risk.

than 10 years ago Wilson and Su proposed that "fine and coarse particulate are separate classes of pollutants and should be measured separately in research and epidemiological studies" [25].

Table 2. Pearson's correlations of particulate matter (PM) concentrations, considered as a whole (PM$_{10}$) or subdivided in the finest (PM$_{2.5}$) and the coarse component (PM$_{10-2.5}$), with activated partial thromboplastin time (aPTT) and prothrombin time (PT).*

		aPTT	PT
PM$_{10}$	R	−0.119	−0.187
	P	0.274	0.060
PM$_{2.5}$	R	−0.089	−0.129
	P	0.416	0.198
PM$_{10-2.5}$	R	−0.112	−0.247
	P	0.303	0.012

*: This analysis was performed in a subgroup of patients admitted to Emergency Department for mild respiratory symptoms, without active thrombosis and not taking anticoagulant therapy, for whom data about coagulation times were available (n = 102).

The association between coarse PM and hospital referral for VTE implies some aspects of novelty. The first observation that particulate air pollution can enhance venous thrombosis risk was done in 2008 by Baccarelli and colleagues, who demonstrated that the exposure to PM$_{10}$ over a long period (1 year) increased the risk of VTE [10]. A recent surveillance study in a Chilean population confirmed these data, emphasizing the effect of the finest component PM$_{2.5}$ [11]. Moreover, in another study the risk of VTE was increased for subjects living near a major traffic road, a recognized major determinant of exposure to particulate pollution, compared with those living farther away and the increase in VTE risk was approximately linear over the observed distance range [26]. On the other hand, these results on VTE risk were not replicated by two recent prospective cohort studies [27,28], indicating the need of further studies addressing such issue. Nonetheless, the biological plausibility of the association between particulate air pollution and VTE was supported by the observation of an inverse correlation between PM$_{10}$ and PT, thus suggesting that air pollution-related changes favour a prothrombotic condition in the global coagulation balance [9]. PM has been associated also with transient increases in plasma viscosity, acute-phase reactants (including fibrinogen), and D-dimer [29,30], as well as with reduced release of tissue plasminogen activator, impairing endogenous fibrinolysis [31]. Remarkably, inhalation of diluted diesel exhaust – an efficacious model for simulation of PM exposure – has been found to increase thrombus formation within 2 hours from exposure, supporting the hypothesis that also short-term exposure to PM at near-ambient levels may favour prothrombotic diathesis [12]. Noteworthy, it is not clear so far whether such changes in haemostatic balance are due to direct interactions of (circulating) PM or secondary to systemic inflammatory response (triggered by pulmonary stress).

Our results are substantially consistent with these previous works on PM-induced VTE risk, but with some remarkable differences. First, while the previous reports investigated a long-term effect of PM exposure, our results on hospital referrals point towards a direct association between PM exposure and VTE risk in the short-term period. Noteworthy, in our analysis long-term exposure for more consecutive days did not result in a more increased VTE risk. The latter result may appear unexpected at a first glance, considering that venous thrombosis is usually a process that takes time to develop and to manifest itself clinically. Nonetheless, given that acute peaks of PM exposure contributes to systemic inflammation and hypercoagulability, a link between short-term PM peaks and VTE events may be even more biologically plausible [9,13].Considering VTE as a multifactorial disease with different steps of progression from initial and asymptomatic phases to late and clinically evident manifestations, we are tempting to speculate that peaks of coarse PM exposure may act as short-term triggers of acute VTE clinical manifestations requiring hospital referral.

Second, the most intriguing result of our analysis was that the coarse and not the finest component of PM was associated with VTE risk. $PM_{2.5}$ and ultrafine particles, on the basis of their capacity to spread into blood circulation throughout the alveolar-capillary wall, were so far considered in experimental models as the most likely culprits influencing haemostatic balance. However, population studies have not clearly defined so far which PM components actually influence VTE. Our findings may contribute to reduce such uncertainty, providing evidence in favour of a role of coarse component $PM_{10-2.5}$, rather than fine $PM_{2.5}$. In our study PM_{10} tended to be associated with VTE, thus our data do not contrast substantially with the previous reports [9,10] but extend their observations to the specific fraction $PM_{10-2.5}$. Baccarelli et al. showed an inverse correlation between PM_{10} and PT, while they founded no association between aPTT and air pollutant level [9]. Thus, they proposed that air pollutants might alter blood coagulation through the induction of tissue factor (TF). Noteworthy, a recent study demonstrated that, although PM-driven long lasting thrombogenic effects were predominantly mediated via formation of activated FXII, PM promoted its early procoagulant actions mostly through the TF-driven extrinsic pathway [32]. Consistently with this hypothesis TF expression is enhanced during systemic inflammatory response, as well as in macrophages exposed to PM [33]. Experimental evidence supports a pro-oxidative and inflammatory role also for coarse PM [21–23], even much stronger than that of fine PM. On this basis, we speculate that short-term exposure to $PM_{10-2.5}$may trigger a systemic inflammatory response through pulmonary stress, which in turn may activate blood coagulation through TF induction. Further studies are undoubtedly needed to address in depth such hypothesis.

Our study has some important limitations. First, the diagnosis classification in ED did not allow a complete characterization of the clinical phenotypes. In particular, we cannot discriminate between idiopathic or secondary VTE events. Moreover, we cannot adjust our analysis for other confounding factors, like socio-economic class. Coarse PM was calculated indirectly by subtracting $PM_{2.5}$ from PM_{10}, so that $PM_{10-2.5}$ measurement may be affected by two measurement errors rather than just one, and ambient air pollution was used as surrogate for personal exposure. However, this approximation is considered to determine only a modest underestimation of PM effects [34]. Certainly, the biological significance of assumptions made on an arithmetic difference should be corroborated by further studies, but the consistency of $PM_{10-2.5}$ association with both PT and VTE was impressive. Another limitation is represented by the lack of assessment of other potential air pollutants, including carbon monoxide, sulphur dioxide, nitrogen oxides, and ozone. On the other hand, one strength of our study is the full adjustment for all the common atmospheric parameters, like temperature, humidity, and pressure. Noteworthy, a recent study supports the presence of seasonal and monthly variability of VTE, to whom PM air pollution may contribute, with a significantly higher risk in winter (absolute increase of risk of about 12%), particularly in January (absolute increase of risk of about 20%) [35].

In summary, our analysis shows that coarse PM air pollution may be a significant predictor of ED admissions for VTE in a urban area in northern Italy, in particular by considering peaks of $PM_{10-2.5}$ exposure in the short-term period. Our results support the claim that not only fine but also coarse PM may have substantial implications on human health [5,25]. The identification of the specific PM fraction/s responsible for increased VTE risk may be useful in view of future large-scale preventive strategies against environmental air pollution. Indeed, although the individual risk of VTE triggered by PM is relatively small, the public health burden could be dramatically larger, owing to the massive exposure of the populations to this risk [36]. Moreover, as hypothesized by Mannucci, if "TVE episodes that we might call unprovoked only because none of the established risk factors are identified are perhaps provoked by air pollution" [37], then information on the local rate of daily PM should be usefully released to ED physicians facing with patients referring for possible VTE.

Materials and Methods

From July 1st, 2007 to June 30th, 2009, all subjects who were examined in Emergency Unit of the University Hospital of Verona for medical (not surgical) problems were initially considered. Of them (n = 18,841 subjects), according to the software "FirstAid" used to classify the referring patients and on the basis of days with the availability of both $PM_{10}$and $PM_{2.5}$data on the admission date (n = 640 days), were subsequently selected 302 cases who were recognized to be affected by VTE (coded according to the International Classification of Diseases(ICD) as ICD-10 I26 and I82). In particular VTE was diagnosed based on evidence from objective methods such as D-Dimer increase associated with radiological demonstration of VTE obtained by means of either compression ultrasonography or lung computed tomographic angiography.

To further investigate the potential influence of PM on thrombophilic diathesis we selected a subgroup of subjects without active thrombosis, within a group of patients admitted for mild and aspecific respiratory symptoms, and for whom data of coagulation times were available (n = 102). In such group the correlations between PM air pollution and coagulation times, both prothrombin time (PT) and activated partial thromboplastin time (APTT), were analysed.

For this epidemiological study only data from the public and anonymous registry about clinical activity of the Emergency Unit of the University Hospital of Verona were used. Therefore, specific approval by the Ethic Committee of our institution

(Azienda Ospedaliera Universitaria Integrata, Verona) was not needed, nor an informed consent was obtained from all the subjects who were examined in Emergency Unit of the University Hospital of Verona from July 1st, 2007 to June 30th, 2009.

Air pollution data

Recordings of air pollution data, measured from July 1st, 2007 to June 30th, 2009, were obtained from the Regional Environmental Protection Agency (ARPAV). The records were by permanent monitors and provided data on air pollution (24-hours averages) of the Verona district; they resulted by standardized quality control procedures for the measurement of the concentration of the different pollution agents. Data about other atmospheric parameters, like temperature, barometric pressure and humidity, were also provided.

The analyzers used by ARPAV for both PM_{10} and $PM_{2.5}$ were based on a filtering pre-treatment of the air sample, aspirated by an appropriate pump system, which was then evaluated by a gravimetric method, as otherwise specified in ARPAV website [38].

The $PM_{10-2.5}$ value was obtained by the difference between the measured values of PM_{10} and $PM_{2.5}$.

Statistical analysis

All the statistical calculations were performed with SPSS 17.0 statistical package (SPSS Inc., Chicago, IL). Because of the skewed distribution, PM values were log-transformed and geometric means with 95% confidence intervals (95%CI) were reported.

Quantitative data were assessed using the Student's t-test or by analysis of variance (ANOVA), with polynomial contrasts for linear trend when indicated. Correlations between quantitative variables were assessed using both Pearson's correlation test and linear regression models. Qualitative data were analyzed with the χ^2-test and with χ^2 for linear trend when indicated. Odds ratios (OR) with 95%CI were estimated by multiple logistic regression models. In order to dissect for possible different correlations between the number of ED admissions for any group of considered disease and PM_{10}, $PM_{2.5}$, and $PM_{10-2.5}$ values were analysed by means of log-linear model of Poisson regression, at first univariate, then adjusted for all the other atmospheric parameters (temperature, barometric pressure, and humidity). A value of $P<0.05$ was considered statistically significant.

The association between hospital referral for VTE and $PM_{10-2.5}$ concentrations at different lags was investigated using distributed lag non-linear models (DLNM) [39]. The DLNM used in the analysis was fitted through a generalized linear model with quasi-Poisson family and a canonical log-link. The lagged effect of $PM_{10-2.5}$ up to 30 days of lag was modelled by a quadratic B-spline. The concentrations of $PM_{10-2.5}$ were not centred in order to describe the effect versus a reference value of 0 mcg/m^3. Such model was also adjusted for atmospheric parameters, i.e. temperature, atmospheric pressure, and humidity. Finally, in order to evaluate the stability of the estimated effects, we fitted a series of DLNM with different configurations of bases in the predictor-lag spaces, different values of the maximum lag and different subsets of the three atmospheric parameters. DLNM estimation was performed with software R, version 2.13.1 [40], using R package dlnm, version 1.5.2 [41].

Author Contributions

Conceived and designed the experiments: NM DC. Performed the experiments: NM DG DC G. Ricci G. Rocca. Analyzed the data: NM DC MS. Contributed reagents/materials/analysis tools: NM DG G. Ricci G. Rocca OO. Wrote the paper: NM DG OO. Co-ordination and critical review of the paper: OO.

References

1. Brook RD, Rajagopalan S, Pope CA, 3rd, Brook JR, Bhatnagar A, et al. (2010) Particulate Matter air pollution and cardiovascular disease. An update to the scientific statement from the American Heart Association. Circulation 121: 2331–78.
2. Brook RD, Franklin B, Cascio W, Hong Y, Howard G, et al. (2004) Expert panel on population and prevention science of the American Hearth Association. Air pollution and cardiovascular disease : a statement for healthcare professionals from the Expert Panel on Population and Prevention Science of the American Heart Association. Circulation 109: 2655–71.
3. Nemmar A, Hoylaerts MF, Nemery B (2006) Effects of particulate air pollution on hemostasis. Clin Occup Environ Med 6: 865–81.
4. Franchini M, Mannucci PM (2007) Short-term effects of air pollution on cardiovascular diseases: outcomes and mechanisms. J Thromb Haemost 5: 2169–74.
5. Brunekreef B, Forsberg B (2005) Epidemiological evidence of effects of coarse airborne particles on health. Eur Respir J 26: 309–18.
6. Venkataraman C, Kao AS (1999) Comparison of particle lung doses from the fine and coarse fractions of urban PM-10 aerosols. Inhal Toxicol 11: 151–69.
7. Dockery DW, Pope CA, 3rd, Xu X, Spengler JD, Ware JH, et al. (1993) An association between air pollution and mortality in six U.S. cities. N Engl J Med 329: 1753–9.
8. Pope CA, 3rd, Thun MJ, Namboodiri MM, Dockery DW, Evans JS, et al. (1995) Speizer FE, Heath CW Jr. Particulate air pollution as a predictor of mortality in a prospective study of U.S. adults. Am J RespirCrit Care Med 151: 669–74.
9. Baccarelli A, Zanobetti A, Martinelli I, Grillo P, Hou L, et al. (2007) Effects of exposure to air pollution on blood coagulation. J ThrombHaemost 5: 252–60.
10. Baccarelli A, Martinelli I, Zanobetti A, Grillo P, Hou L-F, et al. (2008) Exposure to particulate air pollution and risk of deep vein thrombosis. Arch Int Med 168: 920–7.
11. Dales RE, Cakmak S, Vidal CB (2010) Air pollution and hospitalization for venous thromboembolic disease in Chile. J ThrombHaemost 8: 669–74.
12. Lucking AJ, Lundback M, Mills NL, Faratian D, Barath SL, et al. (2008) Diesel exhaust inhalation increases thrombus formation in man. Eur Heart J 29: 3043–51.
13. Franchini M, Mannucci PM (2011) Thrombogenicity and cardiovascular effects of ambient air pollution. Blood 118: 2405–12.
14. Dales R, Liu L, Szyszkowicz M, Dalipaj M, Willey J, et al. (2007) Particulate air pollution and vascular reactivity: the bus stop study. Int Arch Occup Environ Health 81: 159–64.
15. Hoffman B, Moebus S, Mohlenkamp S, Stang A, Lehmann N, et al. (2007) Residential exposure ti traffic is associated with coronary atherosclerosis. Circulation 116: 489–96.
16. Diez Roux AV, Auchincloss AH, Franklin TG, Raghunathan T, Barr RG, et al. (2008) Long-term exposure to ambient particulate matter and prevalence of subclinical atherosclerosis in the Multi-Ethnic Study of Atherosclerosis. Am J Epidemiol 167: 667–75.
17. Brook RD, Urch B, Dvonch JT, Bard RL, Speck M, et al. (2009) Insights into the mechanisms and mediators of the effects of air pollution exposure on blood pressure and vascular function in healthy humans. Hypertension 54: 659–67.
18. Chahine T, Baccarelli A, Litonjua A, Wright RO, Suh H, et al. (2007) Particulate air pollution, oxidative stress genes, and heart rate variability in an elderly cohort. Environ Health Perspect 115: 1617–22.
19. Chuang KJ, Chan C-C, Su T-C, Lee C-T, Tang C-S (2007) The effect of urban air pollution on inflammation, oxidative stress, coagulation, and autonomic dysfunction in young adults. Am J RespirCrit Care Med 176: 370–6.
20. Zeka A, Sullivan JR, Vokonas PS, Sparrow D, Schwartz J (2006) Inflammatory markers and particulate air pollution: characterizing the pathway to disease. Int J Epidemiol 35: 1347–54.
21. Ghio AJ, Hall A, Bassett MA, Cascio WE, Devlin RB (2003) Exposure to concentrated ambient air particles alters hematologic indices in humans. Inhal Toxicol 15: 1465–78.
22. Shi T, Knaapen AM, Begerow J, Birmili W, Borm PJ, et al. (2003) Temporal variation of hydroxyl radical generation and 8-hydroxy-2'-deoxyguanosine formation by coarse and fine particulate matter. Occup Environ Med 60: 315–21.
23. Hsiao WL, Mo ZY, Fang M, Shi XM, Wang F (2000) Cytotoxicity of $PM_{2.5}$ and $PM_{2.5-10}$ ambient air pollutants assessed by the MTT and the Comet assays. Mutat Res 471: 45–55.
24. Becker S, Soukup JM, Gallagher JE (2002) Differential particulate air pollution induced oxidant stress in human granulocytes, monocytes and alveolar macrophages. Toxicol In Vitro 16: 209–18.

25. Wilson WE, Suh HH (1997) Fine particles and coarse particles: concentration relationships relevant to epidemiologic studies. J Air Waste Manag Assoc 47: 1238–49.

26. Baccarelli A, Martinelli I, Pegoraro V, Melly S, Grillo P, et al. (2009) Living near major traffic roads and risk of deep vein thrombosis. Circulation 119: 3118–24.

27. Kan H, Folsom AR, Cushman M, Rose KM, Rosamond WD, et al. (2011) Traffic exposure and incident venous thromboembolism in the atherosclerosis risk in communities (ARIC) study. J Thromb Haemost 9: 672–678.

28. Shih RA, Griffin BA, Salkowski N, Jewell A, Eibner C, et al. (2011) Ambient particulate matter air pollution and venous thromboembolism in the Women's Health Initiative Hormone Therapy trials. Environ Health Perspect 119: 326–331.

29. Ghio AJ, Kim C, Devlin RB (2000) Concentrated ambient air particles induce mild pulmonary inflammation in healthy human volunteers. Am J Respir Crit Care Med 162: 981–988.

30. Sun Q, Hong X, Wold LE (2010) Cardiovascular effects of ambient particulate air pollution exposure. Circulation 121: 2755–65.

31. Mills NL, Toernqvist H, Robinson SD, Gonzalez M, Darnley K, et al. (2005) Diesel exhaust inhalation causes vascular dysfunction and impaired endogenous fibrinolysis. Circulation 112: 3930–6.

32. Kilinç E, Van Oerle R, Borissoff JI, Oschatz C, Gerlofs-Nijland ME, et al. (2011) Factor XII activation is essential to sustain the procoagulant effects of particulate matter. J ThrombHaemost 9: 1359–67.

33. Gilmour PS, Morrison ER, Vickers MA, Ford I, Ludlam CA, et al. (2005) The procoagulant potential of environmental particles (PM_{10}). Occup Environ Med 62: 164–71.

34. Zeger SL, Thomas D, Dominici F, Samet JM, Schwartz J, et al. (2000) Exposure measurement error in time-series studies of air pollution: concepts and consequences. Environ Health Perspect 108: 419–26.

35. Dentali F, Ageno W, Rancan E, Donati AV, Galli L, et al. (2011) Seasonal and monthly variability in the incidence of venous thromboembolism. A systematic review and a meta-analysis of the literature. Thromb Haemost 106: 439–47.

36. Nawrot TS, Perez L, Künzli N, Munters E, Nemery B (2011) Public health importance of triggers of myocardial infarction: a comparative risk assessment. Lancet 377: 732–40.

37. Mannucci PM (2010) Fine particulate: it matters. J Thromb Haemost 8: 659–61.

38. ARPAV (Agenzia Regionale per la Prevenzione e Protezione Ambientale del Veneto) website. Available: http://www.arpa.veneto.it/aria_new/htm/metodi_misura.asp?3 Accessed September 2011.

39. Gasparrini A, Armstrong B, Kenward MG (2010) Distributed lag non-linear models. Statistics in Medicine 29: 2224–34.

40. R Development Core Team (2011) R: A language and environment for statistical computing. R Foundation for Statistical Computing, Vienna, Austria. ISBN 3-900051-07-0, URL http://www.R-project.org/.

41. Gasparrini A (2011) Distributed lag linear and non-linear models in R: the package dlnm. Journal of Statistical Software 43: 1–20.

3

Deaths and Medical Visits Attributable to Environmental Pollution in the United Arab Emirates

Jacqueline MacDonald Gibson[1]*, Jens Thomsen[2], Frederic Launay[3], Elizabeth Harder[1], Nicholas DeFelice[1]

1 Department of Environmental Sciences and Engineering, Gillings School of Global Public Health, University of North Carolina–Chapel Hill, Chapel Hill, North Carolina, United States of America, 2 Health Authority–Abu Dhabi, Abu Dhabi, United Arab Emirates, 3 Environment Agency–Abu Dhabi, Abu Dhabi, United Arab Emirates

Abstract

Background: This study estimates the potential health gains achievable in the United Arab Emirates (UAE) with improved controls on environmental pollution. The UAE is an emerging economy in which population health risks have shifted rapidly from infectious diseases to chronic conditions observed in developed nations. The UAE government commissioned this work as part of an environmental health strategic planning project intended to address this shift in the nature of the country's disease burden.

Methods and Findings: We assessed the burden of disease attributable to six environmental exposure routes outdoor air, indoor air, drinking water, coastal water, occupational environments, and climate change. For every exposure route, we integrated UAE environmental monitoring and public health data in a spatially resolved Monte Carlo simulation model to estimate the annual disease burden attributable to selected pollutants. The assessment included the entire UAE population (4.5 million for the year of analysis). The study found that outdoor air pollution was the leading contributor to mortality, with 651 attributable deaths (95% confidence interval [CI] 143–1,440), or 7.3% of all deaths. Indoor air pollution and occupational exposures were the second and third leading contributors to mortality, with 153 (95% CI 85–216) and 46 attributable deaths (95% CI 26–72), respectively. The leading contributor to health-care facility visits was drinking water pollution, to which 46,600 (95% CI 15,300–61,400) health-care facility visits were attributed (about 15% of the visits for all the diseases considered in this study). Major study limitations included (1) a lack of information needed to translate health-care facility visits to quality-adjusted-life-year estimates and (2) insufficient spatial coverage of environmental data.

Conclusions: Based on international comparisons, the UAE's environmental disease burden is low for all factors except outdoor air pollution. From a public health perspective, reducing pollutant emissions to outdoor air should be a high priority for the UAE's environmental agencies.

Editor: Hamid Reza Baradaran, Tehran University of Medical Sciences, Iran (Republic of Islamic)

Funding: This work was funded by the Environment Agency-Abu Dhabi (www.ead.ae). Staff from the funding organization provided environmental data to support this research but had no other role in the study design. One staff member from the funding organization (Dr. Launay) helped to prepare this manuscript and review it for accuracy. Otherwise, the funder had no role in the decision to publish this manuscript.

Competing Interests: The authors have declared that no competing interests exist.

* E-mail: jackie.macdonald@unc.edu

Introduction

Over the past half century, the United Arab Emirates (UAE) has developed at an unprecedented rate. Prior to the discovery of oil in 1958, the UAE (then called the Trucial States) was among the Arab world's poorest nations, with "no electrical grid, indoor plumbing, telephone system, public hospital, or modern school" [1]. Fueled by oil, the UAE has transformed over the past 50 years. Steel-and-glass metropolises tower above the once barren desert, state-of-the-art desalination plants supply fresh water, factories export locally produced goods worldwide, and once-remote airports serve as international hubs. By 2010, the UAE boasted the world's third-largest gross national income per capita: $59,993, trailing only Liechtenstein and Qatar [2].

Overall, development has dramatically improved public health in the UAE. Life expectancy has increased from less than 45 in the 1950s to 75.9 in 2010, and infant mortality rates have plummeted from more than 180 to fewer than 5 per 1,000 births [3]. However, as in every other nation undergoing an industrial transition, development also has created new forms of environmental pollution, leading to new health risks.

As part of an anticipatory project to confront emerging environmental risks and prevent downstream public health impacts, the Environment Agency–Abu Dhabi (EAD) commissioned a study to quantify the UAE's environmental burden of disease (EBD). The EAD is charged with protecting environmental quality in Abu Dhabi, the UAE's largest emirate (comprising 80% of the UAE's land area and 33% of its population) and source of most of the UAE's petroleum wealth [4]. The EAD collaborated with the Health Authority–Abu Dhabi (HAAD), responsible for monitoring and improving public health and regulating the health-care sector in Abu Dhabi emirate. The EAD and HAAD recruited an international research team of environmental and public health

Table 1. Causes of mortality considered in this study.

Exposure route	Cause of mortality	ICD-10 code(s)	Baseline mortality (deaths in 2008)	Pollutants	Exposure estimation method	Relative risk (95% CI)
Outdoor air pollution	All causes (adults>30)	N/A	8,865	$PM_{2.5}$ (average annual concentration, $\mu g/m^3$)	Abu Dhabi outdoor air quality monitors [14]	1.06 (1.02–1.11) (per 10 $\mu g/m^3$); see [25]
	Respiratory disease (children <5)	J00–99	27	PM_{10} (average daily concentration, $\mu g/m^3$)	Same as for $PM_{2.5}$	1.017 (1.0034, 1.03) (per 10 $\mu g/m^3$); see [26]
Indoor air pollution	Cardiovascular disease	I00–79	2,310	Environmental tobacco smoke (ETS), present or absent in home	Household surveys: ETS present in 19% of homes	Male nonsmokers: 1.25 (1.06, 1.47); female nonsmokers: 1.35 (1.11, 1.64) [27]
	Lung cancer	C33–4	120	ETS	Same as previous	Male nonsmokers: 1.1 (0.6, 1.8); female nonsmokers: (1.2 (0.8, 1.6) [28]
				Radon (average daily concentration, Bq/m^3)	Household measurements* (Abu Dhabi City and Sharjah only): Abu Dhabi, lognormal (mean = 14.4, sd = 7.37); Sharjah, triangular (8, 50.3, 164); assumed zero elsewhere	1.08 (1.13, 1.16) (per 100 Bq/m^3) [29]
				Incense use (frequency per week)	Household surveys: Daily users = 43.54% of population; intermittent users = 42.86%	Daily users: 1.8 (1.2, 2.6); intermittent users (1–5 times/week): 1.2 (0.9–1.6) [30]
Occupational exposures	Asthma	J45	10	Employment in occupation involving exposure to dusts, fumes	UAE Ministry of Economy data on workforce participation by industry sector and occupation within sector; see [20]	Varies by occupation and gender; see [20]
	Chronic obstructive pulmonary disease	J44	37	Employment in occupation involving exposure to dusts, fumes	Same as previous	Varies by exposure level and gender; see [20]
	Asbestosis	501	0	Asbestos exposure	NA	100% of observed cases
	Malignant mesothelioma	C45	6	Asbestos exposure	NA	90% of observed cases in males and 25% in females; see [20]
	Silicosis	502	0	Silica exposure	NA	100% of observed cases
	Leukemia	C91–5	130	Employment in occupation with exposure to diesel exhaust, benzene, ethylene oxide	UAE Ministry of Economy data on workforce employed by industry sector; Carcinogen Exposure (CAREX) database; see [20]	Low exposure: 1.9 (1.6, 2.2); high exposure: 4 (3.6, 4.4); see [20]
	Lung cancer	C33–4	120	Employment in occupation with exposure to arsenic, asbestos, beryllium, cadmium, chromium, nickel, silica	Same as previous	Low exposure: 1.21 (1.18, 1.24); high exposure: 1.77 (1.71,1.83); see [20]
Climate change	Cardiovascular disease	I00–79	2,310	Increase in ambient temperature attributable to global climate change	100% of population exposed	1.001 (1.000, 1.003) [31]
Drinking water contamination	Bladder cancer	C67, C68	23	Drinking chlorinated water	Citizens: 10.5% consume tap (chlorinated) water; non-citizens: tap water consumption represented as uniform (84%, 96.4%) distribution**	Males: 1.24 (0.97, 1.57); females: 1.17 (1.03, 1.34) [32]
	Colon cancer	C18	80	Same as previous	Same as previous	Males: 1.09 (0.81, 1.48); females: 1.19 (0.93, 1.53) [32]
	Rectal cancer	C19–21	30	Same as previous	Same as previous	Males: 1.24 (0.86, 1.79); females: 1.10 (0.90, 1.36) [32]

Table 1. Cont.

Exposure route	Cause of mortality	ICD-10 code(s)	Baseline mortality (deaths in 2008)	Pollutants	Exposure estimation method	Relative risk (95% CI)
	Gastroenteritis	A00–9	7	Access to regulated drinking water supply and sewage treatment	Population divided into two groups: (1) access to regulated water supply and sanitation (population fraction represented as triangular (0.96, 0.98, 1.0) distribution); (2) access to improved but unregulated water, no sanitation	Group 1: uniform (1, 4); group 2: uniform (7.2, 10.2) [21,33]

*EAD provided measured radon concentrations from 111 Abu Dhabi residential dwellings (202 measurements in total) and a mean, minimum, and maximum value for measurements taken in Sharjah.
**Our survey of 628 citizen households found 10.5% drink tap water, 84.6% bottled water, 3.4% well water, and 1.5% water from undefined other sources [19]. Estimates using bottled water industry sales data suggest noncitizens consume 84%–96.4% of water from taps [34,35].

specialists to carry out the analysis. This article reports on the resulting UAE EBD estimates.

The World Health Organization (WHO) in 2007 released preliminary EBD estimates for nations around the world [5]. WHO's EBD estimates included risks of four categories of environmental pollution: (1) particulate matter in outdoor air, (2) combustion of solid fuels indoors, (3) particulate matter and carcinogens in occupational environments, and (4) pathogens in water. The WHO attributed 13 million deaths each year to these risks, with the magnitude and relative importance of the risks differing considerably by epidemiologic subregion [6]. As a result of the high regional variability, a June 2007 editorial in *The Lancet* noted, "With its release of each country's profile of environmental factors and their impact on health, WHO has made a first, very important step towards facilitating more joined-up thinking by policymakers when planning interventions that have the greatest effect at a population level," but that these estimates "should only serve as a starting point for countries to collect their own data for refinement and validation" [6]. To our knowledge, the UAE is the first country to follow through on completing its own comprehensive EBD assessment, considering all the risk factors included in the WHO estimates and several others as well. As such, this project may serve as a model for similar projects in other countries.

Our research team constructed a computer simulation model, the *United Arab Emirates Environmental Burden of Disease Model*, to link data on environmental pollutant concentrations with new UAE public health data and with epidemiologic studies that estimate the relative risks of various illnesses due to pollutant exposures. To our knowledge, this model is the first to implement a comprehensive, national-scale EBD analysis in a flexible computer simulation platform that reflects uncertainty in the estimates and that can be readily updated in the future as conditions change and new local data are collected. The model is designed to support policy analyses comparing the effectiveness of alternative options for reducing environmental risks to health.

Although "environmental risk" may be defined much more broadly, for this project we focused on environmental risks that are within the mandate and capability of the EAD and HAAD to address. Specifically, we considered six categories of risk, corresponding to six different exposure routes:

1. outdoor air pollution,
2. indoor air pollution,
3. occupational exposures,

4. drinking water contamination,
5. coastal water pollution, and
6. global climate change.

This list of risk factors includes all those in the WHO preliminary estimates except for indoor air pollution due to solid fuel use, since solid fuel is no longer used for cooking in the UAE. We also included pollutants for each exposure route that were relevant to the UAE but were not considered in the previous WHO estimates (see "Pollutants," below). Since the scope of potential pollutant-exposure route combinations is in the thousands, we narrowed the list through a two-step process. First, we conducted preliminary risk assessments for candidate pollutant exposure route combinations identified by the EAD and the WHO Centre for Environmental Health Activities for the Eastern Mediterranean Region. Then, we presented the preliminary risk assessment results at workshops involving government environment and health officials, faculty at local universities, international experts, local industries, and environmental groups. Participants were led through a systematic process, developed through previous research, to prioritize the pollutant-exposure route combinations for consideration [7]. Stakeholder engagement in the selection of risks to consider was essential because the results were intended to inform future UAE strategic planning [8].

Methods

Study Population

The EBD estimates include all residents of the UAE's seven emirates, including citizens and expatriates, during the year 2008 (Figure S1 and Table S1). Expatriates constituted 81% of the population–a result of workers emigrating to the UAE to fill jobs arising from the nation's ambitious development agenda, which requires more manpower than is available from the indigenous population (864,000 in 2008). About 85% of the expatriates are Asians (mostly from India and Pakistan), and another 10% are Arab [9]. The remaining 5% includes Australasians, Europeans, and North Americans.

Health Outcomes

We provide EBD estimates for two health outcome categories: mortality and morbidity, with the latter expressed as number of health-care-facility visits. Health-care facility visits include all patient use of hospitals, doctor's offices, and pharmacies. Table 1

Table 2. Nonfatal illnesses considered in this study.

Exposure route	Illness	ICD-9 code(s)	2008 health-care visits	Pollutants	Exposure estimation method	Relative risk
Outdoor air pollution	Cardiovascular disease	390–448	307,667	PM$_{10}$ (daily average, µg/m³)	Abu Dhabi outdoor air quality monitors [14]	1.003 (1.0024–1.0036) (per 10 µg/m³); see [14]
	Respiratory diseases	480–6; 490–7; 507	176,048	PM$_{10}$ (daily average, µg/m³)	Same as previous	1.008 (1.0047–1.012) (per 10 µg/m³); see [14]
				Ozone (daily average, ppb)	Same as for PM$_{10}$	1.03 (1.02–1.05) (per 10 ppb); see [14]
Indoor air pollution	Asthma (age <18)	493	24,418	Mold (presence in home)	Household surveys: present in 16% of homes [19]	1.35 (1.20, 1.51) [36]
				Environmental tobacco smoke (ETS) in home	Household surveys: present in 19% of homes [19]	1.48 (1.32, 1.65) [37]
	Asthma (age≥18)	493	32,388	Mold (presence in home)	Household surveys: present in 16% of homes [19]	1.54 (1.01, 2.32) [38]
	Asthma (age≤6)	493	13,879	Formaldehyde (daily average, µg/m³)	Household surveys: lognormal (mean = 22.5, sd = 63.6) [19]	1.003 (1.002, 1.004) (per 10 µg/m³) [39]
	Cardiovascular disease	390–448	307,667	ETS	Household surveys: present in 19% of homes [19]	1.25 (1.17, 1.32) [40]
	Lower respiratory tract infection (age≤ 6)	480–92	13,996	ETS	Same as previous	1.57 (1.28, 1.91) [41]
	Leukemia	204–208.9	1,520	ETS	Same as previous	2.28 (1.15, 4.53) [42]
	Lung cancer	162	444	Radon concentration (Bq/m³)	Data from Abu Dhabi Food Control Authority: Abu Dhabi City, lognormal (mean = 14.4, sd = 7.37); Sharjah, triangular (8, 50.3, 164); other emirates = 0	1.08 (1.13, 1.16) (per 100 Bq/m³) [29]
				Incense use (frequency)	Household surveys: Daily users = 43.54% of population; intermittent users = 42.86% [19]	Daily users: 1.8 (1.2, 2.6); intermittent users (1–5 times/week): 1.2 (0.9–1.6) [30]
Occupational exposures	Asthma	493	72,301	Employment in occupation with dusts, fumes	UAE Ministry of Economy data on workforce participation by industry sector and occupation within sector [20]	Varies by occupation and gender; see [20]
	Chronic obstructive pulmonary disease	490–2, 494, 496	27,212	Same as previous	Same as previous	Varies by occupation and gender; see [20]
	Leukemia	204–208.9	1,520	Employment in occupation with diesel exhaust, benzene, ethylene oxide	UAE Ministry of Economy data on workforce employed by industry sector; Carcinogen Exposure (CAREX) database; see [20]	Low exposure: 1.9 (1.6, 2.2); high exposure: 4 (3.6, 4.4) [20]
	Lung cancer	162	443	Employed in occupation with arsenic, asbestos, beryllium, cadmium, chromium, nickel, silica	Same as previous	Low exposure: 1.21 (1.18, 1.24); high exposure: 1.77 (1.71,1.83); see [20]
	Malignant mesothelioma	163	28	Diagnosed mesothelioma	NA	90% of observed cases in males; 25% of cases in females; see [20]
	Asbestosis	501	3	Diagnosed asbestosis	NA	100% of observed cases
	Silicosis	502	8	Diagnosed silicosis	NA	100% of observed cases
Climate change	Cardiovascular disease	390–448	307,667	Increase in annual average ambient temperature	100% of population exposed	1.001 (1.000, 1.003) [31]
Drinking water contamination	Bladder cancer	188	929	Total trihalomethane (TTHM) concentration (µg/l)	10.5% of citizens consume tap water; noncitizen consumption represented as uniform (84%, 96.4%) distribution*	Males: 1.24 (0.97, 1.57); females: 1.17 (1.03, 1.34) [32]
	Colon cancer	153	2,191	TTHM (µg/l)	Same as previous	Males: 1.09 (0.81, 1.48); females: 1.19 (0.93, 1.53) [32]
	Rectal cancer	154	639	TTHM at tap (µg/l)	Same as previous	Males: 1.24 (0.86, 1.79); females: 1.10 (0.90, 1.36) [32]

Table 2. Cont.

Exposure route	Illness	ICD-9 code(s)	2008 health-care visits	Pollutants	Exposure estimation method	Relative risk
	Gastroenteritis	008–9, 558.9	81,110	Availability of regulated drinking water supply and sewage treatment	Population divided into groups: (1) access to regulated water supply and sanitation (population fraction represented as triangular (0.96, 0.98, 1.0) distribution); (2) access to improved but unregulated water, no sanitation	Group 1: uniform (1, 4) Group 2: uniform (7.2, 10.2) [21]
Coastal water pollution	Gastroenteritis[4]	008–9, 558.9	81,110	Enterococci concentration in beach water (number/100 ml)	Water quality samples from Abu Dhabi beaches**; previous surveys of swimming frequency***	1.34 (1.00, 1.75) (per log-10) [43]

*Our survey of 628 citizen households found 10.5% drink tap water; 84.6% bottled water; 3.4% well water; and 1.5% water from undefined other sources [19]. Estimates using bottled water industry sales data suggest noncitizens consume 84%–96.4% of water from taps [34,35].

**Monthly observations were available for two Abu Dhabi beaches for 2006. We therefore represented coastal water quality in each month as a uniform distribution with a lower bound equal to the lowest observed concentration and an upper bound equal to the highest observed concentration. Due to wastewater overflows in Dubai over the time period of this study, we assumed enterococci concentrations at Dubai beaches were twice those observed in Abu Dhabi (while for all other emirates, concentrations were assumed the same as in Abu Dhabi). The uniform distribution parameters for all emirates other than Dubai (in enterococci/100 ml) are as follows: Jan. (2, 8); Feb. (0, 4); Mar (0, 3); Apr (0, 0); May (0, 0); Jun (0, 12); Jul (0, 85); Aug (0, 85); Sept. (0, 43); Oct. (4, 250); Nov. (5, 6); Dec. (3, 12). For Dubai, these parameters were doubled.

***Proportion of citizens swimming in coastal waters in any given month were estimated from Badrinath et al. [44], as follows: males ≤14: 3.8%, males >14, 1.4%; females ≤14, 0.87%; females >14, 0%. Proportion of non-citizens swimming in coastal waters were estimated from the Australian Sports Commission [45], assumed to be 6.2% (both genders, all ages).

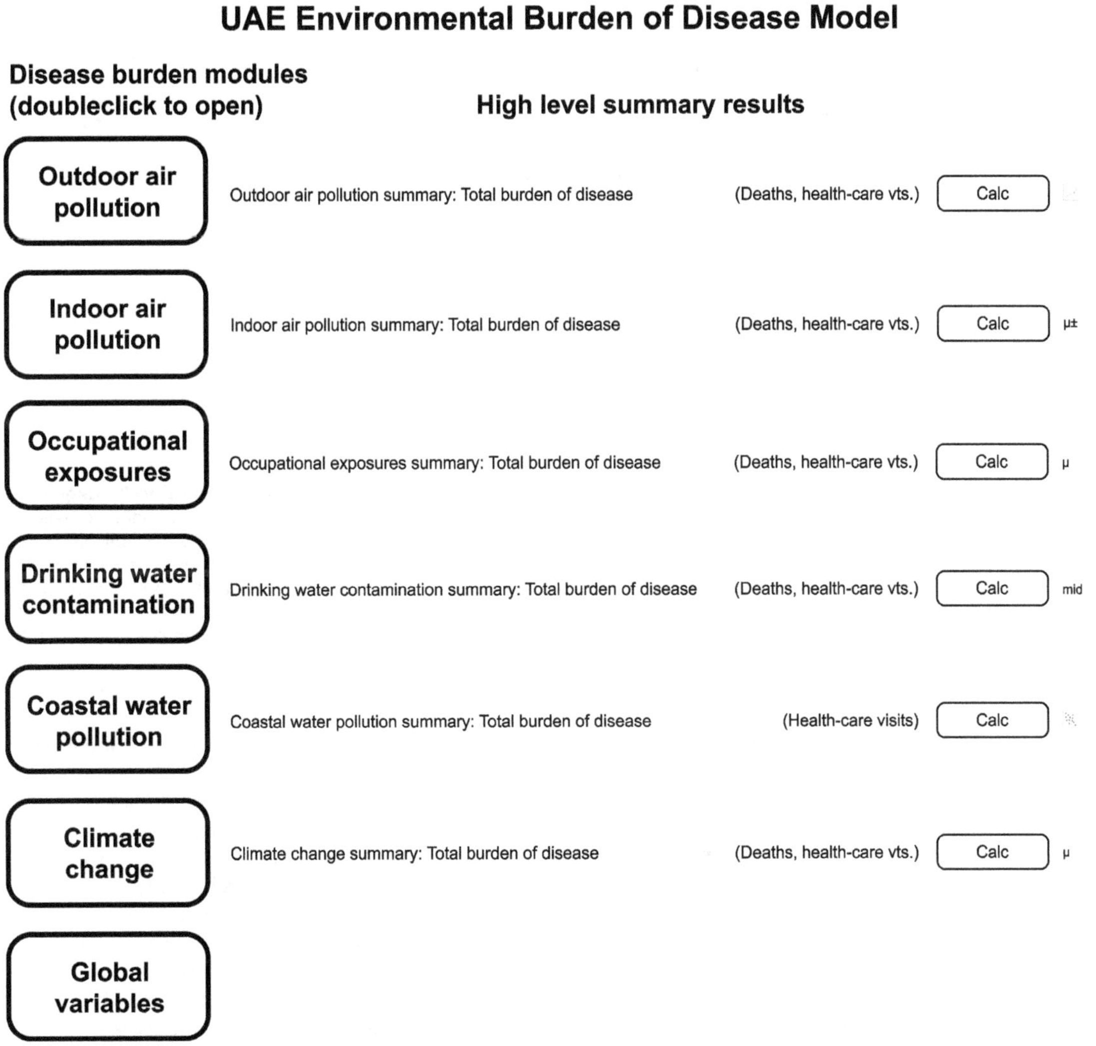

UAE Environmental Burden of Disease Model

Disease burden modules
(doubleclick to open)   **High level summary results**

Outdoor air pollution — Outdoor air pollution summary: Total burden of disease — (Deaths, health-care vts.) — Calc

Indoor air pollution — Indoor air pollution summary: Total burden of disease — (Deaths, health-care vts.) — Calc — μ±

Occupational exposures — Occupational exposures summary: Total burden of disease — (Deaths, health-care vts.) — Calc — μ

Drinking water contamination — Drinking water contamination summary: Total burden of disease — (Deaths, health-care vts.) — Calc — mid

Coastal water pollution — Coastal water pollution summary: Total burden of disease — (Health-care visits) — Calc

Climate change — Climate change summary: Total burden of disease — (Deaths, health-care vts.) — Calc — μ

Global variables

Figure 1. Top level of the UAE EBD model. Double-clicking on any node opens further layers of a module that shows how the EBD for each exposure route is estimated. The "global variables" node contains all health outcome and population distribution data for all modules.

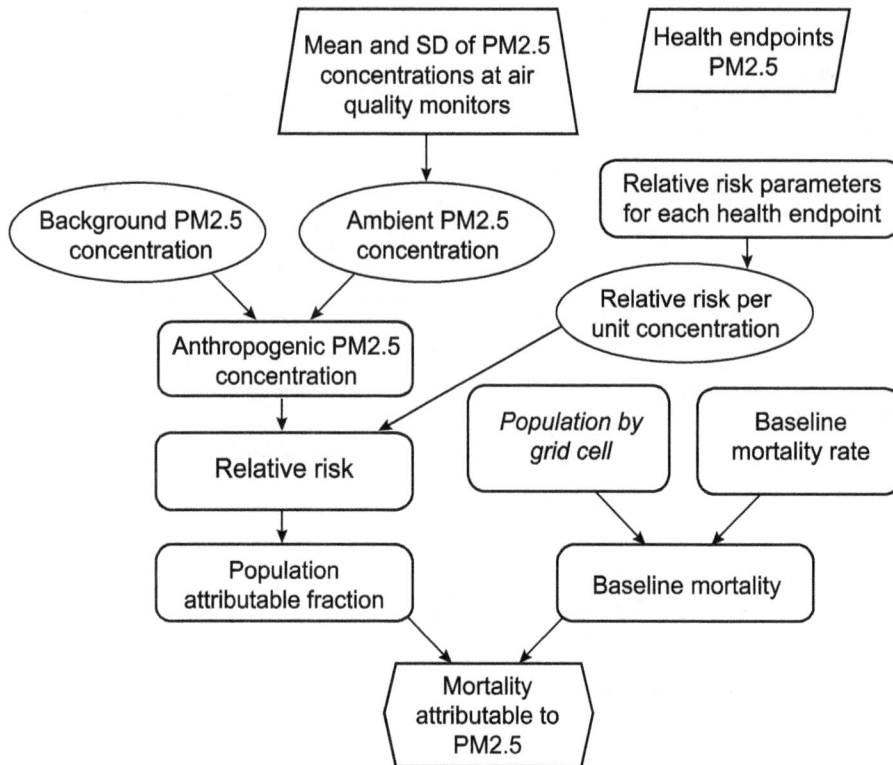

Title: Background PM2.5 concentration

Description: This variable specifies the shape of the probability density function (PDF) of the background PM2.5 concentration as uniform with lower bound 5 and upper bound 35.

Definition: Uniform   (5, 35)

Value: 20

Outputs: Anthropogenic_pm2... Anthropogenic PM2.5 concentration

Source: WHO in Ostro 2004 recommends assuming the natural background level of PM2.5 is 5 μg/m3. However, this value might be too low for the UAE, where dust storms can significantly increase natural background PM levels. Given this, the WHO-recommended level was chosen as a lower bound and 35 μg/m3 was chosen as an upper bound based on previous studies in desert regions in the United States.

Reference: Ostro, B. 2004. Outdoor Air Pollution: Assessing the Environmental Burden of Disease at National and Local Levels. Environmental Burden of Disease Series No. 5. Geneva: World Health Organization.

Figure 2. Layer within the model's outdoor air pollution module and the notecard that opens when clicking on the "Background PM2.5 concentration" node. Trapezoids indicate deterministic variables; ovals indicate random variables; rectangles with rounded corners are variables determined from equations involving higher-level nodes; and the hexagon indicates an objective node. Key nodes are as follows: *Health endpoints PM2.5*: listing of health endpoints associated with PM$_{2.5}$; *Background PM2.5 concentration*: PM$_{2.5}$ concentration in the absence of human activity; *Mean and SD of PM2.5 concentrations at air quality monitors*: mean and SD of a year's worth of measurements at UAE air quality monitors (interpolated from monitors for each of 1,164 cells in a grid used to divide the UAE into subunits for analysis); *Relative risk parameters for each health endpoint*: as shown in Table 1, last column; *Baseline mortality rate*: mortality rate by emirate and citizenship; *Population by grid cell*: population by citizenship in each of the 1,164 geographic grid cells.

lists the causes of mortality, and Table 2 the illnesses considered in this research, along with the total numbers of each in 2008.

To support this analysis, HAAD compiled death records for Abu Dhabi emirate for 2008. The records included 2,949 deaths listed by cause (by ICD-10 code), time, location, age, gender, and nationality. This database includes all deaths reported in Abu Dhabi for the study year. HAAD considers the death notification rate to be 100%, since it is not legal to bury, cremate, or expatriate a body without a death certificate. Hence, death rates estimated from this data set should be very accurate. Comparable information was not available from other emirates. Baseline death information for the other emirates was estimated by calculating death rates for gender-citizenship groups in Abu Dhabi and then

applying those rates to population estimates for those same demographic groups in the other emirates.

HAAD also provided patient encounter records for the diseases of interest in this study from Abu Dhabi's largest health insurance provider, Daman, which covers 73% of the emirate's population. We used these data as a surrogate for morbidity estimates because they provided the most accurate and comprehensive database of incidences of illness available for this research. Prior to 2008, Abu Dhabi lacked standards for medical records coding; hence, the data set employed in this research is the first in Abu Dhabi to be compiled and quality-assured according to international best practices in medical records management [10]. The records included the date of encounter, ICD-9 code for the corresponding

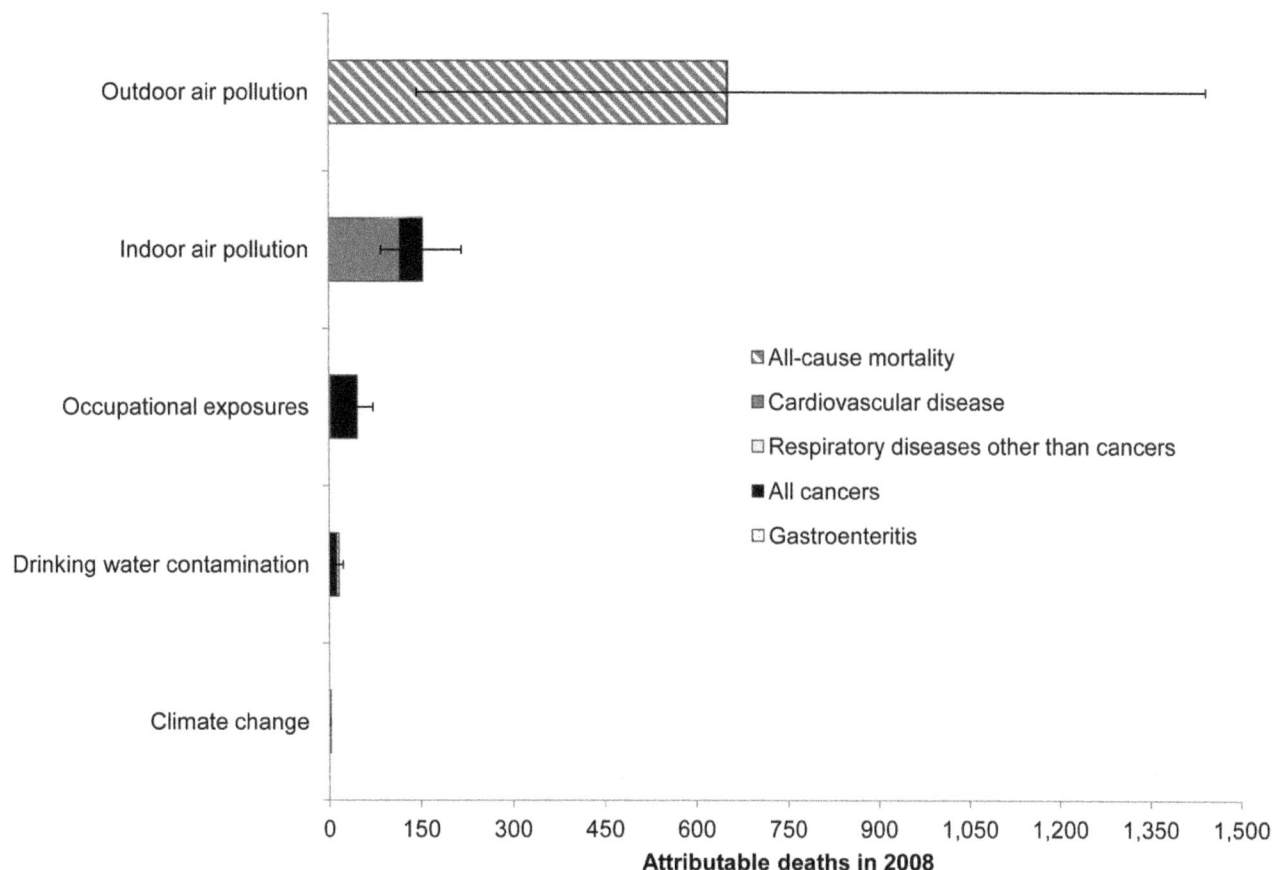

Figure 3. Estimated deaths attributable to environmental pollutants in the UAE. In addition to those easily seen in the chart, two deaths from respiratory diseases other than cancers were attributed to environmental pollution in 2008, both due to outdoor air pollution.

illness, health-care facility name, and patient demographic information (age, gender, citizenship). Patient identifiers were not provided. Noninfectious disease health-care facility visit records were provided for 2008 (162,228 visits recorded by Daman). Gastrointestinal illness records (10,581 in total) were provided for the first half of 2009; we assumed this represented 50% of the visits that would have occurred in 2008 among individuals covered by Daman. Like the mortality data, the health insurance claims data were scaled to cover the entire UAE population.

It is possible that using the Daman data as the basis for estimating health-care facility visit rates may have biased our results because the population insured by Daman may not be representative of the UAE population as a whole. Health insurance is mandatory in Abu Dhabi, and Daman is the major insurance provider, covering all Emiratis, all unskilled expatriate laborers, and many higher skilled expatriates. However, some highly skilled expatriates purchase enhanced insurance from one of dozens of private companies [11]. Hence, it is possible that the Daman data set under-represents skilled expatriate workers. These skilled workers may be healthier than unskilled workers due to their higher socioeconomic status. However, previous analyses have shown that skilled workers visit health-care facilities more often than unskilled workers (3.8 versus 2.7 visits per subscriber per year, respectively) [10]. Hence, the direction of the bias introduced by under-representing skilled workers is unknown, since on the one hand excluding wealthier workers would be expected to bias predicted disease rates upwards while on the other

hand wealthier workers visit doctors more often than their lower-wage counterparts, creating the potential for downward bias by excluding these workers. Nonetheless, data from the other insurance providers, which are private, were not available to support this analysis, and the HAAD data from Daman represent the most comprehensive and most effectively quality-assured health data available for our study year [10].

Pollutants

As in previous EBD studies (see Table S2 for a listing), we focused on pollutants for which exposure is potentially common and strong epidemiologic evidence is available to predict the occurrence of illness due to contaminant exposure. As noted above, preliminary risk assessments by experts and stakeholder engagement sessions narrowed the list of candidates [7]. Tables 1 and 2 list the pollutants for each exposure route-health endpoint combination. In the case of radon, exposure data were available only for a subset of the population (residents of the city of Abu Dhabi and the emirate of Sharjah), so we report those results separately.

EBD Estimation Method

To estimate the burden of disease due to each combination of exposure pathway and pollutant, we used the "population attributable fraction" (PAF) approach. The PAF is the proportion of reduction in disease or mortality that would be expected if exposure to a pollutant were reduced to an alternative (known as

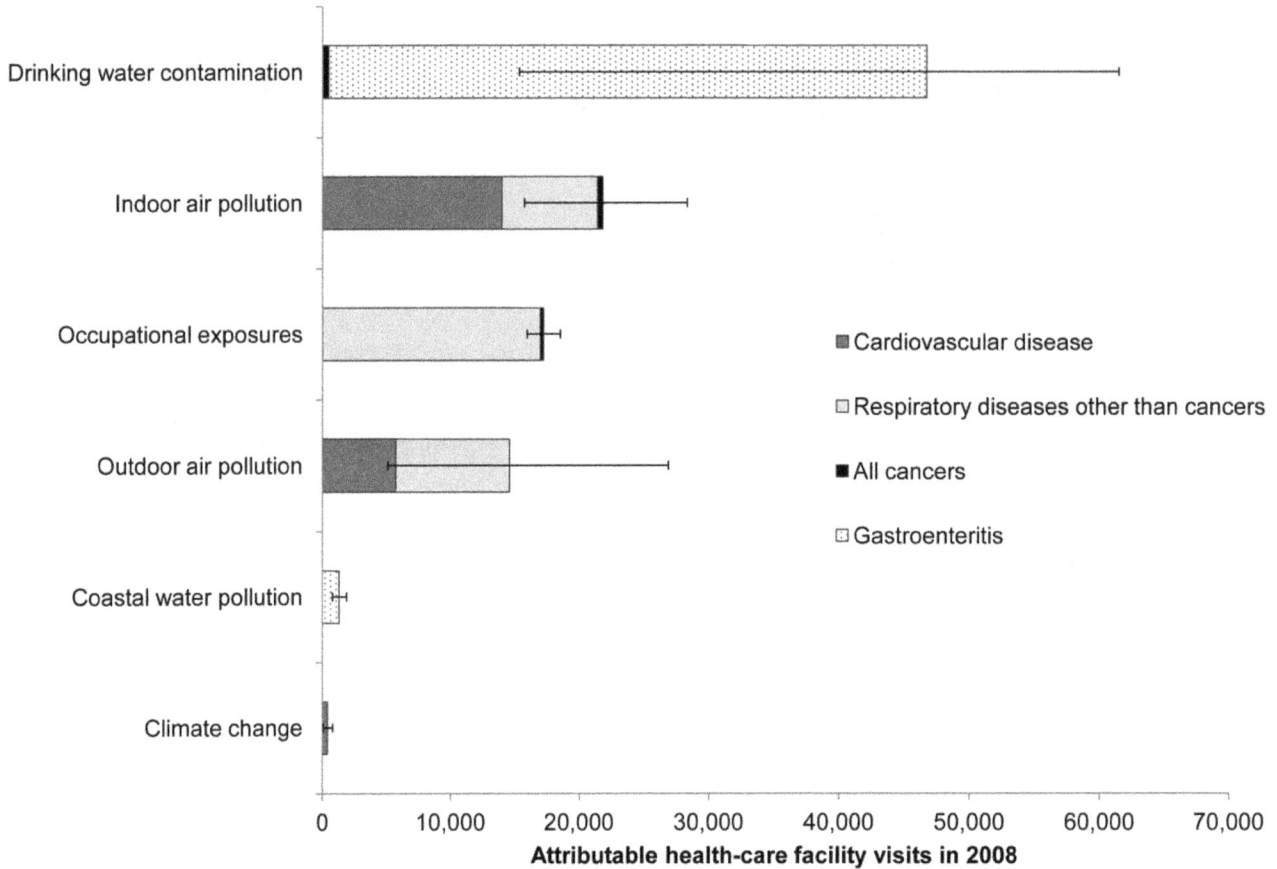

Figure 4. Estimated health-care facility visits attributable to environmental pollutants in the UAE.

"counterfactual") level and can be computed from equation 1 [12,13].

$$PAF = \frac{\int_{x=0}^{m} RR(x)P(x)dx - \int_{x=0}^{m} RR(x)P'(x)dx}{\int_{x=0}^{m} RR(x)P(x)dx}$$

$$= \frac{\sum_{i=1}^{n} RR(x_i)P(x_i) - \sum_{i=1}^{n} RR(x_i)P'(x_i)}{\sum_{i=1}^{n} RR(x_i)P(x_i)} \quad (1)$$

where x is the exposure level; $RR(x)$ is the relative risk at exposure level x; $P(x)$ is the population distribution of exposure; $P'(x)$ is the alternative or counterfactual distribution of exposure; m is the maximum possible exposure level; and n is some finite number of discrete exposure intervals. For this study, the counterfactual exposure level was the elimination of all pollutants to background levels. Background levels were assumed equal to zero for all exposure pathways except for outdoor air, for which background concentrations were represented as uniform distributions with parameters (10, 90) $\mu g/m^3$, (5, 35) $\mu g/m^3$, and (0, 25) ppb for PM_{10}, $PM_{2.5}$, and O_3, respectively, where the first parameter represents the lower bound and the second the upper bound of the distribution [14]. We estimated the EBD separately for each exposure route (i.e., we assumed all other exposure routes were unchanged while exposures via the given route were decreased). For the indoor air exposure route, mold, environmental tobacco smoke, and formaldehyde in indoor air all can exacerbate asthma in children; therefore, we estimated the total PAF for that exposure route from equation 2 [13].

$$PAF = 1 - \prod_{i=1}^{3} (1 - PAF_i) \quad (2)$$

where PAF_i is the PAF for pollutant i (mold, environmental tobacco smoke, or formaldehyde). Equation 2 assumes that the exposures to the three pollutants are uncorrelated and that the pollutants act independently in triggering asthma. The first assumption is reasonable in the case of these three pollutants. For example, a smoker's home is no more or less likely to contain radon or mold than a nonsmoker's home. On the other hand, the second assumption (biological independence) is problematic because evidence exists that these pollutants may act synergistically. For example, previous studies suggest that children exposed to both damp indoor environments (associated with mold growth) and environmental tobacco smoke have a greatly increased risk of developing asthma compared with the risk predicted by considering each of these exposures separately [15,16]. Hence, our estimates of asthma risks associated with indoor air pollutants may be underestimates.

Combining the estimated PAF with the observed total number of cases of the health outcome (D_{total}, from the fourth columns in Tables 1 and 2) gives the number of cases (deaths or health-care facility visits) attributable to the environmental exposure, D_{attrib}, as shown in Equation 3.

Table 3. Deaths attributable to environmental pollution risk factors in 2008.

Exposure route	Cause of mortality	Attributable fraction	Attributable deaths	Confidence interval lower bound	Confidence interval upper bound
Outdoor air pollution	All causes (adults>30)	7.3%	649	143	1,438
	Respiratory disease (children<5)	7.4%	2	0	6
	Total		651	143	1,444
Indoor air pollution	Cardiovascular disease	5.0%	115	50	178
	Lung cancer	31.7%	38[*]	14	55
	Total		153	85	216
Occupational exposures	Asthma	10.0%	1	1	1
	Chronic obstructive pulmonary disease	5.4%	2	NA	NA
	Asbestosis	NA	0	NA	NA
	Silicosis	NA	0	NA	NA
	Malignant mesothelioma	100.0%	6	NA	NA
	Leukemia	9.2%	12	5	22
	Lung cancer	20.8%	25	12	41
	Total		46	26	72
Drinking water contamination	Bladder cancer	7.5%	3	1	5
	Colon cancer	10.0%	6	0	12
	Rectal cancer	57.1%	3	0	6
	Gastroenteritis	12.8%	4	1	5
	Total		15	8	23
Climate change	Cardiovascular disease	0.1%	2	0	2

*An additional two deaths (95% CI 1–3) may be attributable to radon exposure in Abu Dhabi city and Sharjah.

$$D_{attrib} = PAF \times D_{total} \qquad (3)$$

The relative risk columns in Tables 1 and 2 show the parameters for $RR(x)$ in equation 1 (all based on previous international epidemiologic studies). Unless otherwise indicated, relative risks were characterized as lognormally distributed with the mean value and 95% confidence intervals drawn from the referenced studies. Relative risk estimates were selected by expert panels (Table S3) based on WHO guidance documents [17,18] and relevant epidemiologic literature.

The "exposure estimation method" columns in Tables 1 and 2 show how we estimated the population distribution of exposure $(P(x)$ in equation 1). Previously collected pollutant concentration data from EAD's environmental monitoring networks were available for outdoor air and coastal water, and we collected new measurements of indoor air quality [19]. However, we had to impute occupational exposures based on local employment data and previous studies in other regions for similar industrial sectors and job descriptions; details are provided elsewhere [20]. Similarly, we had to impute exposure to drinking water contaminants based on local information about sources of drinking water and access to improved water and sanitation services, using methods established in previous global EBD studies [21].

Pollutant concentrations in outdoor air can vary highly even in relatively small sub-regions. To represent this variability, we subdivided the UAE into 55 km² grid cells, each with different pollutant concentration parameters. Pollutant concentration parameters (mean and standard deviation) in each cell within the grid were estimated from a year's worth of outdoor air monitoring data provided by EAD. Li et al. [14] provide additional details on the monitoring network and statistical estimation techniques.

Due to the complexities of representing variability and uncertainty, previous EBD estimates generally have represented the input variables in equations 1–3 as deterministic–that is, as variables having fixed values. However, this representation can be misleading because these variables may take on any of a variety of values, causing the actual EBD experienced by a specific population to differ considerably from predictions derived only from a single fixed value for each input. For example, the concentration, x, of a contaminant to which an individual is exposed varies with time and location and also may be uncertain due to limitations in pollutant monitoring systems. Similarly, the relative risks associated with exposure $RR(x)$ also may vary by individual and be uncertain due to limitations in the epidemiologic studies from which they were estimated.

Morgan and Henrion [22], in their classic and widely recognized guide for incorporating technical and scientific uncertainty into risk analysis, observe Policies that ignore uncertainty about technology, and about the physical world, often lead in the long run to unsatisfactory technical, social, and political outcomes. By definition, risk involves an 'exposure to a chance of injury or loss' (Random House, 1966). The fact that risk inherently involves chance or probability leads directly to a need to describe and deal with uncertainty).

The EBD analysis presented here follows Morgan and Henrion's recommended protocols for incorporating variability and uncertainty into risk analysis. We represented input variables that are uncertain and/or subject to variability as random

Table 4. Health-care facility visits attributable to environmental pollution risk factors in 2008.

Exposure route	Health outcome	Attributable fraction	Attributable health-care facility visits	Confidence interval lower bound	Confidence interval upper bound
Outdoor air pollution	Cardiovascular disease	1.9%	5,700	1,910	10,500
	Respiratory disease	5.0%	8,850	2,930	17,300
	Total		14,600	5,090	26,900
Indoor air pollution	Asthma (<18) (environmental tobacco smoke and mold)	14.4%	3,510*	964	7,860
	Asthma (age≥18)	7.8%	2,541	63	4,730
	Cardiovascular disease	4.5%	13,940	9,620	18,200
	Lower respiratory tract infection (age≤6)	9.7%	1,360	710	1,970
	Leukemia	19.0%	289	44	477
	Lung cancer	29.3%	130**	38	195
	Total		21,800	15,700	28,300
Occupational exposures	Asthma	16.5%	11,900	10,500	13,100
	Chronic obstructive pulmonary disease	18.4%	5,010		
	Asbestosis	100.0%	3		
	Silicosis	100.0%	8		
	Malignant mesothelioma	89.3%	25		
	Leukemia	9.1%	138	57	255
	Lung cancer	25.3%	112	54	180
	Total		17,200	15,900	18,500
Drinking water contamination	Bladder cancer	16.6%	154	10	296
	Colon cancer	10.6%	232	0	569
	Rectal cancer	15.0%	96	0	219
	Gastroenteritis	57.0%	46,200	14,700	60,900
	Total		46,600	15,300	61,400
Coastal water pollution	Gastroenteritis	1.6%	1,300	792	1,880
Climate change	Cardiovascular disease	0.2%	410	84	802

*Formaldehyde exposure adds another 74 (95% confidence interval 50–99) visits for children under age 6, not included in this total. The total here accounts for the combined risks of ETS and mold exposure and assumes those risks are independent.
**An additional 20 visits (95% confidence interval 12–28) may occur due to radon exposure in Abu Dhabi city and Sharjah.

variables (that is, as probability distributions). We then employed the Monte Carlo simulation technique to propagate the variability and uncertainty in input parameters through the calculations. In brief, equations 1–3 were computed thousands of times, each time using a new selection of input variable values drawn from the appropriate probability distributions; for a further explanation of this method, which is widely employed in risk assessment, see Morgan and Henrion [22] as well as Thomopoulos [23].

To implement the Monte Carlo simulation, we employed *Analytica* software (Lumina Decision Systems, Los Gatos, California), which was developed specifically to enable the modular construction of simulation models. Within *Analytica*, we specified appropriate probability distributions (or deterministic values where appropriate) for the input variables for each combination of contaminant and exposure pathway. We added nodes that carry out the computations represented by equations 1–3. *Analytica* then uses a method known as Latin hypercube sampling in order to compute a probability distribution for the results of equations 1–3 for each contaminant and exposure pathway based on samples drawn at random from the probability distributions for the input

variables. For these estimations, we used 1,000 iterations (i.e., 1,000 samples of each input variable), which yielded stable results. The resulting estimates indicate not only the most likely number of diseases attributable to each contaminant-exposure pathway combination but also the range of plausible values, given the uncertainty and variability in existing information.

Figure 1 shows the front page of the UAE EBD model. From within the *Analytica* program, double clicking on the name of an exposure route opens further layers, such as the layer illustrated in Figure 2. These lower layers further document the input variables and relationships among them. Figure 2 shows the bottom layer of the module for estimating the EBD due to $PM_{2.5}$ in outdoor air. From within *Analytica*, double clicking on any variable icon opens a display that completely specifies the variable (e.g., observational data, parameters of the underlying probability distribution). An advantage of *Analytica* is its ability to handle very large matrix operations, which enabled us to divide the UAE into geographic subunits smaller than a single emirate for purposes of analysis when sufficient exposure data were available; each geographic subunit is represented as a row of a matrix, and population,

Annual attributable deaths per 100,000 population

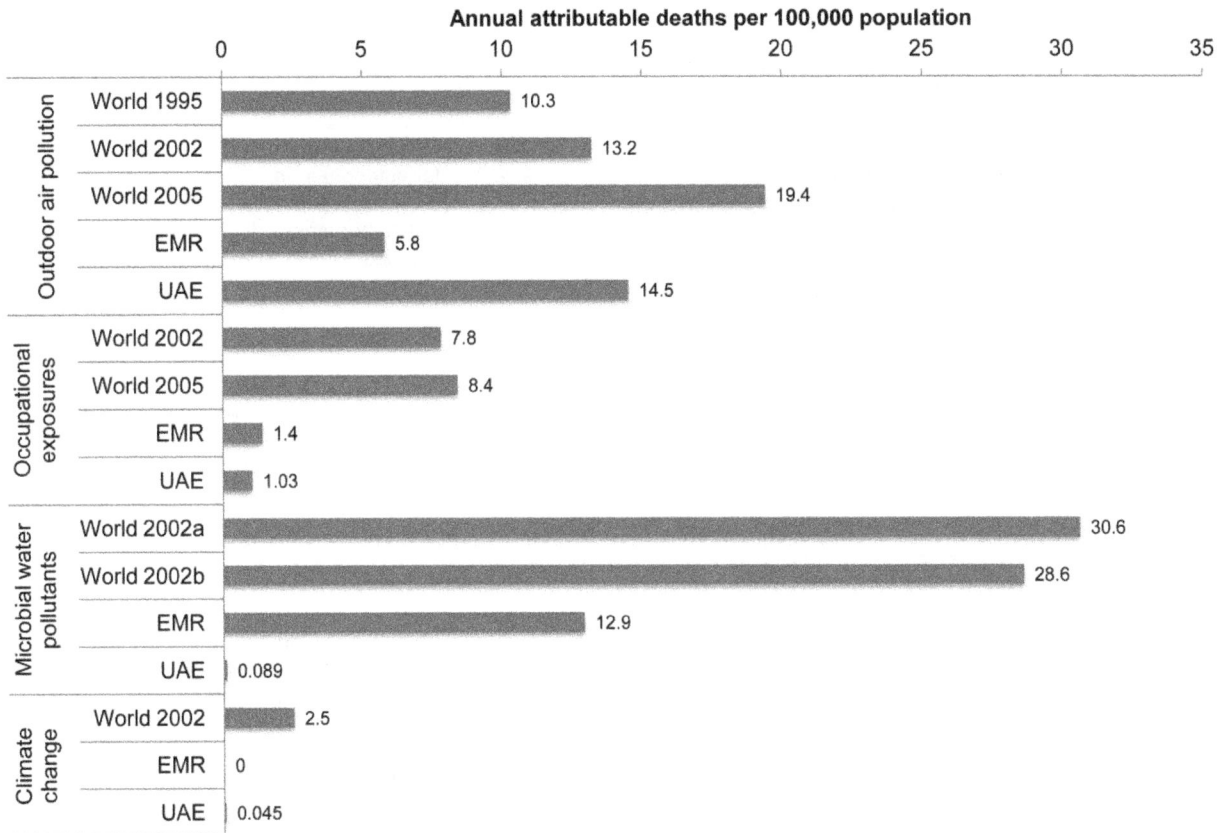

Figure 5. Comparison of annual deaths attributed to environmental pollutants in the UAE with previous global and regional estimates. See Table S2 for sources of international comparison estimates.

baseline disease, and exposure information are assigned accordingly. A further advantage is that the visual construction facilitates communicating the methods to nonspecialists.

Results

Figures 3 and 4 illustrate the estimated deaths and health-care facility visits attributable to the environmental exposure routes considered in this analysis. Tables 3 and 4 provide detailed results, including the PAF for each exposure route–health endpoint combination.

As Figure 3 indicates, outdoor air pollution contributed to 651 deaths (95% CI 143–1,440) in 2008–more than all the other risk factors combined. Indoor air pollution and occupational exposures were the second and third leading contributors to mortality, with 153 (95% CI 85–216) and 46 deaths (95% CI 26–72) attributable to these factors, respectively.

The estimated health-care facility visits attributable to environmental pollution show a different prioritization (Figure 4). Drinking water pollution was the leading contributor, with an estimated 46,600 (95% CI 15,300–61,400) attributable visits (approximately 15% of the health-care facility visits for all the illnesses considered in this study). Indoor air pollution, occupational exposures, and outdoor air pollution contributed to similar numbers of visits (15,000–20,000 each). While these results present a somewhat different ordering than those shown for deaths in Figure 3, it is important to note that the severity of the illnesses prompting these health-care facility visits varies considerably. For example, a heart attack (to which indoor and outdoor air pollution

are important risk contributors) is much more severe than a case of mild gastroenteritis associated with drinking water contamination, especially in a wealthy nation such as the UAE, where child mortality is very low and medical assistance is readily available. Previous EBD studies have typically reported results as disability-adjusted life years (DALYs), representing the number of healthy life years lost due to living in a state of less than perfect health or dying prematurely [17], but the data necessary to translate health-care facility visits into DALYs were unavailable in this case. Future studies using DALYs as a unit of measure would account for such differences and provide a single prioritized ranking.

Figure 5 compares our estimates of the per-capita attributable deaths to previous global estimates and the WHO's estimates for the Eastern Mediterranean Region for environmental risk factors for which comparable information is available. As shown, the UAE's environmental burden of disease is low for all the factors for which relevant comparisons were available except for outdoor air pollution (see Table S2 for additional details on previous studies).

Discussion

These results provide an initial indication of the public health benefits possible in the UAE through new interventions to reduce environmental pollution. Reducing pollutant concentrations in outdoor and indoor air and protecting workers from occupational pollutant exposures could prevent hundreds of premature deaths and tens of thousands of health-care facility visits each year. In addition, although the risk of gastrointestinal illnesses due to water pollution is very low in the UAE by global standards, additional

Limited number of pollutants
While capturing pollutants that are important exposure indicators for each exposure route, this analysis nonetheless neglects the vast array of other anthropogenic pollutants to which people may be exposed.

Limited pollutant concentration data
At the time this study was conducted, pollutant measurements were available only for outdoor air and coastal water, and even these measurements were extremely limited in number. We collected additional data on indoor air quality. Lacking were drinking water quality data as measured at consumer taps and measured occupational exposure data; hence, exposures for these routes had to be imputed from other information (see Tables 1 and 2). The estimates presented here should be updated once measured exposure data are available.

Lack of local relative risk estimates
Relative risks were estimated from international epidemiologic studies since such studies are lacking in the UAE. Although the studies we used are those commonly employed in international EBD studies, UAE-specific environmental epidemiologic studies would indicate whether the UAE population response to pollutants is similar to that observed elsewhere in the world.

Extrapolation of mortality and morbidity estimates from Abu Dhabi emirate to the rest of the UAE
We were able to obtain detailed mortality and medical records only for Abu Dhabi emirate. We extrapolated death and health-care facility visit rates to other emirates, accounting for differences in proportions of nationals and expatriates and in genders by emirate. However, if population characteristics other than nationality and gender differ substantially in other emirates from those in Abu Dhabi, then baseline death and health-care facility visit rates estimated for the other emirates could be biased. A comparison of detailed medical records for the different emirates with those of Abu Dhabi would reveal whether such biases exist, but such a comparison was not possible for the analysis presented here.

Lack of epidemiologic evidence of synergistic effects of exposures to multiple contaminants
While the relative risk estimates we used control for smoking, they generally do not control for the potential for exposure to multiple contaminants through multiple pathways. It is possible that contaminants may have synergistic effects, so that, for example, exposure to multiple contaminants increased susceptibility to health endpoints. The epidemiologic evidence base is not yet sufficient to consider such synergistic effects, in most cases. Further, adding such effects to the model would require information about the joint distribution of exposures to all the contaminants considered. If contaminants act in synergy, then the estimates in this analysis may be too low.

Figure 6. Key sources of uncertainty.

measures to improve drinking water and coastal water quality could prevent a large fraction of mild gastrointestinal illnesses.

In many developed countries, a fear of cancer has driven programs to reduce environmental pollution [24]. It is interesting to note that environmental pollution is associated with a far larger number of deaths for noncancer health endpoints, especially cardiovascular disease, which is the leading cause of death in the UAE.

These results provide one indication of the relative impacts environmental pollutants on health in the UAE and should not be considered as exact. Indeed, the large confidence intervals (Figures 3 and 4) indicate a high level of uncertainty. Figure 6 lists important sources of uncertainty.

In addition to these uncertainties, a further limitation of the study is the inability to combine deaths and health-care facility visits into a single summary measure, such as DALYs, which would facilitate comparisons among risk factors. The information needed to calculate DALYs for the UAE population was unavailable for this study. Most importantly, calculating DALYs requires that each illness experienced by a member of the population is counted only once. However, the data available to us listed each visit to a health-care facility regardless of whether the visit was a repeat encounter for a previously diagnosed illness. To protect patient confidentiality and due to the evolving nature of the UAE's medical record systems, we were not provided with information that would have enabled us to screen out these multiple visits. Hence, one respiratory illness case would be counted four times in the database if the same patient returned to the doctor three times after the initial visit. Hence, had we attempted to calculate DALYs, the results would have been inflated due to such repeat visits. Although it was not possible to compute DALYs, the indicators used in this research (deaths and health-care facility visits) nonetheless provide valuable information about the relative impacts of different types of environmental pollution in the UAE from a public health perspective.

Despite these limitations, this study builds on methods that the WHO has advocated over the past decade in order to support national efforts to reduce environmental risks to health [6,18]. The results yield information about opportunities to prevent illnesses and reduce health-care spending through new environmental interventions. Indeed, in part as a result of this study, Abu Dhabi is already pursuing a number of new programs to reduce environmental pollution exposures, including upgrading its outdoor air monitoring system in order to plan for air quality improvements, establishing a new occupational health management system, implementing an aggressive marine water quality monitoring program, and instituting new anti-smoking campaigns.

Supporting Information

Table S1 Population distribution of the UAE by emirate and gender (2008).

Table S2 Previous estimates of deaths attributable to environmental pollution.

Table S3 Experts contributing to burden of disease analyses for each exposure route.

Acknowledgments

Special thanks are due to H.E. Majid Al Mansouri, Dr. Amir Johri, and Dr. M. Z. Ali Khan for being the original champions of this project and to H.E. Razan Al Mubarak for her continued support. This report represents research and data contributions from a large number of scientists, including the following: Saravanan Arunachalam, Gregory W. Characklis, Leslie Chinery, Christopher A. Davidson, Zeinab Farah, Mohammed Zuber Farooqui, Tiina Folley, Mejs Hasan, Prahlad Jat, Leigh-Anne H. Krometis, Ying Li, Joseph N. LoBuglio, Leena A. Nylander-French, Gavino Puggioni, Marc Serre, Kenneth G. Sexton, Uma Shankar, William Vizuete, and J. Jason West. This manuscript was copy edited and prepared for publication by Angela S. Brammer, a freelance editor and layout designer.

Author Contributions

Conceived and designed the experiments: JMG. Performed the experiments: JMG EH ND. Analyzed the data: JMG EH ND. Contributed reagents/materials/analysis tools: FL JT. Wrote the paper: JMG JT FL EH.

References

1. Walters TN, Kadragic A, Walters LM (2006) Miracle or mirage: Is development sustainable in the United Arab Emirates? Middle East Rev 10: 77–91.
2. United Nations Development Program (2011) Human development report. New York.
3. World Bank (2001) World Databank. Available: http://data.worldbank.org/country/united-arab-emirates. Accessed 2011.
4. Davidson C (2005) The United Arab Emirates: A study in survival. Boulder, Colo.: Lynne Rienner Publishers.
5. World Health Organization (2009) Country profiles of environmental burden of disease: United Arab Emirates. Geneva: World Health Organization. Available: http://www.who.int/quantifying_ehimpacts/national/countryprofile/unitedarabemirates.pdf. Accessed 31 January 2013.
6. The Lancet (2007) The environment's impact on health. Lancet 369: 2052.
7. Willis HH, MacDonald Gibson J, Shih R, Geschwind S, Olmstead S, et al. (2010) Prioritizing environmental health risks in the UAE. Risk Anal 30: 1842–1856.
8. MacDonald Gibson J, Farah ZS (2012) Environmental risks to public health in the United Arab Emirates: A quantitative assessment and strategic plan. Environ Health Perspect 120: 681–686.
9. Kapiszewski A (2001) Nationals and expatriates: Population and labor dilemmas of the Gulf Cooperation Council states. Reading, United Kingdom: Ithaca Press and Garnet Publishing Ltd.
10. Vetter P, Boecker K (2012) Benefits of a single payment system: Case study of Abu Dhabi health system reforms. Health Policy (Amsterdam, Netherlands) 108: 105–114.
11. Koornneef EJ, Robben PBM, Al Seiari MB, Al Siksek Z (2012) Health system reform in the emirate of Abu Dhabi, United Arab Emirates. Health Policy (Amsterdam, Netherlands) 108: 115–21.
12. Murray CJL, Ezzati M, Lopez AD, Rodgers A, Van der Hoorn S (2003). Comparative quantification of health risks: Conceptual framework and methodological issues. Popul Health Metr 1: 1–20.
13. Ezzati M, Van der Hoorn S, Rodgers A, Lopez AD, Mathers CD, et al. (2003) Estimates of global and regional potential health gains from reducing multiple major risk factors. Lancet 362: 271–280.
14. Li Y, MacDonald Gibson J, Jat P, Puggioni G, Hasan M, et al. (2010) Burden of disease attributed to anthropogenic air pollution in the United Arab Emirates: Estimates based on observed air quality data. Sci Total Environ 408: 5784–5793.
15. Andrae S, Axelson O, Bjorksten B, Fredriksson M, Kjellman NI (1988) Symptoms of bronchial hyperreactivity and asthma in relation to environmental factors. Arch Dis Child 63: 473–478.
16. Lindfors A, Wickman M, Hedlin G, Pershagen G, Rietz H, et al. (1995) Indoor environmental risk factors in young asthmatics: A case-control study. Arch Dis Child 73: 408–412.
17. Mathers CD, Vos T, Lopez AD, Salomon J, Ezzati M (2001) National burden of disease studies: A practical guide. Geneva: World Health Organization.
18. Prüss-Üstün A, Mathers CD, Corvalan C, Woodward A (2003) Introduction and methods: Assessing the environmental burden of disease at national and local levels. Geneva: World Health Organization.
19. Yeatts KB, El-Sadig M, Leith D, Kalsbeek W, Al-Maskari F, et al.(2012) Indoor air pollutants and health in the United Arab Emirates. Environ Health Perspect 120: 687–694.
20. Folley TJ, Nylander-French LA, Joubert DM, MacDonald Gibson J (2012) Estimated burden of disease attributable to selected occupational exposures in the United Arab Emirates. Am J Ind Med doi:10.1002/ajim.22043.
21. Fewtrell LA, Prüss-Üstün A, Bartram J, Bos R (2007) Water, sanitation and hygiene: Quantifying the health impact at national and local levels in countries with incomplete water supply and sanitation coverage. Public Health 15: 1–15.
22. Morgan MG, Henrion M, Small M (1990) Uncertainty: A guide to dealing with uncertainty in quantitative risk and policy analysis. Cambridge, New York: Cambridge University Press.
23. Thomopoulos NT (2013) Essentials of Monte Carlo simulation: Statistical methods for building simulation models. New York: Springer.
24. Trumbo CW, McComas K, Kannaovakun P (2007) Cancer anxiety and the perception of risk in alarmed communities. Risk Anal 27: 337–350.
25. Pope CA III, Burnett RT, Thun MJ, Calle EE, Krewski D, et al. (2002) Lung cancer, cardiopulmonary mortality, and long-term exposure to fine particulate air pollution. JAMA 287(9): 1132–1141.
26. Ostro B (2004) Outdoor air pollution: Assessing the environmental burden of disease at national and local levels. Environmental Burden of Disease Series, No. 5. Geneva: World Health Organization.
27. Hill SE, Blakely T, Kawachi I, Woodward A (2007) Mortality among lifelong nonsmokers exposed to secondhand smoke at home: cohort data and sensitivity analyses. Am J Epidemiol 165: 530–540.
28. Cardenas VM, Thun MJ, Austin H, Lally CA, Clark WS, et al. (1997) Environmental tobacco smoke and lung cancer mortality in the American Cancer Society's Cancer Prevention Study. II. CCC 8: 57–64.
29. Darby S, Hill D, Deo H, Auvinen A, Barros-Dios JM, et al. (2006) Residential radon and lung cancer–detailed results of a collaborative analysis of individual data on 7148 persons with lung cancer and 14,208 persons without lung cancer from 13 epidemiologic studies in Europe. Scand J Work Environ Health 32 Suppl 1: 1–83.
30. Friborg JT, Yuan JM, Wang R, Koh WP, Lee HP, et al. (2008) Incense use and respiratory tract carcinomas: A prospective cohort study. Cancer 113: 1676–1684.
31. McMichael AJ, Campbell-Lendrum D, Kovats S, Edwards S, Wilkinson P, et al. (2004) Global climate change. In: Ezzati M, Lopez AD, Rodgers A, Murray CJL, eds. Comparative quantification of health risks: Global and regional burden of disease attributable to selected major risk factors, vol. 1. Geneva, World Health Organization. 1543–1649.
32. Morris RD, Audet AM, Angelillo IF, Chalmers TC, Mosteller F (1992) Chlorination, chlorination by-products, and cancer: A meta-analysis. Am J Public Health 82: 955–963.
33. Prüss-Üstün A, Bos R, Gore F, Bartram J (2008) Safer water, better health: Costs, benefits and sustainability of interventions to protect and promote health. Geneva: World Health Organization.
34. Wilk R (2006) Bottled water: The pure commodity in the age of branding. J Consumer Culture 6: 303–325.
35. (2009) UAE leads bottled water industry. Food and Drink Insight. Available http://www.foodanddrinkinsight.com/file/77640/uae-leads-bottled-water-industry.html. Accessed 25 May 2012.
36. Antova T, Pattenden S, Brunekreef B, Heinrich J, Rudnai P, et al. (2008) Exposure to indoor mould and children's respiratory health in the PATY study. J Epidemiol Community Health 62: 708–714.
37. Vork KL, Broadwin RL, Blaisdell RJ (2007) Developing asthma in childhood from exposure to secondhand tobacco smoke: insights from a meta-regression. Environ Health Perspect 115: 1394–1400.
38. Jaakkola MS, Nordman H, Piipari R, Uitti J, Laitinen J, et al. (2002) Indoor dampness and molds and development of adult-onset asthma: A population-based incident case-control study. Environ Health Perspect 110: 543–547.

39. Rumchev K, Spickett J, Bulsara M, Phillips M, Stick S (2004) Association of domestic exposure to volatile organic compounds with asthma in young children. Thorax 59: 746–751.

40. He J, Whelton PK (1999) Passive cigarette smoking increases risk of coronary heart disease. Eur Heart J 20: 1764–1765.

41. Li JS, Peat JK, Xuan W, Berry G (1999) Meta-analysis on the association between environmental tobacco smoke (ETS) exposure and the prevalence of lower respiratory tract infection in early childhood. Pediatr Pulmonol 27: 5–13.

42. Kasim K, Levallois P, Abdous B, Auger P, Johnson KC (2005) Environmental tobacco smoke and risk of adult leukemia. Epidemiol 16: 672–680.

43. Wade TJ, Calderon RL, Sams E, Beach M, Brenner KP, et al. (2006) Rapidly measured indicators of recreational water quality are predictive of swimming-associated gastrointestinal illness. Environ Health Perspect 114: 24–28.

44. Badrinath P, Al-Shboul QA, Zoubeidi T, Gargoum AS, El-Rufaie OE (eds.) (2002) Measuring the health of the nation: United Arab Emirates health and lifestyle survey 2000. Al Ain: UAE University, Faculty of Medicine and Health Sciences and College of Business and Economics.

45. Australian Sports Commission (2011) Participation in exercise, recreation, and sport: 2010 annual report. Sydney: Government of Australia, Standing Committee on Recreation and Sport.

4

Acute and Chronic Effects of Particles on Hospital Admissions in New-England

Itai Kloog[1]*, Brent A. Coull[2], Antonella Zanobetti[1], Petros Koutrakis[1], Joel D. Schwartz[1]

1 Exposure, Epidemiology and Risk Program, Department of Environmental Health, Harvard School of Public Health, Boston, Massachusetts, United States of America, 2 Department of Biostatistics, Harvard School of Public Health, Boston, Massachusetts, United States of America

Abstract

Background: Many studies have reported significant associations between exposure to $PM_{2.5}$ and hospital admissions, but all have focused on the effects of short-term exposure. In addition all these studies have relied on a limited number of $PM_{2.5}$ monitors in their study regions, which introduces exposure error, and excludes rural and suburban populations from locations in which monitors are not available, reducing generalizability and potentially creating selection bias.

Methods: Using our novel prediction models for exposure combining land use regression with physical measurements (satellite aerosol optical depth) we investigated both the long and short term effects of $PM_{2.5}$ exposures on hospital admissions across New-England for all residents aged 65 and older. We performed separate Poisson regression analysis for each admission type: all respiratory, cardiovascular disease (CVD), stroke and diabetes. Daily admission counts in each zip code were regressed against long and short-term $PM_{2.5}$ exposure, temperature, socio-economic data and a spline of time to control for seasonal trends in baseline risk.

Results: We observed associations between both short-term and long-term exposure to $PM_{2.5}$ and hospitalization for all of the outcomes examined. In example, for respiratory diseases, for every10-$\mu g/m^3$ increase in *short-term* $PM_{2.5}$ exposure there is a 0.70 percent increase in admissions (CI = 0.35 to 0.52) while concurrently for every10-$\mu g/m^3$ increase in *long-term* $PM_{2.5}$ exposure there is a 4.22 percent increase in admissions (CI = 1.06 to 4.75).

Conclusions: As with mortality studies, chronic exposure to particles is associated with substantially larger increases in hospital admissions than acute exposure and both can be detected simultaneously using our exposure models.

Editor: Mike B. Gravenor, University of Swansea, United Kingdom

Funding: The research was supported by the Harvard Environmental Protection Agency (EPA) Center, Grants R-832416 and RD 83479801, NIH grants ES00002 and ES012044. The funders had no role in study design, data collection and analysis, decision to publish, or preparation of the manuscript.

Competing Interests: The authors have declared that no competing interests exist.

* E-mail: ekloog@hsph.harvard.edu

Introduction

Short-term variations in air pollution have been associated with hospital admissions for various causes in cities all over the world [1,2,3,4,5,6]. These associations include admissions for respiratory disease [7,8,9], ischemic heart disease-IHD [10,11], cardiovascular disease-CVD [7,12], myocardial infarction-MI [13,14], congestive heart failure-CHF [15,16], pneumonia [17,18], and diabetes [19,20].

For $PM_{2.5}$ in particular Dominici and colleagues [21] reported associations with hospitalizations for multiple diseases, using single day average $PM_{2.5}$. Zanobetti and colleagues [22] estimated the association between two-day mean $PM_{2.5}$ and emergency hospital admissions for CVD,MI,CHF, respiratory disease, and diabetes in 26 US communities, and reported larger effect sizes than those reported in Dominici et al. [21]. There are currently, to the best of our knowledge, no published studies on the effects of long-term (chronic) particulate matter (PM) exposure and hospital admissions. There are however some studies that provide general evidence for long-term associations of air pollution with hospital admissions, although not specifically focusing on $PM_{2.5}$

[23,24,25,26]. For example, Oudin and colleagues [25] investigated whether the effects of major risk factors for ischemic stroke were modified by long-term exposure to air pollution in Scania, southern Sweden. They found that in low level air pollution areas, the risk for ischemic stroke associated with diabetes seemed to increase with long-term exposure to air pollution. Hruba and colleagues [24] studied the effects of long-term exposure to air pollution on respiratory symptoms and respiratory hospitalization in a cross-sectional study of children. They showed found a significant increase in hospital admissions for asthma, bronchitis or pneumonia associated with increasing air pollution. Andersen and colleagues [26] studied the association between chronic exposure to traffic-related air pollution (NO_2) and incidence of diabetes. They found that chronic exposure to NO_2 may contribute to the development of diabetes, especially in individuals with a healthy lifestyle, nonsmokers, and physically active individuals.

All previous studies have been limited by the lack of high resolution daily exposure data. Many early studies had only 1 in 3 day measurements, and locations without nearby monitors could not be analyzed at all. In addition all previous studies focused on short-term PM exposure and not long term (chronic) exposure or both.

We have recently presented a new method of assessing temporally-and spatially-resolved $PM_{2.5}$ exposures for epidemiological studies which is an extension of existing land use models [27,28]. In this paper, we use our model predictions to study the association between $PM_{2.5}$ exposure and hospital admissions among elderly (aged 65 and older from Medicare data) across New England, and to investigate the effects of both short term (acute) and long-term (chronic) exposure on these outcomes for the first time concurrently. In addition our study investigates the entire population of a region, rather than selected locations near monitoring sites as commonly done in previous studies.

Methods

Study domain

The presented study's spatial domain included the New-England region comprising the states of Connecticut, Maine, Massachusetts, New Hampshire, Rhode Island and Vermont, (Figure 1). The total area of New England is 186,460 km^2. The total population in New-England as of 2010 is 14,444,865. The average size of population in New-England zip codes for the general population is 8130 and 1105 for people 65 and over. The median population is 3535 for the general population and 430 for people 65 and over [29].

Data

Exposure data. Land use regression (LUR) models provide good estimates of spatially resolved long term exposures, but are poor at capturing short term exposures. Due to its large spatial coverage and reliable repeated measurements, satellite remote sensing, provides another important tool for monitoring aerosols, particularly for areas and exposure scenarios where surface $PM_{2.5}$ monitors are not available [30,31,32,33]. Using satellite derived aerosol optical depth (AOD) measurements allowed us to predict daily $PM_{2.5}$ concentration levels across New England for 2000–2008 at a 10×10 km spatial resolution [34]. This published model has been slightly updated to include nested regions in the yearly models and weights to account for non-random missingness in AOD.

In brief, we used day-specific calibrations of AOD data, using ground $PM_{2.5}$ measurements from 78 monitoring sites in the EPA (Environmental Protection Agency) and IMPROVE (Interagency Monitoring of Protected Visual Environments) monitoring network to avoid prediction error due to changes in planetary boundary layer etc. previously noted by Paciorek et al. [28]. We also incorporated land use regression and meteorological variables (temperature, wind Speed, visibility, elevation, distance to major roads, percent of open space, point emissions and area emissions). To estimate $PM_{2.5}$ concentrations in each grid cell on each day we start by calibrating the AOD-$PM_{2.5}$ relationship for each day using grid cells with both monitors and AOD values using mixed models with random slopes for day and nested regions. To validate our first model, the dataset was repeatedly randomly divided into 90% and 10% splits. Predictions for the held-out 10% of the data were made from the model fit of the remaining 90% of the data. This "out of sample" process was repeated ten times and cross-validated (CV) R^2 values were computed. The first stage calibrations resulted in high out-of-sample R^2 (mean out-of-sample $R^2 = 0.85$). Later, we used a second model to address days when *AOD measures are not available* (due to cloud coverage, snow etc...). We thus fit a model with a smooth function of latitude and longitude and a random intercept for each cell (similar to universal kriging) that takes advantage of the association of grid cells AOD values with $PM_{2.5}$ monitoring located elsewhere, and the association with available AOD values in neighboring grid cells. Even for location-day combinations without AOD data our model performance was still excellent (mean out-of-sample $R^2 = 0.81$). Importantly, these R^2 are for daily observations, rather than monthly or yearly, values. By averaging our estimated daily exposures at each location we generated long term exposures. This enabled us to study both the short term and long term effects of ambient particles, respectively.

$PM_{2.5}$ exposure data were generated by our prediction models. The New-England exposure dataset contains daily $PM_{2.5}$ concentrations at a 10×10 km spatial resolution across New-England for the whole study period (Figure 1). This data was matched to zipcodes using ArcGIS and SAS based on spatial location and date.

Hospital Admittance data. Individual hospital admittance records were obtained from the US Medicare program and covers hospitalization for all residents aged 65 and older, for all available years (2000–2006). There were around 3000 hospitals under the study area. We defined cases as those with an emergency admission and a primary discharge diagnosis of all respiratory (ICD 9 460–519), CVD (ICD 9 390–429), stroke (ICD 9 430–436) and diabetes (both primary and secondary admission cause) (ICD 9 250).

We choose broader areas of admissions, since one would expect broader areas of admission to produce less noisy estimates for two reasons. First, the counts are higher and therefore there is more power to examine CVD admissions than IHD admissions. Secondly, studies of misdiagnosis in hospital administrative records show that the broader the categories, the less misclassification there is, which would also eliminate noise and produce more stable results. For diabetes, which is a chronic condition, we looked at the rate of admission of subjects for any primary cause with diabetes as a secondary cause, as well as the small number of admissions with diabetes listed as the primary cause of admission. This allows us to examine whether long term exposure to particles is associated with higher rates of hospitalization of diabetics, as well as whether diabetics have higher rates of acute hospitalizations on high air pollution days.

These records included information such as age, sex, date of admission, race/ethnicity, and zipcode of residence. From this data, we constructed daily counts for each admission cause for each zip code. This allows us to examine the effects of both day-to-day contrasts within residential area, as well as long term contrasts across locations.

Covariates. Temperature data were obtained through the National Climatic Data Center (NCDC) [35]. Only continuous operating stations with daily data running from 2000–2006 were used. Zipcodes were matched to the closest weather station for meteorological variables. All *Socioeconomic* variables were obtained through the U.S. Census Bureau Census from the 2000 social, economic and housing characteristics datasets [36]. *Socio-economic* variables used included the following zipcode level information: Percent of minorities, age, education (people with no high school education) and median income.

Statistical Methods

The admission counts by zip code were matched with our exposure estimates for each 10×10 km grid cell it fell into. While short-term effects of air pollution are traditionally studied using Poisson log-linear models and long-term effects are estimated using the Cox proportional hazard model, we make use of the equivalence between Poisson regression and the piecewise constant

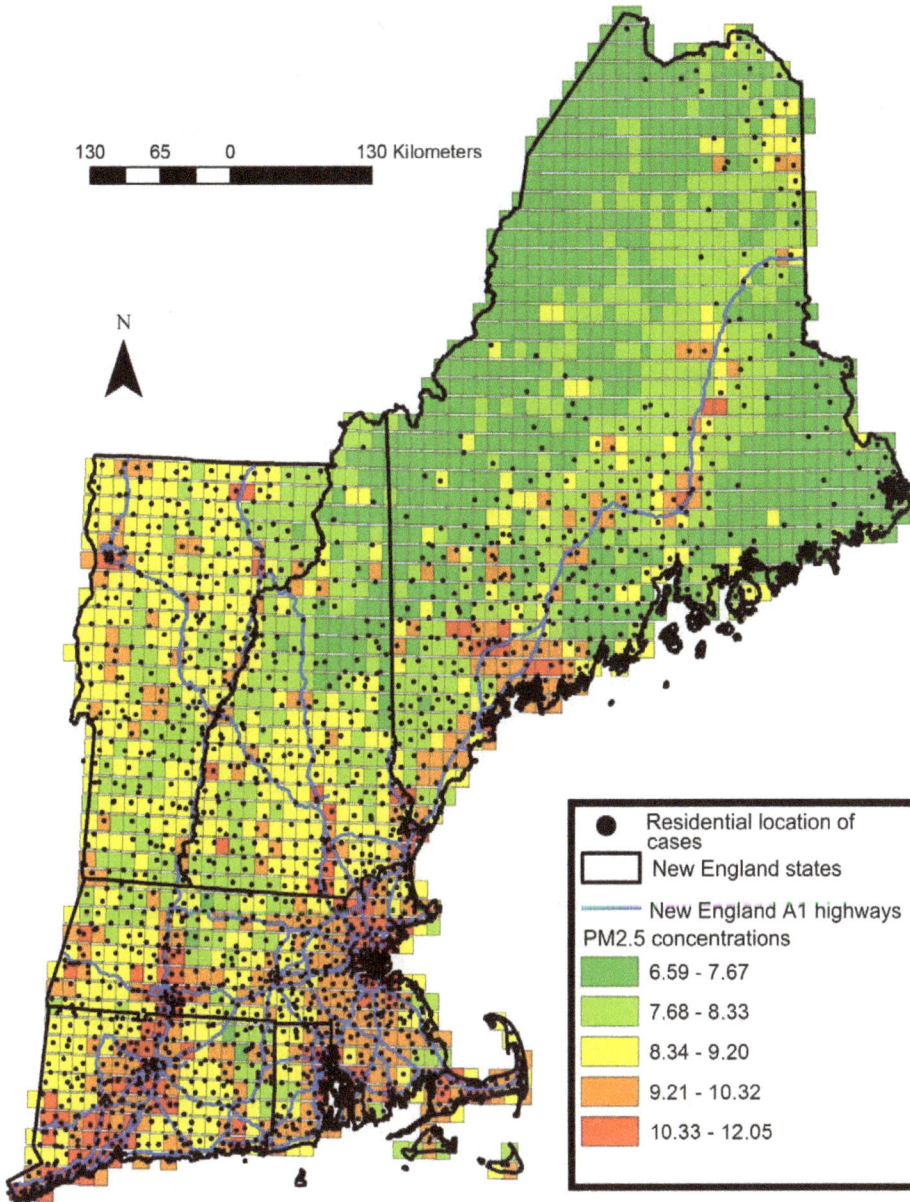

Figure 1: Map of the study area showing the residential location of admission cases juxtaposed over a sample PM2.5 10×10 km pollution grid for 01/07/2001.

proportional hazard model noted by Laird and Oliver [37]. This approach allows us to model the time to a hospital admission as a function of both long term and short term exposure simultaneously. Most time series studies have reported stronger associations with mean $PM_{2.5}$ taken over the current and previous day as compared to same day exposure [38]; therefore for the short-term exposure we used the mean of $PM_{2.5}$ on the day of admission and day before admission in all models. Long-term exposure was calculated as the mean exposure in each zip-code across the whole study period (7 years). Short term exposure was defined as the difference between the two day average and the long-term average. To check the linearity of main effects investigated we fit a piecewise linear model estimating the effect of PM for levels below and above the median of short and long term $PM_{2.5}$. We did not find a significantly different effect between the two slopes

above and below the median which suggests a linear relationship of these variables.

The basic model takes advantage of the fact that a hierarchical mixed Poisson regression can capture both acute and chronic effects. Specifically, we assume that the admission rate λ_{it} in the ith cell on the tth day can be modeled as follows:

$$\log(\lambda_{it}) = \lambda_i + \beta_1 \Delta PM_{it} + \lambda(t) + temporal \; \mathrm{co}variates$$

$$where$$

$$\lambda_i = \delta + \gamma PM_{i.} + spatial \; \mathrm{co}variates + e_i$$

where λ_i is the long term admission rate in grid cell i, ΔPM_{it} is the deviation of the $PM_{2.5}$ concentration in cell i from its long term average on day t, $\lambda(t)$ is a smooth function of time, temporal

Table 1. Descriptive statistics stratified by long term exposure: Hospital admissions by type of admission across New-England for the years 2000–2006.

Characteristic	All Respiratory	CVD	Stroke	Diabetes
	No. (%)	No. (%)	No. (%)	No. (%)
Low pollution				
Sex				
Male	89241 (44.63)	131234(45.52)	24066(41.71)	77553(43.59)
Female	11073 (55.37)	157039(54.48)	33638(58.29)	100382(56.41)
Race				
White	192257 (94.41)	277404(96.23)	55112(95.51)	165174(92.83)
Black	3321 (1.66)	4885(1.69)	1186(2.06)	6339(3.56)
other	4395 (2.20)	5984(2.08)	1406(2.44)	6422(3.61)
Age	79.55	79.24	80.30	77.24
High pollution				
Sex				
Male	101629(44.52)	148566(44.55)	27516(40.66)	93918(42.56)
Female	126658 (55.48)	184948(55.45)	40162(59.34)	126743(57.44)
Race				
White	213519 (93.53)	312202(93.61)	62741(92.71)	194360(88.08)
Black	7672 (3.36)	11920(3.57)	2920(4.31)	15682(7.11)
other	7096 (3.11)	9392(2.82)	2017(2.98)	10619(4.81)
Age	79.64	79.31	80.27	77.26

covariates are temperature and day of the week, PM_i is the long term average PM concentration in cell i, spatial covariates are socioeconomic factors defined at the zipcode level, and e_i is the remaining unexplained difference in admission rate between cell i and other cells, which is treated as a mean zero normal random effect with variance estimated from the data. This model combines the usual Poisson time series analysis with a Poisson representation of a piecewise-constant proportional hazards model. The resulting model specifies that each time interval defining the constant hazards has a separate intercept, and an offset representing person-time at risk. Since the entire population is being analyzed, and specific admissions cases by type are rare events, the person-time at risk varies slowly and smoothly across time. In the limit as the time interval gets small, the time-period specific intercept also approaches a smooth function of time, and hence both can be replaced with the smooth function of time, $\lambda(t)$.

The specific covariates we used were temperature with the same moving average as $PM_{2.5}$, age, percent minorities, median income

and percent of people with no high school education. $\lambda(t)$ was estimated with a natural cubic spline with 35 degrees of freedom (5 df per year).

To investigate the robustness of our results various sensitivity analysis were run on the all respiratory admission as a sample group. We analyzed other averaging periods (lag1,lag2 vs lag0) and the addition of the land use and temporal variables (percent of, percent of house owners living in owned house, Percent of occupied housing units with more than one person per room and median home value, absolute humidity).

We also wanted to compare the results from our novel prediction models with an analysis of this data base using a traditional time series approach. We ran the analysis for the Boston area (Suffolk, Norfolk and Middlesex counties) using daily admission counts for all respiratory admissions, $PM_{2.5}$ from a central PM monitor (Countway monitor at Harvard school of public health) and temperature data from Logan airport as commonly done in time series analysis.

Table 2. Descriptive statistics for short term $PM_{2.5}$ exposure, long term $PM_{2.5}$ exposure and temperature in New-England for 2000–2006.

Covariate	Mean	Min	Max	Median	SD	Range	IQR	Q1	Q3	Days of data available
Lag0 $PM_{2.5}$ (acute PM)	9.60	0.01	72.59	8.55	4.90	72.59	5.32	6.35	11.67	2557
1 year $PM_{2.5}$ (Chronic PM)	9.65	3.54	17.79	9.65	0.81	14.25	0.98	9.16	10.14	2557
Temperature	46.52	−23.80	90.10	47.90	18.73	113.90	29.30	33.00	62.30	2557

Note: Q1 and Q3 are quartiles.

Results

Descriptive statistics stratified by long term pollution (high and low split by the mean) are presented in Table 1. The majority of people included in our analyses which were admitted to hospital were white (84.92%–96.23% across all admission causes) while the average age was 76.65–80.30 years.

Table 2 contains a summary of the predicted exposures for both the acute exposure (0 day lag) and the chronic exposure (365 day moving average) across all grid cells in the analysis.

Table 3 presents the estimated percent increase in hospital admissions for a 10 $\mu g/m^3$ increase for both short term and long term $PM_{2.5}$ by cause of admission and associated 95% confidence intervals. For all respiratory, for every 10-$\mu g/m^3$ increase in short term $PM_{2.5}$ exposure there is a 0.70 percent increase in admissions (95% CI = 0.35 to 0.52) while concurrently for every 10-$\mu g/m^3$ increase in long term $PM_{2.5}$ exposure there is a 4.22 percent increase in admission (95% CI = 1.06 to 4.75). For CVD, for every10-$\mu g/m^3$ increase in short term $PM_{2.5}$ exposure there is a 1.03 percent increase in admission rate (95% CI = 0.69 to 0.45) while concurrently for every10-$\mu g/m^3$ increase in long term $PM_{2.5}$ exposure there is a 3.12 percent increase in admission (95% CI = 0.30 to 4.29). For strokes, for every 10-$\mu g/m^3$ increase in short term $PM_{2.5}$ exposure there is a 0.24 percent increase in admissions (95% CI = -0.13 to 0.56) while concurrently for every10-$\mu g/m^3$ increase in long term $PM_{2.5}$ exposure there is a 3.49 percent increase in admissions (95% CI = 0.09 to 5.18). Finally for diabetes, for every10-$\mu g/m^3$ increase in short term $PM_{2.5}$ exposure there is a 0.96 percent increase in admissions (95% CI = 0.62 to 0.51) while concurrently for every10-$\mu g/m^3$ increase in long term $PM_{2.5}$ exposure there is a 6.33 percent increase in admissions (CI = 3.22 to 4.59).

The results from the sensitivity analysis are presented in table 4. In general the results of the sensitivity analysis were consistent with the primary analysis for the added spatial variables and added temporal variable as well as for the different lags (excluding the acute $PM_{2.5}$ exposure in lag02).

The results from the classic times series analysis were similar to the main model (1.51 vs 0.72 percent change) albeit with higher standard error (0.002 vs 0.001) and much larger CI (0.42–1.65 Vs. 0.35–0.52).

The crude and final estimates as well as the estimates for the model covariates are presented in appendix S1 and S2.

Discussion

In this paper we report, for the first time, that long term exposure to $PM_{2.5}$ is associated with increased hospital admissions of the elderly (aged 65 and older) for all respiratory, CVD, stroke, and diabetes. As with mortality studies, this long term impact is higher than the acute effects. Importantly, we continue to see acute effects independent of the chronic effects. In addition, this analysis covers all zip codes in New England, not just subset zipcodes locations near $PM_{2.5}$ monitors. This represents an important extension of previous Medicare analyses, since we now have estimates that include suburban, small town, and rural populations. Finally, the use of a spatiotemporal model reduces exposure misclassification that exists in, for example time series studies that use a single exposure metric for daily exposure in an entire metropolitan area. Such error is a mixture of classical exposure error, which likely biases the effect estimates downward, and Berkson error, which increases the confidence interval [39]. The results from our novel method presented much tighter confidence intervals compared to the classic time series analysis, indicating that our method could potentially reduce measurement error. Another advantage our method adds is the ability to include population that lives far from monitor compared to the traditional methods

One of the key components of this study is that we showed that by using our prediction models (which produce daily $PM_{2.5}$ predictions) we are able to simultaneously examine short term and long term association with hospital admissions and to do it for the entire population of New England, avoiding issues of selection or non-representative samples, and accounting for small area measures of potential confounders.

The putative biological mechanisms linking both short term and long term exposure to air pollution and CVD involve direct effects of pollutants on the cardiovascular system, blood, and lung receptors, and/or indirect effects mediated through pulmonary oxidative stress and inflammatory responses [40]. The biological mechanisms linking both short term and long term exposure to air pollution and respiratory diseases include reduced lung function, pulmonary inflammation and oxidative stress [41]. Further, an intervention trial of air filtration for elderly adults reduced particles levels and reported improved endothelial function [42]. Similarly, a trial comparing blood pressure when subjects were walking in Beijing with our without a particle filter reported blood pressure was lower when wearing the filter208.

These studies are also supported by toxicologic indicators of mechanism. For example, a recent study of mice genetically prone to atherosclerosis and on a high fat western diet exposed to concentrated particles from the outside air showed that the particle exposure lead to more atherosclerotic plaque, and increased macrophages and tissue factor in the plaques, which reduce plaque stability and increase the risk of a heart attack [43]. Another study, using a different mouse model of atherosclerosis, documented that particle exposure increased oxidation of LDL, increased the thickness of the arterial wall, and promoted plaque growth and instability [44]. A number of studies have directly linked particle

Table 3. Estimated percent increase in hospital admissions for a 10 $\mu g/m^3$ increase for both short term and long term $PM_{2.5}$ by cause of admission.

$PM_{2.5}$ exposure type	All Respiratory	CVD	Stroke	Diabetes
	Percent increase [a]	Percent increase [a]	Percent increase [a]	Percent increase [a]
Short term $PM_{2.5}$ exposure	0.70 (0.35–0.52)	1.03 (0.69–0.45)	0.24(−0.13–0.56)	0.96(0.62– 0.51)
Long term $PM_{2.5}$ exposure	4.22(1.06–4.75)	3.12(0.30–4.29)	3.49 (0.09–5.18)	6.33(3.22– 4.59)

Note: [a]Values are percent.

Table 4. Sensitivity analysis (estimated percent increase in hospital admissions for a 10 µg/m^3 increase in chronic PM$_{2.5}$ exposure) for all respiratory admission causes.

PM$_{2.5}$ exposure type	All Respiratory-acute PM$_{2.5}$	All Respiratory-chronic PM$_{2.5}$
	Percent increase [a]	Percent increase [a]
Baseline	0.70(0.35–0.52)	4.22(1.06–4.75)
Added SES variable	0.70(0.35–0.52)	3.84(0.67–4.74)
Added temporal variable lag 0	0.58(0.23–0.52)	4.40(1.36–4.56)
Added temporal variable lag 01	0.35(0.01–0.52)	4.65(1.60–4.57)
Added temporal variables lag 02	−0.18(−0.52 – −0.52)	5.32(2.25–4.57)

Note: [a]Values are percent.

exposure with ischemia. Wellenius exposed dogs to either filtered air or concentrated air particles, followed by a temporary occlusion of the coronary artery. The animals exposed to particles experienced greater ischemia than those exposed to filtered air [45,46].

Several studies suggested an enhanced susceptibility of people with diabetes to exposure to air pollution partly due to inflammatory mechanisms [19,20,47]. In addition there are reports associating air pollution with incidence of diabetes [48].

Our estimated associations between short term exposure to PM$_{2.5}$ and hospital admissions revealed results qualitatively similar to those studies previously published analyzing short term PM$_{2.5}$ and hospital admissions [10,21,49]. To the best of our knowledge there are no studies on exposure to long term PM$_{2.5}$ and hospital admissions. However our long term exposure results for CVD are in agreement with those reported by Miller et al. [50], who studied postmenopausal women without previous CVD in 36 U.S. metropolitan areas from 1994 to 1998. They estimated that each 10 µg/m^{3}increase in PM$_{2.5}$ was associated with a 24% increase in the risk of a cardiovascular event (hazard ratio, 1.24; 95% CI = 1.09 to 1.41). Those events would almost certainly have resulted in hospitalizations.

These findings also clarify a previous apparent inconsistency. Cohort studies of the association of PM2.5 and deaths from CVD or stroke have reported much larger effect sizes than the time series studies of PM2.5 and admissions from those causes [51,52,53,54]. This seems implausible since many of those events result in hospitalizations. However, these chronic mortality estimates are also much larger than the time series estimates of the acute effects of recent PM2.5 exposure on deaths from those causes. The usual explanation is that chronic exposure produces greater effects because it leads to cumulative damage, such as atherosclerosis etc. [55,56,57,58,59]. Those arguments would be equally applicable to the effects of long term exposure on chronic rates of admissions for these causes. In this paper we show that such larger effects in fact are seen.

A major limitation of this study is our limited ability to control for individual level potential confounders, such as socio-economic factors, diet, exercise, etc. We have used area-based measures of socio-economic factors. To test the potential for confounding, we used data from the Normative Aging Study [60,61], a population based study of an aging cohort, resident in Maine, Massachusetts, New Hampshire and Rhode Island As a general population of subjects eligible for Medicare, we think this is a reasonable test of the potential for confounding. We assigned the same 365 day average exposure to those participants from our model, and examined the association with packyears, with physical activity (METS), and with dietary fish intake. In no case was there a significant association.

Another limitation of the present study is the relatively coarse spatial resolution of 10×10 km. However, as satellite remote sensing evolves and progresses, higher spatial resolution data (3×3 km and 1×1 km) should become available which will further reduce exposure error. Such finer resolution should enable us to assess more precise estimated daily individual exposure as they relate to different location such as residence, work place etc.

In conclusion, we have demonstrated how our prediction models perform well in assessing short term and long term human exposures. Our findings indicate that hospital admission were associated with both short term and long term exposure to PM$_{2.5}$. These findings present new opportunities to study the effects of both the long and short term exposure and human health.

Acknowledgments

The authors want to thank Steven J. Melly, Department of Environmental Health, Harvard School of Public Health, Harvard University.

Author Contributions

Conceived and designed the experiments: IK JS. Performed the experiments: IK AZ BC. Analyzed the data: IK PK BC AZ JS. Contributed reagents/materials/analysis tools: BC. Wrote the paper: IK JS.

References

1. Brunekreef B, Holgate ST (2002) Air pollution and health. The lancet 360: 1233–1242.
2. Pope CA, Dockery DW, Schwartz J (1995) Review of epidemiological evidence of health effects of particulate air pollution. Inhalation toxicology 7: 1–18.
3. Schwartz J (1994) Air pollution and hospital admissions for the elderly in Birmingham, Alabama. Am J Epidemiol 139: 589.
4. Spix C, Anderson HR, Schwartz J, Vigotti MA, Letertre A, et al. (1998) Short-term effects of air pollution on hospital admissions of respiratory diseases in Europe: a quantitative summary of APHEA study results. Archives of Environmental Health: An International Journal 53: 54–64.

5. Zanobetti A, Franklin M, Koutrakis P, Schwartz J (2009) Fine particulate air pollution and its components in association with cause-specific emergency admissions. Environmental Health 8: 58.

6. Zanobetti A, Schwartz J, Dockery DW (2000) Airborne particles are a risk factor for hospital admissions for heart and lung disease. Environ Health Perspect 108: 1071.

7. Dominici F, Peng RD, Bell ML, Pham L, McDermott A, et al. (2006) Fine particulate air pollution and hospital admission for cardiovascular and respiratory diseases. Jama 295: 1127.

8. Fusco D, Forastiere F, Michelozzi P, Spadea T, Ostro B, et al. (2001) Air pollution and hospital admissions for respiratory conditions in Rome, Italy. European respiratory journal 17: 1143.

9. Schwartz J (1996) Air pollution and hospital admissions for respiratory disease. Epidemiology 7: 20–28.

10. Mann JK, Tager IB, Lurmann F, Segal M, Quesenberry CP Jr., et al. (2002) Air pollution and hospital admissions for ischemic heart disease in persons with congestive heart failure or arrhythmia. Environ Health Perspect 110: 1247.

11. Schwartz J, Morris R (1995) Air pollution and hospital admissions for cardiovascular disease in Detroit, Michigan. Am J Epidemiol 142: 23–35.

12. Schwartz J (1997) Air pollution and hospital admissions for cardiovascular disease in Tucson. Epidemiology 8: 371–377.

13. D'Ippoliti D, Forastiere F, Ancona C, Agabiti N, Fusco D, et al. (2003) Air pollution and myocardial infarction in Rome: a case-crossover analysis. Epidemiology 14: 528.

14. Zanobetti A, Schwartz J (2005) The effect of particulate air pollution on emergency admissions for myocardial infarction: a multicity case-crossover analysis. Environ Health Perspect 113: 978–982.

15. Symons J, Wang L, Guallar E, Howell E, Dominici F, et al. (2006) A case-crossover study of fine particulate matter air pollution and onset of congestive heart failure symptom exacerbation leading to hospitalization. Am J Epidemiol 164: 421.

16. Wellenius GA, Bateson TF, Mittleman MA, Schwartz J (2005) Particulate air pollution and the rate of hospitalization for congestive heart failure among Medicare beneficiaries in Pittsburgh, Pennsylvania. Am J Epidemiol 161: 1030.

17. Ilabaca M, Olaeta I, Campos E, Villaire J, Tellez-Rojo MM, et al. (1999) Association between levels of fine particulate and emergency visits for pneumonia and other respiratory illnesses among children in Santiago, Chile. JOURNAL-AIR AND WASTE MANAGEMENT ASSOCIATION 49: 154–163.

18. Medina-Ramon M, Zanobetti A, Schwartz J (2006) The effect of ozone and PM10 on hospital admissions for pneumonia and chronic obstructive pulmonary disease: a national multicity study. Am J Epidemiol 163: 579.

19. O'Neill MS, Veves A, Zanobetti A, Sarnat JA, Gold DR, et al. (2005) Diabetes enhances vulnerability to particulate air pollution-associated impairment in vascular reactivity and endothelial function. Circulation 111: 2913.

20. Zanobetti A, Schwartz J (2001) Are diabetics more susceptible to the health effects of airborne particles? Am J Respir Crit Care Med 164: 831.

21. Dominici F, Peng RD, Bell ML, Pham L, McDermott A, et al. (2006) Fine particulate air pollution and hospital admission for cardiovascular and respiratory diseases. Jama 295: 1127–1134.

22. Zanobetti A, Franklin M, Koutrakis P, Schwartz J (2009) Fine particulate air pollution and its components in association with cause-specific emergency admissions. Environ Health 8: 58–70.

23. Andersen ZJ, Bønnelykke K, Hvidberg M, Jensen SS, Ketzel M, et al. (2011) Long-term exposure to air pollution and asthma hospitalisations in older adults: a cohort study. Thorax.

24. Hruba F, Fabianova E, Koppova K, Vandenberg JJ (2001) Childhood respiratory symptoms, hospital admissions, and long-term exposure to airborne particulate matter. Journal of Exposure Analysis and Environmental Epidemiology 11: 33.

25. Oudin A, Strömberg U, Jakobsson K, Stroh E, Lindgren AG, et al. (2010) Hospital Admissions for Ischemic Stroke: Does Long-Term Exposure to Air Pollution Interact with Major Risk Factors. Cerebrovascular Diseases 31: 284–293.

26. Andersen ZJ, Raaschou-Nielsen O, Ketzel M, Jensen SS, Hvidberg M, et al. (2012) Diabetes Incidence and Long-Term Exposure to Air Pollution. Diabetes Care 35: 92–98.

27. Gryparis A, Paciorek CJ, Zeka A, Schwartz J, Coull BA (2009) Measurement error caused by spatial misalignment in environmental epidemiology. Biostatistics 10: 258–274.

28. Paciorek CJ, Yanosky JD, Puett RC, Laden F, Suh HH (2009) Practical large-scale spatio-temporal modeling of particulate matter concentrations. The Annals of Applied Statistics 3: 370–397.

29. USCB (2000) United States Census Bureau of 2000. US Census Bureau.

30. Engel-Cox J, Holloman C, Coutant B, Hoff R (2004) Qualitative and quantitative evaluation of MODIS satellite sensor data for regional and urban scale air quality. Atmospheric Environment 38: 2495–2509.

31. Gupta P, Christopher S, Wang J, Gehrig R, Lee Y, et al. (2006) Satellite remote sensing of particulate matter and air quality assessment over global cities. Atmospheric Environment 40: 5880–5892.

32. Koelemeijer R, Homan C, Matthijsen J (2006) Comparison of spatial and temporal variations of aerosol optical thickness and particulate matter over Europe. Atmospheric Environment 40: 5304–5315.

33. Liu Y, Park R, Jacob D, Li Q, Kilaru V, et al. (2004) Mapping annual mean ground-level PM2. 5 concentrations using Multiangle Imaging Spectroradiometer aerosol optical thickness over the contiguous United States. J Geophys Res 109: 3269–3278.

34. Kloog I, Koutrakis P, Coull BA, Lee HJ, Schwartz J (2011) Assessing Temporally and Spatially Resolved PM2. 5 Exposures for Epidemiological Studies Using Satellite Aerosol Optical Depth Measurements. Atmospheric Environment.

35. NCDC (2010) The national climatic data center data inventories.

36. Census US (2000) U.S. Census of Population and Housing. Government Printing Office.

37. Laird N, Olivier D (1981) Covariance analysis of censored survival data using log-linear analysis techniques. Journal of the American Statistical Association. pp 231–240.

38. Schwartz J, Dockery DW, Neas LM (1996) Is daily mortality associated specifically with fine particles? Journal of the Air & Waste Management Association (1995) 46: 927.

39. Zeger SL, Thomas D, Dominici F, Samet JM, Schwartz J, et al. (2000) Exposure measurement error in time-series studies of air pollution: concepts and consequences. Environ Health Perspect 108: 419–426.

40. Brook RD, Franklin B, Cascio W, Hong Y, Howard G, et al. (2004) Air pollution and cardiovascular disease. Circulation 109: 2655–2671.

41. Pope CA, Young B, Dockery D (2006) Health effects of fine particulate air pollution: lines that connect. Journal of the Air & Waste Management Association 56: 709–742.

42. Brauner EV, Forchhammer L, Moller P, Barregard L, Gunnarsen L, et al. (2008) Indoor particles affect vascular function in the aged: an air filtration-based intervention study. Am J Respir Crit Care Med 177: 419–425.

43. Sun Q, Yue P, Kirk RI, Wang A, Moatti D, et al. (2008) Ambient air particulate matter exposure and tissue factor expression in atherosclerosis. Inhal Toxicol 20: 127–137.

44. Soares SR, Carvalho-Oliveira R, Ramos-Sanchez E, Catanozi S, da Silva LF, et al. (2009) Air pollution and antibodies against modified lipoproteins are associated with atherosclerosis and vascular remodeling in hyperlipemic mice. Atherosclerosis 207: 368–373.

45. Bartoli CR, Wellenius GA, Coull BA, Akiyama I, Diaz EA, et al. (2009) Concentrated ambient particles alter myocardial blood flow during acute ischemia in conscious canines. Environ Health Perspect 117: 333–337.

46. Wellenius GA, Coull BA, Godleski JJ, Koutrakis P, Okabe K, et al. (2003) Inhalation of concentrated ambient air particles exacerbates myocardial ischemia in conscious dogs. Environ Health Perspect 111: 402–408.

47. Baja ES, Schwartz JD, Wellenius GA, Coull BA, Zanobetti A, et al. (2010) Traffic-related air pollution and QT interval: modification by diabetes, obesity, and oxidative stress gene polymorphisms in the normative aging study. Environ Health Perspect 118: 840–846.

48. Kramer U, Herder C, Sugiri D, Strassburger K, Schikowski T, et al. (2010) Traffic-related air pollution and incident type 2 diabetes: results from the SALIA cohort study. Environ Health Perspect 118: 1273–1279.

49. Wellenius GA, Schwartz J, Mittleman MA (2005) Air pollution and hospital admissions for ischemic and hemorrhagic stroke among Medicare beneficiaries. Stroke 36: 2549.

50. Miller KA, Siscovick DS, Sheppard L, Shepherd K, Sullivan JH, et al. (2007) Long-term exposure to air pollution and incidence of cardiovascular events in women. New England Journal of Medicine 356: 447–458.

51. Krewski D, Jerrett M, Burnett RT, Ma R, Hughes E, et al. (2009) Extended follow-up and spatial analysis of the American Cancer Society study linking particulate air pollution and mortality. Res Rep Health Eff Inst: 5–114; discussion 115–136.

52. Laden F, Neas LM, Dockery DW, Schwartz J (2000) Association of fine particulate matter from different sources with daily mortality in six US cities. Environ Health Perspect 108: 941.

53. Ostro B, Lipsett M, Reynolds P, Goldberg D, Hertz A, et al. (2010) Long-term exposure to constituents of fine particulate air pollution and mortality: results from the California Teachers Study. Environ Health Perspect 118: 363.

54. Puett R, Hart J, Yanosky J, Paciorek C, Schwartz J, et al. (2009) Chronic fine and coarse particulate exposure, mortality, and coronary heart disease in the Nurses' Health Study. Environ Health Perspect 117: 1702.

55. Adar SD, Klein R, Klein BEK, Szpiro AA, Cotch MF, et al. (2010) Air pollution and the microvasculature: a cross-sectional assessment of in vivo retinal images in the population-based Multi-Ethnic Study of Atherosclerosis (MESA). PLoS Medicine 7: e1000372.

56. Bauer M, Moebus S, Mohlenkamp S, Dragano N, Nonnemacher M, et al. (2010) Urban Particulate Matter Air Pollution Is Associated With Subclinical Atherosclerosis:: Results From the HNR (Heinz Nixdorf Recall) Study. Journal of the American College of Cardiology 56: 1803–1808.

57. Hansen CS, Sheykhzade M, Moller P, Folkmann JK, Amtorp O, et al. (2007) Diesel exhaust particles induce endothelial dysfunction in apoE-/-mice. Toxicology and applied pharmacology 219: 24–32.

58. Künzli N, Jerrett M, Garcia-Esteban R, Basagaña X, Beckermann B, et al. (2010) Ambient air pollution and the progression of atherosclerosis in adults. PLoS One 5: e9096.

59. Sun Q, Wang A, Jin X, Natanzon A, Duquaine D, et al. (2005) Long-term air pollution exposure and acceleration of atherosclerosis and vascular inflammation

in an animal model. JAMA: the journal of the American Medical Association 294: 3003.

60. Halonen JI, Zanobetti A, Sparrow D, Vokonas PS, Schwartz J (2010) Associations between outdoor temperature and markers of inflammation: a cohort study. Environmental Health 9: 42.

61. Madrigano J, Baccarelli A, Wright RO, Suh H, Sparrow D, et al. (2010) Air pollution, obesity, genes and cellular adhesion molecules. Occup Environ Med 67: 312–317.

5

Assessing the Impact of Water Filters and Improved Cook Stoves on Drinking Water Quality and Household Air Pollution: A Randomised Controlled Trial in Rwanda

Ghislaine Rosa[1]*, **Fiona Majorin**[1], **Sophie Boisson**[1], **Christina Barstow**[2], **Michael Johnson**[3], **Miles Kirby**[1], **Fidele Ngabo**[4], **Evan Thomas**[5], **Thomas Clasen**[1,6]

1 Department of Disease Control, London School of Hygiene and Tropical Medicine, London, United Kingdom, 2 Department of Civil, Environmental and Architectural Engineering, University of Colorado at Boulder, Boulder, Colorado, United States of America, 3 Berkeley Air Monitoring Group, Berkeley, California, United States of America, 4 Ministry of Health, Government of Rwanda, Kigali, Rwanda, 5 Department of Mechanical and Materials Engineering, Portland State University, Portland, Oregon, United States of America, 6 Department of Environmental Health, Emory University, Atlanta, Georgia, United States of America

Abstract

Diarrhoea and respiratory infections remain the biggest killers of children under 5 years in developing countries. We conducted a 5-month household randomised controlled trial among 566 households in rural Rwanda to assess uptake, compliance and impact on environmental exposures of a combined intervention delivering high-performance water filters and improved stoves for free. Compliance was measured monthly by self-report and spot-check observations. Semi-continuous 24-h $PM_{2.5}$ monitoring of the cooking area was conducted in a random subsample of 121 households to assess household air pollution, while samples of drinking water from all households were collected monthly to assess the levels of thermotolerant coliforms. Adoption was generally high, with most householders reporting the filters as their primary source of drinking water and the intervention stoves as their primary cooking stove. However, some householders continued to drink untreated water and most continued to cook on traditional stoves. The intervention was associated with a 97.5% reduction in mean faecal indicator bacteria (Williams means 0.5 *vs.* 20.2 TTC/100 mL, $p<0.001$) and a median reduction of 48% of 24-h $PM_{2.5}$ concentrations in the cooking area ($p=0.005$). Further studies to increase compliance should be undertaken to better inform large-scale interventions.
Trial registration: Clinicaltrials.gov; NCT01882777; http://clinicaltrials.gov/ct2/results?term = NCT01882777&Search = Search

Editor: Robert K. Hills, Cardiff University, United Kingdom

Funding: This study was funded by DelAgua Health, a for-profit company that implements the intervention in Rwanda in conjunction with the Rwanda Ministry of Health. The funders had no role in study design, data collection and analysis, decision to publish, or preparation of the manuscript.

Competing Interests: We have the following interests: This study was funded by DelAgua Health, a for-profit company that implements the program in Rwanda in conjunction with the Rwanda Ministry of Health. Evan Thomas and Christina Barstow, co-authors of this article, are compensated consultants to DelAgua Health, and are in charge of overseeing the implementation of the program in Rwanda. Michael Johnson, co-authors of this article, is employed by the commercial company, Berkeley Air Monitoring Group, which was contracted to provide advice on household air pollution monitoring.

* E-mail: ghislaine.rosa@lshtm.ac.uk

Introduction

Environmental contamination at the household level is a major cause of death and disease, particularly among rural populations in low-income countries. Unsafe drinking water, together with poor sanitation, account for an estimated 0.9% of the global burden of disease and 0.3 million deaths [1]. Much of this disease burden is associated with diarrhoea, which alone accounts for 10.5% of deaths in children under 5 years in low-income countries [2]. Household air pollution (HAP) from biomass fuel smoke has been linked to increased risk of respiratory tract infections, low birth weight, exacerbations of inflammatory lung conditions, cardiac events, stroke, eye disease, tuberculosis, cancer and nutritional deficiencies [3]. The Global Burden of Disease (GBD) 2010 project found HAP from solid fuels to be responsible for 3.5 million premature deaths globally [1]. In this same assessment, smoke from household cooking fuels was also responsible for another half a million premature deaths due to contributions to outdoor air pollution [1]. These environmental hazards are aggravated among rural inhabitants of sub-Saharan Africa who are more likely to rely on unsafe water supplies and cook using biomass fuels on inefficient stoves [4–6].

Inefficient cookstoves also present substantial economic, developmental and environmental costs. At the household level, poverty is exacerbated and time spent at school is reduced by the burden of collecting more fuel for boiling drinking water and cooking [7]. Individuals, households and governments bear the cost of expenditures for seeking treatment of enteric and respiratory infections. Cookstove emissions also contribute to greenhouse gas and black carbon emissions, and in some cases the fuel harvesting can result in denuding of forests [8,9].

With a population of 10.5 million and a density of 412 persons per sq. km, Rwanda is the most densely populated country in East Africa [10]. Eighty per cent of the population of Rwanda lives in rural areas and is engaged in agriculture [11]. Despite significant progress over the last decade, 57% of the population is living below the poverty line, 37% of them living in extreme poverty [11]. While a large proportion of the rural population has access to improved water sources (71.2%), mainly through protected springs, only 2.2% of rural areas have water on their premises [12], resulting in an increased risk for drinking water contamination during transport and storage [13]. Almost all of rural Rwanda (99.0%) relies on biomass for their cooking needs [12]. Morbidity and mortality are largely dominated by communicable diseases, including HIV/AIDS, acute respiratory infections, diarrhoeal diseases, intestinal parisitoses, and malaria [14]. Among deaths of children under 5, pneumonia accounts for 20% and diarrhoea for 12% [15].

In an effort to reduce the disease burden in rural Rwanda, decrease poverty associated with expenditures for fuel, and minimize the impact of greenhouse gases from inefficient combustion of biomass in low-efficiency stoves, the Rwanda Ministry of Health (MiniSante) and the Rwanda Environmental Management Authority (REMA) have partnered with DelAgua Health (implementer) to design, deploy and evaluate the impact of a project that will deliver and promote the use of advanced water filters and high efficiency cookstoves to lower-income households in Rwanda. Prior to initiating the full campaign, the implementer with the Ministry of Health undertook a pilot distribution of filters and cookstoves to approximately 2200 households in 15 villages in 11 of the country's 30 districts. We conducted this study in three of those villages in order to assess the uptake of the intervention and its impact on drinking water quality and household air pollution.

Methods

Study setting

The study was conducted from September 2012 to April 2013 in three rural villages, Nyarutovu and Kabuga located in Muhanga district, Southern province; and Rubona, located in Gakenke district, Northern province. These villages were purposely selected from the 15 villages comprising the pilot distribution phase. The sites were changed from the original protocol, Karongi and Ngororero districts in Western province, to accommodate access to better microbiology laboratory facilities in Kigali.

Study design and sample size

The study employed a parallel, household-randomised, control trial design with a 1:1 ratio. This trial followed a non-blinded design because previous attempts to blind an earlier version of the LifeStraw Family filter in the Democratic Republic of Congo were unsuccessful [16]. The objectives of the study were to assess (i) uptake and use of the intervention by the target population when delivered programmatically, and (ii) the impact of the intervention on the microbiological quality of household drinking water and air quality near the self-reported cooking area over the 5-month follow-up period. Our primary outcomes were (i) to assess levels of faecal contamination (measured by thermotolerant coliforms, TTC) in stored water in the home that householders used for drinking, and (ii) to determine average 24-h concentrations of $PM_{2.5}$ in the main cooking area as identified by participants. Our secondary outcome was to assess use of the intervention filters and stoves based on self-report and spot-check observations.

The sample size calculation was based on $PM_{2.5}$ emissions reductions rather than TTC reductions in drinking water as the former was determined to require a larger sample. Assuming a 50% reduction in $PM_{2.5}$ emissions, 80% power, $\alpha = 0.05$ and a coefficient of variation (COV) of 1, we estimated a sample size of 63 households per arm.

The protocol of this trial and CONSORT checklist are available as supporting information; see Text S1 and S2.

Intervention

Each intervention household received one LifeStraw Family 2.0 filter and one EcoZoom Dura improved wood burning stove. The filter is the second-generation of a gravity-based water purifier that uses ultrafiltration in the form of a hollow-fibre cartridge to remove pathogens from drinking water. The first generation device has been shown in field studies to be highly effective in improving water quality and to achieve consistent (though not exclusive) use [16]. The second-generation version used in this study employs a table-top design and an integrated safe storage vessel. Untreated water is poured through a 20-μm pre-filter plastic mesh into a 6.0 L container; over time, gravity forces the water through the cartridge comprised of hollow-fibres with a 20-nm pore size. The water then passes into a 5.5 L storage vessel where it can be dispensed via a plastic tap. The device is cleaned daily by backwashing the cartridge using a squeeze-pump mounted on the back of the storage container. The device is designed to treat 18,000 L of water [17] with a flow rate of approximately 3 L per hour. In the laboratory, the filter cartridge was found to meet the USEPA standards for microbiological water purifiers by reducing bacteria by 6 logs, viruses by 5 logs and protozoa by 4 logs [18]. The filter meets the "highly protective" World Health Organization (WHO) rating for household water treatment technologies [19].

The intervention stove is based on the 'rocket' concept that uses an internal 'chimney' in the stove that directs air through the burning fuel (usually biomass), and encourages the mixing of gases and flame above it. Precise internal stove dimensions are used to achieve high combustion efficiency and transfer heat to the cooking pot. Two additional components are included with the stove, a "stick support" onto which fuel wood is placed to promote airflow and a "pot skirt" which increases fuel efficiency. A study comparing cookstoves in Uganda, Kenya and Tanzania reported that the EcoZoom (aka StoveTec) stove saved 39% to 54% of fuel compared to open fires, cooked meals faster, and was participants' most preferred stove during controlled cooking of local dishes [20–22]. In the intervention group, householders were encouraged to cook outdoors on the EcoZoom stove and to use dry wood only to increase the efficiency of the stove. Further details on the messaging used in the pilot distribution can be found elsewhere [23].

Houses that were allocated to the intervention group also received a poster with illustrations and instructions in *Kinyarwanda*, the local language, on filter and stove use, maintenance, and contact names and phone numbers for the implementer. Most households had easy access to a cellular phone for contacting the implementer. Intervention households received one-to-one training on use and maintenance in their homes by community health workers (CHWs) who were previously trained by trainers who themselves had been trained by the filter and stove manufacturers and implementer. Intervention households were then visited periodically at approximately one-month intervals by CHWs to refresh health messaging and encourage use. Households allocated to the control group were instructed to continue usual practices throughout the study. At the end of the study in April 2013, these control households received their own filters, stoves, posters and training.

Enrolment, baseline survey, randomisation and deployment of devices

Households were eligible to participate in the study if (i) they were registered as being members of the village, (ii) the head of the household was over 18 years, and (iii) no members of the household worked as a CHW. The last criterion was included after the original protocol was drafted as at the time of the design the researchers were not aware that the CHW that would deliver the intervention resided in the villages selected for the study. It was explained that while all participating households would receive filters and stoves, half would receive them at the outset of the study and the balance at the conclusion of the 5-month follow-up period. After obtaining consent from the heads of participating households, a baseline survey was undertaken in September-November 2012 to collect information on demographics, socio-economic characteristics, water, hygiene and sanitation practices as well as fuel and cooking practices. Data collection tools were translated into *Kinyarwanda* and piloted before use.

Following the baseline survey, a public lottery was organised by the implementer and research teams during a village meeting to randomly allocate an approximately equal number of households from each village to intervention or control groups. Local authorities and village chiefs were extensively engaged to assess the suitability of this randomisation approach. After the lottery, members of control households were invited to leave the venue while those of intervention households attended a demonstration on the use and maintenance of the filter and stove, collected their devices, and carried them to their homes.

Outcome assessment

Compliance. Monthly cross-sectional surveys were conducted by trained field investigators (the evaluation team) working independently of the implementation team at unannounced visits among each household. At each visit participants were asked to identify the main drinking water container in the household, whether it was the intervention filter or another container; the surveyor also recorded whether the filter contained water at the time of the visit, a possible objective indicator of filter use. The field investigators also observed the cookstove and if cooking was taking place at the time of the unannounced visit, recorded where and whether such cooking was on the intervention stove or the traditional stove. If no cooking was taking place, field investigators noted the presence of smoke marks on the intervention stove, a possible objective indicator of use. Reported measures of stove use were also collected by asking participants what stove had been used the last time cooking took place in their home.

Independent to our study, the implementers undertook a separate survey, conducted by Environmental Health Officers (EHOs), to assess use and acceptability of the intervention for their own monitoring and evaluation purposes. The details of this assessment have been presented elsewhere [23].

Additionally, to assess use of the intervention in a more objective manner, remotely reporting electronic sensors were mounted onto 23 intervention filters and 27 intervention stoves and deployed in a randomly selected sub-sample of intervention households for a two-week period. The details of the implementation of this nested study, data handling and analysis, and results are presented elsewhere [24].

Water quality. During each of the five monthly visits, field investigators took a sample from the water container identified by the householder as being used mainly for drinking by children under 5 years of age, or adults if no under 5 s resided in the household. If this was other than directly from the intervention filter, a second sample was taken directly from the filter if it

contained water. All water samples were collected in sterile Whirl-Pak bags (Nasco, Fort Atkinson, WI) containing a tablet of sodium thiosulphate to neutralize any halogen disinfectant. Samples were placed on ice and processed within 6 h of collection to assess levels of TTC. Microbiological assessment was performed using the membrane filtration technique [25] on membrane lauryl sulphate medium (Oxoid Limited, Basingstoke, Hampshire, UK) using a DelAgua field incubator (Robens Institute, University of Surrey, Guilford, Surrey, UK).

Household air pollution. Monitoring of particulate matter with an aerodynamic diameter $<2.5\ \mu m$ ($PM_{2.5}$) in the main cooking area took place between November 2012 and March 2013. 126 households (63 control and 63 intervention households) were randomly selected for semi-continuous 24-h $PM_{2.5}$ monitoring. Households were numbered and selected by using a computerised random number generator. Upon arrival at the participant's home, the family member mainly responsible for cooking was identified and a short survey was employed to identify the area in the household where cooking primarily took place. "Stacking" of stoves (using different stoves, often in different locations) [26] was a common scenario, both in control and intervention households, though more common in the latter. In cases where the participant reported cooking equally in two locations or with two or more stoves, we sampled from indoor rather than outdoor locations and from traditional rather than intervention stove. UCB-PATS $PM_{2.5}$ monitors (described below) were placed 1.5 m above the ground and 1 m away from the stove and, whenever possible, at least 1.5 m from windows and doors by suspending the monitors from the roof beams. When cooking was reported to take place outdoors, the $PM_{2.5}$ monitor was mounted onto a vertical wooden stand and placed at the same distance and height from the stove. The location of the stand was marked on the floor and participants were advised not to touch or move the equipment.

$PM_{2.5}$ was measured using the University of California, Berkeley Particle and Temperature Sensor (UCB-PATS[TM]), (Berkeley Air Monitoring Group, USA), a semicontinuous (1-min averages), light-scattering nephelometer [27,28]. Laboratory and field validations of the UCB-PATS have been described previously [27–30]. To take into account that nephelometer sensitivity is a function of an aerosol's specific optical properties such as size, colour, and shape [31], calibration of the UCB response with the target aerosol was undertaken by conducting 24-h $PM_{2.5}$ gravimetric co-location measurements in a sub-sample of homes (n = 30). Five field blanks were obtained, resulting in an adjustment of subtracting 5 µg to the final filter masses ($<1\%$ of the mean mass deposition). The UCB-PATS response was then linearly regressed against the gravimetric samples (n = 27, $R^2 = 0.86$), with the resulting equation then used to adjust the UCB-PATS response to the gravimetric measures (Figure S1 of supporting information). Three gravimetric samples were omitted due to incomplete sampling durations.

Gravimetric $PM_{2.5}$ samples were collected using standard air sampling pumps (PXR8, SKC Inc., USA) with $PM_{2.5}$ cyclones (SCC 1.062, BGI, USA) using a flow rate of 1.5 L/min. Flow rates were measured before and after installation of the sampling equipment in the home with a rotameter (Matheson Trigas, Montgomeryville, PA, USA) that had been calibrated using a TSI Flow Calibrator 4146 (TSI, Inc., USA). $PM_{2.5}$ was collected on 37-mm Teflon filters (Pall, USA). Filters were stored at $4°C$ until shipment to Berkeley Air Monitoring Group in California, USA for weighing. Filters were equilibrated for 24 h at $22\pm3°C$ and $40\pm5\%$ relative humidity before being weighed on a 0.1

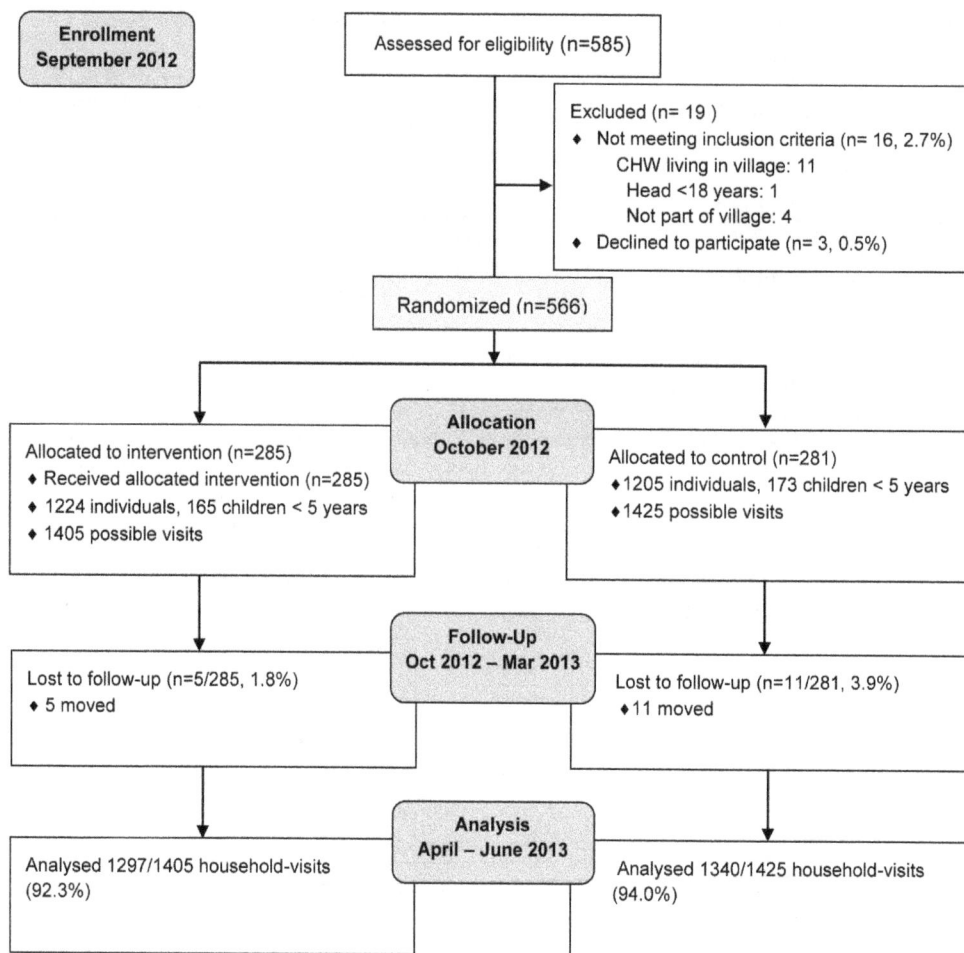

Figure 1. CONSORT diagram showing the flow of participants through the trial.

microgram resolution electro microbalance (XP2U, Mettler Toledo, USA).

Data analysis

All data were analysed using Stata 12 (Stata Corporation, College Station, TX, USA). Because both $PM_{2.5}$ concentrations and TTC counts in drinking water followed non-normal distributions, medians, geometric means and Williams means are presented together with arithmetic means. The Williams mean is calculated by adding 1 to all the data values, then taking the geometric mean, then subtracting 1 again [32]. Categorical data were compared using a Chi square or a Fisher's exact test where appropriate. The non-parametric Wilcoxon rank sum test was used to compare $PM_{2.5}$ concentrations in the main cooking area between intervention and control groups. To assess the effect of the intervention on water quality, TTC counts during follow-up were compared using random effects negative binomial regression as describe elsewhere [33] to account for (i) repeated observations within households and, (ii) the skewed distribution of the TTC counts. Model comparison was assessed by using the Bayesian information criterion (BIC), which is a well-established measure of goodness of fit that also applies to non-nested models [33,34]. For the purpose of analysis, plates that yielded coliform forming units (CFUs) that were too numerous to count (TNTC) were assigned a value of 300 TTC/100 mL. Data were analysed in an intention-

to-treat basis in order to estimate the effect of the intervention regardless of compliance. Only those households with complete follow-up data were analysed.

Ethics

The study was reviewed and approved by the ethics committee at the London School of Hygiene and Tropical Medicine (No. 6239, as amended) and the Rwanda National Ethics Committee (No. 328 RNEC/2012). Written informed consent to participate in the research was obtained from the male or female head or the wife of each participating household.

Results

Study population

The three villages participating in the study comprised 585 households, all of which were screened to participate in the study, 16 (2.7%) were ineligible and 3 (0.5%) refused to participate (Figure 1). A total of 566 households with 2429 individuals were enrolled in the study. Of those 281 (49.7%) were assigned to the control group and 285 (50.4%) were assigned to receive the intervention filter and cookstove. Household loss-to-follow-up was 2.8%, primarily due to participants moving out of the study area. A total of 2737 household-visits were completed during the follow-

Table 1. Filter and stove use among intervention households: Evaluator's survey.

	All visits	
	N	**%**
Filter use[1]		
Reported drinking container		
Intervention filter	1210	89.2
Other container	146	10.8
Water stored in other container treated	57	39.0
Method of treatment: Intervention filter	56	98.2
No water in intervention filter among households identifying filter as drinking container	12	1.0
No water in intervention filter among households not identifying filter as drinking container	83	56.8
Stove use[2]		
Observation data on use		
Intervention household cooking at time of visit	280	26.9
Stove in current use		
Intervention stove only	152	54.3
Both stoves simultaneously	12	4.3
Traditional stove only	116	41.4
Currently cooking outdoors	59	21.1
Intervention stove users cooking outdoors[3]	49	32.2
Reported data on use		
Reported last stove used[4]		
Intervention stove only	593	78.0
Both stoves simultaneously	15	19.3
Traditional stove only	147	2.0
Reported using intervention stove In last three follow-up visits	130	47.5

[1]Based on households that completed the visits and allowed enumerators to observe the container, 1356/1393 = 97.3%.
[2]Data only available from mid follow-up 2 onwards (1040/1393 = 74.7%).
[3]Among those households cooking only on the intervention stove.
[4]Excludes those households cooking at time of home visit.

up period (96.7%) and data on one of the primary outcomes (water quality) was collected for 2637 households-visits (93.2%).

Baseline characteristics

Baseline characteristics were distributed evenly between the trial arms, with the exception of availability of soap among households with a designated hand washing area and boiling or chlorination of drinking water (see Table S1 of supporting information). At baseline, drinking water samples were obtained from 551 (97.3%) households. The median and Williams mean of drinking water was 14 and 20.2 TTC/100 mL (95% CI: 15.0–27.0 TTC/100 mL) and 22 and 30.3 TTC/100 mL (95% CI: 22.8–40.2 TTC/100 mL) for control and intervention groups, respectively.

Filter and improved stove use and compliance

Most households used the filter throughout the study period (Table 1). Intervention households identified the filter as the main drinking source in 89.2% of all household visits where drinking water was available. Visual inspection at the time of the unannounced visit was consistent with reported use, with 99% of the filters containing water. Of the 10.8% of intervention households that stored their drinking water elsewhere, overall only 39.0% of them reported that the water had been treated with the intervention filter. Over the course of the study, however, only

62.9% of intervention households identified the filter as the main drinking water storage container in all five follow-up visits with available water (n = 240, 84.2%). Of the remainder, 11.2% reported treating it and storing it elsewhere at least once during the 5-month follow-up, 25.0% reported drinking untreated water at least once during follow-up and 0.8% did not know the status of their water in at least one of the visits. During the last follow-up visit, the major reasons for not having filtered water at the time of the visit were (i) forgetting to fill the filter (48.1%), (ii) drinking mainly locally produced beer instead of water (22.2%), or (iii) having a broken or not properly functioning filter (18.5%).

The intervention stoves were also used throughout the study, though most householders also continued to use their traditional stoves. Field investigators observed actual cooking on about a quarter (26.9%) of their unannounced visits. Of these, 54.3% were cooking only with the intervention stove and 4.3% were using both the intervention and traditional stoves (Table 1). Reported use was higher, with householders claiming they last cooked solely on the intervention stove on 78.0% of visits. Use of the intervention stove was not consistent, with 47.5% of intervention households reporting to have used the intervention stove during the last cooking event at all three home visits (data not collected during initial phases of follow-up). Likewise, of the households that were cooking at all three unannounced visits (n = 8), or at two of the

Table 2. Filter and stove use among intervention households: Implementer's survey.

	N	%
Filter use		
Filter presence confirmed in households[1]	283	99.7
Tap accessible to <5 s	267	94.4
Water present in filter	269	95.1
Stove use		
Observation data on use		
Intervention household cooking at time of visit[2]	78	27.7
Stove in current use		
Intervention stove only	50	64.1
Both stoves simultaneously	4	5.1
Traditional stove only	24	30.8
Currently cooking outdoors	30	38.5
Intervention stove users cooking outdoors[3]	28	17.2
Reported data on use		
Reported last stove used[4]		
Intervention stove only	163	79.9
Both stoves simultaneously	2	1.0
Traditional stove only	39	19.1
Primary stove in current use is intervention stove[4]	253	89.1
Use intervention stove ≥7/week	236	93.3
Use intervention stove ≥14/week	137	54.2
Continue using traditional stove	217	76.4
Use traditional stove ≥7/week	58	26.7
Reported cooking less indoors	175	61.6
Reported main cooking is outdoors	163	57.4
Tend more the fire with the intervention stove[5]	212	83.8

[1]Observation not allowed in one household.
[2]Of those households that allowed the observation (n = 282, 99.3%).
[3]Among those households cooking only on the intervention stove.
[4]Excludes those households cooking at time of home visit.
[5]Among those households identifying intervention stove as main cooking stove.

three unannounced visits (n = 52), only 50.0% and 34.6% were using the intervention stove at all three or two visits, respectively. During the last follow-up visit, the major reasons for not using the intervention stove during the last cooking event were (i) having no time to tend the fire (34.1%), (ii) not having dry (30.7%) or the right-size wood (10.2%) or, (iii) cooking beans for which a traditional stoves was regarded most appropriate (10.2%).

Data on use of the intervention from the implementer's survey was very similar to our assessment (Table 2). A similar percentage of intervention households (27.7%) were cooking at the time of the visit. Of these, just over two thirds (64.1%) were exclusively cooking on the intervention stove, but only 17.2% of these were cooking outdoors, a figure just slightly lower than the one observed on our independent follow-up. Data from the implementer's more extensive survey confirmed the stacking of stoves in intervention households, with 76.4% of intervention households reporting to continue using their traditional stoves. Of these, 26.7% reported using it ≥7 times per week. Of interest was the fact that 83.8% of intervention households identifying the intervention stove as their primary cookstove reported that the intervention stove required

more active tending of the fire as compared to the traditional stove.

In the last round of our follow-up, 5.4% and 5.1% of intervention households reported having problems with their filter or stove at the time of the visit, respectively. Data collected from the implementer's repair team indicates that 24.9% of filters and 6.7% of stoves had to be repaired during the study. No devices had to be fully replaced, though some repairs involved the replacement of individual components. The main reasons for filters being repaired were (i) filters being clogged (48.6%) and, (ii) tubes being damaged by rodents (27.0%). The main reasons for the intervention stoves being repaired included (i) pot skirts melting (65%), and (ii) stick supports breaking (10%).

Overall, only 1.0% of water samples collected from control households were reported to have been treated with a neighbour's intervention filter, showing low levels of cross-contamination between groups.

Water quality

The microbiological quality of the stored drinking water was significantly higher in intervention households than control households (Williams means 0.5 vs. 20.2 TTC/100 mL, respectively, $p < 0.001$). Overall, 86.8% (95% CI: 84.9%–88.6%) of drinking water samples from intervention households were free of TTC compared to 22.4% (95% CI: 20.1%–24.6%) of control household samples ($p < 0.001$) (Figure 2). The proportion of samples that had >100 TTC/100 mL was 3.6% (95% CI: 2.6%–4.6%) for intervention households and 31.9% (95% CI: 29.4%–34.5%) for control households. Overall, 96.6% of drinking water samples collected directly from filters were free of TTC. In intervention households, water quality was significantly higher in water samples collected directly from the filter (Williams mean 0.14 TTC/100 mL; 95% CI: 0.10–0.18) than water stored in another container (Williams mean 13.8 TTC/100 mL; 95% CI: 9.0–20.7) (see Table S2 of supporting information). The quality of the drinking water stored in other containers did not differ significantly between control and intervention households ($p = 0.07$). However, among intervention households, water that was stored in another container and was reportedly treated with the intervention filter was significantly of higher quality than reportedly non-treated stored water (Williams means 5.4 vs. 23.2 TTC/100 mL, respectively, $p < 0.001$). Throughout the duration of the study, only 2.5% of control households had drinking water free of TTC on all follow-up visits as opposed to 56.5% of intervention households. Overall 15.2% of samples from control households and 5.1% of samples from intervention households yielded plates that were TNTC.

Air quality

A total of 121 households (60 intervention and 61 control) completed the 24-h PM$_{2.5}$ monitoring of the main cooking area. 66.7% of intervention households identified the intervention stove as their main cooking stove. However, only 23.3% of intervention households reported that their main cooking area was outdoors as promoted by the intervention. Of these, all households reported cooking with the intervention stove. Among the control households, the three stone fire was identified as the main cooking stove in 65.6% of cases, followed by the locally made *rondereza* stove (24.6%). Only one control household reported cooking outdoors.

Table 3 shows the PM$_{2.5}$ concentrations of the main cooking area for control and intervention households on an aggregate level and stratified by reported main area of cooking. Overall, mean and median 24-h PM$_{2.5}$ concentrations in intervention households were 0.485 mg/m^3 and 0.267 mg/m^3, respectively, compared to

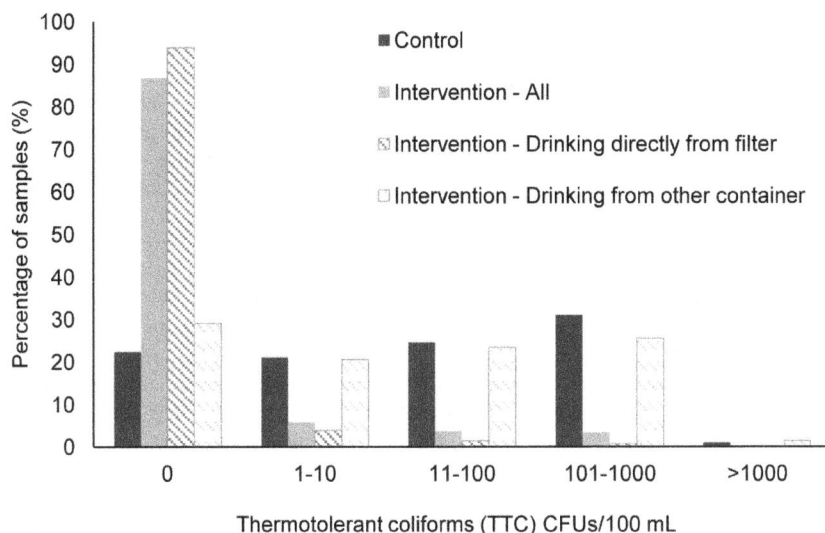

Figure 2. Percentage of water samples by level of contamination (TTC/100 mL).

0.905 mg/m^3 and 0.509 mg/m^3 for control households. This represents a 48% reduction in median 24-h concentrations ($p = 0.005$). Compared to control households that predominantly cooked indoors, intervention homes that reported indoor cooking showed a reduction in median concentrations of 37%, which was only borderline significant, possibly due to the smaller sample size ($p = 0.08$). Outdoor cooking in the intervention was associated with a median reduction of 73% when compared to control households ($p<0.001$) and 57% reduction when compared to indoor-cooking intervention homes ($p = 0.02$).

Discussion

We report on a randomised controlled trial to independently evaluate a pilot implementation program distributing free water filters and improved cooking stoves to rural homes in Rwanda. We found high reported use of the intervention filter, which was associated with significantly higher microbiological quality of drinking water when consumed directly from the filter. Nevertheless, such use was not exclusive; a sizable proportion of householders continued to drink untreated water. We also found improved household air quality among intervention households despite continued use of the traditional stove.

Filter uptake among the intervention population was high, with filters being reportedly used in 89.2% of all household visits. Similar levels of uptake of filter-based interventions have been reported elsewhere [16,35,36]. Nevertheless, we found that 25% of intervention householders were reporting untreated water in at least one of the five follow-up visits. The nested study within this RCT using remotely reporting electronic sensors that collected objective data on use of the intervention devices (mainly times and volumes of water filtered for the intervention filter and times and duration of use for the intervention stove) corroborated our findings, showing that the filters and stoves were not used in a consistent and exclusive manner [24]. Epidemiological modelling based on quantitative microbial risk assessment suggests that even occasional consumption of untreated water can vitiate the health benefits associated with improved water quality interventions [37–39]. However, the intervention did significantly improve the microbiological quality of the drinking water when the filter was used as the main storage container. Since 96.6% of drinking water samples collected directly from filters were free of TTC, the conditions for achieving health gains may be achieved with better messaging.

Exclusive use was more problematic for the intervention stove. Only half of the intervention households reported that the last

Table 3. Summary statistics for 24-h PM$_{2.5}$ concentrations in the reported main cooking area.

PM$_{2.5}$ (mg/m^3)	N	Mean	SD	Min	Median	Max	Geometric mean	% Mean reduction	% Median reduction	Wilcoxon RST[1] p-value
Control	61	0.905	1.05	0.06	0.509	4.69	0.51	-	-	
Intervention	60	0.485	0.53	0.04	0.267	2.28	0.28	46%	48%	0.005
Reported cooking location										
Control- Indoor cooking	60	0.910	1.06	0.06	0.506	4.69	0.51	-	-	
Intervention- Indoor cooking	46	0.558	0.56	0.04	0.321	2.28	0.33	39%	37%	0.08
Intervention- Outdoor cooking	14	0.243	0.34	0.05	0.139	1.40	0.16	73%	73%	<0.001
[1]Wilcoxon rank-sum (Mann-Whitney) test										

cooking event was performed with the intervention stove in the last three monthly follow-up visits. Likewise, only a third of those households that were visited twice at times that cooking was taking place were using the intervention stove at both instances, showing that among the intervention arm, households continued to rely on their traditional stove. Results from the implementers' survey showed similar results, with 76.4% of households reporting the continued use of their traditional stove, 26.7% of them using it more than 7 times per week. This is consistent with other studies that have shown that the introduction of a new stove often results in "stacking" rather than an immediate complete substitution [40–43].

Households reported continuing the use of their traditional stove because the intervention stove required more tending, unavailability of the adequate fuel or personal preferences for cooking traditional dishes. Context-specific issues regarding a community's cooking needs and preferences have been commonly cited in the literature as reasons for not achieving higher uptake and/or exclusive sustained use of improved cookstoves [42,44]. Thus re-considerations of the promoted stove or more active messaging addressing each of the main barriers may be required if a switching of the stove as opposed to an addition of the intervention stove to the current cooking system is to be achieved. This is not only going to affect the potential health impact of the intervention but also its environmental impact.

The assessment of HAP among control and intervention households showed an overall reduction of 48% of 24-h $PM_{2.5}$ among intervention households, which was comparable to reductions in household air pollution for rocket stove interventions in Ghana (52%) and Kenya (33%) [22,45]. Indoor cooking with the intervention stoves as opposed to the traditional stove was associated with a 37% reduction in 24-h $PM_{2.5}$, which was of borderline significance. However, we cannot rule out that this association may be due to residual bias by comparing sub-groups. Likewise, cooking outdoors, as recommended by the implementer, doubled the reduction in 24-h $PM_{2.5}$ from 37% to 73% as compared to indoor cooking on traditional stoves. Future studies, randomising participants not only to stove technology but also to cooking location (indoors vs. outdoors) would be advisable. More effective messaging may increase the levels of outdoor cooking expected by the intervention, as only 57.4% of households reported that their main cooking area was outdoors. Nevertheless, both the indoor and outdoor concentrations in the cooking area were well over even the initial interim 24-h WHO target for $PM_{2.5}$ (75 µg/m^3) [46]. At the same time, it will be important to monitor personal exposure directly, as most householders that identified the intervention stove as their primary cooking stove (83.8%) reported that the intervention stove required more tending than their traditional one, which could mitigate some of the impact from the household level reductions in $PM_{2.5}$. Indeed, many studies have found that reductions in personal exposure tend to be lower than reductions of emissions in the cooking area [47,48]. Given that a recent RCT study suggested that personal exposure reductions exceeding 50% may be required to achieve meaningful health impacts [47], further assessments of the intervention stove maybe be needed to determine whether the use of the intervention stove translates into meaningful health benefits.

This study has certain limitations. First, the villages included in the RCT were not selected randomly and should not be viewed as representative of any larger population. Second, we cannot rule out the potential for reactivity due to repeated monthly follow-up visits [49]. Third, while we attempted to collect objective indicators of use, by both undertaking visual observations of the filter and stoves and cooking events, the study relied heavily on reported data, which is susceptible to reporting bias. Furthermore, in this study we failed to collect data on reported supplementation of treated water with untreated water, which would have further implications for the health impact of the study. Previous studies with the earlier version of the LifeStraw Family filter have found quite varied results. A study in the Democratic Republic of Congo showed substantial supplementation despite high levels of filter use [16]. On the other hand, a study among HIV-positive mothers, who may be more aware of their health and their children's health, reported almost no supplementation [36]. However, in the latter storage containers were provided. Fourth, budget constraints allowed only the main cooking area, as identified by the participant, to be monitored for HAP. Given the potential for reporting bias and that stacking was commonly reported among the study population, it is very likely that cooking events may have taken place during the monitoring period in areas other than the one being monitored, thus giving a misleading and probable underestimate of the actual total HAP. Likewise, budget constraints did not permit personal $PM_{2.5}$ assessment, a more reliable metric for exposures associated with health outcomes [50]. Fifth, we did not collect any self-reported or other measures of diarrhoea or respiratory infections in our study communities. Finally, the follow-up period of this evaluation was limited to 5 months. This represents under a fraction of the lifespan of both the filter and stove and provided little opportunity to assess the impact of seasonal variations that are common in water quality and HAP. It also provided no opportunity to assess long-term patterns of use, which have been shown to diminish or vary over time for both water filters and improved cookstoves [35,43]. We are endeavouring to address some of these shortcomings in a longer-term follow-up study, currently underway, that will focus on health outcomes and sustained use.

Notwithstanding these limitations, this study suggests that a combined filter/stove intervention accompanied by consistent follow-up to promote use has the potential to significantly improve drinking water quality and household air pollution among a vulnerable population in Rwanda. If the longer-term follow-up study demonstrates sustained use with more exclusive reliance on the intervention hardware and lower personal exposure to HAP, then a large-scale roll out in Rwanda could significantly reduce exposures linked to much of the country's disease burden.

Acknowledgments

We thank all of the participants who contributed to this study. We also thank the staff for their diligent efforts in collecting the data. We thank too Wolf-Peter Schmidt and Alexandra Huttinger for their contribution to the design of the study. Michael Johnson, senior scientist at Berkeley Air Monitoring Group was responsible for providing advice on HAP monitoring and acts as lead author for the group, mjohnson@berkeleyair.com.

Author Contributions

Conceived and designed the experiments: GR SB FN TC. Performed the experiments: GR FM MK. Analyzed the data: GR MJ. Contributed reagents/materials/analysis tools: CB ET. Wrote the paper: GR TC.

References

1. Lim SS, Vos T, Flaxman AD, Danaei G, Shibuya K, et al. (2012) A comparative risk assessment of burden of disease and injury attributable to 67 risk factors and risk factor clusters in 21 regions, 1990–2010: a systematic analysis for the Global Burden of Disease Study 2010. Lancet 380: 2224–2260.

2. Liu L, Johnson HL, Cousens S, Perin J, Scott S, et al. (2012) Global, regional, and national causes of child mortality: an updated systematic analysis for 2010 with time trends since 2000. Lancet 379: 2151–2161.

3. Ezzati M, Kammen DM (2001) Quantifying the effects of exposure to indoor air pollution from biomass combustion on acute respiratory infections in developing countries. Environ Health Perspect 109: 481.

4. Rehfuess E, Mehta S, Pruss-Ustun A (2006) Assessing Household Solid Fuel Use: Multiple Implications for the Millenium Development Goals. Environ Health Perspect 114.

5. WHO/UNICEF (2012) Progress on drinking water and sanitation: 2012 update. Geneva: WHO.

6. Bonjour S, Adair-Rohani H, Wolf J, Bruce NG, Mehta S, et al. (2013) Solid Fuel Use for Household Cooking: Country and Regional Estimates for 1980–2010. Environ Health Perspect 121: 784–790.

7. Barnes DF, Openshaw K, Smith KR, van der Plas R (1994) What Makes People Cook with Improved Biomass Stoves? A Comparative International Review of Stove Programs. World Bank.

8. Johnson M, Edwards R, Ghilardi An, Berrueta V, Gillen D, et al. (2009) Quantification of Carbon Savings from Improved Biomass Cookstove Projects. Environ Sci Technol 43: 2456–2462.

9. Bond TC, Doherty SJ, Fahey DW, Forster PM, Berntsen T, et al. (2013) Bounding the role of black carbon in the climate system: A scientific assessment. J. Geophys Res: Atmos 118: 5380–5552.

10. NISR (2012) Rwanda Ministry of Health Annual Health Statistics Booklet 2011. National Institute of Statistics of Rwanda.

11. NISR (2008) Rwanda in Statistics and Figures 2008. National Institute of Statistics of Rwand.

12. National Institute of Statistics of Rwanda (NISR) [Rwanda], Ministry of Health (MOH) [Rwanda], International I (2012) Rwanda Demographic and Health Survey 2010. Calverton, Maryland, USA: NISR, MOH, and ICF International.

13. Wright J, Gundry S, Conroy R (2004) Household drinking water in developing countries: a systematic review of microbiological contamination between source and point-of-use. Trop Med Int Health 9: 106–117.

14. RMOH (2011) Rwanda Health Statistics Booklet 2011. Rwanda Ministry of Health.

15. UNICEF (2012) Committing to Child Survival: A Promise Renewed. Progress Report 2012. New York: United Nations Children's Fund.

16. Boisson S, Kiyombo M, Sthreshley L, Tumba S, Makambo J, et al. (2010) Field Assessment of a Novel Household-Based Water Filtration Device: A Randomised, Placebo-Controlled Trial in the Democratic Republic of Congo. PLoS One 5: e12613.

17. Clasen T, Naranjo J, Frauchiger D, Gerba C (2009) Laboratory Assessment of a Gravity-Fed Ultrafiltration Water Treatment Device Designed for Household Use in Low-Income Settings. Am J Trop Med Hyg 80: 819–823.

18. Naranjo J, Gerba C (2011) Assessment of the LifeStraw Family Unit using the World Health Organization Guidelines for "Evaluating Household Water Treatment Options: Health-based Targets and Performance Specifications". University of Arizona, Department of Soil, Water and Environmental Science.

19. WHO (2011) Evaluating household water treatment options: health-based targets and performance specifications. Geneva: WHO.

20. Adkins E, Tyler E, Wang J, Siriri D, Modi V (2010) Field testing and survey evaluation of household biomass cookstoves in rural sub-Saharan Africa. Energy Sustain Dev 14: 172–185.

21. USAID (2011) In-Home Emissions of Greenhouse Pollutants from Rocket and Traditional Biomass Cooking Stoves in Uganda.

22. Pennise D, Brant S, Agbeve SM, Quaye W, Mengesha F, et al. (2009) Indoor air quality impacts of an improved wood stove in Ghana and an ethanol stove in Ethiopia. Energy Sustain Dev 13: 71–76.

23. Barstow CK, Thomas EA, Ngabo Fidele, Ghislaine R, Majorin F, et al. (2013) Environmental health product delivery and education in rural Rwanda through a public-private partnership. Unpublished.

24. Thomas EA, Barstow CK, Rosa G, Majorin F, Clasen T (2013) Use of Remotely Reporting Electronic Sensors for Assessing Use of Water Filters and Cookstoves in Rwanda. Environ Sci Technol 47: 13602–13610.

25. APHA (2001) Standard Methods for the Examination of Water and Wastewater. 21st Edition. Washington, (DC): APHA.

26. Masera OR, Saatkamp BD, Kammen DM (2000) From Linear Fuel Switching to Multiple Cooking Strategies: A Critique and Alternative to the Energy Ladder Model. World Development 28: 2083–2103.

27. Edwards R, Smith K, Kirby B, Allen T, Litton C, et al. (2006) An inexpensive dual-chamber particle monitor: laboratory characterization. J Air Waste Manag Assoc 56: 789–799.

28. Edwards RD, Liu Y, He G, Yin Z, Sinton J, et al. (2007) Household CO and PM measured as part of a review of China's National Improved Stove Program. Indoor Air 17: 189–203.

29. Chowdhury Z, Edwards RD, Johnson M, Naumoff Shields K, Allen T, et al. (2007) An inexpensive light-scattering particle monitor: field validation. J Environ Monit 9: 1099–1106.

30. Dutta K, Shields KN, Edwards R, Smith KR (2007) Impact of improved biomass cookstoves on indoor air quality near Pune, India. Energy Sustain Dev 11: 19–32.

31. Cynthia AA, Edwards RD, Johnson M, Zuk M, Rojas L, et al. (2008) Reduction in personal exposures to particulate matter and carbon monoxide as a result of the installation of a Patsari improved cook stove in Michoacan Mexico. Indoor Air 18: 93–105.

32. Alexander N (2012) Review: analysis of parasite and other skewed counts. Trop Med Int Health 17: 684–693.

33. McElduff F, Cortina-Borja M, Chan S-K, Wade A (2010) When t-tests or Wilcoxon-Mann-Whitney tests won't do. Adv Physiol Educ 34: 128–133.

34. Kuha J (2004) AIC and BIC: Comparisons of Assumptions and Performance. Sociol Methods Res 33: 188–229.

35. Hunter PR (2009) Household Water Treatment in Developing Countries: Comparing Different Intervention Types Using Meta-Regression. Environ Sci Technol 43: 8991–8997.

36. Peletz R, Simunyama M, Sarenje K, Baisley K, Filteau S, et al. (2012) Assessing Water Filtration and Safe Storage in Households with Young Children of HIV-Positive Mothers: A Randomized, Controlled Trial in Zambia. PLoS One 7: e46548.

37. Hunter PR, Zmirou-Navier D, Hartemann P (2009) Estimating the impact on health of poor reliability of drinking water interventions in developing countries. Sci Total Environ 407: 2621–2624.

38. Brown J, Clasen T (2012) High Adherence Is Necessary to Realize Health Gains from Water Quality Interventions. PLoS One 7: e36735.

39. Enger KS, Nelson KL, Rose JB, Eisenberg JNS (2013) The joint effects of efficacy and compliance: A study of household water treatment effectiveness against childhood diarrhea. Water Res 47: 1181–1190.

40. Ruiz-Mercado I, Masera O, Zamora H, Smith KR (2011) Adoption and sustained use of improved cookstoves. Energy Policy 39: 7557–7566.

41. Burwen J, Levine DI (2012) A rapid assessment randomized-controlled trial of improved cookstoves in rural Ghana. Energy Sustain Dev 16: 328–338.

42. Hanna R, Duflo E, Greenstone M (2012) Up in smoke: The influence of household behaviour on the long-run impact of improved cooking stoves. NBER Working Paper No. 18033. Cambridge, MA: National Bureau of Economic Research.

43. Pine K, Edwards R, Masera O, Schilmann A, Marrón-Mares A, et al. (2011) Adoption and use of improved biomass stoves in Rural Mexico. Energy Sustain Dev 15: 176–183.

44. Manibog FR (1984) Improved Cooking Stoves in Developing Countries: Problems and Opportunities. Annu Rev Energy 9: 199–227.

45. Ochieng CA, Vardoulakis S, Tonne C (2013) Are rocket mud stoves associated with lower indoor carbon monoxide and personal exposure in rural Kenya? Indoor Air 23: 14–24.

46. WHO (2006) Global update 2005: Particulate matter, ozone, nitrogen dioxide and sulfur dioxide, air quality guidelines. Copenhagen: WHO.

47. Smith KR, McCracken JP, Weber MW, Hubbard A, Jenny A, et al. (2011) Effect of reduction in household air pollution on childhood pneumonia in Guatemala (RESPIRE): a randomised controlled trial. Lancet 378: 1717–1726.

48. Fitzgerald C, Aguilar-Villalobos M, Eppler AR, Dorner SC, Rathbun SL, et al. (2012) Testing the effectiveness of two improved cookstove interventions in the Santiago de Chuco Province of Peru. Sci Total Environ 420: 54–64.

49. Zwane AP, Zinman J, Van Dusen E, Pariente W, Null C, et al. (2011) Being surveyed can change later behavior and related parameter estimates. Proc Natl Acad Sci U S A.

50. Naeher LP, Brauer M, Lipsett M, Zelikoff JT, Simpson CD, et al. (2007) Woodsmoke health effects: A review. Inhal Toxicol 19: 67–106.

6

Personal Exposure to Household Particulate Matter, Household Activities and Heart Rate Variability among Housewives

Ya-Li Huang[1,2], Hua-Wei Chen[3], Bor-Cheng Han[2], Chien-Wei Liu[4], Hsiao-Chi Chuang[5,6], Lian-Yu Lin[7], Kai-Jen Chuang[1,2]*

1 Department of Public Health, School of Medicine, College of Medicine, Taipei Medical University, Taipei, Taiwan, 2 School of Public Health, College of Public Health and Nutrition, Taipei Medical University, Taipei, Taiwan, 3 Department of Cosmetic Application and Management, St. Mary's Junior College of Medicine, Nursing and Management, Yilan, Taiwan, 4 Department of Information Management, St. Mary's Junior College of Medicine, Nursing and Management, Yilan, Taiwan, 5 Division of Pulmonary Medicine, Department of Internal Medicine, Shuang Ho Hospital, Taipei Medical University, Taipei, Taiwan, 6 School of Respiratory Therapy, College of Medicine, Taipei Medical University, Taipei, Taiwan, 7 Department of Internal Medicine, Division of Cardiology, National Taiwan University Hospital, Taipei, Taiwan

Abstract

Background: The association between indoor air pollution and heart rate variability (HRV) has been well-documented. Little is known about effects of household activities on indoor air quality and HRV alteration. To investigate changes in HRV associated with changes in personal exposure to household particulate matter (PM) and household activities.

Methods: We performed 24-h continuous monitoring of electrocardiography and measured household PM exposure among 50 housewives. The outcome variables were log_{10}-transformed standard deviation of normal-to-normal (NN) intervals (SDNN) and the square root of the mean of the sum of the squares of differences between adjacent NN intervals (r-MSSD). Household PM was measured as the mass concentration of PM with an aerodynamic diameter <2.5 μm ($PM_{2.5}$). We used mixed-effects models to examine the association between household $PM_{2.5}$ exposure and log_{10}-transformed HRV indices.

Results: After controlling for potential confounders, an interquartile range change in household $PM_{2.5}$ with 1- to 4-h mean was associated with 1.25–4.31% decreases in SDNN and 0.12–3.71% decreases in r-MSSD. Stir-frying, cleaning with detergent and burning incense may increase household $PM_{2.5}$ concentrations and modify the effects of household $PM_{2.5}$ on HRV indices among housewives.

Conclusions: Indoor $PM_{2.5}$ exposures were associated with decreased SDNN and r-MSSD among housewives, especially during stir-frying, cleaning with detergent and burning incense.

Editor: Rudolf Kirchmair, Medical University Innsbruck, Austria

Funding: This study was supported by grants (NSC 101-2314-B-038-053-MY3 and TMU101-AE1-B08) from the National Science Council of Taiwan and Taipei Medical University. The funders had no role in study design, data collection and analysis, decision to publish, or preparation of the manuscript.

Competing Interests: The authors have declared that no competing interests exist.

* E-mail: kjc@tmu.edu.tw

Introduction

Air pollution exposure, particularly particulate matter (PM), has been associated with increased cardiovascular mortality and morbidity [1,2]. These associations have been partially supported by the association of PM with heart rate variability (HRV) changes, and previous panel studies have reported this association as a possible mechanism linking PM to increased risk for cardiovascular diseases [3]. Recently, several studies have reported that the association of cardiovascular endpoints with personal exposure to PM [4–6]. It is also known that people spend 87% of their time in enclosed buildings [7]. These findings imply that exposure to indoor PM may increase the cardiovascular effects of PM exposure. Moreover, the World Health Organization (WHO) considers indoor air pollution as the 3rd most important risk factor, responsible for 4.3% of the global burden of disease [8].

There are many household sources of PM, such as cooking, cleaning and tobacco smoke [7,9,10], which worsen indoor air quality and induce cardiovascular effects among housekeepers who spend most of their time in houses. However, studies of the adverse effect on HRV of personal exposure to household PM among housekeepers are lacking. The effect modification of household activities on the association of PM exposure with HRV changes remains unclear. The aim of this study was to investigate the association between personal exposure to household PM and HRV indices among housewives using four 24-h visits to monitor each participant's household PM exposure, HRV indices and household activity pattern in their private homes.

Materials and Methods

Ethics approval

The study was approved by the ethics committee of St. Mary's Medicine Nursing and Management College. Written informed consent was obtained from each participant before the study began.

Study participants and design

This panel study was designed to simultaneously and continuously monitor changes in household PM levels and HRV indices as well as household activities among housewives in their own homes. The selection criteria for volunteer participants were as follows: no history of smoking or drinking; no medication that might affect cardiac rhythm; and no cardiovascular diseases, such as coronary artery disease, arrhythmia, hypertension, diabetes mellitus and dyslipidemia. Ninety-two housewives responded to our recurring advertisement; 50 of them (54%) living in 50 homes in the Taipei metropolitan area met the criteria and were willing to participate in this study after our protocols had been explained.

The protocol included four home visits that entailed continuous 24-h monitoring of electrocardiography (ECG), household PM, noise, meteorological conditions and time-activity patterns at approximately one-season intervals in the years 2010 to 2012. Each of the 50 housewives had four home visits, for a total of 200 home visits. During their first visits, age, sex, and household characteristics were recorded using a questionnaire. Height and weight were measured and used to calculate body mass index (BMI). Information on indoor environmental measurements and time-activity patterns during the study periods were collected at all visits.

Household particulate matter, meteorological conditions, noise and time-activity patterns

We conducted 24-h continuous monitoring of household PM, meteorological conditions, noise and time-activity patterns during each visit for each housewife. Household PM less than 2.5 μm in diameter (PM$_{2.5}$), temperature and relative humidity were measured continuously using a personal dust monitor (DUSTcheck Portable Dust Monitor, model 1.108; temperature and humidity sensor, model 1.153FH; Grimm Labortechnik Ltd., Ainring, Germany), which measured and recorded 1-min mass concentrations of PM$_{2.5}$ as well as temperature and relative humidity. Noise level was measured using a portable noise dosimeter (Logging Noise Dose Meter Type 4443, Brüel & Kjær, Nærum, Denmark), which reported 1-min continuous equivalent sound levels (Leq) and the time-weighted-averages (TWAs) of noise doses. A range of 30–100 dBA was used to measure noise exposure with 1-min readings over 24 hours.

We asked each participant to carry the dust monitor personally from 0700 hr to 2200 hr to measure personal household PM$_{2.5}$ and noise exposure during the participants' household activities. Participants themselves recorded their time-activity patterns, such as indoor tobacco smoke exposure, indoor chemical dispersion, burning incense, stir-frying and cleaning with detergent every hour from 0700 hr to 2200 hr during the study periods. After sampling, the raw data for 1-min household PM$_{2.5}$, temperature, relative humidity and noise measurements were matched with the sampling time of HRV monitoring and then calculated as 1-, 2-, 3- and 4-h means if 75% of the data were present.

Heart rate variability monitoring

We performed continuous ambulatory ECG monitoring using a PACERCORDER 3-channel device (model 461A; Del Mar Medical Systems LLC, Irvine, CA, USA) with a sampling rate of 250 Hz (4 msec) from 0700 hr to 2200 hr (15 hours) during the study periods. ECG tapes were analyzed using a Delmar Avionics model Strata Scan 563 (Irvine, CA). A complete 5-min segment of the normal-to-normal (NN) interval was taken for HRV analysis, including the standard deviation of NN intervals (SDNN) and the square root of the mean of the sum of the squares of differences between adjacent NN intervals (r-MSSD). Each participant obtained approximately 720 successful 5-min HRV measurements during the four visits (12 measurements for each hour, 180 measurements for each visit) for data analysis. We obtained approximately 32,432 5-min measurements of HRV indices for 50 participants in our data analysis (missing data rate = 9.9%).

Statistical analysis

Mixed-effects models were used to examine the association between household PM$_{2.5}$ and log$_{10}$-transformed HRV indices by running R statistical software version 2.15.1. The independent variables were the 1-, 2-, 3- and 4-h mean household PM$_{2.5}$, whereas the dependent variables were SDNN and r-MSSD. We treated participant age, BMI, hour of day, temperature, relative humidity, noise, household PM$_{2.5}$ and household activity periods (yes vs. no) including indoor tobacco smoke exposure, indoor chemical dispersion, burning incense, stir-frying and cleaning with detergent as fixed effects and fitted the participant identity number as a random intercept term in our mixed-effects models.

Effect modification by indoor tobacco smoke exposure, burning incense, stir-frying and cleaning with detergent were assessed in separate mixed-effects models by including interaction terms between household PM$_{2.5}$ effects and each potential effect modifier. Household PM$_{2.5}$ effects are expressed as percent changes by interquartile range (IQR) changes as $[10^{(\beta \times IQR)}-1] \times 100\%$ for log$_{10}$-transformed HRV indices, where β is the estimated regression coefficient. Power analysis and sample sizes calculation were performed with power analysis and sample size (PASS) (NSCC, Kaysville, UT, USA). A significance level of 0.05 was used to determine statistical significance in our models.

Results

Thirty-six thousand 5-min measurements of indoor environmental variables and 32,423 5-min measurements of HRV indices were included in the data analyses. As shown in Table 1, the age range of the 50 housewives was 25–64 years. The mean BMI was 23.2 kg/m^2. Of the 50 participants, 44 with 157 home visits cooked by stir-frying, 38 participants with 129 home visits cleaned with detergent, and 24 participants with 96 home visits burned incense. Only 2 participants with 6 home visits went out shopping, and 50 participants with 194 home visits stayed home during the study periods. All participants used gas for cooking during the study period.

The participants' household PM$_{2.5}$ exposure, meteorological conditions, noise exposure and HRV indices are summarized in Table 2. When the participants' HRV indices were measured during the study periods of the years 2010 to 2012, they demonstrated relatively normal PM levels (WHO air quality guidelines for 24-h mean PM$_{2.5}$: 25 μg/m^3) [11], with a household PM$_{2.5}$ of 23.5 μg/m^3 (SD = 19.4 μg/m^3). The mean noise level was under 50 dBA, which may not enhance sympathetic activity [12]. The mean values (SD) of the log$_{10}$-transformed HRV indices were 1.62 msec (0.32) for SDNN and 1.11 msec (0.28) for r-MSSD.

The associations between household PM$_{2.5}$ and log$_{10}$-transformed HRV indices estimated by the mixed-effects models are

Table 1. Basic characteristics of the 50 participants (Mean ± SD).

Variables	
Age, years	
Mean	38.0±10.5
Range	25–64
Body mass index, kg/m^2	
Mean	23.2±2.5
Range	19.0–31.0
Household activities among the 50 participants, no (%)	
Indoor tobacco smoke exposure	7 (14)
Incense burning	24 (48)
Cooking with stir-fry	44 (88)
Cleaning with detergent	38 (76)
Indoor chemical dispersion	5 (10)
Shopping	2 (4)
Household activities during 200 home visits, no (%)	
Indoor tobacco smoke exposure	14 (7)
Incense burning	96 (48)
Cooking with stir-fry	157 (78.5)
Cleaning with detergent	129 (64.5)
Chemical indoor dispersion	8 (4)
Shopping	6 (3)

Table 2. Summary statistics of 5-min household PM$_{2.5}$, meteorological conditions, noise and HRV indices for the 50 participants experienced during 200 visits.

Variables	No.	Mean ± SD	Range
Household PM$_{2.5}$, μg/m^3	36,000	23.5±19.4	12.0–121.0
Noise, dBA	36,000	47.5±22.5	26.0–78.0
Temperature, °C	36,000	24.7±3.6	14.0–33.2
Relative humidity, %	36,000	69.5±3.0	65.1–78.2
Log$_{10}$ SDNN, msec	32,423	1.62±0.32	1.01–2.00
Log$_{10}$ r-MSSD, msec	32,423	1.11±0.28	0.55–1.65

interaction was found between indoor tobacco smoke exposure and household PM$_{2.5}$ for HRV indices.

Discussion

To our knowledge, this is the first study to evaluate the impact of personal exposure to household PM$_{2.5}$ and the impact of household activities on acute changes in HRV indices among housewives. In general, our findings suggest that personal exposure to household PM$_{2.5}$ may impair autonomic function and result in decreased HRV indices. Few studies have investigated the relationship between personal PM$_{2.5}$ exposure and HRV indices [5,13,14], and the majority of those have examined effects of ambient PM$_{2.5}$ exposure on autonomic function [15–17] in human subjects. Our PM$_{2.5}$-induced HRV reductions are in agreement with previous findings [13–17]. The findings support the statement of the American Heart Association's expert panel regarding the biological mechanisms of the effects of PM$_{2.5}$ on cardiovascular events, which are thought to occur through a neural mechanism, altering central nervous system functions [3].

We found that stir-frying and burning incense increased indoor PM$_{2.5}$ levels and that the increase may modify the association between household PM$_{2.5}$ and HRV indices. Epidemiological studies have reported that individuals exposed to indoor cooking oil fumes have a high risk of respiratory diseases [18] and lung cancer [19]. Few panel studies have reported the association between air pollution due to cooking and cardiopulmonary endpoints. In a panel of 387 nonsmoking Chinese restaurant workers, exposure to cooking oil fumes in kitchens was associated with increased urinary 8-OHdG levels. Female workers had a greater oxidative stress response to cooking oil fumes than male

shown in Table 3. After adjusting the models for age, BMI, hour of day, temperature, relative humidity, noise and household activity periods including indoor tobacco smoke exposure, indoor chemical dispersion, burning incense, stir-frying and cleaning with detergent, household PM$_{2.5}$ exposures significantly decreased SDNN and r-MSSD. Interquartile increases in the 1-, 2-, 3- and 4-h mean household PM$_{2.5}$ (19.8, 17.4, 16.5, and 16.2 μg/m^3, respectively) were associated with 1.25–4.31% decreases in SDNN. The 2-, 3- and 4- means were associated with 1.96–3.71% decreases in the r-MSSD. The greatest decreases in log$_{10}$-transformed HRV indices occurred at the 4-h mean. Age, BMI, temperature and household activity periods including indoor tobacco smoke exposure, burning incense, stir-frying and cleaning with detergent were significantly associated with decreased SDNN and r-MSSD. No association was observed between relative humidity, noise, indoor chemical dispersion and decreased HRV indices.

We found a consistent effect modification for household PM$_{2.5}$ by different household activity periods, including stir-frying, cleaning with detergent and burning incense (Table 4). Participants showed changes of −4.52% and −2.94% in SDNN and r-MSSD, respectively, associated with increased household PM$_{2.5}$ during stir-frying, whereas participants showed no significant change in HRV indices during study periods without cooking. We also found relatively stronger effects of household PM$_{2.5}$ on participants during cleaning with detergent compared to study periods without cleaning. A similar result was observed in a model including an interaction term between burning incense and household PM$_{2.5}$, although the statistical significance of the interaction was weaker than those in models evaluating effect modifications by cooking and cleaning. However, no significant

Table 3. Percentage changes (95% CI)[a] in HRV indices for interquartile range changes in household PM$_{2.5}$.

Outcome	1-hr mean	2-hr mean	3-hr mean	4-hr mean
Log$_{10}$ SDNN	−1.25	−2.15	−3.02	−4.31
	(−2.00, −0.50)	(−2.77, −1.53)	(−3.87, −2.17)	(−6.50, −2.12)
Log$_{10}$ r-MSSD	−0.12	−1.96	−2.64	−3.71
	(−2.78, 2.54)	(−3.01, −0.91)	(−4.80, −0.48)	(−5.11, −2.30)

[a]Coefficients are expressed as % changes for interquartile range changes in household PM$_{2.5}$ exposure in models adjusting for age, BMI, hour of day, temperature, relative humidity, noise and household activity periods including indoor tobacco smoke exposure, indoor chemical dispersion, burning incense, stir-frying and cleaning with detergent.

Table 4. Effect modification[a] of the association of HRV indices with interquartile range increases in 4-h mean household $PM_{2.5}$ by household activities.

	Log$_{10}$ SDNN (95% CI)	Log10 r-MSSD (95% CI)
Indoor tobacco smoke exposure		
Yes	−0.99 (−2.22, 0.24)	−1.49 (−2.84, −0.14)
No	0.31 (−1.50, 2.12)	1.52 (−3.07, 6.11)
P-value, interaction[b]	0.458	0.227
Burning incense		
Yes	−2.25 (−4.02, −0.48)	−1.99 (−3.42, −0.56)
No	1.89 (−0.45, 4.23)	0.65 (−1.00, 2.30)
P-value interaction	0.025	0.087
Stir-frying		
Yes	−4.52 (−5.37, −3.67)	−2.94 (−3.86, −2.02)
No	−1.15 (−3.08, 0.78)	0.42 (−1.58, 2.42)
P-value interaction	<0.001	<0.001
Cleaning with detergent		
Yes	−3.38 (−4.78, −1.98)	−2.44 (−4.31, −0.57)
No	−0.68 (−1.25, −0.11)	2.68 (1.57, 3.79)
P-value interaction	0.012	<0.001

[a]Coefficients are expressed as % changes for interquartile range changes in household $PM_{2.5}$ exposure in models adjusting for age, BMI, hour of day, temperature, relative humidity, noise and interaction terms.
[b]P-value is for effect modification.

workers [20]. Another panel study observed the association between household wood smoke exposure and ST-segment depression in a panel of 70 women using open wood fires for cooking [21]. The present study showed the effect modification of stir-frying on the association between household PM and HRV reduction. Burning incense is a long-standing Asian tradition used to give respect to ancestors. It has been reported that Taiwan households worship twice per day and are exposed to high levels of particulate air pollution [22]. A recent epidemiological study has reported that exposure to incense smoke in the home may increase the risk of lung cancer among smokers [23]. An *in vitro* study showed that exposure of human coronary artery endothelial cells to burning incense particles induced cytokine production and reduced nitric oxide formation [24]. Studies evaluating incense PM-induced autonomic dysfunction are lacking. Our study suggested some caution in the use of incense for housewives due to incense PM-induced decreases in HRV indices. Overall, our findings add to the growing evidence that air pollutants from stir-frying and incense burning can induce autonomic dysfunction in human subjects similar to those from vehicle and industrial emissions [13–17]. The public health implication is grave because high levels of exposure to indoor air pollution from stir-frying and burning incense are common in Asian countries.

Another interesting finding in our study was that use of detergents when cleaning appeared to modify the effects of household $PM_{2.5}$ on HRV indices; greater household $PM_{2.5}$ effects on HRV indices were observed when participants cleaned with detergent. A recent cross-sectional study used the indoor air quality checklist published by the Department of Occupational Health and Safety to evaluate the health risk of 102 building occupants in a nonindustrial workplace setting. The results showed that the main factors influencing the high number of complaints regarding indoor air quality included indoor detergent and chemical dispersion. Cleanliness led to high pollutant levels and

complaints from occupants due to health risks when working inside [10]. Although a limited understanding of the indoor dispersion of detergents and chemicals can make them the primary source of indoor air pollution, odor-related complaints are an example of the human sense of the existence of indoor chemical pollutants. The present study indicated that cleaning with detergent increased the levels of household $PM_{2.5}$ and modified the association between household $PM_{2.5}$ and HRV reduction. These findings have important implications for the feasibility of reliably investigating the associations between cleanliness, indoor air quality and health effects in large-scale epidemiological and intervention studies of household PM. We recommend further studies to investigate the clinical significance of the association between household particle control and cardiovascular health improvement.

Some possible limitations may confound our findings of HRV reduction by household $PM_{2.5}$, including unavailable data on associations with outcomes and some key physiologic and environmental information. First, we could not adjust for respiration-modulated autonomic activity in our study because we were unable to measure key respiration parameters, such as nasal and mouth airflow, chest wall movement and abdominal movement, during the study periods. Second, medication and comorbidity among older housewives could still confound our findings for household $PM_{2.5}$ effects on HRV reduction even though we used strict criteria to exclude cases with chronic cardiopulmonary diseases and specific medication from our study. Third, other unmeasured indoor air pollutants, such as ozone, carbon monoxide and total volatile organic compounds, may have confounded our findings. Fourth, non-randomized recruitment may result in selection bias and confound the association of HRV with household $PM_{2.5}$. Last, the effects of noise and household activities on HRV require further clarification because the sample size of our study may not be large enough to adjust for their effects completely.

Conclusions

We believe our findings generally indicate that household $PM_{2.5}$ was associated with autonomic function in housewives. Household activities including stir-frying, burning incense and cleaning with detergent were associated with increased levels of household $PM_{2.5}$ and modify its effects on HRV reduction.

Author Contributions

Conceived and designed the experiments: LL KC. Performed the experiments: YH HC BH CL. Analyzed the data: HC KC. Contributed reagents/materials/analysis tools: LL HC KC. Wrote the paper: LL KC.

References

1. Pope CA 3rd, Burnett RT, Thurston GD, Thun MJ, Calle EE, et al. (2004) Cardiovascular mortality and long-term exposure to particulate air pollution: epidemiological evidence of general pathophysiological pathways of disease. Circulation 109:71–77.
2. Pope CA 3rd, Dockery DW (2006) Health effects of fine particulate air pollution: lines that connect. J Air Waste Manag Assoc 56:709–742.
3. Brook RD, Rajagopalan S, American Heart Association Council on Epidemiology, Prevention, Council on the Kidney in Cardiovascular Disease, Council on Nutrition, Physical Activity, Metabolism, et al. (2010) Particulate matter air pollution and cardiovascular disease: An update to the scientific statement from the American Heart Association. Circulation 121:2331–2378.
4. Chan CC, Chuang KJ, Shiao GM, Lin LY (2004) Personal exposure to submicrometer particles and heart rate variability in human subjects. Environ Health Perspect 112:1063–1067.
5. Chuang KJ, Chan CC, Chen NT, Su TC, Lin LY (2005) Effects of particle size fractions on reducing heart rate variability in cardiac and hypertensive patients. Environ Health Perspect 113:1693–1697.
6. Lanki T, Ahokas A, Alm S, Janssen NA, Hoek G, et al. (2007) Determinants of personal and indoor PM2.5 and absorbance among elderly subjects with coronary heart disease. J Expo Sci Environ Epidemiol 17:124–133.
7. Klepeis NE, Nelson WC, Ott WR, Robinson JP, Tsang AM, et al. (2001) The National Human Activity Pattern Survey (NHAPS): a resource for assessing exposure to environmental pollutants. J Expo Sci Environ Epidemiol 11:231–252.
8. Lim SS, Vos T, Flaxman AD, Danaei G, Shibuya K, et al. (2012) A comparative risk assessment of burden of disease and injury attributable to 67 risk factors and risk factor clusters in 21 regions, 1990–2010: a systematic analysis for the Global Burden of Disease Study 2010. Lancet 380:2224–2260.
9. Polidori A, Turpin B, Meng QY, Lee JH, Weisel C, et al. (2006) Fine organic particulate matter dominates indoor-generated PM2.5 in RIOPA homes. J Expo Sci Environ Epidemiol 16:321–331.
10. Syazwan A, Rafee BM, Juahir H, Azman A, Nizar A, et al. (2012) Analysis of indoor air pollutants checklist using environmetric technique for health risk assessment of sick building complaint in nonindustrial workplace. Drug Healthc Patient Saf 4:107–126.
11. WHO-Europe (2006) Air Quality Guidelines, Global Update 2005: Particulate Matter, Ozone, Nitrogen Dioxide, and Sulfur Dioxide: World Health Organization Europe.
12. Lee GS, Chen ML, Wang GY (2010) Evoked response of heart rate variability using short-duration white noise. Auton Neurosc 155:94–97.
13. Magari SR, Hauser R, Schwartz J, Williams PL, Hauser R, et al. (2002) Association between personal measurements of environmental exposure to particulates and heart rate variability. Epidemiology 13:305–310.
14. Magari SR, Hauser R, Schwartz J, Williams PL, Smith TJ, et al. (2001) Association of heart rate variability with occupational and environmental exposure to particulate air pollution. Circulation 104:986–991.
15. Gold DR, Litonjua A, Schwartz J, Lovett EG, Larson AC, et al. (2000) Ambient pollution and heart rate variability. Circulation 101:1267–1273.
16. Chuang KJ, Chan CC, Su TC, Lee CT, Tang CS (2007) The effect of urban air pollution on inflammation, oxidative stress, coagulation, and autonomic dysfunction in young adults. Am J Respir Crit Care Med 176:370–376.
17. Chuang KJ, Chan CC, Su TC, Lin LY, Lee CT. (2007) Associations between particulate sulfate and organic carbon exposures and heart rate variability in patients with or at risk for cardiovascular diseases. J Occup Environ Med 49:610–617.
18. Svendsen K, Sjaastad AK, Siverstsen I (2003) Respiratory symptoms in kitchen workers. Am J Ind Med 43:436–439.
19. Behera D, Balamugesh T (2006) Dose-response relationship between cooking fumes exposures and lung cancer among Chinese nonsmoking women. Cancer Res 66:4961–4967.
20. Pan CH, Chan CC, Wu KY (2008) Effects on Chinese restaurant workers of exposure to cooking oil fumes: a cautionary note on urinary 8-hydroxy-2'-deoxyguanosine. Cancer Epidemiol Biomarkers Prev 17:3351–3357.
21. McCracken J, Smith KR, Stone P, Díaz A, Arana B, et al. (2011) Intervention to lower household wood smoke exposure in Guatemala reduces ST-segment depression on electrocardiograms. Environ Health Perspect 119:1562–1568.
22. Lung SC, Mao IF, Liu IJ (1999) Community air quality monitoring and resident's personal exposure assessment in large metropolitan areas in Taiwan. Taiwan Environmental Protection Administration.
23. Tse LA, Yu IT, Qiu H, Au JS, Wang XR (2011) A case-referent study of lung cancer and incense smoke, smoking, and residential radon in Chinese men. Environ Health Perspect 119:1641–1646.
24. Lin LY, Lin HY, Chen HW, Su TL, Huang LC, et al. (2012) Effects of temple particles on inflammation and endothelial cell response. Sci Total Environ 414:68–72.

7

Ligia italica (Isopoda, Oniscidea) as Bioindicator of Mercury Pollution of Marine Rocky Coasts

Guglielmo Longo[1], Michelanna Trovato[1], Veronica Mazzei[1], Margherita Ferrante[2]*, Gea Oliveri Conti[2]

1 Dipartimento di Scienze Biologiche, Geologiche e Ambientali, Università di Catania, Catania, Italy, 2 Dipartimento di Anatomia, Biologia e Genetica, Medicina Legale, Neuroscienze, Patologia Diagnostica, Igiene e Sanità Pubblica "G. F. Ingrassia", Università di Catania, Catania, Italy

Abstract

In this study, we evaluated the possible role of *Ligia italica* as a bioindicator for the monitoring of heavy metals pollution in the supralittoral zone of marine rocky coasts. Between 2004 and 2011 specimens of *L. italica* were collected along the Eastern Sicilian coasts from sites known for their high pollution levels as they are near to an area where in September 2001 a refinery plant discharged into the sea some waste containing Hg. Other specimens were collected from the Vendicari Natural Reserve located about 30 miles from the polluted sites and used as control area. On a consistent number of animals, the concentration *in toto* of As, Cd, Cr, Hg, Ni, Pb, V, was determined by Atomic Absorption Spectrometry. On other animals, investigations were carried out in order to check for ultrastructural alterations of the hepatopancreas, that is the main metals storage organ in isopods. Results revealed the presence, in the animals collected in 2004 from the polluted sites, of considerable concentrations of Hg and of lower concentrations of other metals such as As, Pb and V. The Hg bioaccumulation resulted in remarkable ultrastructural alterations of the two cellular types (B and S cells) in the epithelium of the hepatopancreas. Surprisingly, a moderate amount of Hg was also found in specimens collected in 2004 from the Vendicari Natural Reserve, proving that the Hg pollution can also spread many miles away. Animals collected from the polluted sites in the following years showed a progressively decreasing Hg content, reaching very low levels in those from the last sampling. Also, the ultrastructural alterations found in the hepatopancreas of the animals from the last sample were quite irrelevant. In conclusion, *Ligia italica* can represent a good bioindicator and the ultrastructure of the hepatopancreas could be used as ultrastructural biomarker of heavy metals pollution in the supralittoral zones.

Editor: Philippe Archambault, Université du Québec à Rimouski, Canada

Funding: The authors have no funding or support to report.

Competing Interests: The authors have declared that no competing interests exist.

* E-mail: marfer@unict.it

Introduction

The coast is the frontier between land and sea. On the rocky shores the supralittoral zone is the emerged area regularly reached by the sea spray and/or splash. In the supralittoral zone of the sea rocky coasts the dynamic influence of the marine water interacts with the terrestrial environment. The supralittoral, for this reason, is an ecotone particularly subject to pollution because it receives contaminants coming from both environments [1–2]; it is characterized by strong thermal range, intense light radiation and significant changes in salinity, depending on evaporation and rainwater inputs (trends in rainfall or rainwater regime). Therefore, its wildlife shows features that exhibit affinity with both environments. The most numerous animal group is represented by crustaceans, particularly by Isopoda, Amphipoda and Decapoda [3–4].

In the last years, the development of sensible techniques and sophisticated assays has improved the environmental monitoring programs by using valid bioindicators of marine ecosystems, particularly for monitoring heavy metals in the sea water [5–8]. A biological indicator is a organism used to monitor the environmental quality of a particular ecosystem that provides quantitative information on the quality of the environment around it. Therefore, a good bioindicator will indicate the presence and the amount of a pollutant and will provide additional information

about the intensity of its exposure [9–11]. A bioindicator specie is considered valid if some basic characteristics are satisfied, in particular: accessibility, bio-ecological suitability, reliability and representativeness.

The marine heavy metals pollution is an issue for many Eastern Sicilian coasts where several industrial and petrochemical plants have discharged massive quantities of polluting gases, organic contaminants and heavy metals, thus causing dramatic levels of pollutants in air, water and land with consequent harmful outcome to human health [12–15]. In 2001, for example, one refinery plant dumped into the Sicilian sea a considerable amount of mercury (Hg) and other polluting heavy metals; in fact, analyses requested by the Public prosecutor's office of Syracuse demonstrated that the amount of Hg present in the sea was twenty thousand times higher than the upper limit established by law. The consequences of contamination by Hg [16–17], and by other metals as well, are known for many years, as they pose a worldwide problem [12–14]. However, despite the seriousness of that episode of pollution, until now, the knowledge of the environmental status is lacking because studies aimed at evaluating its effects on biological communities present in the concerned area are numerically very poor, probably due also to the lack of a validated biondicator for studying the supralittoral zone of the rocky coasts.

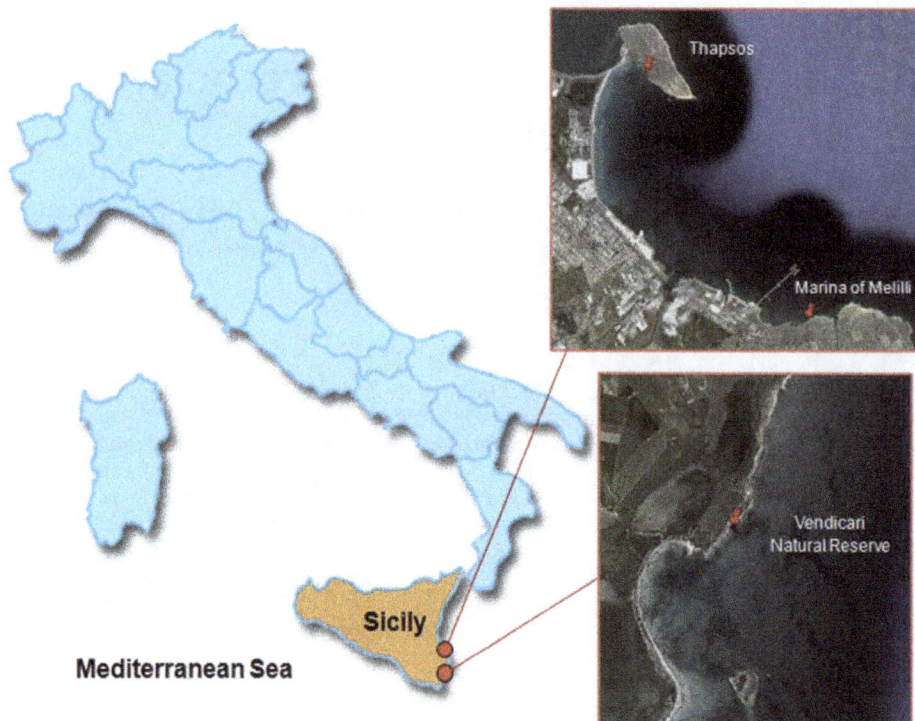

Figure 1. Map of sampling sites.

Ligia italica Fabricius 1798 (one of many coastal *Ligia*) is an oniscidean isopod particularly abundant in this Sicilian ecotone and it is being closely bound to the marine environment from which leaves only temporarily and only for few meters [4;7]. 'The low vagility and high levels of isolation observed among populations of coastal *Ligia* make this isopod a potential good bioindicator, free from cross contamination, as they appear to be highly constrained throughout their life cycle to the same rocky beach [4]. The effectiveness of Isopoda as excellent bioindicators and bioaccumulators of heavy metals in biomonitoring programs is claimed by scientific literature [4;18–27]. The main metal storage organ in isopods is the hepatopancreas [28–29]. Specifically, Hg accumulates mainly in the exoskeleton while cadmium, lead and chromium can accumulate preferably into the hepatopancreas [29]. In the light of these considerations, it is extremely important to verify the potential for bioaccumulation and biological responses of *L. italica* Fabricius 1798 in order to assess their health status and, indirectly, to evaluate the quality of the environment in which they live.

In the present research we evaluated the possible role of *Ligia italica*, as bioaccumulator and bioindicator, in monitoring Hg and other heavy metals pollution in the supralittoral zone of marine rocky coasts. The validation of this species as bioindicator has been carried out by calculating the amount of Hg and other heavy metals, such as arsenic (As), cadmium (Cd), chromium (Cr), nickel (Ni), lead (Pb), and vanadium (V), in animals in toto and by observing their effects on hepatopancreas ultrastructure.

Materials and Methods

The research was carried out on a consistent number of sexually mature male and female *L. italica*, samples were collected twice in 2004, 2007 and 2011 from three different sites of the district of Syracuse (Sicily): Thapsos (Penisola Magnisi), along the rocky

shore in front of the petrochemical pole (where in 2001 the discharge of a large amount of Hg was registered); Marina di Melilli, another very polluted coastline near Thapsos, and the Vendicari Natural Reserve (see Fig. 1). We chose the last site, being non-polluted, because we wanted to utilize it for collection of control samples. In order to calculate metals concentrations, from each of the three research sites, we took three groups of ten animals (five males and five females); they were weighed, beheaded and immediately immersed in 10 ml vials containing 3 ml of HNO_3 (Suprapur) +1 ml of H_2O_2 (mineralization solution). Analysis were carried out in the Laboratory of Environmental and Food Hygiene of the Hygiene and Public Health Department "*G. F. Ingrassia*", University of Catania, through the following steps:

1) the digestion solution was adjusted to a final volume of 6 ml of HNO_3 and 1 ml of H_2O_2;

2) samples were mineralized in a Milestone 1200 microwave digestor, in teflon dedicated vessels, using the specific program;

3) then, samples were transferred into 10 ml flasks and were brought to volume with ultrapure metal free water for determination of As, Cd, Cr, Ni, Pb, V, by the atomic absorption spectrometer (AAS) Perkin Elmer Analyst 800.

For Hg determination, we used the FIAS-AAS (Cold vapour Technique or Hydride Generation System) with the same Analyst 800. The instrument was properly calibrated for each metal by means of certified Perkin Elmer calibration standards. The values of metals contents were reported in $\mu g\ g^{-1}$ dry weight. Statistical analysis was performed using the Kruskal-Wallis test on SPSS-18.0 software.

The ultrastructural investigations were carried out on the hepatopancreas, the organ primarily involved in the bioaccumu-

Table 1. Heavy metal content (μg g^{-1} dry weight) of *Ligia italica* in the different samples.

	THAPSOS			MARINA OF MELILLI			VENDICARI NATURAL RESERVE		
METALS	**2004**	**2007**	**2011**	**2004**	**2007**	**2011**	**2004**	**2007**	**2011**
Hg	8.13±1.78	3.24±0.42	0.33±0.07	12.84±2.34	0.75±0.18	0.03±0.01	4.29±0.55	0.75±0.11	0.02±0.01
As	5.46±1.08	6.75±1.50	15.30±3.19	3.99±0.81	5.10±0.91	15.54±3.11	4.50±0.78	3.75±0.60	4.09±0.63
Cd	0.24±0.07	0.36±0.11	1.05±0.22	0.27±0.09	0.22±0.06	1.26±0.17	0.33±0.07	0.25±0.04	0.22±0.07
Cr	0.27±0.09	1.68±0.27	1.41±0.22	0.18±0.05	0.67±0.14	1.50±0.31	0.21±0.04	1.01±0.13	0.39±0.06
Ni	0.24±0.05	1.80±0.37	3.99±0.62	0.03±0.01	0.81±0.18	4.26±0.59	0.03±0.01	1.20±0.22	0.47±0.09
Pb	4.92±1.70	4.02±0.73	4.17±0.73	7.83±1.62	1.74±0.45	9.30±1.56	2.37±0.40	2.37±0.34	2.04±0.22
V	3.75±0.82	4.20±0.99	0.63±0.12	1.65±0.43	0.70±0.13	0.57±0.06	1.26±0.12	0.78±0.09	0.91±0.11

Means ± SD.

lation of heavy metals. Some males and females were beheaded and immediately immersed in Ringer's saline solution modified for land isopods acc. Legrand cited by Besse [30]; then the hepatopancreas were removed. Fixation was carried out through 2.5% glutaraldehyde in 0.1 M phosphate buffer, pH 7.3 for 4 h at 4°C; after repeated washing in the same buffer the specimens were post fixed in 1% OsO_4, in the same buffer, for 1 h at room temperature. The samples were dehydrated in ethanol followed by propylene oxide and embedded in Embed 812 (EMS). For light microscopy, semi-thin sections were cut on a Ultracut Leica ultramicrotome with diamond blades, then they were stained with 0.5% blue toluidine in 0.1 M phosphate buffer, pH 7.3 or with the polychromatic staining method of Sato and Shamoto [31]. For electron transmission microscopy, ultra-thin sections were collected on Cu/Rh grids of 200 mesh, were stained with uranyl acetate and lead citrate acc. Reynolds [32] and examined in a Philips CM 10 electron microscope, at 60 or 80 kV.

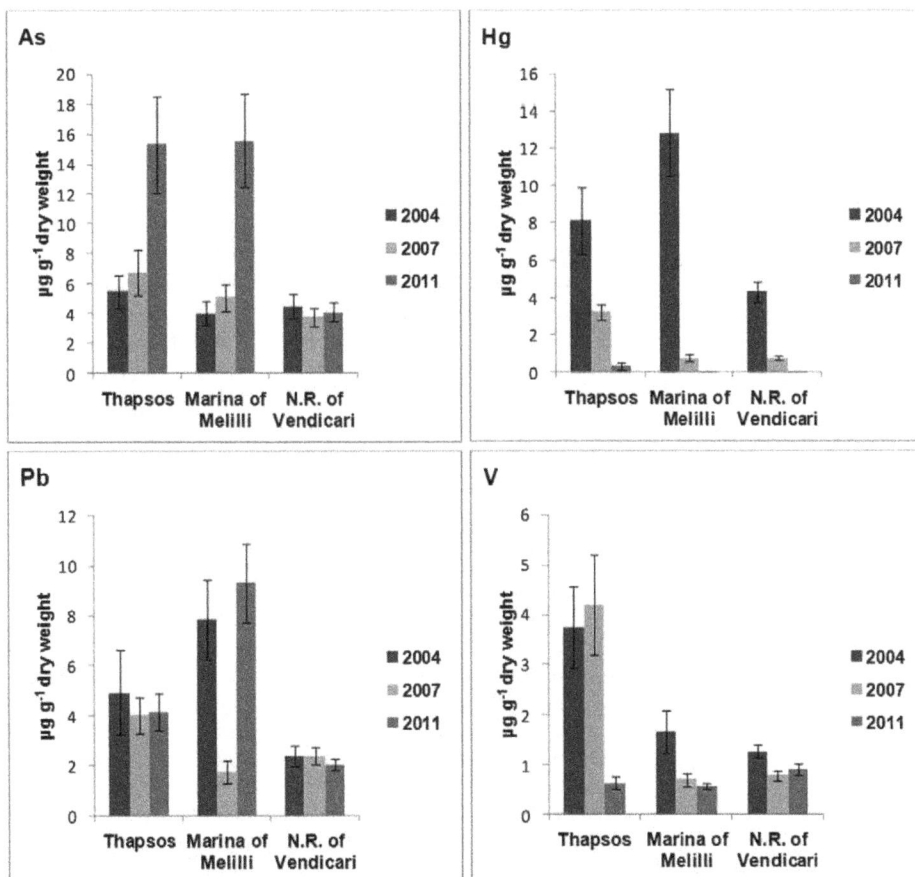

Figure 2. As, Hg, Pb and V content (μg g^{-1} dry weight) in the different samples.

Table 2. Kruskal-Wallis test.

YEARS	As	Hg	Pb	V
2004	2.76	7.20**	6.49*	5.96*
2007	5.07	5.53	6.49*	5.80*
2011	5.42	7.32	7.20**	1.69

Comparison of the different sites for the same year.
($H_{0.05} = 5.60 - *: p<0.05$, $H_{0.01} = 7.20 - **: p<0.01$).

Results

Metal concentration

The metals concentrations determined in the hepatopancreas are reported in Table 1. The content of heavy metals was generally higher in the animals collected from the polluted sites compared to those from the Vendicari Natural Reserve and, moreover, it was significantly different in relation to the year in which sampling was done. In the animals collected in 2004 from the two polluted sites (Table 1) the most significant data were those on the content of Hg and, to a lesser extent, of Pb, As and V,. In particular, the highest content of Hg – 12.84 $\mu g\ g^{-1}$ dry weight – was found in animals from Marina di Melilli, whereas in those from Thapsos, which is located in front of the petrochemical plants responsible for the release in the sea of large quantities of waste containing Hg, the content of this metal was lower (8.13 $\mu g\ g^{-1}$). However, the most striking feature was the presence of Hg (4.29 $\mu g\ g^{-1}$) even in animals collected from the Vendicari Natural Reserve, site theoretically not involved in heavy metals pollution (Fig. 2). The animals collected in 2007 showed a Hg content still appreciable only in those collected from Thapsos, whereas those collected from Marina di Melilli and from the Vendicari Natural Reserve showed a very modest content of this metal (Table 1, Fig. 2). Seven years after the first sampling, the Hg content had decreased to very low levels in all animals collected from the three different sites (Table 1, Fig. 2).

The amount of other tested metals, such as Cd, Cr and Ni, was rather inconsistent, whereas low but appreciable concentrations of As, Pb and V, were detected in the animals collected from all sites (Table 1, Fig. 2). Statistically highly significant differences in the content of Hg, Pb and As were found in samples collected in the same year from the different sites, as well as in samples collected in different years from the same site (Table 2, Table 3).

Morphology and ultrastructure of hepatopancreas

In the animals collected from the Vendicari Natural Reserve in 2004, the hepatopancreas showed morphological and ultrastructural features very similar to those already known from other oniscidean isopods; however, while in the terrestrial isopods the

Table 3. Kruskal-Wallis test.

SITES	As	Hg	Pb	V
Thapsos	5.96*	7.20**	1.07	5.96*
Marina of Melilli	5.96*	7.26**	5.96*	5.42
Vendicari N.R.	0.80	7.26**	0.36	2.49

Comparison of the different years for the same site.
($H_{0.05} = 5.60 - *: p<0.05$, $H_{0.01} = 7.20 - **: p<0.01$).

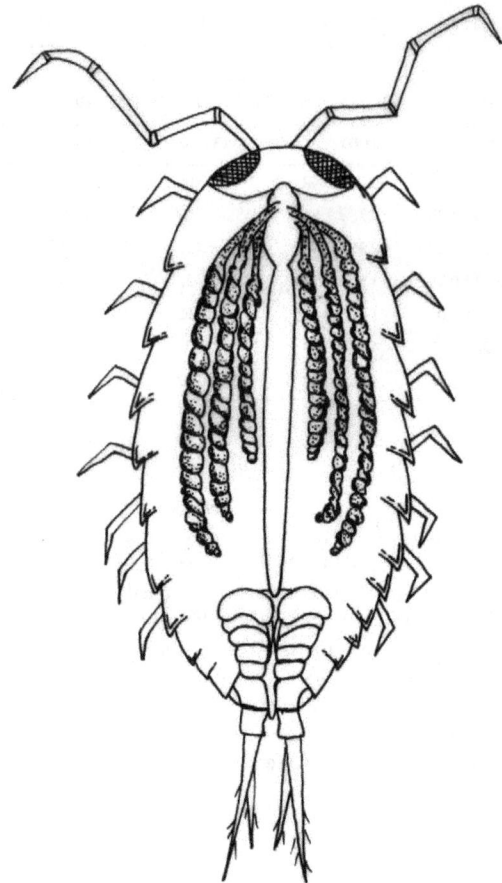

Figure 3. Schematic drawing of *in situ* hepatopancreas tubules of *Ligia italica*.

hepatopancreas consists of two pairs of spiral blind-ending tubules that run adjacent to the hindgut throughout the length of the pereion, in the *L. italica*, as in the marine isopods, a third pair of shorter and thinner tubules is present (Figs. 3 and 4A). The histological structure of the tubules is very simple and almost homogeneous throughout their length with exception of the proximal and distal regions; their wall consists of a monolayered epithelium that lies on a thin basal lamina surrounded by a net of myocites (Fig. 4B). As in other species so far studied, two types of cells are present even in the epithelium of *L. italica*: the large basophilic cells (B cells) and the small cells (S cells), that alternate almost regularly in the organ cross sections (Fig. 4C). The B cells are dome-shaped and their apical portions protrude considerably into the organ's lumen (Fig. 4C). The wide basal plasmalemma of the B cells makes a typical membranous labyrinth, characterized by a relevant number of tubular invaginations (Fig. 4D). The large nucleus of the B cells is mostly located in a basal position; in some cells it is euchromatic, in others it shows a more condensed chromatin (Figs. 4C and 5A). The cytoplasm is slightly dense and contains abundant lipid droplets of various sizes, rough reticuloendothelial elements (RER) and vesicles in addition to numerous free ribosomes and glycogen granules (Fig. 5B). The mitochondria are numerous and prevalently localized in the subnuclear region (Fig. 5A); most of them appear swollen and show a great reduction of their cristae (Fig. 5C). The rough endoplasmic reticulum is extended above and below the nucleus; in some B cells it is arranged in small cisternae that show some fenestrations while in

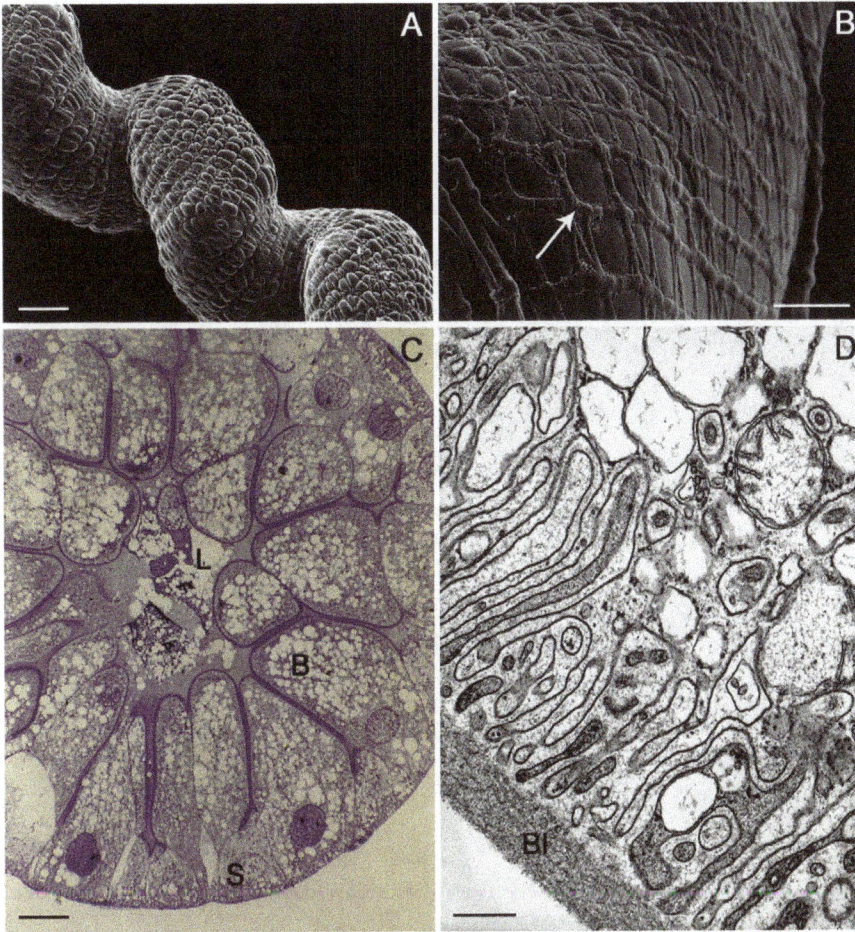

Figure 4. Tubule of the hepatopancreas of *Ligia italica* from Vendicari Natural Reserve. A–B: SEM, C: LM, D: TEM. A) The tubule shows a regular spiral trend and it is possible to observe the basal portions of S and B cells that protrude from its surface. B) A network of miocytes (arrow) surrounds the thin basal lamina that covers the surface of the tubules. C) Cross section of the proximal region of a tubule. B, dome-shaped large cells; S, small cells. The body of the B cells is filled with numerous lipid droplets. L, lumen of the organ (Semithin section; Sato and Shamoto stain). D) The basal plasmalemma of the B cells makes a typical membranous labyrinth characterized by a relevant number of tubular invaginations; Bl, basal lamina. Scale bar: A = 200 µm; B = 30 µm; C = 20 µm; D = 0.5 µm.

most of them it is formed by small vesicles or large vacuoles (Figs. 5B–C). The apical regions of the B cells show a thick border of microvilli (Fig. 6A) that in some cells undergo vesiculation and subsequent disorganization (Fig. 6B). The S cells are wedge-shaped and are much lower than the B cells; their lateral surface is closely adjacent to that of the B cells, so, only a small extent of their apical surface overlooks the lumen of the hepatopancreas (Fig. 4C). Many of the characteristics of the S cells are comparable to those of the B cells but their cytoplasm is less dense and the nucleus has sometimes an irregular shape. The RER is well developed and consists mostly of rounded vesicles (Fig. 6C); golgian structures and lipid droplets are rarely observed. Small and rounded mitochondria showing a reduction of their cristae are scattered in the cytoplasm as small vesicles with an electron-dense content (Fig. 6C). The S cells contain, moreover, some vesicles that store metals and others with a heterogeneous content, most likely of lysosomal derivation (Fig. 6D).

In the specimens of *L. italica* collected in 2004 from the two polluted sites, the hepatopancreas does not show considerable morphological modifications but rather it undergoes more marked alterations of the cellular ultrastructure that have affected the majority of cells present along the entire wall of the organ. Some of these alterations are common to the two types of cells, some are more specific. The most important alterations common to both types of cells relate to:

♦ damage of the microvillous border; this alteration does not affect all cells equally as in some of them the microvilli retain an aspect almost normal, in others they show an irregular profile like "sausage", in others they undergo a process of vesiculation that results in their complete disorganization (Fig. 7A);

♦ partial or total disappearance in many cells of the basal membranous labyrinth (Fig. 7B);

♦ a more marked swelling of the mitochondria that are also almost completely devoid of cristae and frequently present an open cavitation of the matrix (Fig. 7C);

♦ a notable reduction of the nuclear volume consequent to a massive condensation of chromatine, sometimes associated to some large dilatations of perinuclear cisterna (Fig. 7D).

Other alterations regarding B cells are:

♦ further increased cytoplasmic density;

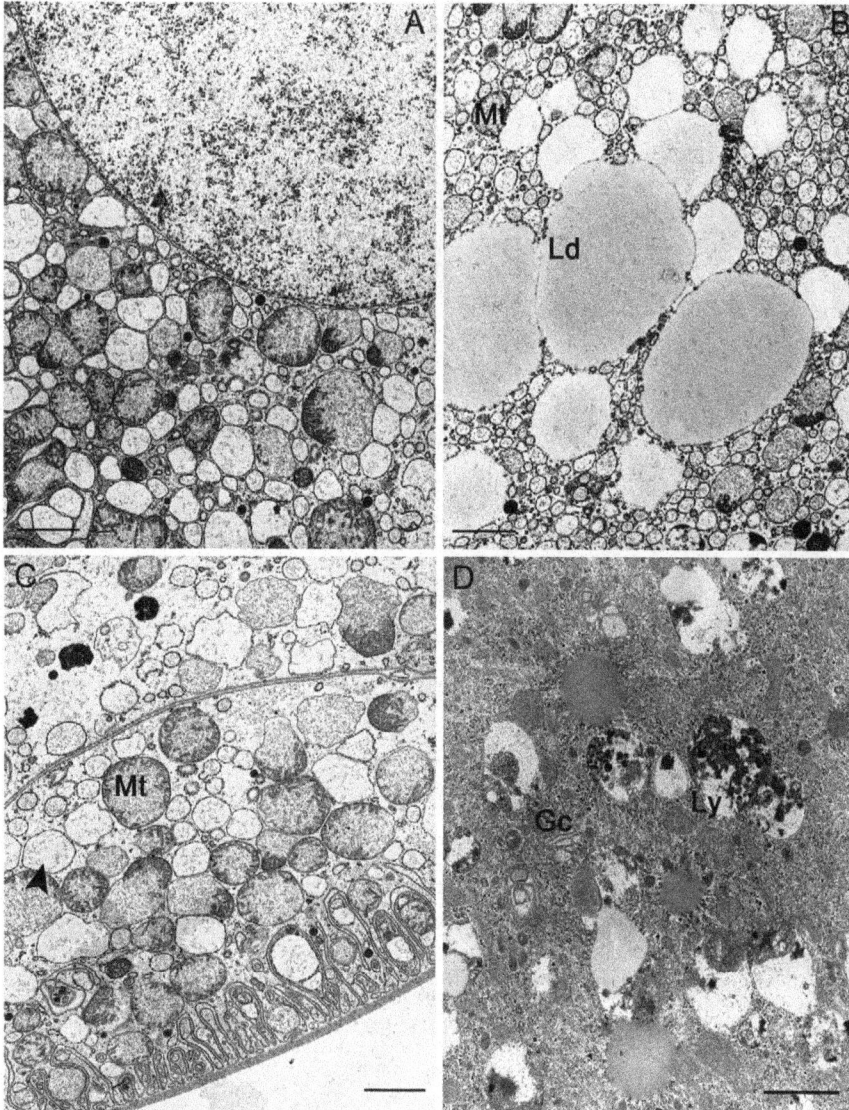

Figure 5. Tubule of the hepatopancreas of *Ligia italica* from Vendicari Natural Reserve. A–D TEM. A) Euchromatic nucleus of B cell. B) The cytoplasm of B cells contains abundant lipid droplets (Ld) of various sizes, mitochondria (Mt), RER elements and vesicles. C) Mitochondria (Mt) and RER elements (arrow head), particularly abundant in the subnuclear region of the B cells; Mt are swollen and show a great reduction of their cristae. D) Several saccules of the Golgi complex (Gc) and lysosomes (Ly) are present in the supranuclear region of the B cells. Scale bar: A–D = 2 μm.

♦ considerable swelling and vesiculation of the RER; large vacuoles, probably derived from the RER, are present around the nucleus in some B cells (Fig. 8A);

♦ presence in their cytoplasm, near the lipid droplets, of small scattered accumulations of electron-dense material that in some cells become much more large and abundant (Fig. 8B).

The major ultrastructural alterations concerning the S cells are represented primarily by an increase in the number of vesicles that accumulate heavy metals (Fig. 8C) and of other vesicles containing an heterogeneous electron-dense material, most likely of lysosomal derivation (Fig. 8D).

In the specimens collected in the subsequent samplings the ultrastructural alterations highlighted in the hepatopancreas were quite irrelevant.

Discussion

Our research, based on a series of sampling carried out in two different sites of Eastern Sicily, has shown that three years after the discharge into the sea of a massive amount of Hg by a petrochemical industry located in the immediate vicinity of the study areas, a significant amount of this metal as well as modest concentrations of other metals, such as As, Pb, and V, were still present in tissues of *Ligia italica*, an oniscidean isopod, frequent and abundant in the rocky supralittoral zone of those sea coasts. Data get more significant value if we consider that Hg, after its discharge into the marine environment, is in part linked to suspended sediments (therefore it is temporarily buried) and then undergoes various processes of transformation and biodegradation including chemical modifications [33–37], so that its concentration in the upper layers of marine waters (that come into contact with *L. italica*) is significantly reduced. Surprisingly, a fair amount of Hg

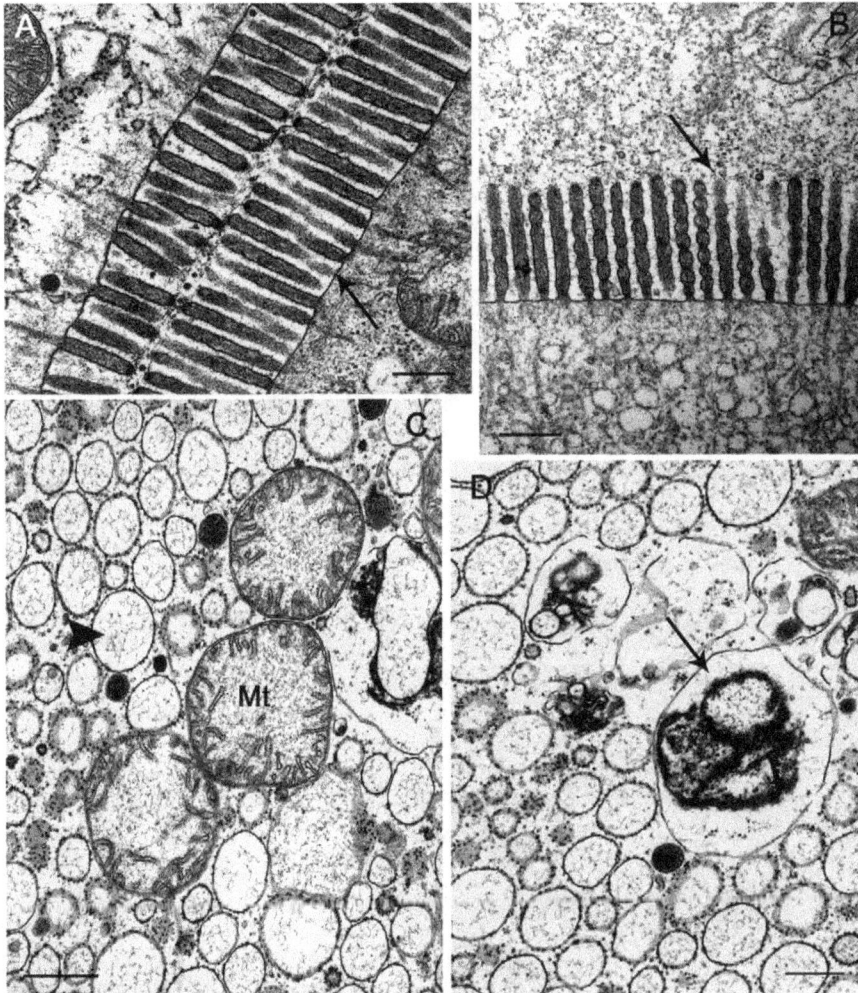

Figure 6. Tubule of the hepatopancreas of *Ligia italica* from Vendicari Natural Reserve. A–D TEM. A–B) Apical surface of B cells bearing a thick border of microvilli that in some cells undergo a progressive vesiculation (arrows). C) The cytoplasm of S cells is less dense and has a well developed rough endoplasmic reticulum in the form of rounded vesicles of various sizes (arrow head); also in the S cells the mitochondria (Mt) are swollen and with a clear reduction of their cristae. D) In the cytoplasm of S cells numerous vesicles (arrow) with a heterogeneous content, most likely of lysosomal derivation, are present. Scale bar: A, B = 0.5 μm; C = 1 μm; D = 1.5 μm.

was also found in some *L. italica* collected from the Vendicari Natural Reserve, the environment chosen as a control area (because presumably free from Hg pollution), located about 30 miles away from the industrial plant responsible for the Hg pollution. This result would indicate that the discharge could propagate, before its immobilization, even at a considerable distance from the release site thanks to the sea currents and the atmospheric spread, as well as to the marine sediments of the primarily polluted site made up of coarse-grained material that could have scavenged only a limited portion of the total Hg [38], thus helping the spreading of the Hg pollution.

In animals collected from the same sites, three and seven years after the first sampling respectively, concentrations of As, Cd, Cr, Ni, Pb and V showed more or less modest changes, probably related to their variable discharge into the marine environment from chemical plants present in the area under examination; the amount of Hg found in the animals of the two polluted sites has, instead, undergone a progressive and marked decrease, until reaching values close to zero in the samples from Marina di Melilli and from the Vendicari Natural Reserve. This would suggest that

industries have not performed a further discharge of waste containing Hg and/or that the significant amount of Hg released into the sea in 2001 and in previous years has been subject of an outflow represented, in descending order, by the sedimentation of particles containing Hg, by the offshore exports and by the dilution in the atmosphere as a result of volatilization [35].

The ultrastructural investigations carried out after the first sampling on the hepatopancreas of animals collected from the two heavily polluted sites, showed a considerable set of cellular alterations that affected both cell types present in the wall of the organ; most probably, these alterations are ascribable to the bioaccumulation of mercury even if cannot be entirely excluded an additional influence resulted from the contemporaneous presence of other metals such as As, Pb and V. Some of these cellular alterations, although less marked and more sporadic, have also been observed in the hepatopancreas of the animals collected from the control area of the Vendicari Natural Reserve. These alterations have not equally involved the entire wall of the organ but have variably involved a more or less relevant number of both types of its epithelial cells. The most significant cellular effect was

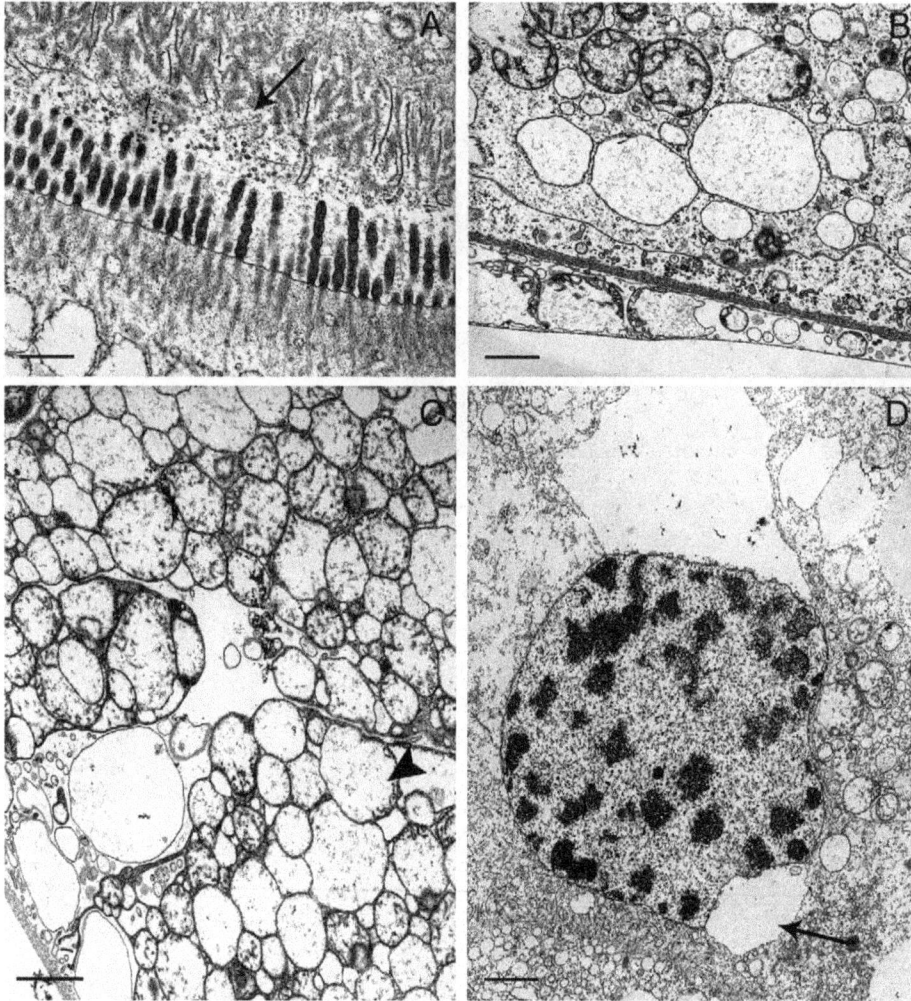

Figure 7. Tubule of the hepatopancreas of *Ligia italica* from the two polluted sites. A–D TEM. A) Most of the B and S cells show a marked alteration of the microvillous border that, in some cells, undergoes a total disorganization (arrow). B) The B cells show a partial or total disappearance of the basal membranous labyrinth. C) The mitochondria of B cells undergo a more marked swelling and are also almost completely devoid of cristae; many show a frank cavitation of matrix (arrow head). D) The nuclei of B cells exhibit a considerable reduction of their volume, consequent to a massive condensation of chromatin, sometimes associated with some large dilatations (arrow) of their perinuclear cisterna. Scale bar: A = 0.5 μm; B = 1 μm; C = 1.5 μm; D = 2 μm.

certainly represented by the disorganization of the microvillous border that is, undoubtedly, the first target of the action exerted by the metal. Some B and S cells have only shown a moderate decrease in the number of microvilli that appear shorter and slightly deformed; in other cells the microvilli undergo a progressive vesiculation or total destruction associated with the disappearance of their actin cytoskeleton and of the below terminal net. A similar effect has already been described by Köhler et al. [28] for the hepatopancreas of *Porcellio scaber* contaminated with high concentrations of Cd, Pb an Zn and represents, therefore, the result of a mechanism common to many heavy metals; however, the same changes have also been reported by Storch e Lehnert-Moritz [39] for the hepatopancreas of animals of *Ligia oceanica* subjected to a prolonged starvation period. The reduced functionality of the microvillous border has an obvious negative effect on the absorption of nutrients by the hepatopancreas [40].

Another main target of the metal is the labyrinth formed by the plasma membrane in the basal region that undergoes a remarkable

reduction in most of the B cells. In their research on the toxic effect of Hg in *Crangon crangon*, Andersen and Baatrup [41] reported that most of the metal accumulated in the hepatopancreas was absorbed from the haemolymph; this finding would explain the severe damage of the plasma membrane also in the basal region of the B cells. The damage induced by Hg on the basal membranous labyrinth do not seem to be induced by other heavy metals; Köhler et al. [28], in fact, in specimens of *P. scaber* contaminated with high concentrations of Cd and Pb never observed a significant alteration of the basal labyrinth.

Other alterations consequential to the Hg accumulation involve:

a) the RER that undergoes a more or less pronounced fragmentation and consequent swelling and vesiculation in many B and S cells;

b) the mitochondria, that show in a large extent a significant swelling and a marked reduction or even a total disappearance of their cristae;

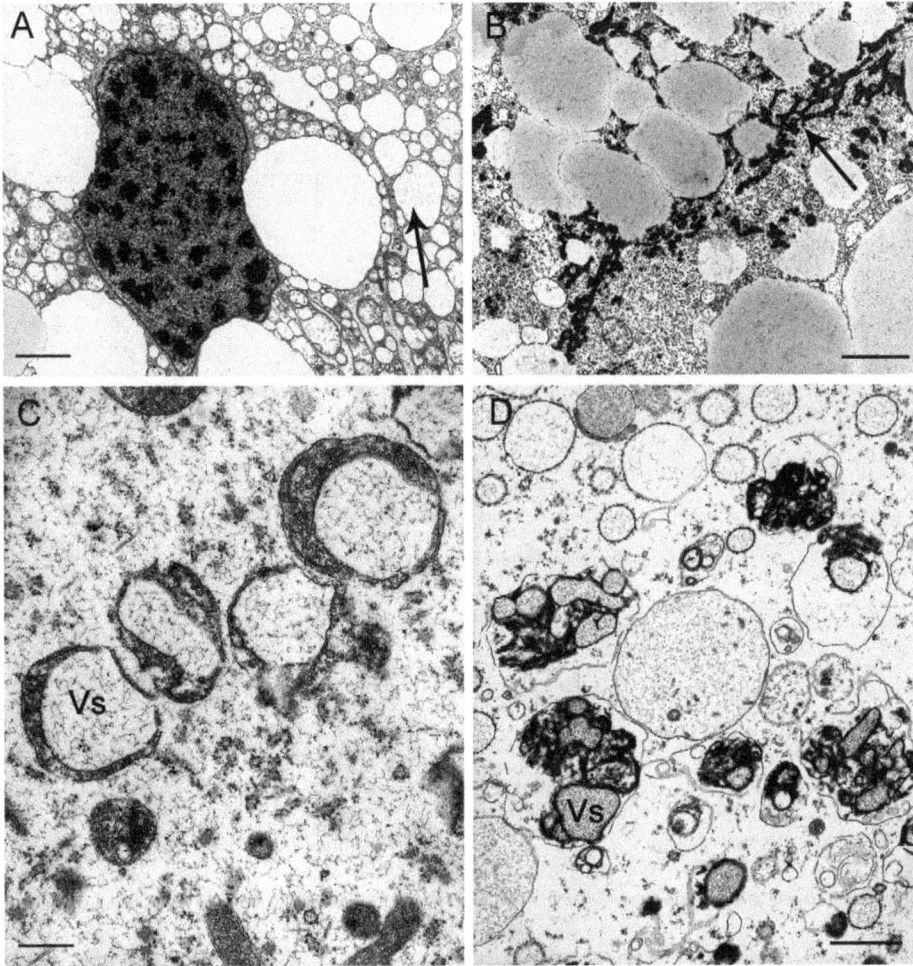

Figure 8. Tubule of the hepatopancreas of *Ligia italica* from the two polluted sites. A–D TEM. A) In many B cells the rough endoplasmic reticulum undergoes a marked swelling and vesiculation (arrow); large vacuoles (Lv), probably derived from the rough endoplasmic reticulum, are present around the nucleus. B) The lipid droplets of B cells are surrounded by scattered accumulations of electron-dense material that in some cells are particularly abundant (arrow). C) In many S cells the number of vesicles of heavy metals accumulation increases moderately. D) The major significant ultrastructural alteration concerning the S cells is, however, represented by the presence of numerous vesicles containing heterogeneous electron-dense material (Vs), that are most likely of lysosomal derivation. Scale bar: A, B = 2 μm; C = 0.5 μm; D = 1 μm.

c) a decay of lipid droplets often showing at the periphery the presence of clusters of electron-dense material probably corresponding to lipofuscin;

d) a marked condensation of nuclear chromatine often associated with one or more large dilatations of perinuclear cisterna;

e) an increased number of vesicles of lysosomal derivation filled with heterogeneous electron-dense materialIn other organisms too, lysosomes have been identified as the major site of Hg accumulation [42–44].

Many of the effects that we have observed in *L. italica* are largely similar to those described in *P. scaber* by Köhler et al. [28] who argue that these ultrastructural alterations could be used as biomarkers for monitoring the subchronic exposure to the heavy metals present in the environment. Nevertheless, as stated by the same aforementioned authors, a pattern of similar alterations were also found in the hepatopancreas of animals of different species of oniscideans subjected to prolonged starvation [39,42–47]; therefore, in our opinion, the ultrastructural effects of Hg or other

heavy metals accumulation could be used as biomarkers only if supported by a qualitative and quantitative analysis of the heavy metals content.

As a conclusion, based on what we said in the introduction about a bioindicator, we can say that *L. italica* is a good indicator for Hg pollution, as it is accessible, because easily available and easy to sample, and it has a very accessible threshold of analytical detection with standard techniques of analysis. *L. italica* is suitable from the bio-ecological point of view because it shows a wide distribution in the study area; its identification is easy, and there is an adequate knowledge on its anatomy, physiology and ecology; moreover, it has a sufficient life cycle and, even more important, it has poor mobility and it is easily available in all seasons. Our results, also, show its reliability and representativeness.

Despite all these strong points we have to highlight that in our study there are still some important weak points that are the lack of data about the content of the metal in the water and soil of the study area, and, in addition, the little knowledge about genetic of *L. italica* that not disclose its possible allopatric differentiation, therefore they lack information about its adaptability to constantly changing environmental conditions. These weak points, however, are easily overcome with appropriate studies, that we are planning in a near future.

Acknowledgments

We wish to thank, for sampling site of Vendicari Natural Reserve, Dr. Filadelfo Brogna, Director U.O.B. n° 2 Management of Protected Areas of the Regional Department of State Forest Company, Service XVI – Provincial Office of Syracuse, which allowed the sampling of the animals for conducting this research in the Natural Oriented Reserve, Faunistic Oasis of Vendicari.

Thanks also to the precious collaboration of Dr. Pasquale Di Mattia for helping in critically reviewing the English grammar of the manuscript.

Author Contributions

Conceived and designed the experiments: MF GL. Performed the experiments: GL MF. Analyzed the data: MF GL. Contributed reagents/materials/analysis tools: GOC MT VM. Wrote the paper: GL MF VM.

References

1. Bianchi C, Morri C (2000) Marine biodiversity of the Mediterranean Sea: situation, problems and prospects for future research. Marine Pollution Bulletin 40: 367–376.

2. Boero F (2003) State of knowledge of marine and coastal biodiversity in the Mediterranean Sea. UNEP, SPA-RAC.

3. Bazairi H, Ben Haj S, Boero F, Cebrian D, De Juan S, et al. (2010) The Mediterranean Sea Biodiversity: state of the ecosystems, pressures, impacts and future priorities. UNEP. Tunis. 1–100. Ed. RAC/SPA.

4. Hurtado LA, Mateos M, Santamaria CA (2010) Phylogeography of Supralittoral Rocky Intertidal *Ligia* Isopods in the Pacific Region from Central California to Central Mexico. PLoS ONE 5(7): e11633. doi:10.1371/journal.pone.0011633.

5. Copat C, Brundo MV, Arena G, Grasso A, Oliveri Conti G, et al. (2012) Seasonal variation of bioaccumulation in Engraulis encrasicolus (Linneaus, 1758) and related biomarkers of exposure. Ecotoxicol Environ Saf 86: 31–7.

6. Copat C, Maggiore R, Arena G, Lanzafame S, Fallico R, et al. (2012) Evaluation of a temporal trend heavy metals contamination in Posidonia oceanica (L.) Delile, (1813) along the western coastline of Sicily (Italy). J Environ Monit. 14(1): 187–92.

7. Conti GO, Copat C, Ledda C, Fiore M, Fallico R, et al. (2012) Evaluation of heavy metals and polycyclic aromatic hydrocarbons (PAHs) in Mullus barbatus from Sicily channel and risk-based consumption limits. Bull Environ Contam Toxicol 88(6): 946–50.

8. Tomasello B, Copat C, Pulvirenti V, Ferrito V, Ferrante M, et al. (2012) Biochemical and bioaccumulation approaches for investigating marine pollution using Mediterranean rainbow wrasse, Coris julis (Linneaus 1798). Ecotoxicol Environ Saf 86: 168–75.

9. Iserentant R, De Sloover J (1976) Le concept de bioindicateur. Mem Soc Roy Bot Belg, 7: 15–24.

10. National Research Council (1987) Biological Markers in Environmental Health Research. Environ. Health Pers. 74: 3–9.

11. Hunsaker CN, Carpenter DE (1990) Environmental Monitoring and Assessment Program –Ecological Indicators. EPA/600/3-060.

12. Ramistella EM, Bellia M, Di Mare S, Rotiroti G, Duscio D (1990) Inquinamento ambientale di origine industriale e mortalità per tumore. Revisione della situazione di Augusta e Priolo. Boll Acc Gioenia Sci Nat Catania 23: 437–462.

13. Bianca S, Li Volti G, Caruso-Nicoletti M, Ettore G, Barone P, et al. (2003) Elevated incidence of hypospadias in two sicilian towns where exposure to industrial and agricultural pollutants is high. Repr Toxicol 17: 539–45.

14. Bianchi F, Bianca S, Linzalone N, Madeddu A (2004) Sorveglianza delle malformazioni congenite in Italia: un approfondimento nella provincia di Siracusa. Epidemiologia e Prevenzione 28: 87–93.

15. Nicosia E (2007) Cause di rischio e sostenibilità urbana nella città di Augusta. Atti delle Giornate della Geografia "Rischi e territorio nel Mondo Globale", Udine, 24–26 maggio 2006 (Available in CD-R).

16. Sciacca S, Oliveri Conti G (2009) Mutagens and carcinogens in drinking water. Mediterr J Nutr Metab 2: 157–162.

17. Mazzariol S, Di Guardo G, Petrella A, Marsili L, Fossi CM, et al. (2011) Sometimes sperm whales (Physeter macrocephalus) cannot find their way back to the high seas: a multidisciplinary study on a mass stranding. PLoS One6(5): e19417.

18. Carefoot TH, Taylor BE (1995) Ligia: a prototypal terrestrial isopod. In: Alikhan MA, editor. Terrestrial isopod biology. Rotterdam: Balkema AA Publishers. 47–60.

19. Wieser W, Busch G, Bijchel L (1976) Isopods as indicators of the copper content of soil and litter. Oecologia 23: 107–114.

20. Martin MH, Coughtrey PJ (1982) Biological monitoring of heavy metal pollution: land and air. Elsevier, London, 475 p.

21. Hopkin SP, Hardisty GN, Martin MH (1986) The woodlouse Porcellio scaber as a biological indicator of zinc, cadmium, lead and copper pollution. Environm Pollut 11B: 271–290.

22. Hopkin SP (1989) Terrestrial isopods as biological indicators of zinc pollution in the Reading area, south east England. Monit Zool Ital Monogr 4: 477–488.

23. Dallinger R, Berger B, Birkel S (1992) Terrestrial isopods: useful bioindicators of urban metal pollution. Oecologia 89: 32–41.

24. Belashov E, Andrusishina I, Shadrin N (1998) Supralittoral crustaceans as bioindicators of quality of the environment. In: Colery PH, Brätter P, Negretti de Brätter V, Khassanova L, Etienne JC, editors. Metal Ions in Biology and Medicine. Paris: John Libbey Eutotext. 317–321.

25. Paoletti MG, Hassall M (1999) Woodlice (Isopoda: Oniscidea): their potential for assessing sustainability and use as bioindicators. Agric Ecosyst Environ 74: 157–165.

26. Souty-Grosset C, Badenhausser I, Reynolds JD, Morel A (2005) Investigations on the potential of woodlice as bioindicators of grassland habitat quality. Eur J Soil Biol 41: 109–116.

27. De Domenico E, Maucieri A, Giordano D, Maisano M, Gioffrè G, et al. (2011) Effect of in vivo exposure to toxic sediments on Juveniles of Sea Bass (Dicentrarchus labrax). Aquatic Toxicology 105(3–4): 688–697.

28. Köhler HR, Hüttenrauch K, Berkus M, Gräff S, Alberti G (1996) Cellular hepatopancreatic reactions in Porcellio scaber (Isopoda) as biomarkers for the evaluation of heavy metal toxicity in soils. Applied Soil Ecology, 3: 1–15.

29. Kouba A, Buřič M, Kozák P (2010) Bioaccumulation and Effects of Heavy Metals in Crayfish: A Review. Water, Air and Soil Pollution, 211: 5–16.

30. Besse G (1976) Contribution a l'étude expérimentale de la physiologie sexuelle femelle chez les Crustacés Isopodes terrestres. PhD dissertation, Université de Poitiers, CNRS, n° AO13017.

31. Sato T, Shamoto M (1973) A simple rapid polychrome stain for epoxy-embedded tissue. Stain Technol 49: 223–227.

32. Reynolds E (1963) The use of lead citrate at high pH as an electron opaque stain in electron microscopy. J Cell Biol 17: 208–212.

33. Covelli S, Faganeli J, Horvat M, Brambati A (2001) Mercury contamination of coastal sediments as the result of a long-term cinnabar mining activity (Gulf of Trieste, Northern Adriatic sea). Appl Geochem16: 541–558.

34. Faganeli J, Horvat M, Covelli S, Fajon V, Logar M, et al. (2003) Mercury and methylmercury in the Gulf of Trieste (northern Adriatic sea). Sci Total Environ 304: 315–326.

35. Fitzgerald WF, Lamborg CH (2005) Geochemistry of mercury in the environment. In: Lollar BS, editor. Environmental geochemistry. Oxford: Elsevier-Pergamon. 107–148.

36. Žagar D, Petkovšek G, Raja R, Sirnik N, Horvat M, et al. (2007) Modelling of mercury transport and transformations in the water compartment of the Mediterranean Sea. Mar Chem 107: 64–88.

37. Youn S, Donkor AK, Attibayeba A, Bonzongo JCJ (2010) Mercury in Aquatic Systems of the Gulf Islands National Seashore, Southeastern USA. J Appl Sci Environ Manage 14: 71–83.

38. Di Leonardo R, Bellanca A, Capotondi L, Cundy A, Neri R (2007) Possible impacts of Hg and PAH contamination on benthic foraminiferal assemblages: an example from the Sicilian coast, central Mediterranean. Sci Total Environ 388: 168–183.

39. Storch V, Lehnert-Moritz K (1980) The effects of starvation on the hepatopancreas of the isopod Ligia oceanica. Zool Anz 204: 137–146.

40. Clifford B, Witkus ER (1971) The fine structure of the hepatopancreas of the woodlouse, Oniscus asellus. J Morphol 135: 335–350.

41. Andersen JK, Baatrup E (1988) Ultrastructural localization of mercury accumulations in the gills, hepatopancreas, midgut and antennal glands of the brown shrimp, Crangon crangon. Aquatic Toxicol 13: 309–324.

42. Raes H, De Coster W (1991) Storage and detoxication of anorganic and organic mercury in honeybees: histochemical and ultrastructural evidence. Prog Histochem Cytoc 23: 316–320.

43. Marigómez I, Soto M, Kortabitarte M (1996) Tissue-level biomarkers and biological effect of mercury on sentinel slugs, Arion ater. Arch Environ Con Tox 31: 54–62.

44. Braeckman B, Raes H (1999) The ultrastructural effect and subcellular localization of mercuric chloride and methylmercuric chloride in insect cells (Aedes albopictus C6/36). Tissue & Cell 31: 223–232.

45. Storch V (1982) Der einfluss der ernährung auf die ultrastruktur der grossen zellen in den Mitteldarmdrüsen terrestrischer Isopoda (Armadillidium vulgare, Porcellio scaber). Zoomorphology 100: 131–142.

46. Storch V (1984) The influence of nutritional stress on the ultrastructure of the hepatopancreas of terrestrial isopods. Symp Zool Soc London 53: 167–184.

47. Štrus J, Burkhardt P, Storch V (1985) The ultrastructure of the midgut glands in Ligia italica (Isopoda) under different nutritional conditions. Helgoländer Meeresunters 39: 367–374.

8

Coal Fly Ash Impairs Airway Antimicrobial Peptides and Increases Bacterial Growth

Jennifer A. Borcherding[1], Haihan Chen[3], Juan C. Caraballo[1], Jonas Baltrusaitis[4], Alejandro A. Pezzulo[1], Joseph Zabner[1], Vicki H. Grassian[2], Alejandro P. Comellas[1]*

1 Department of Internal Medicine, Carver College of Medicine, University of Iowa, Iowa City, Iowa, United States of America, 2 Department of Chemistry, University of Iowa, Iowa City, Iowa, United States of America, 3 Department of Chemical and Biochemical Engineering, Iowa City, Iowa, United States of America, 4 Central Microscopy Research Facility, University of Iowa, Iowa City, Iowa, United States of America

Abstract

Air pollution is a risk factor for respiratory infections, and one of its main components is particulate matter (PM), which is comprised of a number of particles that contain iron, such as coal fly ash (CFA). Since free iron concentrations are extremely low in airway surface liquid (ASL), we hypothesize that CFA impairs antimicrobial peptides (AMP) function and can be a source of iron to bacteria. We tested this hypothesis in vivo by instilling mice with Pseudomonas aeruginosa (PA01) and CFA and determine the percentage of bacterial clearance. In addition, we tested bacterial clearance in cell culture by exposing primary human airway epithelial cells to PA01 and CFA and determining the AMP activity and bacterial growth in vitro. We report that CFA is a bioavailable source of iron for bacteria. We show that CFA interferes with bacterial clearance in vivo and in primary human airway epithelial cultures. Also, we demonstrate that CFA inhibits AMP activity in vitro, which we propose as a mechanism of our cell culture and in vivo results. Furthermore, PA01 uses CFA as an iron source with a direct correlation between CFA iron dissolution and bacterial growth. CFA concentrations used are very relevant to human daily exposures, thus posing a potential public health risk for susceptible subjects. Although CFA provides a source of bioavailable iron for bacteria, not all CFA particles have the same biological effects, and their propensity for iron dissolution is an important factor. CFA impairs lung innate immune mechanisms of bacterial clearance, specifically AMP activity. We expect that identifying the PM mechanisms of respiratory infections will translate into public health policies aimed at controlling, not only concentration of PM exposure, but physicochemical characteristics that will potentially cause respiratory infections in susceptible individuals and populations.

Editor: Olivier Neyrolles, Institut de Pharmacologie et de Biologie Structurale, France

Funding: This publication was made possible by Grant Number UL1RR024979 from the National Center for Research Resources (NCRR), a part of the National Institutes of Health (NIH). Its contents are solely the responsibility of the authors and do not necessarily represent the official views of the CTSA or NIH. This publication was also supported by Center for Health Effects of Environmental Contamination (CHEEC) Seed Grants: FY 2010 and NIH Grant: KO1HL080966, and NIH P30 ES005605. The funders had no role in study design, data collection and anlysis, decision to publish, or preparation of the manuscript.

Competing Interests: The authors have declared that no competing interests exist.

* E-mail: alejandro-comellas@uiowa.edu

Introduction

Coal is one of the most abundant sources of energy production globally and continues to grow on an annual basis. In 2010, U.S. coal consumption was 1,048.3 million short tons, an increase of 50.8 short tons from the previous year [1]. Coal Fly Ash (CFA), a byproduct of coal combustion, is considered a poorly soluble particle comprised of various transition metals such as iron, and aluminum silicate as classified by ACGIH (American Conference of Industrial Hygienists) [2]. The majority of CFA (99%) is collected and deposited in landfills, therefore providing a potential source of transition metals into the water supply and redistributing itself into the atmosphere [3]. Due to the increased global demand and the limited regulations in growing economies such as China, CFA released into the atmosphere continues to be a large anthropogenic source of air pollution.

Epidemiological studies show a strong correlation between respiratory infections and $PM_{2.5}$ [4]. Ambient air pollution is associated with cystic fibrosis (CF) and chronic obstructive pulmonary disease (COPD) exacerbations [4,5,6]. The majority

of these exacerbations are infectious in nature [7]. In addition, a correlation between biomass fuels used for indoor cooking, including coal, and acute respiratory infections in children has been reported [8]. Therefore, due to the association between respiratory exacerbations and increased pollution, further investigation needs to be conducted in order to understand the mechanism of PM induced respiratory infections.

PM which is rich in iron [9] can increase iron bioavailability to microorganisms [10,11], such as Pseudomonas Aeruginosa (PA01). The amount of soluble, and therefore potentially bioavailable iron in PM, specifically CFA, has been correlated with particle size [12], source of CFA [12] and amount of aluminum silicate present within the particle [13]. Therefore, CFA can be an exogenous iron source for bacteria in biological fluids, such as the airway surface liquid (ASL), that the body maintains at low iron concentrations ($<10^{-18}$ M) [14] and thus become potentially detrimental to human health. Although there have been significant studies of the effects of PM on the lung epithelium, there is a paucity of data on the effects of PM induced bacterial growth and pathogenicity that can lead to respiratory infections.

We hypothesize that CFA will impair airway bacterial clearance by both promoting bacterial growth and impairing airway epithelial antimicrobial peptide function. To test this hypothesis we set out to determine the effects of CFA on *Pseudomonas Aeruginosa* (PA01) *in vivo* and *in vitro*. Three CFA particles with different iron content, that were previously characterized for iron solubility and mobilization, were used for this study (Table 1) [11].

Methods

Ethics Statement

All animals (mice) used in this study were according to protocols approved by the University of Iowa Institutional Animal Care and Use Committee (IACUC). Animals were anesthetized prior to instillations and harvest in order to reduce animal distress.

X-Ray Photoelectron Spectroscopy (XPS)

Surface composition for all fly ash particles was performed using a custom-designed Kratos Axis Ultra X-Ray photoelectron spectrometer with a monochromatic Al Ka X-Ray source as previously described [15]. The fly ash particles were pressed onto indium foil which was fixed on a stainless steel bar or copper stub for further analysis. The pressed particles were then transferred into the XPS analysis chamber, which had a pressure that was maintained in the 10^{-9} Torr range during analysis. Charging was prevented by using the following instrumental parameters: energy range from 1200 to -5 eV, pass energy of 160 eV, step size of 1 eV, dwell time of 200 ms, and an X-ray spot size of $\sim 700 \times 300$ μm. Survey spectra were collected at three different locations on the sample stub, and reported elemental compositions represent the average and one standard deviation of the three analyses. XPS data collected were analyzed using CasaXPS data processing software.

Energy Dispersive X-Ray Spectroscopy (EDX)

The morphology and total elemental composition of fly ash particles were examined using a Hitachi S-3400 N scanning electron microscopy (SEM) coupled with energy dispersive X-ray spectroscopy system. Particles were sprinkled onto carbon tape that had been mounted on an aluminum stub and were subsequently carbon coated. Elemental analyses used an integrat-

ed IXRF System Inc. X-ray microanalysis system and an accelerating voltage of 10 kV with a detection limit of 1 wt%. SEM/EDX elemental maps were collected as well to examine the distribution of Fe in fly ash particles. A resolution of 256×200 pixels, and a dwell time of 1 second were used.

Experimental Preparation of Particles

Characterized coal fly ash from the National Institute of Standards and Technology (NIST) were suspended in an iron deficient media [BD Difco Minimal media (M9) with 2.2 mM glucose, 0.002 M magnesium sulfate ($MgSO_4$), 0.001 M calcium chloride ($CaCl_2$) and 25 mM sodium succinate]. The particle suspension was sonicated for ten minutes immediately prior to conducting experiments.

Bacteria

Pseudomonas Aeruginosa (PA01) was chosen as a model in our studies due to its prevalence and importance in disease such as COPD and CF.

In Vivo Mouse Instillation

Six to eight week Harlan C57/BL6 males (20–25 g) were intranasally instilled with 50 μl $OD_{600} = 0.03$ PA01 with or without 10 μg/mL freshly dissolved and sonicated CFA. PA01 was exposed to CFA for a minimal amount of time ($\sim$10 minutes). After 24 hours, BAL was performed or lungs were removed and homogenized in 2 ml PBS. In BAL performed mice, BALF samples were used to determine cell count and differential by using Wright-Geisma staining. Non-lavaged samples were plated on lauria broth agar (LB) plates and CFUs of PA01 were recorded.

TNF-α and IL-1β

R& D DuoSet ELISAs were conducted according to manufacturer's instructions to determine TNF-α and IL-1β production in BAL fluid.

Cell Culture

Briefly cells were isolated from donor lungs and plated on cell culture inserts in an air liquid interface. Human airway epithelial cells were obtained from the University of Iowa cell culture core and changed to antibiotic free USG media two weeks prior to

Table 1. Coal Fly Ash Particles.

Total Iron Content, Dissolved Iron %, and Elemental Composition of Coal Fly Ash [11]

	FA 2689	FA 2690	FA 2691
Source	Stilesboro, GA	Criag, CO	Iatan, MO
Size	1–50 μm	1–50 μm	1–50 μm
Specific Surface Area (m² g⁻¹)	0.8±0.1	3.8±0.1	2.2±0.1
Total Fe Content (%)	9.32±0.06	3.57±0.06	4.42±0.03
Dissolved Fe (%) pH 7.5	0.028	0.032	0.057
XPS/EDX Feᵃ	1.3±0.2	1.4±0.3	0.6±0.2
XPS/EDX Alᵃ	0.6±0.1	0.6±0.1	0.4±0.2
XPS/EDX Siᵃ	1.0±0.1	0.7±0.1	0.7±0.3

Total iron content, aluminum silicate content, dissolved iron, particle size, and specific surface area of three different coal fly ash particles (FA 2689, FA 2690 and FA 2691).
XPS was used to determine surface composition.
EDX was used to determine bulk composition.
ᵃXPS/EDX Ratio: High ratio of XPS/EDX indicates elemental enrichment at the surface, low ratio content (<1) indicates enrichment of element at inner core.

experiments [16]. Cells were washed with PBS three times and media was changed to antibiotic free USG media two weeks prior to experiments. Media was changed every four days and experiments were conducted day four post media change in order to ensure adequate airway surface liquid levels. Sterility of cell culture was determined as previously published (Phil H. Karp 2002). Briefly, a dose response of PA01 was conducted to determine inoculum of complete bacterial killing. 0.1 μL with 12 CFU PA01 and 10 μg/mL CFA was placed on apical surface. Growth was determined by washing epithelial apical surface with 50 ul PBS and growing in LB media for eight hours.

p-hydroxyphenylacetate Assay (pHPA)

H_2O_2 was measured by adding 1.6 mM pHPA (Sigma), 95 μg/mL Horseradish Peroxidase (HRP) (Sigma), 1 mM Hepes, 6.5 mM glucose and 6 mM $NHCO_3$ in Hanks balanced salt solution (HBSS). Solution was added to cell culture and fluorescence was measured over one hour. A standard H_2O_2 curve was generated and pHPA dimer concentration was determined.

Transepithelial Electrical Conductance (Gt)

Airway cells were submerged in 500 μl of media and transepithelial electrical resistance (Rt) was measured with Millicell Electrical Resistance System (ERS) (Millipore Corporation, Bedford, MA) and Gt was calculated as the reciprocal of Rt [17].

Antimicrobial Peptide Activity

PA01 was grown overnight in M9 media, subcultured and diluted to $OD_{600} = 0.45$. The culture was then diluted and 13500 PA01 was added to start experiment. Sodium phosphate buffer at pH 7.8 was used and a cocktail of antimicrobial peptides (600 μg/mL Lysozyme, 200 μg/mL Lactoferrin and 100 ng/mL β-Defensins 1&2) equaling 400 μl were added to a 96 deep well plate. 10 μg/mL CFA was added with AMPs and PA01. Mixture was incubated for one hour at 37°C and 300 rpm. ¼ diluted Lauria Broth (LB) media was added to mixture and grown overnight. OD_{600} was measured to determine level of antimicrobial peptide activity. CFUs were determined by conducting the above experiment, serially diluting and plating cultures on LB agar plates at beginning and endpoints to determine exact colony count.

Growth Experiments

PA01 was grown overnight in an iron deficient media BD Difco Minimal media (M9) with 2.2 mM glucose, 0.002 M magnesium sulfate ($MgSO_4$), 0.001 M calcium chloride ($CaCl_2$) and 25 mM sodium succinate. 10 μg/mL CFA was added to three hour subcultured overnight cultures to equal a volume of 250 μl and growth was observed by measuring OD_{600} at 37°C for nine hours. 25 μM iron chloride ($FeCl_3$) [18], a soluble source of iron was used as a positive control. CFU experiments were conducted in a 5 mL volume in same conditions as above. Samples were taken at time T = 0 and T = 18, serially diluted and plated on LB plates to determine exact colony count.

Fe-Dissolution

An inductively coupled plasma optical emission spectrometer (ICP/OES) (Varian, 720-ES) was used to determine the concentration of dissolved iron in PA01 media (≥5 ppb). Suspensions of M9 media with particles were spun at 2950 rpm for 15 minutes and filtered with 0.2 μm filters to remove particles before ICP/OES analysis. The concentration of dissolved iron in solution was calculated from the working calibration curve generated from iron standard solution data. Blank samples were also analyzed using ICP/OES to ensure that no significant iron was detected as ions in blank solution. Gamble's buffer was used for measurement of pH 7.5 experiments and autosomal lysosomal fluid (ALF) was used for pH 4.0 experiments.

Statistical Analysis

Data are presented as means ± SEM. The program used for data analysis was GraphPad Prism 5.00 (San Diego, CA). The following information provides the analysis method for each figure and panel. Fisher's analysis of a contingency table of sterility was used to determine significance in Figure 1A. In Figures 1B and 2A–D, One-way ANOVA using Dunnett's Multiple Comparison Test was used. Figure 3A, Fisher's analysis of a contingency table of sterility was used to determine significance. Figures 3B–C and Figures 4A–B, One-way ANOVA using Dunnett's Multiple Comparison Test was used to determine significance. In Figure 5, non-linear regression (curve-fit) with variable slope from three independent experiments was used for statistical analysis. Data was compared for all parameters of the growth curve using the extra sum of squares F-test to detect differences throughout the entire growth curve. A p value of <0.05 was considered statistically significant.

Results

CFA Increases PA01 Growth *in vivo*

In order to test our hypothesis of CFA induced PA01 growth, we used a mouse model to determine the effects on PA01 clearance in the presence of CFA. Three CFA particles from different sources that have been well characterized for size, surface area and elemental composition were used in these experiments. Before experiments were conducted, CFA particles were sonicated for ten minutes in order to reduce aggregates. These three sources are standard reference materials (SRM) from the National Institute of Standards and Technology (NIST) and include FA 2689, FA 2690 and FA 2691 (Table 1) [11]. Six to eight week old Harlen C57BL/6 male mice were instilled with PA01 (4.5 10^6 PA01/mouse) in the presence and absence of CFA (10 μg/mL) in M9 media. According to the ACGIH, insoluble or poorly soluble particles Threshold Limit Value (TLV) is 3 mg/m³ for respirable particles and 10 mg/m³ for inhalable particles [2]. Therefore, this CFA dose is at the TLV for particles of this composition, which translate our results into relevant daily human exposures.

Twenty four hours later, mice were sacrificed and lungs harvested. None of the mice died over this time period nor were there any significant weight changes (data not shown). After 24 hours, under control conditions, 62% of the mice were sterile. Conversely, FA 2689 only exhibited 41% bacterial clearance (p = 0.0045) and FA 2690 and FA 2691 exhibited 33% bacterial clearance (p<0.0001) (Figure 1A). Among the non-sterile mice, there was a significant amount of bacteria recovered in mice instilled with PA01 and FA 2689 and FA 2690 (p<0.05) (Figure 1B). Therefore, it appears that CFA decreases or delays PA01 clearance in lungs of healthy mice provides a source of iron for bacterial growth or allows the bacteria to persist in the lungs.

CFA-induced Increased PA01 Recovery is not due to Inflammatory Response *in vivo*

PM has been linked to acute exacerbations of COPD (AECOPD) through neutrophil recruitment and cytokine release [5,6,19,20]. In conjunction with this, air pollution has been shown to inhibit bacterial clearance by increasing inflammation. More

A.

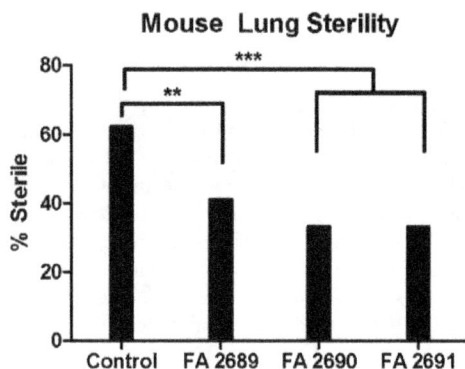

Mouse Lung Sterility

B.

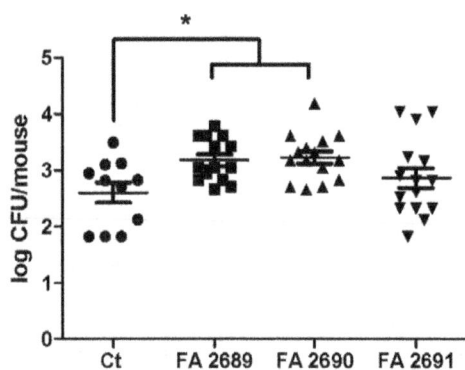

Figure 1. CFA increases bacterial grown *in vivo*. Panel A. Total mouse lung sterility *in vivo*. After 24 hours, CFUs were determined after homogenizing lungs and plating to determine growth. In the presence of CFA, lung sterility was decreased by 20–30% **p<0.01, ***p<0.0001. PA01 alone N = 26, PA01+ FA 2689 N = 22, PA01+ FA 2690 N = 21, and PA01+ FA 2691 N = 24. **Panel B.** Growth in non-sterile mice. Log CFU/mouse was determined. FA 2689 and FA 2690 increased growth more than control (p<0.05).

specifically, Harrod et al. reported that diesel exhaust particles (DEP) increase infection through an inflammatory response [21]. In order to determine whether the effect of decreased PA01 clearance in our mouse model was due to an increased inflammatory response, we instilled five mice in each condition as stated above and measured neutrophil recruitment and cytokine release. Although the PA01 groups had higher bronchoalveolar lavage (BAL) cell count, the presence of CFA (10 µg/mL, FA2690) did not affect BAL total cell count in the control and PA01 treated group (Figure 2A). In addition, PA01 induced a higher amount of neutrophils (p<0.05) when compared to control, but as shown in Figure 2B, the presence of CFA did not change the neutrophil count significantly more than PA01 alone. After measuring two inflammatory cytokines, TNF-α and IL-1β, it appears that PA01 in the presence of FA 2690 does not increase IL-1β nor TNF-α when compared with PA01 alone (Figures 2C&D).

CFA on Human Airway Epithelia Increases PA01 Growth without Disrupting Epithelial Barrier Integrity

Since it appears that the decreased airway clearance of PA01 in the presence of CFA *in vivo* is not due to an inflammatory response, we set out to determine if bacterial proliferation in the presence of CFA was due to structural abnormalities at the cellular level. Reactive oxygen species (ROS) production in the airway had been linked to the presence of PM [22] which can generate ROS through oxidation on the surface of the particles, including CFA, which is shown to elicit ROS damage to DNA [23]. It has been reported that in cell culture, at concentrations of 100 µg/mL, CFA with LPS increases ROS [24]. Therefore, using the same cultures for all of the following experiments, we exposed primary human airway epithelial cells cultured in an air-liquid interface [16] to PA01 (12 CFU in 0.1 µL) in the presence of 10 µg/mL CFA and tested growth, trans-epithelial electrical conductance (Gt) and hydrogen peroxide (H_2O_2) production. As shown in Figure 3A, human airway epithelial cells were 100% sterile after 24 hours of incubation with PA01, while the presence of CFA impaired airway bacterial clearance or increase bacterial growth (p<0.0001). Specifically, the percentage of growth in cell cultures treated with FA 2689 and FA 2691 was 11% (95% CI 0 to 37%) whereas in the presence of FA 2690, growth was present in 22% (95% CI 0 to 56%) of the cultures (n = 3 in triplicates from two different human donors).

After 24 hours, H_2O_2 production was measured by determining p-hydroxyphenylacetate (pHPA) oxidation [25]. CFA (10 µg/ml) in the presence of PA01 does not increase H_2O_2 production when compared with PA01 alone, which is consistent with other reports [24] (Figure 3B). Also, Gt was measured to determine the effect on the epithelial barrier integrity. Exposing primary human airway epithelial cells to CFA and PA01 at the above concentrations did not disrupt airway epithelial barrier integrity (Figure 3C), nor increased cell death as determined by propidium iodide staining (data not shown).

CFA Decreases Antimicrobial Peptide Activity

The above results led to the hypothesis that CFA impairs airway innate immunity mechanisms. The lung has various mechanisms to protect itself against pathogens and one of the primary defense systems are AMPs, which are present in the airway surface liquid (ASL). This is comprised primarily of lactoferrin, lysozyme and β-Defensins 1&2, which behave synergistically but also have specific functions. Specifically, lysozyme degrades the bacterial cell walls via its muramidase activity, lactoferrin sequesters iron and inhibits microbial respiration, therefore limiting iron availability, and β-defensins have broad antibacterial activity [26].

In order to test whether CFA inhibits AMP activity, we combined physiologically relevant concentrations of AMP present in the lung [27] (600 µg/mL Lysozyme, 200 µg/mL Lactoferrin and 100 ng/mL B Defensins 1&2) with 10 µg/mL CFA and determined the effect on PA01 growth by measuring OD_{600} after 18 hours. As shown in Figure 4A, PA01 grew in the absence of AMP. In the presence of a positive control, $FeCl_3$ (25 µM), and CFA (10 µg/mL) there was an increase in bacterial growth, although not statistically significant. When PA01 was treated with AMP, there was significant growth inhibition. Conversely, as shown in Figures 4B & C, AMP activity was impaired when PA01 cultures (determined by measuring CFU and OD_{600}) were treated with all three forms of CFA and $FeCl_3$.

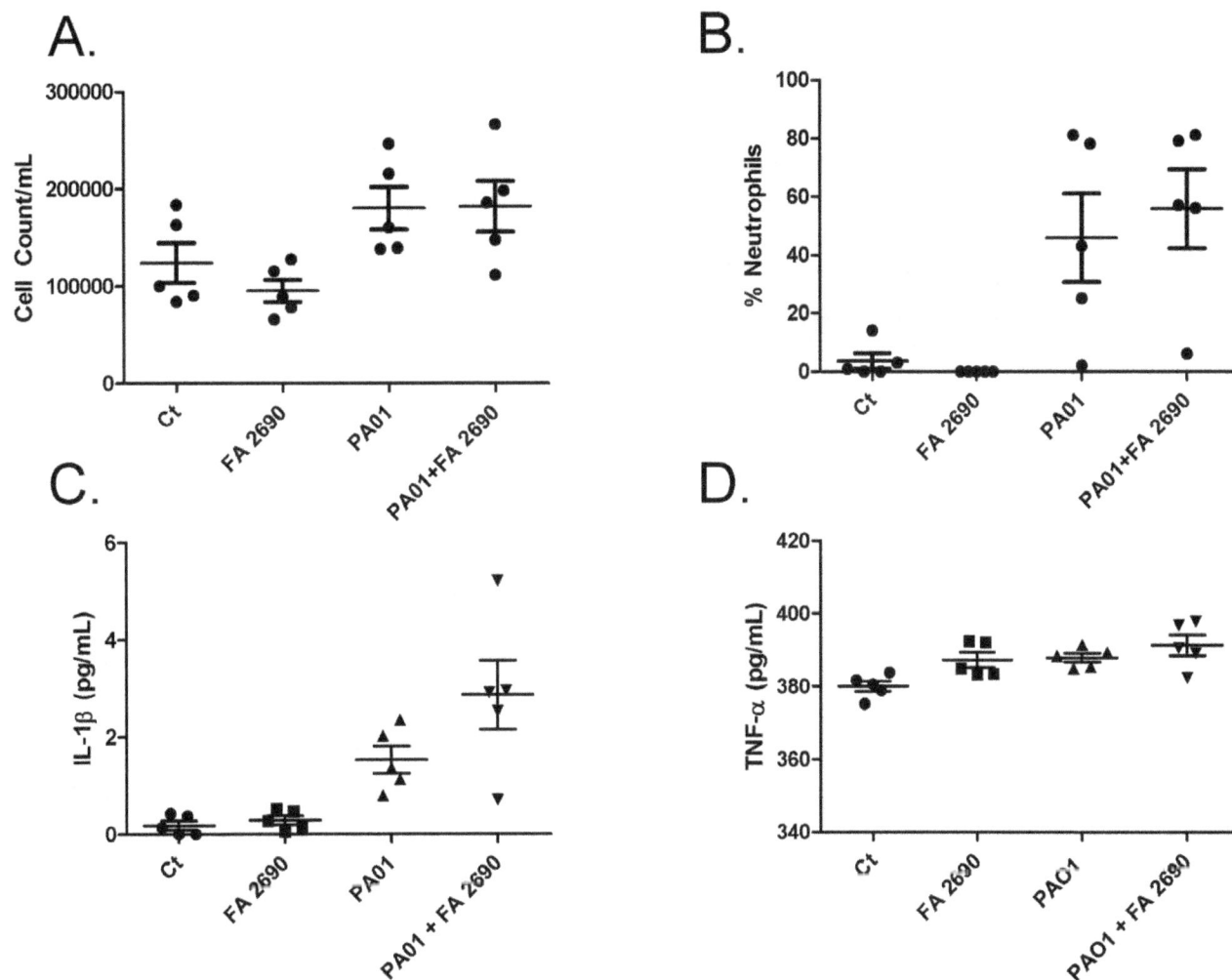

Figure 2. CFA (10 µg/mL) in the presence of PA01 does not significantly increase neutrophil recruitment or cytokine production in BAL. Panel A. When C57/Bl6 mice were exposed to 10 µg/mL CFA in the presence and absence of PA01 (4.5 10^6 PA01/mouse), total BAL cell count was not significantly different between Ct and PA01 but there was an increase in neutrophil percentage in the presence of PA01. However, there were no significant cell count differences between PA01 and PA01 with FA2690. **Panel B.** Although neutrophil recruitment in mice BAL was higher, there was no statistically significant difference between control and PA01 nor PA01 and PA01 with FA2690. **Panel C.** IL-1β in the presence of PA01 and CFA increased production more than PA01 alone but was not significantly different. **Panel D.** TNF-α in the presence of PA01 and CFA does not significantly increase compared with PA01 alone. N = 5.

CFA Provides a Bioavailable Source of Iron for PA01

Due to the increase in bacterial growth (Figure 4A), we set out to determine if there were growth differences between the three different CFA particles. Before conducting our experiments, we set out to create a media with non-detectable iron levels (<5 ppb; M9) in order to mimic as much as possible the ASL iron content. To test our hypothesis of CFA induced bacteria growth and the role of dissolved iron, 10 µg/mL of CFA particles (FA 2689, 2690, 2691) were added to three hour sub-cultured PA01 cultures in M9 media and growth was observed by measuring OD_{600} at 37°C for nine hours while correcting for particle absorbance effects. $FeCl_3$ (25 µM), a soluble source of iron, was used as a positive control (data not shown). All CFA particles induced bacteria growth compared to control (p<0.0001). In addition, FA 2691 appeared to contribute to PA01growth more than FA 2689 and FA 2690 (Figure 5) (p<0.0001). When iron dissolution in M9 media was measured, using an inductively coupled plasma optical emission spectrometer (ICP-OES), FA 2691 had 0.057 mg/L of dissolved

iron compared with 0.028 and 0.032 mg/L in FA 2689 and 2690 respectively (Table 1).

Discussion

The World Health Organization (WHO) reports that acute respiratory infections (ARIs) are the leading cause of acute illnesses worldwide and remain one of the most important causes of death, especially in the very young, the elderly, and the immunocompromised. In addition, the WHO and the Environmental Protection Agency (EPA) recognize ambient air pollution as an important risk factor for ARIs. Despite the magnitude of this problem, the ambient air pollution mechanism responsible for the development of respiratory infections is not well known. One of the main components of ambient air pollution is particulate matter (PM), thus PM must play an important role in this mechanism. Since respiratory infections are in part the consequence of mechanisms that will promote bacterial growth and will impair innate immunity, we initially hypothesize that PM will increase

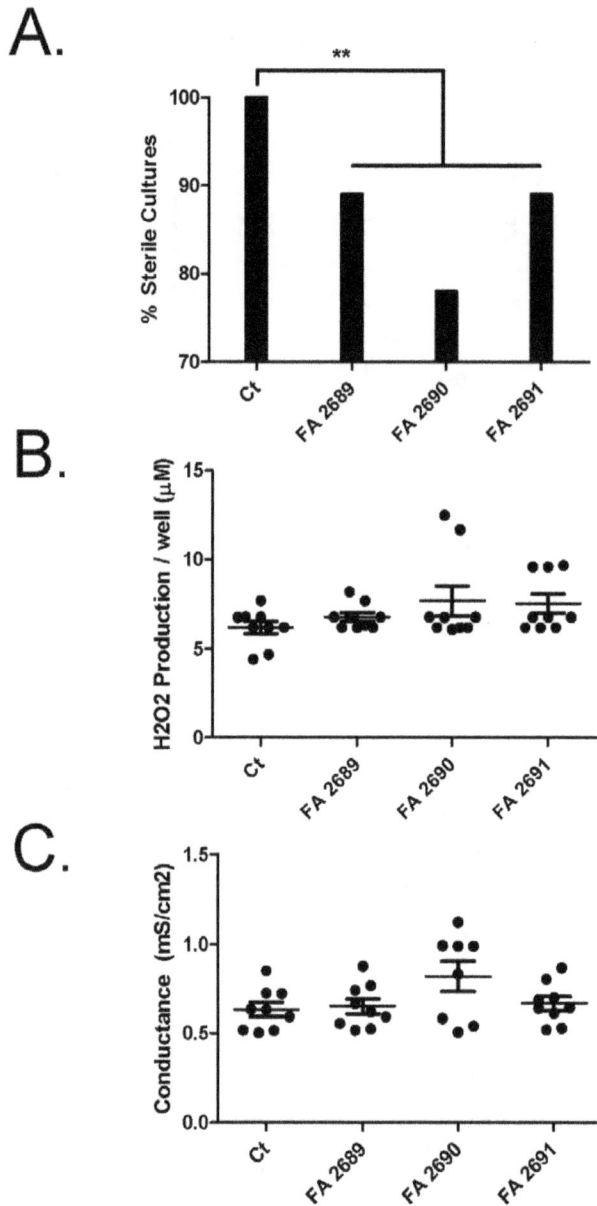

Figure 3. CFA increases PA01 growth in isolated human airway epithelia. Panel A. After 24 hours, PA01 (12 CFU) growth in the presence of CFA (10 µg/mL) on isolated human airway epithelia was measured. CFA increases PA01 growth in cell culture. FA 2689 and FA 2691 increased PA01 growth more than CT by 11% and FA 2690 increased PA01 growth 22% **$p < 0.0001$. **Panel B.** Hydrogen peroxide (H_2O_2) production in the presence of PA01 and CFA does not significantly increase more than PA01 alone. **Panel C.** Transepithelial electrical conductance (Gt) across primary human airway epithelia does not significantly decrease in the presence of PA01 and CFA when compared with PA01 alone. $N = 3$ in triplicates from two different human donors.

nutrient bioavailability for bacteria, and will impair airway antimicrobial peptide function.

Our results show that CFA reduces or delays bacterial clearance *in vivo* and *in vitro* as well as provides a source of iron for bacterial growth. The reduced bacterial clearance is consistent with reports of rats and mice exposed to PM [21,28], where one mechanism implicated in bacterial clearance impairment is an increase

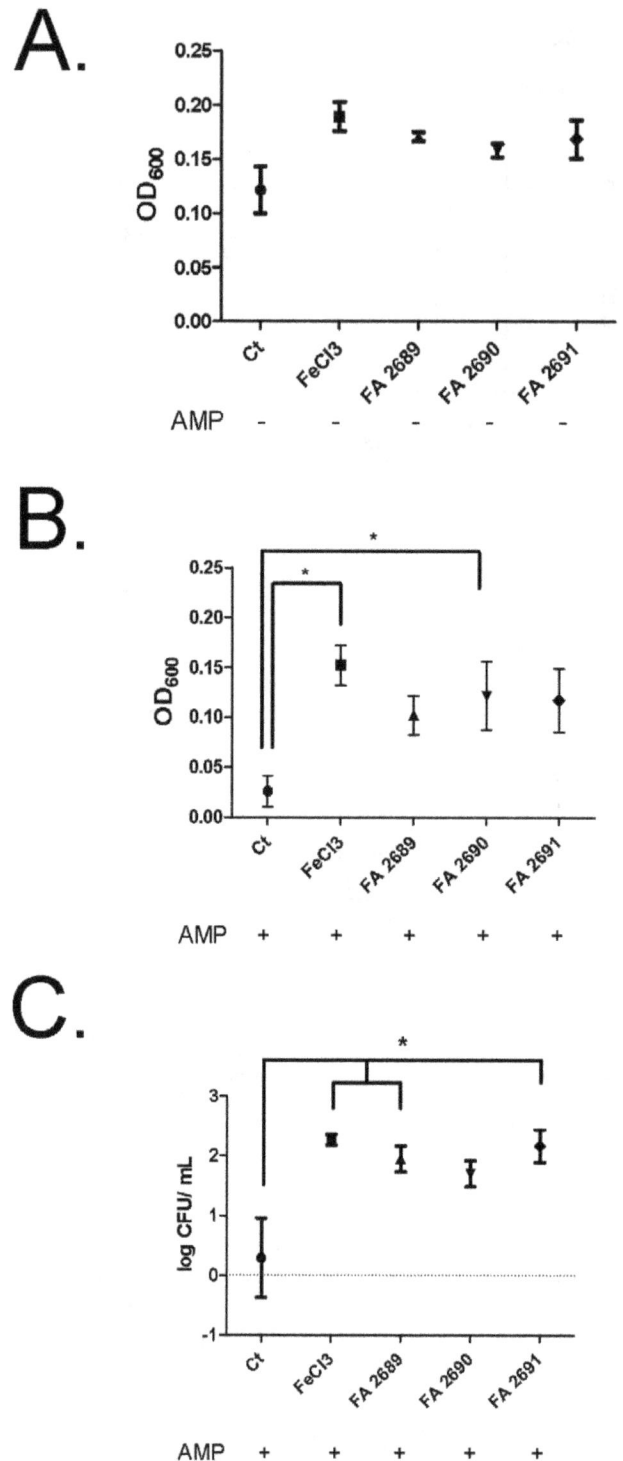

Figure 4. CFA inhibits antimicrobial peptide activity. Panel A. PA01 in the presence of $FeCl_3$ (25 µM) and three different types of CFA (FA 2689, FA 2690 and FA 2691) increased growth at 10 µg/mL without AMP cocktail more than control, however the growth increase is not statistically significant. **Panel B.** PA01 growth in the presence of AMP cocktail (600 µg/mL Lysozyme, 200 µg/mL Lactoferrin and 100 ng/mL β-Defensin 1&2) inhibits PA01 growth. $FeCl_3$ (25 µM) and FA 2690 (10 µg/mL) inhibit AMP activity *$p \leq 0.05$, **$p < 0.0001$. **Panel C.** CFU count of PA01 after 18 hours in the presence of AMPs. $N = 3$ in triplicates. SEM reported.

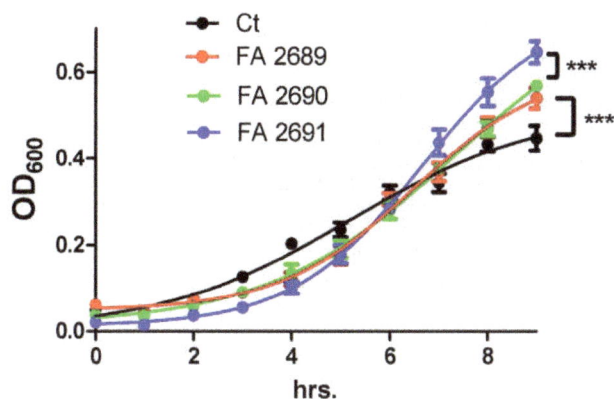

Figure 5. CFA increases PA01 growth. Subcultured PA01 was grown in M9 with $FeCl_3$ (25 μM), three different CFA particles (10 μg/mL) or no particles (CT). Growth was recorded over nine hours. CFA increased growth more than CT (p<0.0001 for all three CFAs). FA 2691 increased PA01 growth more than FA 2689 or FA 2690 ***p<0.0001. N = 3 in triplicates.

inflammatory response in the lung. However, the overall inflammatory response in the presence of CFA and PA01 was not significantly increased over PA01 alone. This discrepancy when compared with previous studies reporting a correlation between neutrophil recruitment and increased infection is perhaps due to differences in dose, 50 μg reported versus 500 ng in our study [29,30]. Another potential mechanism of reduced bacterial clearance could be due to macrophage function impairment. This mechanism has been shown in models where PM exposure, at high doses, inhibits phagocytosis. However, the relevance of this inhibition in people exposed to ambient air pollution has been raised recently in a review by Miyata et al. [31]. Part of the argument lies in regard to the experimental doses that show this effect, and its relevance to actual ambient air pollution exposures. In contrast, our *in vivo* and *in vitro* models, with much lower PM doses, suggest other mechanisms of increased infection susceptibility, specifically the impairment of AMP function. Our results of reduced AMP activity in the presence of CFA provide insight into diseases with persistent colonization, which is consistent with a recent report that showed AMP activity impairment in a cystic fibrosis model [32]. Also, Parameswaran et al. reported that AMP levels in COPD subjects likely affect pathogen clearance and clinical outcomes of infection [33].

Determining the mechanism of CFA impairment on AMP function is challenging since these particles are physicochemically complex. However, it seems that PM could potentially decrease lysozyme activity. For example, a small cohort of peat dust exposed workers showed decreased lysozyme positive macrophages, indicating increased macrophage phagocytosis and a potential effect on lysozyme activity [34]. Also Noble et al. reported that cigarette smoke and dust decreased human nasal lysozyme concentrations [35]. Other studies that have attempted to understand the mechanism of lysozyme inhibition has shown that lysozyme activity is inhibited by cations [36] thus CFA particles could leach certain cations, such as Fe (III), Fe (II) or Al (III) and

inhibit AMP activity. PM not only can affect lysozyme, but β-Defensins, as it has been reported that oil fly ash, a byproduct of oil-fired power plants with a composition of carbon, silicates, and iron oxides can impair β-Defensin synthesis in epithelial cells [37]. Furthermore, lactoferrin can be inhibited by its complete iron saturation, which in turns, impairs it ability to sequester iron. Therefore, several mechanisms impairing AMP function can potentially play a role in PM induced decrease bacterial clearance.

CFA not only decreased AMP activity, but increased bacterial growth. CFA is known to be an iron containing particle, thus CFA can be an important nutrient for bacteria growth. In addition, a recent report correlated iron mobilization in CFA with iron associated within aluminosilicate glass phases [13]. Therefore, it appears the effect of CFA on bacterial growth is more complex than just total iron content alone, since FA 2689 had the largest amount of iron, but it did not translate into the highest dissolved iron [11], nor in the highest growth curve. As shown in Table 1, the propensity for iron to be mobilized and thus dissolved is due in part to its enrichment within the aluminum silicate phase, specifically iron in FA 2691 is to a large extent associated with the aluminum silicate content (XPS/EDX ratio range: 0.4–0.8 for Al, Fe and Si) (Table 1) compared with FA 2689 (XPS/EDX ratio range: 0.6–1.3) and FA 2690 (XPS/EDX ratio range: 0.6–1.4). CFA spheres commonly contain aluminosilicate-phase iron in the inner core with iron oxide aggregate on the surface (see Chen, Laskin et al. 2012) Therefore, due to the decreased XPS/EDX ratio and thus high iron content in aluminosilicate phase, we propose that one mechanism of PM induced bacteria growth is dependent on the iron dissolution from the aluminum silicate glass content (Table 1).

In summary, our results show the following: i) the CFA concentrations used in this study are potentially very relevant to human daily exposures, thus posing a potential public health risk for susceptible subjects living in urban areas and for those exposed to Fe-containing anthropogenic particles; ii) although CFA provides a source of bioavailable iron for bacteria, not all CFA particles have the same biological effects, and their propensity for iron dissolution can be an important factor on susceptible subjects and populations; iii) CFA impairs lung innate immune mechanisms of bacterial clearance, specifically AMP activity.

These results provide a potential mechanism to explain the epidemiological data that associates ambient air pollution and bacterial infections. However, we recognize that PM is very complex and requires the design of experiments that will control for different physicochemical characteristics such as size, shape, presence of other transition metals, aluminum silicate content, and iron species. We expect that identifying the PM mechanisms of respiratory infections will translate into public health policies aimed at controlling, not only concentration of PM exposure, but physicochemical characteristics that will potentially cause respiratory infections in susceptible individuals and populations.

Author Contributions

Conceived and designed the experiments: JAB HC JCC JB AP AC VG JZ. Performed the experiments: JAB JB JCC HC. Analyzed the data: JAB JB AC VG HC. Contributed reagents/materials/analysis tools: AC JZ VG. Wrote the paper: JAB VG AC.

References

1. Watson W, Paduano N, Raghuveer T, Thapa S (2010) U.S. Coal Supply and Demand: 2010 Year in Review. In: Administration USEI, editor. Washington, DC: Environmental Protection Agency.

2. Hygienists ACGIH (2009) The Documentation of the Threshold Limit Values and Biological Exposure Indicies. Cincinnati: ACGIH. 74–75.

3. Giere R, Carleton LE, Lumpkin GR (2003) Micro- and nanochemistry of fly ash from a coal-fired power plant. American Mineralogist 88: 1853–1865.

4. Goss CH, Newsom SA, Schildcrout JS, Sheppard L, Kaufman JD (2004) Effect of ambient air pollution on pulmonary exacerbations and lung function in cystic

fibrosis. American journal of respiratory and critical care medicine 169: 816–821.

5. Ling SH, Van Eeden SF (2009) Particulate matter air pollution exposure: role in the development and exacerbation of chronic obstructive pulmonary disease. Int J Chron Obstruct Pulmon Dis 4: 233–243.

6. Arbex MA, de Souza Conceicao GM, Cendon SP, Arbex FF, Lopes AC, et al. (2009) Urban air pollution and chronic obstructive pulmonary disease-related emergency department visits. Journal of epidemiology and community health 63: 777–783.

7. Gilligan PH (1991) Microbiology of airway disease in patients with cystic fibrosis. Clinical microbiology reviews 4: 35–51.

8. Smith KR, Samet JM, Romieu I, Bruce N (2000) Indoor air pollution in developing countries and acute lower respiratory infections in children. Thorax 55: 518–532.

9. Ghio AJ, Carraway MS, Madden MC (2012) Composition of air pollution particles and oxidative stress in cells, tissues, and living systems. Journal of toxicology and environmental health Part B, Critical reviews 15: 1–21.

10. Shi Z, Krom MD, Bonneville S, Baker AR, Bristow C, et al. (2011) Influence of chemical weathering and aging of iron oxides on the potential iron solubility of Saharan dust during simulated atmospheric processing. Global Biogeochemical Cycles 25.

11. Chen H, Laskin A, Baltrusaitis J, Gorski CA, Scherer MM, et al. (2012) Coal Combustion Fly Ash as a Source of Iron in Atmospheric Dust. Environmental science & technology 46(4): 2112–2120.

12. Veranth JM, Smith KR, Hu AA, Lighty JS, Aust AE (2000) Mobilization of iron from coal fly ash was dependent upon the particle size and source of coal: analysis of rates and mechanisms. Chemical research in toxicology 13: 382–389.

13. Veranth JM, Smith KR, Huggins F, Hu AA, Lighty JS, et al. (2000) Mossbauer spectroscopy indicates that iron in an aluminosilicate glass phase is the source of the bioavailable iron from coal fly ash. Chemical research in toxicology 13: 161–164.

14. Bullen JJ, Rogers HJ, Spalding PB, Ward CG (2005) Iron and infection: the heart of the matter. FEMS immunology and medical microbiology 43: 325–330.

15. Baltrusaitis J, Usher CR, Grassian VH (2007) Reactions of sulfur dioxide on calcium carbonate single crystal and particle surfaces at the adsorbed water carbonate interface. Physical chemistry chemical physics : PCCP 9: 3011–3024.

16. Karp PH, Moninger TO, Weber SP, Nesselhauf TS, Launspach JL, et al. (2002) An In Vitro Model of Differentiated Human Airway Epithelia: Epithelial Cell Culture Protocols. Methods in molecular biology 188: 115–137.

17. Caraballo JC, Yshii C, Butti ML, Westphal W, Borcherding JA, et al. (2011) Hypoxia increases transepithelial electrical conductance and reduces occludin at the plasma membrane in alveolar epithelial cells via PKC-zeta and PP2A pathway. American journal of physiology Lung cellular and molecular physiology 300: L569–578.

18. Kaneko Y, Thoendel M, Olakanmi O, Britigan BE, Singh PK (2007) The transition metal gallium disrupts Pseudomonas aeruginosa iron metabolism and has antimicrobial and antibiofilm activity. The Journal of clinical investigation 117: 877–888.

19. Finnerty K, Choi JE, Lau A, Davis-Gorman G, Diven C, et al. (2007) Instillation of coarse ash particulate matter and lipopolysaccharide produces a systemic inflammatory response in mice. Journal of toxicology and environmental health Part A 70: 1957–1966.

20. Nurkiewicz TR, Porter DW, Barger M, Millecchia L, Rao KM, et al. (2006) Systemic microvascular dysfunction and inflammation after pulmonary particulate matter exposure. Environmental health perspectives 114: 412–419.

21. Harrod KS, Jaramillo RJ, Berger JA, Gigliotti AP, Seilkop SK, et al. (2005) Inhaled diesel engine emissions reduce bacterial clearance and exacerbate lung disease to Pseudomonas aeruginosa infection in vivo. Toxicological sciences : an official journal of the Society of Toxicology 83: 155–165.

22. Mukherjee B, Dutta A, Roychoudhury S, Ray MR (2011) Chronic inhalation of biomass smoke is associated with DNA damage in airway cells: involvement of particulate pollutants and benzene. Journal of applied toxicology 33(4): 281–289.

23. Jones T, Brown P, BeruBe K, Wlodarczyk A, Longyi S (2010) The physicochemistry and toxicology of CFA particles. Journal of toxicology and environmental health Part A 73: 341–354.

24. Voelkel K, Krug HF, Diabate S (2003) Formation of reactive oxygen species in rat epithelial cells upon stimulation with fly ash. Journal of biosciences 28: 51–55.

25. Panush D, Fulbright R, Sze G, Smith RC, Constable RT (1993) Inversion-recovery fast spin-echo MR imaging: efficacy in the evaluation of head and neck lesions. Radiology 187: 421–426.

26. Wiesner J, Vilcinskas A (2010) Antimicrobial peptides: the ancient arm of the human immune system. Virulence 1: 440–464.

27. Ganz T (2002) Antimicrobial polypeptides in host defense of the respiratory tract. The Journal of clinical investigation 109: 693–697.

28. Roberts JR, Young SH, Castranova V, Antonini JM (2009) The soluble nickel component of residual oil fly ash alters pulmonary host defense in rats. Journal of immunotoxicology 6: 49–61.

29. Sigaud S, Goldsmith CA, Zhou H, Yang Z, Fedulov A, et al. (2007) Air pollution particles diminish bacterial clearance in the primed lungs of mice. Toxicology and applied pharmacology 223: 1–9.

30. Hatch GE, Boykin E, Graham JA, Lewtas J, Pott F, et al. (1985) Inhalable particles and pulmonary host defense: in vivo and in vitro effects of ambient air and combustion particles. Environmental research 36: 67–80.

31. Miyata R, van Eeden SF (2011) The innate and adaptive immune response induced by alveolar macrophages exposed to ambient particulate matter. Toxicology and applied pharmacology 257: 209–226.

32. Pezzulo AA, Tang XX, Hoegger MJ, Alaiwa MH, Ramachandran S, et al. (2012) Reduced airway surface pH impairs bacterial killing in the porcine cystic fibrosis lung. Nature 487: 109–113.

33. Parameswaran GI, Sethi S, Murphy TF (2011) Effects of bacterial infection on airway antimicrobial peptides and proteins in COPD. Chest 140: 611–617.

34. Sandstrom T, Kolmodin-Hedman B, Ledin MC, Bjermer L, Hornqvist-Bylund S, et al. (1991) Exposure to peat dust: acute effects on lung function and content of bronchoalveolar lavage fluid. British journal of industrial medicine 48: 771–775.

35. Noble RE (2002) Effect of environmental contaminants on nasal lysozyme secretions. The Science of the total environment 284: 263–266.

36. Travis SM, Conway BA, Zabner J, Smith JJ, Anderson NN, et al. (1999) Activity of abundant antimicrobials of the human airway. American journal of respiratory cell and molecular biology 20: 872–879.

37. Klein-Patel ME, Diamond G, Boniotto M, Saad S, Ryan LK (2006) Inhibition of beta-defensin gene expression in airway epithelial cells by low doses of residual oil fly ash is mediated by vanadium. Toxicological sciences : an official journal of the Society of Toxicology 92: 115–125.

9

Is Particle Pollution in Outdoor Air Associated with Metabolic Control in Type 2 Diabetes?

Teresa Tamayo[1]*, Wolfgang Rathmann[1], Ursula Krämer[2], Dorothea Sugiri[2], Matthias Grabert[3], Reinhard W. Holl[3]

1 Institute of Biometrics and Epidemiology, German Diabetes Center, Leibniz Center for Diabetes Research at Heinrich Heine University Düsseldorf, Düsseldorf, Germany, 2 Institute for Environmental Medicine (IUF), Leibniz Center at Heinrich Heine University Düsseldorf, Düsseldorf, Germany, 3 Institute for Epidemiology and Medical Biometry, University of Ulm, Ulm, Germany

Abstract

Background: There is growing evidence that air pollutants are associated with the risk of type 2 diabetes. Subclinical inflammation may be a mechanism linking air pollution with diabetes. Information is lacking whether air pollution also contributes to worse metabolic control in newly diagnosed type 2 diabetes. We examined the hypothesis that residential particulate matter (PM_{10}) is associated with HbA_{1c} concentration in newly diagnosed type 2 diabetes.

Methods: Nationwide regional levels of particulate matter with a diameter of ≤ 10 µm (PM_{10}) were obtained in 2009 from background monitoring stations in Germany (Federal Environmental Agency) and assigned to place of residency of 9,102 newly diagnosed diabetes patients registered in the DPV database throughout Germany (age 65.5 ± 13.5 yrs; males: 52.1%). Mean HbA_{1c} (%) levels stratified for air pollution quartiles (PM_{10} in µg/m^3) were estimated using linear regression models adjusting for age, sex, BMI, diabetes duration, geographic region, year of ascertainment, and social indicators.

Findings: In both men and women, adjusted HbA_{1c} was significantly lower in the lowest quartile of PM_{10} exposure in comparison to quartiles Q2–Q4. Largest differences in adjusted HbA_{1c} (95% CI) were seen comparing lowest quartiles of exposure with highest quartiles (men %: -0.42 (-0.62; -0.23)/mmol/mol: -28.11 (-30.30; -26.04), women, %: -0.28 (-0.47; -0.09)/mmol/mol: -0.28 (-0.47; -0.09)).

Interpretation: Air pollution may be associated with higher HbA_{1c} levels in newly diagnosed type 2 diabetes patients. Further studies are warranted to examine this association.

Editor: Florian Kronenberg, Innsbruck Medical University, Austria

Funding: Data are part of the BMBF Diabetes Meta-Database project which was funded by the Competence Network for Diabetes mellitus of the Federal Ministry of Education and Research (support codes 01GI1110D/01GI1106). The funders had no role in study design, data collection and analysis, decision to publish, or preparation of the manuscript.

Competing Interests: The authors have declared that no competing interests exist.

* E-mail: teresa.tamayo@ddz.uni-duesseldorf.de

Background

The associations between exposure to traffic-related air pollution and cardiovascular disease, cardiovascular hospital admission rates, and all-cause or cardiovascular mortality are well established [1,2]. Patients with type 2 diabetes are more susceptible to these adverse effects [3]. Besides, traffic-related air pollution was associated with diabetes-associated mortality in a current study [4]. Recently, evidence is growing that air pollutants (nitrogen oxides (NOx), particulate matter (PM) with a diameter of ≤ 10 µm or 2.5 µm) may also be associated with type 2 diabetes prevalence and incidence [5–9]. Diabetes risk was increased by 4%–15% per interquartile range (IQR) of particulate matter with a diameter of ≤ 10 µm (PM10) [5,7], by 25% per IQR increase in nitrogen oxides (NOx) [9], and by 11% for living in short distance (<50 m) to a major road [8].

Furthermore, in a cross-sectional Taiwanese study, higher HbA1c levels were observed with increased traffic-related air pollution in the general population [10]. To our knowledge, the association with HbA1c has not been examined in patients with type 2 diabetes so far. HbA1c is mainly used as an indicator for metabolic control in persons with type 2 diabetes. Guidelines stress the importance of a good metabolic control in most patients in order to prevent complications [11]. Even small increase in HbA1c due to worse metabolic control could affect long-term cardiovascular risk and mortality [12,13]. However, air pollution may impede an optimal metabolic control by increased subclinical inflammation [14]. In addition, inflammatory processes may also increase the vulnerability to cardiovascular health effects (e.g. myocardial infarction) in persons with type 2 diabetes who are exposed to air pollution [5,15].

Thus, we examined HbA_{1c} concentration in individuals with newly diagnosed type 2 diabetes and its association with residential air pollution in a large German cohort based on the DPV documentation system (Diabetessoftware für Prospektive Verlaufsbeobachtung) using data assessed in routine care [16].

Table 1. Characteristics of participants with type 2 diabetes*.

	Total	Women	Men
Number of participants (N)	9,102	4,356	4,746
Age at baseline examination (years)	65.5 (13.5)	67.3 (14.0)	63.9 (12.9)
Diabetes duration (years)	1.5 (0.6)	1.5 (0.6)	1.5 (0.6)
HbA$_{1c}$ (%)	7.2 (1.9)	7.1 (1.8)	7.3 (2.0)
Body mass index (kg/m^2)	30.6 (6.4)	30.9 (6.9)	30·4 (5.9)
Mean PM10 year 2009 (µg/m^3)	19.6 (4.3)	19.7 (4.2)	19·6 (4.3)
Hypertension/antihypertensive treatment (%)	65.2	64.7	65.6
Dyslipidaemia/lipid-lowering treatment (%)	66.8	67.2	66.5
Insulin treatment (%)	35.3	33.9	36.6
Oral antihyperglycaemic drugs (%)	57.1	55.6	58.5

*Results are numbers (N), frequencies in % or means (SD). HbA1c levels in % (NGSP) can be converted to mmol/mol (IFCC) by application of the following formula: IFCC = (10.93*NGSP) − 23.50.

Methods

Ethics Statement

Informed consent was obtained from every patient at each participating center (more than 300 GP practices, hospitals, rehabilitation clinics). The consent procedure and documentation (either verbal or written depending on institution) was approved by local institutional review boards or the responsible commissioners for data protection of participating centers. The locally collected study data was anonymized before transferal to the data management center at Ulm University. The DPV study and the consent and data collection procedures were approved by local data control authorities and the institutional review board at Ulm University.

Study Population

Patients with newly diagnosed type 2 diabetes aged 18 years or older who were registered between 2005 and 2009 in the DPV database were selected for the analysis. The DPV database covers anonymized data on more than 200,000 patients from 336 participating health facilities such as diabetologists, primary care practices, hospitals and rehabilitation clinics in Germany [16]. Nationwide, physicians from participating centers document each patient with diabetes diagnosis and data e.g. on age, sex, diabetes duration, HbA$_{1c}$, laboratory measurements and medication. After informed consent, patient data are transferred electronically twice a year to the documentation center in Ulm. For this purpose, a computer software was installed in participating centers that serves for medical documentation in routine care as well as for data collection in the DPV study with considerable overlap between both functions.

For the present study, the data was analyzed cross-sectionally. Only patients with a diabetes duration of a maximum of 2.5 years (range: 0.5–2.5 years) were included in order to examine a homogeneous patient group with doctor's visits in a comparable time frame in which similar treatment options were available. Of 14,042 patients, 1,984 individuals had missing information on body mass index (BMI) or HbA$_{1c}$. Furthermore, we restricted the sample to persons treated in ambulatory care units of hospitals to examine a more homogeneous patient group with better documentation. Thus, 9,102 participants were available for analysis.

Measurements

HbA$_{1c}$ measurements were measured locally and adjusted to the Diabetes Control and Complications Trial (DCCT) normal range using the multiple of the mean method based on the reference range for healthy subjects in each laboratory [16,17].

Nationwide regional levels of particulate matter with a diameter of ≤10 µm (PM$_{10}$) were obtained for Germany based on a raster with a cell size of eight kilometer × eight kilometer. These maps were generated by the environmental agency of Germany "Umweltbundesamt II 4.2" (monitoring of air quality) using the chemical REM-CALGRID^2 (RCG) model into which PM10 measured at background monitoring stations was integrated [18]. The REM-CALGRID^2 itself has been used since 1999. It is fitted with meteorological and PM time-series data of 150 German monitoring stations and additional data from other European countries. The integration of PM10 for 2009 from the monitoring sites was done using the optimal interpolation method (OI). This model includes inhomogeneous spatial auto-covariance between PM10 from monitoring stations and the broad scale background information for representative (reference) areas. Different models of covariance were applied for each calculated rasterpoint depending on monitored and modeled PM$_{10}$. Suburban sites were overrepresented in monitoring stations. Therefore, prior to integration, PM$_{10}$ measurements from monitoring stations were corrected for this suburban/rural bias. Based on this approach, we calculated the annual PM$_{10}$ for each five-digit postcode area (100 areas) by intersection of the German PM$_{10}$ raster with the German postcode map. Each postcode area obtained an area-weighted mean of PM10 of included rastercells. Intersection was done with ArcGIS version 9, Environmental Systems Research Institute (ESRI), California, USA.

Data of the interpolation raster are leveled to the measurement range of background monitoring stations. The measurements are further leveled out by integration by interpolation. In other words, data smoothing is needed to characterize the average pollution level in each raster area which includes measurements from e.g. urban and rural sites with heterogeneous levels of PM$_{10}$. This was accomplished by interpolation over the raster cells and furthermore by integration to calculate the mean pollution of each raster.

Additionally, patient's residency was available on postcode level only. Depending on population density, postcode areas may extend to dozens of km^2 in regions with low population density. As a consequence, the interpolation raster is further coarsened to

Table 2. Characteristics of participants with type 2 diabetes per geographic region (Nielsen area).

	1 (N)	2 (W)	3 (SW)	4 (S)	5 (B)	6 (NE)	7 (E)	P-value
Number of participants (N)	451	2729	3050	1451	412	850	159	
Women (N)	176	1344	1435	706	215	409	71	0.002
Age at baseline examination (years)	64.8 (12.4)	67.6 (12.8)	64.1 (14.0)	66.2 (13.2)	63.0 (13.2)	64.9 (13.9)	63.8 (15.4)	<0.0001
Diabetes duration (years)	1.5 (0.6)	1.5 (0.6)	1.5 (0.6)	1.5 (0.6)	1.4 (0.6)	1.5 (0.6)	1.5 (0.6)	0.438
HbA$_{1c}$ (%)	7.4 (2.7)	7.1 (1.8)	7.2 (1.9)	7.2 (1.9)	7.4 (1.9)	7.6 (2.0)	7.1 (1.7)	<0.0001
Body mass index (kg/m^3)	30.4 (5.7)	30.2 (6.2)	30.8 (6.4)	30.3 (6.5)	31.5 (6.8)	31.2 (6.5)	31.6 (6.3)	<0.0001
Mean PM10 year 2009 (µg/m^3)	15.9 (2.1)	23.5 (3.8)	17.5 (2.9)	17.7 (2.1)	25.0 (1.7)	17.1 (2.5)	19.1 (1.5)	<0.0001
Hypertension/antihypertensive treatment (%)	73.4	53	68.7	76.6	68.4	67.1	61	<0.0001
Dyslipidaemia/Lipid-lowering treatment (%)	60.5	66.6	67.4	71.3	51.9	70.7	54.7	<0.0001
Insulin treatment (%)	25.1	29.8	38.7	28.5	41.7	54.5	39	<0.0001
Oral antihyperglycaemic drugs (%)	56.3	55.6	54.4	71.1	54.6	49.1	60.4	<0.0001

Results are numbers (N), frequencies in % or means (SD). Abbreviation of Nielsen areas: 1 (N) = Hamburg, Bremen, Schleswig-Holstein, Lower Saxony (North); 2 (W) = North Rhine-Westfalia (West); 3 (SW) = Hesse, Rhineland-Palatinate, Saarland, Baden-Württemberg (Southwest); 4 (S) = Bavaria (South); 5 (B) = Berlin (Northeast); 6 (NE) = Mecklenburg-Vorpommern, Brandenburg, Saxony-Anhalt (Northeast); 7 (E) = Thuringia, Saxony (East). HbA1c levels in % (NGSP) can be converted to mmol/mol (IFCC) by application of the following formula: IFCC = (10.93*NGSP) − 23.50.

area-weighted averages for each postcode area. Therefore measurements are not precise for the place of residence but leveled to the area around it. However, people usually do not stay at their place of residence throughout the day but change their position for work or leisure time. Hence, this approach yields a rather valid estimation of annual background exposure for patients who mostly stay within the range of some kilometers around their place of residence. Further information on the application of the REM-CALGRID^2 model has been described elsewhere [19].

Height and weight were measured during doctors' visits. Hypertension and dyslipidemia were defined according to doctor's diagnosis or disease-specific medication.

As social indicators we included formal schooling (no high school diploma, yes/no) or immigrant status (yes/no).

Geographic Location of Nielsen Areas

Nielsen areas were first used in market research to determine geographic regions sharing characteristics of federal economy and consumer behavior (The Nielsen Company, NY, USA). In Germany, seven Nielsen areas are distinguished each of which includes one or more coherent federal states. Germany consists of sixteen federal states with autonomous jurisdiction in many aspects of administrative law affecting e.g. health politics and education. Some of the German federal states are rather small (Bremen and Hamburg). In these smaller federal states, only some physicians participated in the DPV study, so that numbers in some regions were low and protection of data privacy was not guaranteed on federal state level. Therefore, Nielsen areas were used to allow for adjustment of regional disparities in health care and of geographic features. Because Nielsen areas are characterized by economic factors they also function as an area-based social indicator reflecting e.g. unemployment rate of a region.

Nielsen area 1: Hamburg, Bremen, Schleswig-Holstein, Lower Saxony (North)

Nielsen area 2: North Rhine-Westfalia (West)

Nielsen area 3 Hesse, Rhineland-Palatinate, Saarland, Baden-Württemberg (Southwest)

Nielsen area 4 Bavaria (South)

Nielsen area 5 Berlin (Northeast)

Nielsen area 6 Mecklenburg-Vorpommern, Brandenburg, Saxony-Anhalt (Northeast)

Nielsen area 7 Thuringia, Saxony (East)

Statistical Analysis

For descriptive analyses, mean (SD) were calculated for continuous variables and proportions for categorical variables. Quartiles of exposure to PM$_{10}$ (µg/m^3) in 2009 were calculated using the distribution in the study population. Mean (adjusted) HbA$_{1c}$ levels stratified for air pollution quartiles (PM$_{10}$ in µg/m^3) and mean difference in HbA$_{1c}$ levels between quartiles and corresponding 95% confidence intervals were estimated using generalized linear regression modeling adjusting for age, sex, BMI, diabetes duration, geographic region (Nielsen area), year of ascertainment, and the social indicator (no high school diploma or immigration background). In addition, analyses stratified by sex were performed. For sensitivity analyses, models were fitted including both patients treated in ambulatory care units of hospitals and those treated in outpatient practices of general practitioners and diabetologists. Furthermore, the linear association of continuously measured PM$_{10}$ (µg/m^3) with HbA1c measurements was analyzed. Moreover, age- and sex- adjusted models were fitted, additionally adjusting for types of medication (oral anti-diabetic medication (OAD) or insulin) and co-morbidities (hypertension or dyslipidaemia). The limit of statistical

Table 3. Adjusted mean HbA$_{1c}$ and difference in HbA$_{1c}$ comparing quartiles of particulate matter (PM$_{10}$) exposure in type 2 diabetes patients*.

Quartiles of exposure	Total sample	Women	Men
	N = 9,102	N = 4,356	N = 4,746
Mean adjusted HbA$_{1c}$ in %			
Q1 < 16.40 µg/m³	6.9	6.7	7.2
Q2 16.40 − < 18.05 µg/m³	7.1	6.9	7.4
Q3 18.05 − < 21.10 µg/m³	7.1	6.9	7.4
Q4 ≥ 21.10 µg/m³	7.3	6.9	7.6
Difference in HbA$_{1c}$ levels in %			
Q1 vs. Q2	**−0.20 (−0.33, −0.08)**	**−0.22 (−0.38, −0.05)**	**−0.20 (−0.37, −0.02)**
Q1 vs. Q3	**−0·21 (−0·32, −0·09)**	**−0·24 (−0·39; −0·08)**	**−0·18 (−0·34; −0·02)**
Q1 vs. Q4	**−0·36 (−0·49, −0·22)**	**−0·28 (−0·47, −0·09)**	**−0·42 (−0·62, −0·23)**
Q2 vs. Q3	0·00 (−0.13, 0.12)	−0·02 (−0.19; 0.15)	0·02 (−0.16, 0.20)
Q2 vs. Q4	**−0.15 (−0.30, −0.01)**	−0.07 (−0.26; 0.13)	**−0.23 (−0.44, −0.02)**
Q3 vs. Q4	**−0.15 (−0.28, −0.02)**	−0.04 (−0.21, 0.13)	**−0.25 (−0.43, −0.06)**

*Results are adjusted means for HbA$_{1c}$ in % calculated from generalized linear regression models. Models were fitted adjusting for age, sex, body mass index, duration of diabetes, geographic region, year of treatment, and social indicators (low education, immigration background). Furthermore, difference in HbA$_{1c}$ levels in % (95% CI) comparing quartiles of PM$_{10}$ exposure also derived from linear regression models are presented. Group differences are considered as significant (highlighted in bold) if corresponding 95% confidence intervals do not include 0. HbA1c levels in % (NGSP) can be converted to mmol/mol (IFCC) by application of the following formula: IFCC = (10.93*NGSP) − 23.50.

significance was set at p < 0.05. Statistical analyses were carried out with SAS for Windows version 9.3. (SAS Institute, Cary, NC, USA).

Results

The studied sample comprised 9,102 patients (4,356 women, 4,746 men) with newly diagnosed type 2 diabetes whose mean diabetes duration was 1.5 years (SD 0.6 years).

Of all patients, 49% were registered in the South (Nielsen regions 3 and 4), 30% in the West (Nielsen 2), 16% in the North-East (Nielsen 5–7) and 5% in the North of Germany (Nielsen 1). In Table 1 characteristics of the study participants are shown. On average, the sample was mostly elderly and obese. Numbers of participants who had left school without a high school diploma or who had immigrant status (social indicator) were low (n = 182). The mean annual HbA$_{1c}$ (%) was 7.2 (SD: 1.9%). Overall, more than one third was treated with insulin, and more than half of the sample received oral glucose-lowering drugs. Hypertension or antihypertensive drug prescriptions were found in approximately two thirds. Patients from different geographic Nielsen areas differed in several aspects (Table 2). Patients from the South and West were slightly older on average, had lower HbA1c levels and a lower mean BMI especially in comparison to patients from the East, Northeast and Berlin. Exposure to PM$_{10}$ was higher in the densely populated areas of Berlin and the Rhine-Ruhr-Area (West). Medical treatment also differed considerably between regions with a very high percentage of patients receiving oral anti-hyperglycaemic drugs and a low percentage receiving insulin in the South and a reversed pattern in Berlin and the Northeast.

Table 3 shows adjusted differences in HbA$_{1c}$ levels and corresponding 95% confidence intervals across particle exposures. HbA$_{1c}$ (%) was significantly lower in both men and women in the lowest quartile of PM$_{10}$ exposure in comparison to quartiles with higher levels of exposure (Q1 vs. Q2–Q4). Men in Q2 and Q3 also had substantially lower adjusted HbA$_{1c}$ levels than those in the

highest quartile of PM$_{10}$ exposure (Q2 vs. Q4: −0.23; 95% CI: −0.44, −0.02/Q3 vs. Q4: −0.25; −0.43, −0.06).

In the adjusted model, further variables associated with HbA$_{1c}$ were sex, BMI, age, diabetes duration and geographic area (Nielsen areas). We did not observe any association with the social indicator variable (p = 0.47). Of note, crude mean HbA1c values changed after adjustment. Adjusting for age and sex only, HbA1c levels were significantly lower in quartile 1 than in all other quartiles (e.g. Q1: 7.1%; Q4: 7.3%). With further adjustment for BMI, diabetes duration and region (Nielsen area), differences in HbA1c levels were more pronounced.

While the main analysis (see tables) included only patients treated in ambulatory care units of hospitals, we carried out a sensitivity analysis encompassing both patients treated in hospitals and in practices of general practitioners. In this sample of 12,058 participants, the differences between quartiles of pollution were slightly attenuated in the final model adjusted for age, sex, BMI, diabetes duration (years), Nielsen areas, year of treatment, institution of treatment (GP yes/no) and the social indicator. Overall, mean adjusted HbA1c levels were 0.2–0.4% (2–4 mmol/mol) lower in all quartiles. Difference in Quartiles Q1 and Q3 did not reach statistical significance (−0.12; 95% CI −0.25, 0.01) while the overall tendency of all other group comparisons remained similar.

In further models also adjusted for age, sex, BMI, diabetes duration (years), Nielsen areas, year of treatment and the social indicator but fitted with the continuous measurements of PM10 confirmed a significant association with HbA1c (estimate: 0.025, SE 0.007, p = 0.0001). With further adjustment for clinical information such as treatment with oral anti-hyperglycemic drugs only (yes/no), hypertension or anti-hypertensive drugs, dyslipidaemia or lipid-lowering drugs results hardly changed. Inclusion of the information on treatment with insulin (yes/no) attenuated the association with air pollution, but it remained significant (0.014, SE 0.002, p = 0.02).

Discussion

The novel finding of the present study is that exposure to particulate matter (PM_{10}) is associated with higher HbA_{1c} levels (worse metabolic control) in newly diagnosed type 2 diabetes patients. Our findings are in line with previous results of a population-based Taiwanese study where HbA_{1c} levels increased by 1.4% (95% CI 1.1–1.7) for each inter-quartile range increase in PM_{10} pollution.[10] In our study in type 2 diabetes, HbA_{1c} increase was less pronounced, however, on substantially lower levels of air pollution exposure. Adjusting for insulin attenuated the association but it remained significant. This has two implications: first, patients living in highly polluted areas would possibly require insulin at an earlier stage of their disease, and second, metabolic control is impaired in these patients, even under early medication with insulin and they would possibly require higher dosages of insulin to achieve HbA_{1c} targets of guideline recommendations. However, further studies are warranted to corroborate these hypotheses.

Despite the relatively small increase in HbA_{1c} levels for each quartile increment of particulate matter exposure in our study, this difference might contribute to a considerable long-term increase of micro- and macrovascular complications. In the population-based Rancho Bernardo study, a 1% increase in HbA_{1c} was associated with a hazard ratio for cardiovascular mortality of 1.26 (95% CI 1.03–1.55) even at non-diabetic levels [20].

Further evidence on the importance of comparatively low levels of air pollution comes from another recent study [21]. In 25 otherwise healthy individuals with impaired glucose tolerance from rural areas, even small increases in $PM_{2.5}$ concentrations affected insulin resistance after short periods and at small levels of exposure (5 days). Furthermore, in a Swedish study, nitrogen oxide exposure at levels below current WHO air quality guidelines during pregnancy was associated with gestational diabetes and pre-eclampsia [22].

Rajagopalan and Brook have summarized pathophysiologic pathways which are currently discussed to explain the association between air pollution and type 2 diabetes [15]. Among these, systemic inflammation and oxidative stress play an important role. This mechanistic pathway may also explain the association of air pollution with type 2 diabetes, gestational diabetes and gestational complications (e.g. pre-eclampsia) and moreover the deleterious cardiovascular effects of air pollution in patients with type 2 diabetes [3,15,23]. In response to inhaled pollutants a state of chronic systemic inflammation and oxidative stress occurs which subsequently may aggravate insulin resistance and trigger metabolic disturbances [24].

Data on possible pathophysiologic pathways have mostly been obtained from mouse models. These mouse models also suggested an interaction of $PM_{2.5}$ exposure with high-fat diet during the development of metabolic disturbances [14]. However, mice fed normal chow also showed enlarged visceral fat contents and increased macrophage infiltration in visceral adipose tissue after exposure to $PM_{2.5}$ for 10 weeks [25]. Another recent mouse model suggested further pathways linking air pollution with type 2 diabetes by showing that $PM_{2.5}$ exposure had adverse effects on glycogen storage in the liver which led to the development of a NASH like phenotype [26].

As an example for the inflammatory response in human beings, short-term exposure with concentrated ambient particles induced mild pulmonary inflammation and increased plasma fibrinogen content in healthy volunteers [27]. However, it should also be noted that a recent cross-sectional study in elderly women found no convincing evidence for an association between exposure to PM_{10} and elevated plasma levels of proinflammatory biomarkers [28].

Thus, effects of air pollutants are suggested to affect various organs and systems of the body including glucose metabolism [3]. Given the world-wide burden of traffic- and industry-related air pollution, and of type 2 diabetes, pollution control might be very effective to lower the burden of disease. Strategies to reduce exposure to traffic related air pollution such as urban planning, land-use decisions and individual strategies need to be developed and tested.

Limitations

First, HbA_{1c} levels were not centrally determined. In order to reduce between-laboratory variation, HbA_{1c} values were standardized to the Diabetes Control and Complication Trial Research Group reference range (DCCT) using the multiple of the mean method.[17]. Furthermore, lifestyle factors (physical activity, nutrition) were not assessed. Also, detailed information on socioeconomic circumstances of patients was not available (e.g. schooling degree, income situation, professional career). Therefore, uncontrolled confounding by individual socioeconomic or lifestyle factors may have played a role in the association that we observed. Finally, there may be an uncertainty of measurements especially in some participants with a high mobility or unusually high exposures at work or indoors (e.g. due to open fires). Thus, the association of air pollution with HbA1c levels we found may actually be caused by socioeconomic or lifestyle factors we could not adjust for. However, the strength of the study is the use of a huge nationwide sample covering both rural and urban regions across Germany.

Conclusions

Patients with type 2 diabetes exposed to higher levels of air pollution showed higher HbA_{1c} levels and consequently might be at a higher risk for complications. However, we cannot rule out residual confounding due to those socioeconomic or lifestyle factors that were not available for analysis. Considering the worldwide burden of type 2 diabetes and of air pollution, this association needs further corroboration.

Acknowledgments

The authors thank Martin de Souza (Institute for Epidemiology and Medical Biometry, University of Ulm, Ulm, Germany) for his contributions to the statistical analysis. We especially thank the German Federal Environment Agency, Department II 4.2, monitoring of air quality, for providing us with the air pollution values.

Author Contributions

Conceived and designed the experiments: WR RH UK TT. Analyzed the data: RH MG. Contributed reagents/materials/analysis tools: RH DS. Wrote the paper: TT.

References

1. Pope CA (2007) Mortality effects of longer term exposures to fine particulate air pollution: review of recent epidemiological evidence. Inhal Toxicol 19: 33–38.
2. Brook RD, Rajagopalan S, Pope CA 3rd, Brook JR, Bhatnagar A, et al. (2010) American Heart Association Council on Epidemiology and Prevention, Council on the Kidney in Cardiovascular Disease, and Council on Nutrition, Physical Activity and Metabolism. Particulate matter air pollution and cardiovascular disease: An update to the scientific statement from the American Heart Association. Circulation 121: 2331–2378.
3. Zanobetti A, Schwartz J (2012) Cardiovascular damage by airborne particles: are diabetics more susceptible? Epidemiology 13: 588–92.

4. Raaschou-Nielsen O, Sørensen M, Ketzel M, Hertel O, Loft S, et al. (2013) Long-term exposure to traffic-related air pollution and diabetes-associated mortality: a cohort study. Diabetologia 56: 36–46.

5. Brook RD, Jerrett M, Brook JR, Bard RL, Finkelstein MM (2008) The relationship between diabetes mellitus and traffic-related air pollution. J Occup Environ Med 50: 32–38.

6. Dijkema MB, Mallant SF, Gehring U, van den Hurk K, Alssema M, et al. (2011) Long-term exposure to traffic-related air pollution and type 2 diabetes prevalence in a cross-sectional screening-study in the Netherlands. Environ Health 10: 76.

7. Krämer U, Herder C, Sugiri D, Strassburger K, Schikowski T et al. (2010) Traffic-related air pollution and incident type 2 diabetes: results from the SALIA cohort study. Environ Health Perspect 118: 1273–1279.

8. Puett RC, Hart JE, Schwartz J, Hu FB, Liese AD, et al. (2011) Are particulate matter exposures associated with risk of type 2 diabetes? Environ Health Perspect 119: 384–389.

9. Coogan PF, White LF, Jerrett M, Brook RD, Su JG et al. (2012) Air pollution and incidence of hypertension and diabetes mellitus in black women living in Los Angeles. Circulation 125: 767–772.

10. Chuang KJ, Yan YH, Chiu SY, Cheng TJ (2011) Long-term air pollution exposure and risk factors for cardiovascular diseases among the elderly in Taiwan. Occup Environ Med 68: 64–68.

11. (2013) Executive Summary: Standards of Medical Care in Diabetes-2013. Diabetes Care 2013 36: 4–10.

12. Zoungas S, Chalmers J, Ninomiya T, Li Q, Cooper ME, et al. (2012) ADVANCE Collaborative Group. Association of HbA$_{1c}$ levels with vascular complications and death in patients with type 2 diabetes: evidence of glycaemic thresholds. Diabetologia 55: 636–643.

13. Kowall B, Rathmann W, Heier M, Giani G, Peters A, et al. (2011) Categories of glucose tolerance and continuous glycemic measures and mortality. Eur J Epidemiol 26: 637–645.

14. Sun Q, Yue P, Deiuliis JA, Lumeng CN, Kampfrath T, et al. (2009) Ambient air pollution exaggerates adipose inflammation and insulin resistance in a mouse model of diet-induced obesity. Circulation 119: 538–546.

15. Rajagopalan S, Brook RD (2012) Air pollution and type 2 diabetes: mechanistic insights. Diabetes 61: 3037–3045.

16. Awa WL, Fach E, Krakow D, Welp R, Kunder J et al. (2012) The DPV Initiative and the German BMBF Competence Networks Diabetes mellitus and Obesity. Type 2 diabetes from pediatric to geriatric age: analysis of gender and obesity among 120.183 patients from the German/Austrian DPV database. Eur J Endocrinol 167: 245–254.

17. The Diabetes Control and Complications Trial Research Group (1993) The effect of intensive treatment of diabetes on the development and progression of long-term complications in insulin-dependent diabetes mellitus. N Engl J Med 329: 977–986.

18. Flemming J, Reimer E, Stern R (2002) Data assimilation for CT-modelling based on optimum interpolation, in ITM Air pollution modelling and its applications XXIV, eds. C. Borrego und G. Schayes, NATO CMS, Kluwer Academic/Plenum Publishers, New York.

19. Stern R, Flemming J (2004) Formulation of criteria to be used for the determination of the accuracy of model calculations according to the requirements of the EU Directives for air quality – Examples using the chemical transport model REM-CALGRID. Final report for the environmental agency of Germany (Umweltbundesamt). Available: http://www.umweltbundesamt.de/sites/default/files/medien/publikation/long/3614.pdf. Accessed 2014 Feb 17.

20. Cohen BE, Barrett-Connor E, Wassel CL, Kanaya AM (2009) Association of glucose measures with total and coronary heart disease mortality: does the effect change with time? The Rancho Bernardo Study. Diabetes Res Clin Pract 86: 67–73.

21. Brook RD, Xu X, Bard RL, Dvonch JT, Morishita M, et al. (2013) Reduced metabolic insulin sensitivity following sub-acute exposures to low levels of ambient fine particulate matter air pollution. Sci Total Environ 448: 66–71.

22. Malmqvist E, Jakobsson K, Tinnerberg H, Rignell-Hydbom A, Rylander L (2013) Gestational Diabetes and Preeclampsia in Association with Air Pollution at Levels below Current Air Quality Guidelines. Environ Health Perspect 121: 488–493.

23. Sibai B, Dekker G, Kupferminc M (2005) Pre-eclampsia. Lancet 365: 785–799.

24. Brook RD, Franklin B, Cascio W, Hong Y, Howard G, et al. (2004) Expert Panel on Population and Prevention Science of the American Heart Association. Air pollution and cardiovascular disease: a statement for healthcare professionals from the Expert Panel on Population and Prevention Science of the American Heart Association. Circulation 109: 2655–2671.

25. Xu X, Yavar Z, Verdin M, Ying Z, Mihai G, et al. (2010) Effect of early particulate air pollution exposure on obesity in mice: role of p47phox. Arterioscler Thromb Vasc Biol 30: 2518–2527.

26. Zheng Z, Xu X, Zhang X, Wang A, Zhang C, et al. (2013) Exposure to Ambient Particulate Matter Induces a NASH-like Phenotype and Impairs Hepatic Glucose Metabolism in an Animal Model. J Hepatol 58: 148–154.

27. Salvi S, Blomberg A, Rudell B, Kelly F, Sandström T, et al. (1999) Acute inflammatory responses in the airways and peripheral blood after short-term exposure to diesel exhaust in healthy human volunteers. Am J Respir Crit Care Med 159: 702–709.

28. Teichert T, Vossoughi M, Vierkötter A, Sugiri D, Schikowski T, Schulte T, et al. (2013) Association between Traffic-Related Air Pollution, Subclinical Inflammation and Impaired Glucose Metabolism: Results from the SALIA Study. PLoS One 8: e83042.

10

The Association of Ambient Air Pollution and Physical Inactivity in the United States

Jennifer D. Roberts[1]*, **Jameson D. Voss**[1,2], **Brandon Knight**[1]

1 Department of Preventive Medicine and Biometrics, F. Edward Hebert School of Medicine, Uniformed Services University, Bethesda, Maryland, United States of America,
2 Epidemiology Consult Service, United States Air Force School of Aerospace Medicine, Wright-Patterson Air Force Base, Ohio, United States of America

Abstract

Background: Physical inactivity, ambient air pollution and obesity are modifiable risk factors for non-communicable diseases, with the first accounting for 10% of premature deaths worldwide. Although community level interventions may target each simultaneously, research on the relationship between these risk factors is lacking.

Objectives: After comparing spatial interpolation methods to determine the best predictor for particulate matter ($PM_{2.5}$; PM_{10}) and ozone (O_3) exposures throughout the U.S., we evaluated the cross-sectional association of ambient air pollution with leisure-time physical inactivity among adults.

Methods: In this cross-sectional study, we assessed leisure-time physical inactivity using individual self-reported survey data from the Centers for Disease Control and Prevention's 2011 Behavioral Risk Factor Surveillance System. These data were combined with county-level U.S. Environmental Protection Agency air pollution exposure estimates using two interpolation methods (Inverse Distance Weighting and Empirical Bayesian Kriging). Finally, we evaluated whether those exposed to higher levels of air pollution were less active by performing logistic regression, adjusting for demographic and behavioral risk factors, and after stratifying by body weight category.

Results: With Empirical Bayesian Kriging air pollution values, we estimated a statistically significant 16–35% relative increase in the odds of leisure-time physical inactivity per exposure class increase of $PM_{2.5}$ in the fully adjusted model across the normal weight respondents (p-value<0.0001). Evidence suggested a relationship between the increasing dose of $PM_{2.5}$ exposure and the increasing odds of physical inactivity.

Conclusions: In a nationally representative, cross-sectional sample, increased community level air pollution is associated with reduced leisure-time physical activity particularly among the normal weight. Although our design precludes a causal inference, these results provide additional evidence that air pollution should be investigated as an environmental determinant of inactivity.

Editor: Jonatan R. Ruiz, University of Granada, Spain

Funding: Funding for this study was provided by a Uniformed Services University of the Health Sciences intramural start-up grant for newly appointed faculty. The funders had no role in study design, data collection and analysis, decision to publish, or preparation of the manuscript.

Competing Interests: The authors have declared that no competing interests exist.

* E-mail: jennifer.roberts@usuhs.edu

Introduction

Worldwide, physical inactivity accounts for more than three million annual deaths and 6–10% of major non-communicable diseases, such as coronary heart disease, type-II diabetes and breast and colorectal cancers [1–5]. Similarly, physical inactivity is strongly associated with obesity and a portion of physical inactivity related mortality is attributed to obesity [6–9]. In the U.S., two-thirds of adults are overweight or obese and approximately six percent are extremely obese, which is a body mass index greater than or equal to 40.0 kg/m^2 [10,11]. While a majority of Americans are overweight or obese, sub-populations are disproportionately impacted. For instance, there are racial, ethnic, geographic and economic disparities in the obesity prevalence throughout the U.S. [12,13]. Research into how the built environment may impact these disparities has shown conflicting results. [14,15]. One explanation is that individual determinants interact with one another in a dynamic system, which suggests future research needs to account for the way factors interrelate with one another in the real world by using an ecological perspective.

Granted, modifiable lifestyle factors such as the increased consumption of unhealthy foods and physical inactivity are important independent contributors to the increasing burden of non-communicable disease. Other insidious factors, however, such as poor air quality, may influence physical inactivity, but current research has not adequately established this role. While not yet considered an environmental determinant of inactivity, there is little confusion about the unfavorable effects of acute and chronic air pollution exposure, particularly from particulate matter (PM) and ozone (O_3), on both the respiratory and cardiovascular systems [16–19]. While some harms likely remain uncharacterized, research has shown that exposure to $PM_{2.5}$ (particulate matter <2.5 μm in aerodynamic diameter), PM_{10} (particulate matter <10 μm in aerodynamic diameter) and O_3 is associated with reduced exercise capacity, higher resting blood pressure, lower ventilator function and decrements in exercise performance [20–23]. Although there is abundant research illustrating these effects in a resting, inactive state, among athletes or normal weight subjects, the data examining the effects of poor air quality in real world settings are meager. Thus, the generalizability of these findings is in question particularly when over 60% of the U.S. population is overweight.

Another important gap is the difficulty in determining the geographic pattern of air pollution exposure. Although the U.S. Environmental Protection Agency (U.S. EPA) monitors and reports air pollution levels throughout the U.S., it is challenging to appreciate how these readings translate to air pollution exposures across standard geographic units, such as U.S. counties.

Thus, the overall aim of this study was to assess the association between ambient air pollution and leisure-time physical activity. Additionally, this association was examined after stratifying by body weight category.

Materials and Methods

In this cross-sectional study, our two sources of data consisted of (1) annual summary measurements of 2011 ambient air quality monitoring data from the U.S. EPA Air Quality System (AQS) Data Mart and (2) Behavioral Risk Factor Surveillance System (BRFSS) survey data collected in 2011 throughout the U.S. that provided self-reported levels of leisure-time physical activity, demographic information and residential location. Using these data sources, we compared spatial interpolation methods to determine the best predictor for county-level $PM_{2.5}$, PM_{10} and O_3 exposures throughout the U.S and we evaluated the possible cross-sectional relationship of ambient air pollution with physical inactivity in the full study population and after stratifying by body weight.

Physical Inactivity

BRFSS, a state-based telephone health survey system, collects data on behavioral and other health risk factors. As a whole, BRFSS, uses a methodology to collect a representative sample of the U.S. non-institutionalized adult population. As guided by the Centers for Disease Control and Prevention (CDC), data are collected from all 50 States, District of Columbia, Puerto Rico, U.S. Virgin Islands, and Guam. Although the CDC and other researchers have described the complex survey design in great detail, it should be noted that for many states, BRFSS is implemented through the use of disproportionate stratified random sampling (DSS) [24–26]. In order to account for the relatively recent rise in the proportion of U.S. households without landline telephones, adjustments were actualized during the fielding of the recent 2011 BRFSS to include households that rely on cellular telephones. Additionally, a more sophisticated weighting methodology known as "raking" was implemented. Raking, in contrast to the previously used poststratification method, forms individual variable adjustments in a series of data processing iterations, and thus reduces the risk of potential bias [27,28].

Using BRFSS 2011 data, we assessed leisure-time physical inactivity through responses to the question, "During the past month, other than your regular job, did you participate in any physical activities or exercises such as running, calisthenics, golf, or walking for exercise?" The responses to this question were either

Table 1. Annual means of $PM_{2.5}$, PM_{10} and O_3 Empirical Bayesian Kriging (EBK) interpolated ambient air pollution concentrations by natural breaks classes.

Air Pollution Classes	$PM_{2.5}$ (ug/m³)	PM_{10} (ug/m³)	O_3 (ppb)
Class 1	3.49–6.52	5.00–13.40	26.93–37.83
Class 2	6.53–8.45	13.41–17.59	37.84–42.40
Class 3	8.46–9.85	17.60–21.27	42.41–45.83
Class 4	9.86–10.89	21.28–26.31	45.84–49.89
Class 5	10.90–15.38	26.32–52.88	49.90–56.94

"yes", "no", "don't know/not sure", and "refused". A response of "no" was defined as leisure-time physical inactivity.

In this study, inclusion criteria for the BRFSS data were as follows: (1) geographically located within the contiguous U.S including the District of Columbia; (2) responses from respondents who either were categorized as normal weight, overweight or obese; and (3) respondents from counties with both county and state Federal Information Processing Standard (FIPS) codes. Missing, "refused" and "don't know" responses were also excluded from the analysis.

Air Pollution Exposure

The air quality data collected by the U.S. EPA AQS contains air monitoring measurements for criteria air pollutants from 1957 to present [29]. The database contains several million observations from thousands of monitors throughout the U.S. [29]. In addition to descriptive and geographic information about the monitoring sites, quality assurance information is also available.

For $PM_{2.5}$ and PM_{10}, annual summaries for 2011 were obtained using the standard 1-hour or 24-hour collection periods. Due to the strong seasonal and diurnal patterns that exist for ground level O_3, U.S. EPA requires that monitoring locations collect only during specified months of the year as determined by their geographic location. Thus, the National Ambient Air Quality Standard (NAAQS) for ozone is based on an 8-hour averaging time. For inclusion in this study, the 2011 annual O_3 summaries calculated using daily maximum 8-hour averages over the effective monitoring season were selected.

To confirm geographical accuracy, the monitoring data were mapped along with 2011 U.S. Census counties. County FIPS codes from the monitoring data were compared to county FIPS from the enclosing Census county. Of the 3945 records, four monitoring locations were found not to have a FIPS match. The spatial locations of these four were examined visually and the discrepant cases were located less than 500 meters from the county border, suggesting the discrepancy was due to error introduced during the import and processing of the air pollution data. Thus, no records were excluded from further analysis based on locational accuracy.

In order to ensure the accuracy and reliability of the air pollution concentrations, U.S. EPA inclusion criteria were applied. The inclusion criteria for the AQS data were as follows: (1) for $PM_{2.5}$ and PM_{10}: availability of greater than 75% of observations was required; (2) for O_3: availability of greater than 75% of valid days in the effective monitoring season was required [29].

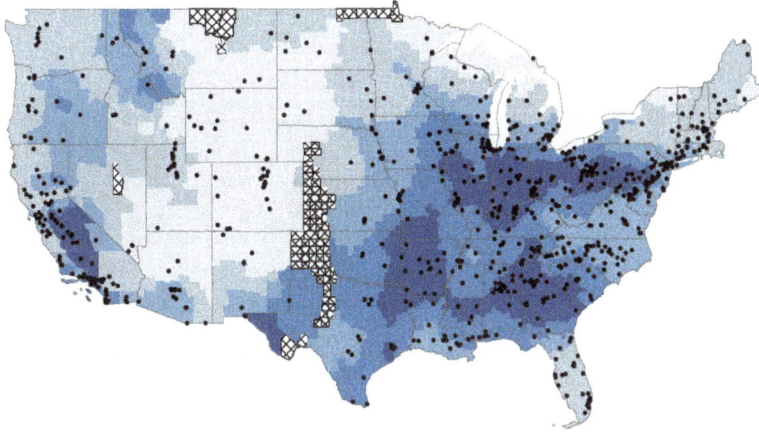

PM$_{2.5}$

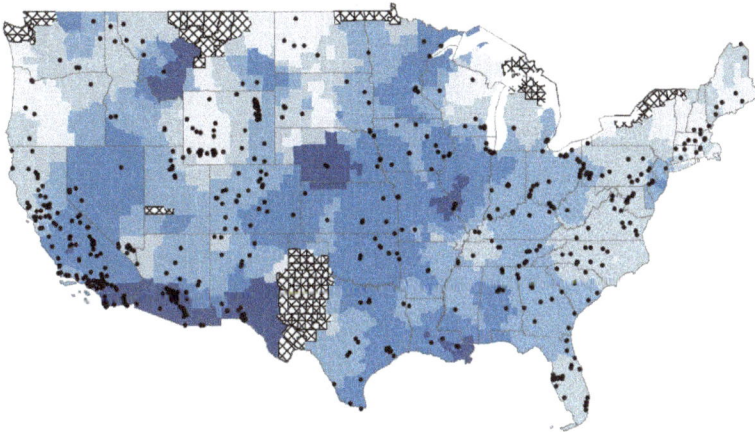

PM$_{10}$

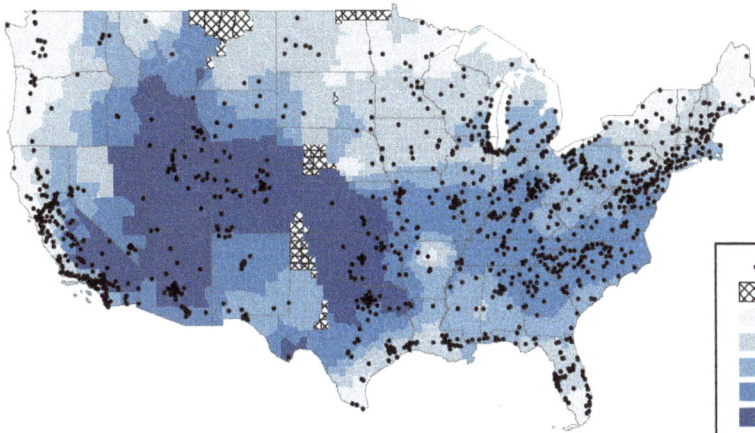

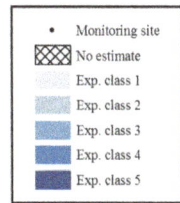

O$_3$

Figure 1. U.S. map of annual mean Empirical Bayesian Kriging (EBK) interpolated ambient air pollution concentrations by natural breaks classes.

Air Pollution Modeling

Since BRFSS data are available at the U.S. county level, the air pollution data provided from the discrete monitoring stations were modeled to estimate county-level average exposures. Studies examining the relationship between air pollution and health outcomes have implemented a variety of techniques to estimate pollution from U.S. EPA monitoring data, including various interpolation and spatiotemporal regression models [30–32]. With the use of ArcGIS 10.1, Inverse Distance Weighting (IDW) and Empirical Bayesian Kriging (EBK) were employed to perform spatial interpolation, creating continuous surfaces for the three air pollution parameters, which were then compared in order to select the best method for inclusion in subsequent analysis. A search window of 250 km was selected in order to ensure that air pollution estimates were generated for a substantial percentage of U.S. counties in the study area, while also maximizing the prediction precision.

The first method, IDW, is a deterministic method that imposes a model of spatial autocorrelation and calculates interpolation weights for each known point (in this case, each monitoring

Table 2. Prevalence of demographic factors by air pollution exposure class.

Demographic Factors	Prevalence (%)[a]					
	$PM_{2.5}$[b]		PM_{10}[b]		O_3[b]	
	Class 1 (n = 32,438)	Class 5 (n = 57,600)	Class 1 (n = 42,539)	Class 5 (n = 16,081)	Class 1 (n = 31,333)	Class 5 (n = 45,901)
Age*						
18–24 years	9.07	10.71	10.07	11.07	9.77	11.29
25–34 years	17.04	17.55	15.06	18.46	16.62	18.92
35–44 years	19.05	19.13	17.13	19.31	19.13	19.54
45–54 years	20.91	20.36	21.80	19.20	20.47	19.92
55–64 years	17.48	15.87	17.79	15.20	16.52	15.04
≥65 years	16.45	16.38	18.16	16.76	17.49	15.29
Sex						
Male	51.71	51.15	51.84	50.12	51.28	51.32
Female	48.29	48.85	48.16	49.88	48.72	48.68
Race/Ethnicity*						
White/non-Hispanic	75.81	64.28	85.48	59.33	62.54	65.18
Black/non-Hispanic	2.93	13.15	3.68	9.56	5.41	7.24
Hispanic	13.62	15.90	5.38	24.29	19.92	20.64
Asian/Pacific Islander	2.52	4.29	2.44	4.25	8.80	3.20
American Indian	3.56	0.87	1.12	0.98	1.19	2.18
Multi/Other	1.56	1.52	1.90	1.60	2.14	1.56
Educational Level*						
Not graduated from high school	10.49	15.19	9.19	15.50	12.02	14.58
High school graduate	27.25	29.59	28.06	25.45	23.85	25.75
Attended college	33.57	29.64	31.66	33.72	32.81	34.26
College graduate or higher	28.70	25.58	31.09	25.33	31.32	25.42
Annual Income Level*						
<$25,000	28.27	32.44	25.21	31.80	28.63	29.91
≥$25,000 to <$50,000	26.84	25.14	25.24	25.79	24.47	26.28
≥$50,000	44.89	42.42	49.55	42.41	46.90	43.81
Marital Status*						
Married/partnered	61.09	57.44	59.35	56.30	58.53	60.55
Divorced/widowed/separated	20.15	19.62	19.24	20.86	19.52	19.27
Never married	18.76	22.93	21.41	22.84	21.95	20.18

[a]Proportions based on frequency-weighted final weight variable rounded to nearest integer.
[b]Air pollution based on five natural breaks air pollution classes (Classes 2-4 not shown).
*p-value <0.0001 Chi-square test of homogeneity – All three pollutants.

Table 3. Prevalence of risk or geographic factors by air pollution exposure class.

Risk/Geographic Factors	Prevalence (%)[a]					
	$PM_{2.5}$[b]		PM_{10}[b]		O_3[b]	
	Class 1 (n = 32,438)	Class 5 (n = 57,600)	Class 1 (n = 42,539)	Class 5 (n = 16,081)	Class 1 (n = 31,333)	Class 5 (n = 45,901)
Smoking*						
Current and former smoker	48.24	44.80	48.27	44.08	43.39	42.92
Never smoked	51.76	55.20	51.73	55.92	56.61	57.08
Body Mass Index[§]						
Normal weight	37.13	33.41	35.63	34.65	36.51	35.37
Overweight	37.58	36.93	37.72	37.27	37.41	36.83
Obese	25.29	29.65	26.65	28.08	26.08	27.81
Disability*						
No	76.32	77.11	75.24	76.57	74.98	77.00
Yes	23.68	22.89	24.76	23.43	25.02	23.00
General Health Status*						
Excellent/good	84.69	81.50	85.38	81.94	82.93	83.52
Fair/poor	15.31	18.50	14.62	18.06	17.07	16.48
Asthma Currently[¶]						
No	91.09	91.72	89.88	91.80	91.03	91.51
Yes	8.91	8.28	10.12	8.20	8.97	8.49
Seasonality*						
Quarter 1 (January to March)	17.13	23.89	27.70	25.40	27.27	20.62
Quarter 2 (April to June)	27.36	23.40	26.88	27.16	25.48	29.27
Quarter 3 (July to September)	30.13	26.86	23.71	21.94	23.80	24.91
Quarter 4 (October to December)	25.39	25.86	21.71	25.49	23.45	25.20
U.S. Geographic Region*						
Northeast	6.79	10.89	68.89	0.00	7.82	0.00
Southeast	0.00	20.73	0.58	1.26	19.51	0.00
Midwest	11.04	33.58	5.59	19.47	4.02	4.04
Southwest	56.50	12.90	1.61	16.26	5.58	59.12
West	25.67	21.89	23.53	62.74	63.07	36.84
Metropolitan County Classification*						
Rural counties	29.84	14.03	21.43	5.37	14.17	12.67
Counties with <250,0000	21.87	9.14	10.12	7.61	8.04	8.49
Counties with 250,000–1 Million	14.86	21.19	23.40	12.59	20.00	20.94
Counties with ≥1 Million	33.43	55.63	45.06	74.44	57.79	57.90

[a]Proportions based on frequency-weighted final weight variable rounded to nearest integer.
[b]Air pollution based on five natural breaks air pollution classes (Classes 2–4 not shown).
*p-value <0.0001 Chi-square test of homogeneity – All three pollutants.
[§]p-value <0.0001 Chi-square test of homogeneity – Only $PM_{2.5}$ and O_3.
[¶]p-value <0.0001 Chi-square test of homogeneity – Only PM_{10} and O_3.

station) as a function of distance between the known points and the predicted points within a specific search window. There is an inverse relationship between the interpolation weights and the distance from the interpolated points to each known point. Hence, the values that are closer to the prediction location have more weight or influence on the predicted values than those farther away. We chose to calculate weights that change linearly in order to create a smooth surface. The second method, EBK, is an implementation of the kriging class of geostatistical methods that allow the development of a statistical autocorrelation model using a sample data set. Common kriging methods such as ordinary,

simple, and universal kriging, require selection of model parameters from an empirical variogram. The empirical variogram is used to calculate interpolating weights such that the mean square error is minimized. EBK automates the parameter selection process through simulation and subsetting and generates accurate results from moderately non-stationary data, indicating that the mean and variance do not differ with geographical position [33,34].

A one square kilometer grid was overlaid on a map of the counties that comprise the contiguous U.S. Each interpolation method produced air pollution estimates at the center of each

Table 4. Association between Empirical Bayesian Kriging (EBK) interpolated ambient $PM_{2.5}$ exposure class and physical inactivity by body weight subset, logistic regression model.

PM2.5 Exposure Class	Obese Weight (n=99,699)			Overweight (n=127,720)			Normal Weight (n=116,927)		
	A[a]	B[b]	C[c]	A[a]	B[b]	C[c]	A[a]	B[b]	C[c]
	Logistic Regression Models*								
	OR (95% CI)			OR (95% CI)			OR (95% CI)		
Class 1	Referent	Referent	Referent	Referent	Referent	Referent	Referent	Referent	Referent
Class 2	1.19 (1.06, 1.33)	1.11 (0.98, 1.25)	1.14 (0.99, 1.31)	1.14 (1.02, 1.28)	1.10 (0.97, 1.25)	1.12 (0.98, 1.29)	1.26 (1.11, 1.42)	1.16 (1.02, 1.32)	1.16 (1.01, 1.34)
Class 3	1.21 (1.09, 1.35)	1.14 (1.02, 1.29)	1.17 (1.02, 1.34)	1.14 (1.02, 1.27)	1.13 (0.99, 1.28)	1.13 (0.98, 1.30)	1.32 (1.18, 1.47)	1.24 (1.09, 1.40)	1.29 (1.11, 1.49)
Class 4	1.33 (1.20, 1.49)	1.15 (1.02, 1.30)	1.17 (1.02, 1.35)	1.40 (1.25, 1.56)	1.24 (1.09, 1.40)	1.24 (1.07, 1.43)	1.59 (1.42, 1.78)	1.30 (1.14, 1.48)	1.35 (1.16, 1.58)
Class 5	1.35 (1.21, 1.50)	1.21 (1.07, 1.36)	1.23 (1.07, 1.42)	1.22 (1.09, 1.37)	1.13 (0.99, 1.28)	1.13 (0.98, 1.31)	1.49 (1.33, 1.67)	1.27 (1.12, 1.44)	1.35 (1.16, 1.58)

[a]Model A unadjusted.
[b]Model B adjusted for age, sex, race/ethnicity, education, annual income, marital status, seasonality and geographic region.
[c]Model C adjusted for age, sex, race/ethnicity, education, annual income, marital status, seasonality, geographic region, general health status, smoking, disability, asthma, urbanization, and the other two air pollutants.
*Wald Chi-square p-value <0.0001 for all three models.

square of the grid. County exposure estimates were calculated by averaging the interpolated values spatially located within the county borders. U.S. counties that did not have a single interpolated estimate within its borders were excluded from further analysis.

Cross-validation was completed for each interpolation method and for each pollutant to test the generalization performance and provide a quantitative comparison of the IDW and EBK methods. This was performed by omitting values for a single monitor and then using the remaining monitors to interpolate the concentration at the removed monitor's location. To identify and select the most appropriate interpolation method to use for further analysis, the cross-validations were compared in terms of prediction root mean square error (RMSE) and prediction mean absolute error (MAE).

Upon this selection, the annual mean of $PM_{2.5}$, PM_{10} and O_3 concentrations were transformed from continuous variables to categorical variables using natural breaks classification method with Jenks optimization. Natural breaks are data-specific classes, which are based on natural groupings inherent in the data and where boundaries are set based on relatively large differences in the data values by reducing the within and maximizing the between class variance. Finally, we linked annual average concentrations of $PM_{2.5}$, PM_{10} and O_3 with the 2011 BRFSS data using FIPS codes as the linking unit.

Statistical Analysis

BRFSS uses a complex survey design with stratification, multistage clustering and sampling weights. Therefore, statistical analysis was performed in STATA MP/12.1 using the -svyset- commands. The weighted prevalence of leisure-time physical inactivity was calculated by physical demographic and risk factor categories. Additionally, the weighted prevalence of demographic and other risk factor variables were calculated by air pollution exposure class. Chi-square tests for homogeneity were performed to investigate the association of the demographic and behavioral risk variables with physical inactivity and air pollution exposure.

We considered whether adults who were exposed to higher levels of $PM_{2.5}$, PM_{10} and O_3 concentrations exhibited higher levels of physical inactivity by performing logistic regression. Additionally, we examined this association after stratifying the data into three subgroups [e.g. (1) normal weight (body mass index (BMI) 18.5 to 24.9); (2) overweight (BMI 25.0 to 29.9), and (3) obese (BMI 30 or higher)] as defined and categorized by the BRFSS data. The air pollution variables were analyzed both as continuous and natural breaks categorical variables using three models for each pollutant. Model A examined the effect on physical inactivity with ambient exposure to $PM_{2.5}$, PM_{10} and O_3 without the adjustment of any confounders. Model B adjusted for age, sex, race/ethnicity, education, annual income, marital status, seasonality and geographic region. Along with the aforementioned confounders, Model C, also adjusted for general health status, smoking, disability, asthma, urbanization and the other two air pollutants. Additionally, we calculated Pearson's correlation coefficients to examine relations between the three pollutant measures.

Ethics Statement

The Uniformed Services University of the Health Sciences, Human Research Protections Program Office, determined that this research was non-human subjects research consistent with 32 CFR 219.102.

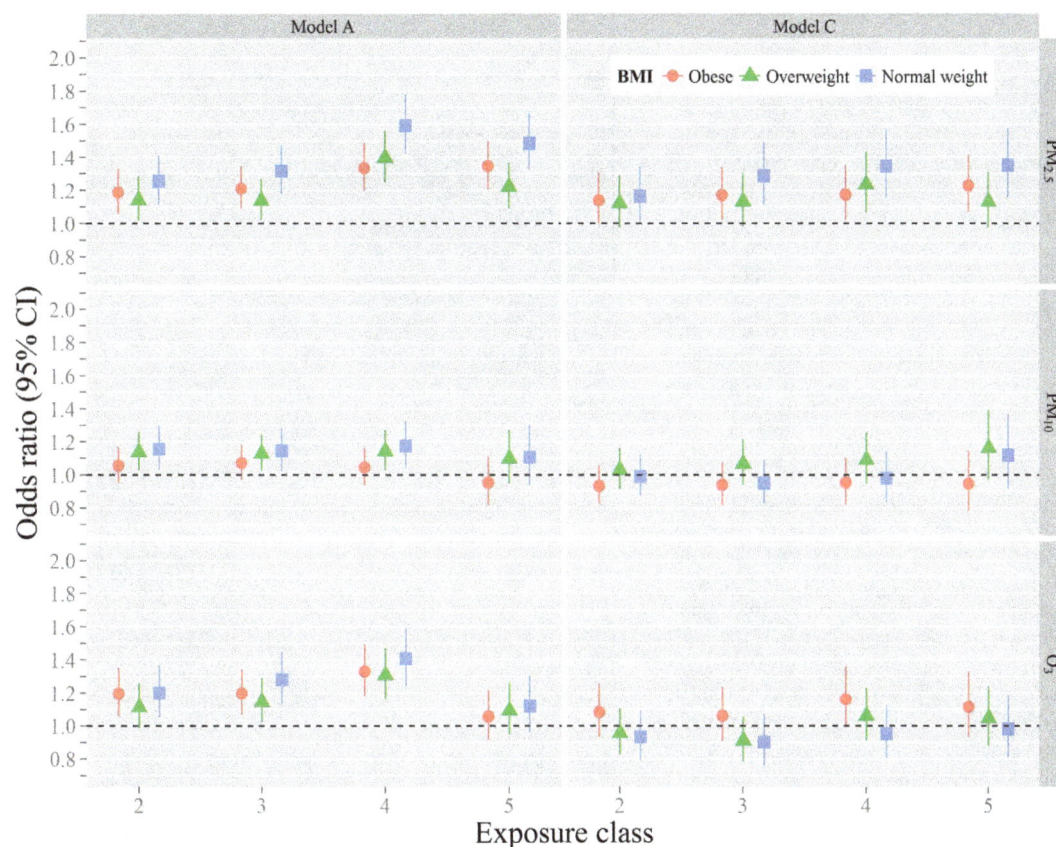

Figure 2. Association between air pollution exposure classes and the odds ratios of physical inactivity [†, ††, †††]. † Model A unadjusted. †† Model C adjusted for age, sex, race/ethnicity, education, annual income, marital status, seasonality, geographic region, general health status, smoking, disability, asthma, urbanization, and the other two air pollutants. ††† Exposure class 1 is referent.

Results

The 2011 BRFSS dataset encompassed 504,408 observations. Based on the inclusion and exclusion criteria, data from 48 states and the District of Columbia were included in the analysis (N = 329,628 subjects from 2249 U.S. counties). A total of 24.5% (n = 80,825) responded "no" to participating in any physical activity during the previous month. The prevalence of leisure-time physical inactivity was higher among older respondents and among females; however with respect to race and ethnicity, Black/ non-Hispanic respondents demonstrated the highest prevalence of physical inactivity, which was closely followed by Hispanic respondents. The highest levels of reported physical inactivity were also observed among the respondents with lower levels of education, those who reported being divorced, widowed or separated, and, those receiving less than $50,000 in annual income. Respondents who reported obesity, disability, asthma, or prior history of smoking were also more likely to report physical inactivity. Physical inactivity was highest during the months of January through March and October through December. In addition, physical inactivity was highest in the Southeastern part of the U.S. while also being lowest in the West. There was also a much higher level of physical inactivity among respondents who resided in rural counties or non-metropolitan counties (Data not shown, p-value <0.001).

The two interpolation methods, IDW and EBK, were evaluated based on accuracy and precision for predicting annual ambient $PM_{2.5}$, PM_{10} and O_3 concentrations. Results showed that the IDW and EBK models were similar in regards to RMSE and MAE with EBK trending toward increased precision for all three air quality parameters. Thus, EBK was identified as the most appropriate method. Furthermore, kriging interpolation methods have been recognized by U.S. EPA as possessing the greatest merit in predicting air pollution concentrations in unknown locations [35].

The mean of $PM_{2.5}$, PM_{10} and O_3 annual mean concentrations were 9.50 ug/m^3 [SD: 1.80], 19.52 ug/m^3 [SD: 4.16], and 45 ppb [SD: 4.48] for the BRFSS counties, respectively. All of the pollutants were positively correlated with $PM_{2.5}$ and PM_{10} having the strongest correlation (r: 0.29) followed by O_3 and PM_{10} (r: 0.23). Furthermore, the air pollution variables were transformed into categorical variables using the Jenks' natural breaks methodology (Table 1). While the highest concentrations of $PM_{2.5}$ were found in the upper Atlantic, Midwest, and the South, along with a small cluster in Southern California, the higher concentration PM_{10} counties were clustered throughout the U.S., particularly in the Southwest (Figure 1). The highest two classes of O_3 were clustered in the middle and western part of the country (Figure 1). When comparing air pollution by natural breaks class, the most evident differences were observed within the race/ethnicity, U.S geographic region and metropolitan county classification categories among all of the air pollutants (Tables 2–3). Generally, White/ non-Hispanic respondents and those living in rural counties or with a population of less than 250,000 were exposed to lower levels of $PM_{2.5}$ and PM_{10} (Tables 2–3).

Association of Physical Inactivity and Air Pollution Exposure by Body Weight

When considering $PM_{2.5}$ as a continuous variable, the odds of leisure-time physical inactivity significantly increased with increasing concentration of $PM_{2.5}$ across all models and strata. For the fully adjusted Model C, we estimated a 2.4% relative increase in the odds of physical inactivity per $\mu g/m^3$ increase of $PM_{2.5}$ exposure among the obese respondents [OR = 1.02 (95% CI: 1.00, 1.05)]. Similarly, increasing concentration of PM_{10} among the normal weight respondents was also associated with higher odds of inactivity [OR = 1.01 (95% CI: 1.00, 1.02)]. We also estimated increases in the odds of physical inactivity per unit increase of each air pollutant for the entire combined dataset and those results were also found to be statistically significant for $PM_{2.5}$ in all models (Data not shown, p-value <0.01).

Alternatively, the associations with inactivity were also modeled using natural breaks exposure classes for each of the three pollutants ($PM_{2.5}$, PM_{10} and O_3). There was a statistically significant 16–35% relative increase in the odds of physical inactivity per increase from the lowest $PM_{2.5}$ exposure class in the fully adjusted Model C among the normal weight respondents (Table 4), which was a stronger association than the other body weight strata. Figure 2 illustrates a relationship between the odds of physical inactivity and increasing dose as represented by exposure class. For O_3, statistical significance was observed only in the unadjusted Model A for all three weight groups (Figure 2). Results found using the full dataset were similar to that of the normal weight stratum (Data not shown).

Lastly, the relationship between air pollution and several other covariates was notable. For instance, odds ratios for physical inactivity increased strongly with increasing age and BMI classes (Table 5). By contrast, higher levels of education and income classes decreased the odds of physical inactivity. Furthermore, respondents in the Western part of the U.S had 35% higher odds of being active than those in the Northeast (Table 6). There was also a 32–33% higher odds of activity found among respondents during the warmer months of the year (Table 6). Finally, the odds of physical inactivity decreased with increasing urbanization (Table 6).

Discussion

In this nationally representative, cross-sectional sample, increased ambient levels of $PM_{2.5}$, PM_{10} and O_3 were associated with reduced physical activity. This association was significant in all models of adjustment for $PM_{2.5}$. Our research demonstrated an association between increasing ordinal air pollution classes and increasing odds of inactivity among adults. Remarkably, the most compelling relationships were evident among normal weight as opposed to overweight or obese respondents. These findings suggest that the presence of air pollution may discourage normal weight individuals from engaging in leisure-time physical activity.

Our research exhibited an inverse relationship between air pollution exposure and leisure-time physical activity among Americans. This is consistent with findings from similar studies that also examined the association of physical inactivity and air pollution and found a direct relationship with increasing $PM_{2.5}$ and the increasing prevalence of physical inactivity (p-value <0.01) [36]. This prior work, however, only examined crude levels of $PM_{2.5}$ and generalized findings across all weight categories. Another study modeled other air pollutants, including O_3 and nitrogen oxides (NO_x), and physical inactivity limited to Southern California, but did not consider air pollution as a determinant of inactivity [37].

Table 5. Association of demographic factors with physical inactivity.

Demographic Factors	OR (95% CI)
Age*	
18–24 years	Referent
25–34 years	1.51 (1.36, 1.67)
35–44 years	1.72 (1.55,1.91)
45–54 years	1.74 (1.57, 1.92)
55–64 years	1.82 (1.64, 2.01)
≥65 years	2.00 (1.80, 2.21)
Sex*	
Male	Referent
Female	1.15 (1.10, 1.19)
Race/Ethnicity*	
White/non-Hispanic	Referent
Black/non-Hispanic	1.19 (1.13, 1.27)
Hispanic	1.22 (1.14, 1.30)
Asian/Pacific Islander	1.80 (1.58.2.05)
American Indian	0.92 (0.79, 1.07)
Multi/Other	0.91 (0.80, 1.03)
Educational Level*	
Not graduated from high school	Referent
High school graduate	0.84 (0.79, 0.90)
Attended college	0.62 (0.58, 0.66)
College graduate or higher	0.42 (0.39, 0.45)
Annual Income Level*	
<$25,000	Referent
≥$25,000 to <$50,000	0.96 (0.91, 1.00)
≥$50,000	0.72 (0.69, 0.76)
Marital Status	
Married/partnered	Referent
Divorced/widowed/separated	1.02 (0.98, 1.06)
Never married	0.97 (0.92, 1.03)

Adjusted for all above variables in the table and all three air pollution variables.
*Joint adjusted Wald test p-value <0.0001.

We add to this literature by using more recent data, sophisticated interpolation methods increasing the coverage and modeling reliability, and clarifying the association across weight strata. Our research revealed a higher prevalence of physical inactivity among obese and overweight, as compared to normal weight, respondents. When stratifying by body weight category, the association between air pollution and leisure-time physical activity varied minimally by body weight category with the exception of $PM_{2.5}$ where the magnitude of association trended higher among the normal weight respondents. One reason for this finding may be due to the fact that normal or healthy weight adults already are more physically active than obese or overweight adults and therefore the reduction in activity is greater. For instance, obesity related disability may create a disparity in discretionary activity - a lean population is active when the conditions are favorable while those who are disabled are inactive regardless.

Although a causal mechanism cannot be elucidated by our study design, the association of physical inactivity and ambient air

Table 6. Association of risk or geographic factors with physical inactivity.

Risk/Geographic Factors	OR (95% CI)
Smoking	
Never smoked*	Referent
Current and former smoker	1.18 (1.14, 1.22)
Body Mass Index*	
Normal weight	Referent
Overweight	1.04 (1.00,1.09)
Obese	1.44 (1.38, 1.50)
Disability*	
No	Referent
Yes	1.42 (1.37, 1.48)
General Health Status*	
Excellent/good	Referent
Fair/poor	1.74 (1.66, 1.82)
Asthma Currently	
No	Referent
Yes	0.98 (0.92, 1.03)
Seasonality*	
Quarter 1 (January to March)	Referent
Quarter 2 (April to June)	0.68 (0.65, 0.72)
Quarter 3 (July to September)	0.67 (0.63, 0.70)
Quarter 4 (October to December)	0.84 (0.80, 0.88)
U.S. Geographic Region*	
Northeast	Referent
Southeast	1.01 (0.96, 1.06)
Midwest	0.90 (0.86, 0.95)
Southwest	0.89 (0.84, 0.96)
West	0.65 (0.61, 0.69)
Metropolitan County Classification*	
Rural counties	Referent
Counties with <250,0000	0.93 (0.89, 0.99)
Counties with 250,000–1 Million	0.90 (0.85, 0.94)
Counties with ≥1 Million	0.88 (0.84, 0.92)

Adjusted for all above variables in the table and all three air pollution variables.
*Joint adjusted Wald test p-value <0.0001.

pollution as mediated by body weight category is plausible. Positive associations have been found with exposure to air pollutants and direct health effects on the respiratory and cardiovascular systems, such as increased blood pressure, asthma exacerbations, cardiac arrhythmia, and decreased lung function [17,38–41]. Therefore, adverse health effects from increasing levels of air pollution could reduce one's capacity for physical activity. In addition to the physiological effects (e.g. difficulty breathing), the possibility of a psychosocial effect (e.g. smog appearance disincentivizing physical activity) could be contributing factor to this association. With readily available information, the U.S. population is likely more aware of the health risk associated with exposure to high levels of air pollution. Hence, this awareness may ultimately discourage individuals from engaging in outdoor physical activity.

While these are important findings, one study limitation was potential misclassification of exposure based on the air pollution modeling, interpolated estimates and annual means. Because we applied the same methods uniformly throughout the entire U.S. we suspect this biases to the null as non-differential measurement error, but we cannot rule out differential misclassification given the geographic variation in our exposure and outcome. Furthermore, the air pollution data used were at the county-level and did not provide information at an individual daily exposure level. Yet, since $PM_{2.5}$ and O_3 are often more homogeneous air pollutants in their distribution over large regions, we believe this misclassification was minimized.

Our study used self-reported leisure-time physical activity and self-reported data are often subject to certain biases. Since physical activity was assessed as a dichotomous variable over a month timeframe, the accuracy of responses may not have been compromised. However, the BRFSS physical activity question provided examples of exercise such as running and calisthenics. Physical activity includes not only the participation of sports and exercise, but also walking or biking on a daily basis by means of an active commute or transport. With the BRFSS examples, respondents may not have considered their less obvious physical activities, such as walking to work because the survey question asked respondents about participation in physical activities outside of their "regular job". Additionally, other built environmental factors, such as neighborhood walkability or safety, may have influenced one's level of physical activity.

The major strength of this study is the use of our novel EBK estimations for air pollution exposures. Unlike other kriging methods, EBK allows for automated model fitting and more accurate and robust predictions. Another strength of this study is the use of a nationally representative sample with extensive demographic, behavioral and risk factor data. The combination of BRFSS data with U.S. EPA air pollution data brought to light novel findings. Lastly, we were able to examine the relationships of three air pollutants and their influence with each other on physical inactivity.

The findings of this research emphasized the phenomenon that there is a complex interplay among many risk factors, behavioral and demographic variables, which are associated with physical activity. Thus, the complexity limits the applications of observational research as it can raise questions of causality and directionality. Because $PM_{2.5}$ is a modifiable exposure with cost effective mitigation strategies, future research could evaluate causality by cluster randomizing the timing of $PM_{2.5}$ reduction interventions and assessing the short-term impact on leisure-time physical inactivity [42,43].

Conclusions

We present evidence that as air pollution concentrations increase, American adults, especially those who are lean, are less likely to be physically active. Given the public health emphasis on community level determinants of inactivity, additional research should determine if environmental air pollution is a modifiable risk factor for inactivity. We postulate those interventions which improve physical activity and reduce air pollution such as transportation interventions will have both primary and secondary benefits.

Acknowledgments

The authors gratefully acknowledge the technical assistance of the U.S. Environmental Protection Agency, Air Quality System (AQS) Team and the biostatistical support provided by Dr. Cara Olsen of Uniformed Services University.

Author Contributions

Conceived and designed the experiments: JDR JDV BK. Performed the experiments: JDR BK. Analyzed the data: JDR BK. Contributed reagents/materials/analysis tools: JDR BK. Wrote the paper: JDR JDV BK.

References

1. Pratt M, Sarmiento OL, Montes F, Ogilvie D, Marcus BH, et al. (2012) The implications of megatrends in information and communication technology and transportation for changes in global physical activity. Lancet 380: 282–293.
2. Lee IM, Shiroma EJ, Lobelo F, Puska P, Blair SN, et al. (2012) Effect of physical inactivity on major non-communicable diseases worldwide: an analysis of burden of disease and life expectancy. Lancet 380: 219–229.
3. Awatef M, Olfa G, Rim C, Asma K, Kacem M, et al. (2011) Physical activity reduces breast cancer risk: a case-control study in Tunisia. Cancer Epidemiol 35: 540–544.
4. Haggar FA, Boushey RP (2009) Colorectal cancer epidemiology: incidence, mortality, survival, and risk factors. Clin Colon Rectal Surg 22: 191–197.
5. Arsenault BJ, Rana JS, Lemieux I, Despres JP, Kastelein JJ, et al. (2010) Physical inactivity, abdominal obesity and risk of coronary heart disease in apparently healthy men and women. Int J Obes (Lond) 34: 340–347.
6. Pietilainen KH, Kaprio J, Borg P, Plasqui G, Yki-Jarvinen H, et al. (2008) Physical inactivity and obesity: a vicious circle. Obesity (Silver Spring) 16: 409–414.
7. Lee IM, Djousse L, Sesso HD, Wang L, Buring JE (2010) Physical activity and weight gain prevention. JAMA 303: 1173–1179.
8. Herring MP, Puetz TW, O'Connor PJ, Dishman RK (2012) Effect of exercise training on depressive symptoms among patients with a chronic illness: a systematic review and meta-analysis of randomized controlled trials. Arch Intern Med 172: 101–111.
9. Motl RW, Birnbaum AS, Kubik MY, Dishman RK (2004) Naturally occurring changes in physical activity are inversely related to depressive symptoms during early adolescence. Psychosom Med 66: 336–342.
10. Ogden C, Carroll M, Kit BK, Flegal KM (2012) Prevalence of obesity in the United States, 2009-2010 - NCHS Data Brief. Centers for Disease Control and Prevention. Available: http://www.cdc.gov/nchs/data/databriefs/db82.pdf. Accessed 2013 February 1.
11. Fryar CD, Carroll MD, Ogden C (2012) Prevalence of overweight, obesity, and extreme obesity among adults: United States, Trends 1960–1962 through 2009-2010 - NCHS Health E-Stat. Centers for Disease Control and Prevention. Available: http://www.cdc.gov/nchs/data/hestat/obesity_adult_09_10/obesity_adult_09_10.htm. Accessed 2013 January 15.
12. Singh GK, Kogan MD, van Dyck PC (2010) Changes in state-specific childhood obesity and overweight prevalence in the United States from 2003 to 2007. Arch Pediatr Adolesc Med 164: 598–607.
13. CDC (2011) Obesity rates among low-income preschool children. Centers for Disease Control and Prevention. Available: http://www.cdc.gov/obesity/childhood/data.html. Accessed 2012 March 12.
14. Casazza K, Fontaine KR, Astrup A, Birch LL, Brown AW, et al. (2013) Myths, presumptions, and facts about obesity. N Engl J Med 368: 446–454.
15. Voss JD, Masuoka P, Webber BJ, Scher AI, Atkinson RL (2013) Association of elevation, urbanization and ambient temperature with obesity prevalence in the United States. Int J Obes (Lond) doi:10.1038/ijo.2013.5 [Online 30 January 2013].
16. Brunekreef B, Holgate ST (2002) Air pollution and health. Lancet 360: 1233–1242.
17. Gent JF, Triche EW, Holford TR, Belanger K, Bracken MB, et al. (2003) Association of low-level ozone and fine particles with respiratory symptoms in children with asthma. JAMA 290: 1859–1867.
18. Laden F, Neas LM, Dockery DW, Schwartz J (2000) Association of fine particulate matter from different sources with daily mortality in six U.S. cities. Environ Health Perspect 108: 941–947.
19. Pope CA, 3rd, Burnett RT, Thurston GD, Thun MJ, Calle EE, et al. (2004) Cardiovascular mortality and long-term exposure to particulate air pollution: epidemiological evidence of general pathophysiological pathways of disease. Circulation 109: 71–77.
20. Cakmak S, Dales R, Leech J, Liu L (2011) The influence of air pollution on cardiovascular and pulmonary function and exercise capacity: Canadian Health Measures Survey (CHMS). Environ Res 111: 1309–1312.
21. Marr LC, Ely MR (2010) Effect of air pollution on marathon running performance. Med Sci Sports Exerc 42: 585–591.
22. Cutrufello PT, Rundell KW, Smoliga JM, Stylianides GA (2011) Inhaled whole exhaust and its effect on exercise performance and vascular function. Inhal Toxicol 23: 658–667.

23. Rundell KW, Caviston R (2008) Ultrafine and fine particulate matter inhalation decreases exercise performance in healthy subjects. J Strength Cond Res 22: 2–5.
24. Mokdad AH, Stroup DF, Giles WH (2003) Public health surveillance for behavioral risk factors in a changing environment. Recommendations from the Behavioral Risk Factor Surveillance Team. Centers for Disease Control and Prevention Morbidity and Mortality Weekly Report (MMWR) 52: 1–12.
25. Nelson DE, Holtzman D, Waller M, Leutzinger CL, Condon K (1998) Objectives and design of the Behavioral Risk Factor Surveillance System. American Statistical Association 1998 Proceedings of the Section on Survey Methods. Alexandria, VAAmerican Statistical Association. pp. 214–218.
26. VDH (2012) BRFSS Methodology. Virginia Department of Health. Available: http://www.vahealth.org/brfss/methodology.htm. Accessed 2012 December 1.
27. CDC (2012) Methodologic changes in the Behavioral Risk Factor Surveillance System in 2011 and potential effects on prevalence estimates. Morbidity and Mortality Weekly Report MMWRCenters for Disease Control and Prevention. pp. 410–413.
28. Battaglia MP, Frankel MR, Link MW (2008) Improving standard poststratification techniques for random-digit-dialing telephone surveys. Survey Research Methods 2: 11–19.
29. EPA (2011) AQS Data Dictionary. U.S. Environmental Protection Agency. Available: http://www.epa.gov/ttn/airs/airsaqs/manuals/AQS Data Dictionary.pdf. Accessed 2012 December 20.
30. Marshall JD, Nethery E, Brauer M (2008) Within-urban variability in ambient air pollution: Comparison of estimation methods. Atmospheric Environment 42: 1359–1369.
31. Son JY, Bell ML, Lee JT (2010) Individual exposure to air pollution and lung function in Korea: spatial analysis using multiple exposure approaches. Environ Res 110: 739–749.
32. Hystad P, Demers PA, Johnson KC, Brook J, van Donkelaar A, et al. (2012) Spatiotemporal air pollution exposure assessment for a Canadian population-based lung cancer case-control study. Environ Health 11,: 22.
33. Krivoruchko K (2012) Empirical Bayesian Kriging: Implemented in ArcGIS Geostatistical Analyst. ESRI. Available: http://www.esri.com/news/arcuser/1012/files/ebk.pdf. Accessed 2012 December 15.
34. Pilz J, Spöck G (2008) Why do we need and how should we implement Bayesian kriging methods. Stoch Environ Res Risk Assess 22: 621–632.
35. EPA (2004) Developing spatially interpolated surfaces and estimating uncertainty. U.S. Environmental Protection Agency. Available: http://www.epa.gov/airtrends/specialstudies/dsisurfaces.pdf. Accessed 2012 February 21.
36. Wen XJ, Balluz LS, Shire JD, Mokdad AH, Kohl HW (2009) Association of self-reported leisure-time physical inactivity with particulate matter 2.5 air pollution. J Environ Health 72: 40–44; quiz 45.
37. Hankey S, Marshall JD, Brauer M (2012) Health impacts of the built environment: within-urban variability in physical inactivity, air pollution, and ischemic heart disease mortality. Environ Health Perspect 120: 247–253.
38. Alexeeff SE, Litonjua AA, Suh H, Sparrow D, Vokonas PS, et al. (2007) Ozone exposure and lung function: effect modified by obesity and airways hyperresponsiveness in the VA normative aging study. Chest 132: 1890–1897.
39. Delfino RJ, Tjoa T, Gillen DL, Staimer N, Polidori A, et al. (2010) Traffic-related air pollution and blood pressure in elderly subjects with coronary artery disease. Epidemiology 21: 396–404.
40. Peters A, Liu E, Verrier RL, Schwartz J, Gold DR, et al. (2000) Air pollution and incidence of cardiac arrhythmia. Epidemiology 11: 11–17.
41. Pope CA, 3rd, Burnett RT, Thun MJ, Calle EE, Krewski D, et al. (2002) Lung cancer, cardiopulmonary mortality, and long-term exposure to fine particulate air pollution. JAMA 287: 1132–1141.
42. Brook RD, Rajagopalan S, Pope CA 3rd, Brook JR, Bhatnagar A, et al. (2010) Particulate matter air pollution and cardiovascular disease: An update to the scientific statement from the American Heart Association. Circulation 121: 2331–2378.
43. EPA (2008) Appendix G: Health-based cost-effectiveness of reductions in ambient PM 2.5 associated with illustrative PM NAAQS attainment strategies. U.S. Environmental Protection Agency. Available: http://www.epa.gov/ttnecas1/regdata/RIAs/Appendix G-Health Based Cost Effectiveness Analysis.pdf. Accessed 2012 March 7.

11

The Inter-Group Comparison – Intra-Group Cooperation Hypothesis: Comparisons between Groups Increase Efficiency in Public Goods Provision

Robert Böhm[1]*, Bettina Rockenbach[2]

1 Center for Empirical Research in Economics and Behavioral Sciences, University of Erfurt, Erfurt, Germany, **2** Department of Economics, University of Cologne, Cologne, Germany

Abstract

Identifying methods to increase cooperation and efficiency in public goods provision is of vital interest for human societies. The methods that have been proposed often incur costs that (more than) destroy the efficiency gains through increased cooperation. It has for example been shown that inter-group conflict increases intra-group cooperation, however at the cost of collective efficiency. We propose a new method that makes use of the positive effects associated with inter-group competition but avoids the detrimental (cost) effects of a structural conflict. We show that the mere comparison to another structurally independent group increases both the level of intra-group cooperation and overall efficiency. The advantage of this new method is that it directly transfers the benefits from increased cooperation into increased efficiency. In repeated public goods provision we experimentally manipulated the participants' level of contribution feedback (intra-group only vs. both intra- and inter-group) as well as the provision environment (smaller groups with higher individual benefits from cooperation vs. larger groups with lower individual benefits from cooperation). Irrespective of the provision environment groups with an inter-group comparison opportunity exhibited a significantly stronger cooperation than groups without this opportunity. Participants conditionally cooperated within their group and additionally acted to advance their group to not fall behind the other group. The individual efforts to advance the own group cushion the downward trend in the above average contributors and thus render contributions on a higher level. We discuss areas of practical application.

Editor: Kevin Paterson, University of Leicester, United Kingdom

Funding: The authors have no funding or support to report.

Competing Interests: The authors have declared that no competing interests exist.

* E-mail: robert.boehm@uni-erfurt.de

Introduction

It is a fundamental question of society how voluntary cooperation in public goods provision can be promoted to counteract society's breakdown as envisioned in Hardin's "tragedy of the commons" [1]. Despite being socially desirable (e.g. reduction of air pollution), provision of public goods (i.e. buying a more environment friendly car) is individually costly and the benefits (i.e. less polluted air) are also shared with non-providers. Therefore, the individual incentives to free-ride on others' contributions lead to a collectively suboptimal outcome. By now a rich literature in biological and social sciences has identified circumstances fostering cooperation in public goods [2–4]. Although the results are promising they are far from satisfactory. In experiments on repeated public goods provision, cooperation typically starts at intermediate levels and declines over time until almost complete free-riding is reached. The most prominent explanation for the observed behavior is based on intra-group comparison in the form of conditional cooperation. Conditional cooperation is the propensity to cooperate as long as others are known (or at least believed) to do the same [5–7]. Although the majority of subjects behave in this manner [6], [8], selfishly-biased conditional cooperation and the adaption to some purely selfish actors may explain the decline in overall cooperation in repeated interactions [5], [9].

Since cooperative groups are – *ceteris paribus* – more successful (e.g. grow faster and endure longer) than less cooperative groups, inter-group conflict may have played an important role in the evolution of human cooperation [10], [11]. It has been shown that cooperation within the own group [12], [13] and (costly) norm enforcement (e.g. punishment of non-cooperative group members) [14], [15] increases if the intra-group conflict is embedded in a structural inter-group conflict; the so-called inter-group conflict – intra-group cooperation/cohesion effect. In a similar vein, self-interested actions by group leaders have been shown to decrease in the presence of a rivaling other group [16]. However, in many cases inter-group conflict is a zero-sum game or may even destroy resources because engaging in inter-group conflict generates costs for both the victorious and the inferior group (for instance casualties, monetary resources, damage, and harm in war). Thus, engaging in inter-group conflict by cooperating with the own group members is often inefficient from the collective perspective [17], [18].

It is an open question how one can make use of the positive effect of inter-group processes on the propensity to cooperate efficiently. In the following, we argue that even in the absence of negative inter-group interdependences (i.e., structural inter-group conflicts) inter-group comparison may increase intra-group cooperation, without sacrificing collective efficiency. Our findings extend previous research on public goods provisions by integrating and testing predictions from intra-group as well as inter-group social comparison processes. By doing so, we propose a new mechanism how to increase human cooperation in public goods provision by the means of mere inter-group comparison.

The Inter-Group Comparison – Intra-Group Cooperation Hypothesis

Adding inter-group comparison to the public good provision problem should not affect players following money-maximizing rationality: They free-ride no matter whether or not another group is present. In the same way, "rational cooperation"[19] should not be affected by the presence of another group: Players cooperate in the repeated game because they believe that others in their own group are altruistic or play a tit-for-tat strategy until an end game effect kicks in. Inequity-averse players [20], [21] decrease cooperation in the case of disadvantageous intra-group comparison (i.e., they contributed more than other group members), and (to a lesser degree) increase their cooperation in the case of advantageous intra-group comparison (i.e. they contributed less than other group members). Hence, players are prone to assimilate to other group members' contributions [22], [23]. If subjects not only compare their payoff to the members of the own group but also to the members of another group, this may influence their intra-group behavior, even if the other group is structurally independent. It is, however, still an open question whether the payoff comparisons to the members of the other group are weighted differently from the payoff comparisons to the own group members.

Research in social psychology has shown that social comparisons may operate on the inter-individual and intra-group level, but also on the inter-group level [24]. Inter-group comparison processes are of utmost importance to social identity theory [25], [26] and self-categorization theory [27]; see [28] for an overview. The social identity perspective of group formation and intergroup relations proposes that group members are prone to increase positive distinctiveness, that is, they positively maximize relative differences between the outcomes of the own and other groups. Thus, inter-group comparisons may activate a comparative focus [29] that motivates group members to increase the relative (joint) outcome of their own group or to decrease the relative disadvantage in comparison with another group. Hence, even in the absence of a material inter-group conflict that may force *realistic competition* for scarce resources [30], [31], group members may engage in *social competition* to boost their social identity [32] or reduce uncertainty [33]. Following this perspective, subjects should increase cooperation in the case of a disadvantageous inter-group comparison (i.e. the own group's outcome is lower than the comparison group's outcome).

What happens if the effects of intra-group and inter-group comparison "pull" in opposite directions? For instance, conditional cooperation may request a high contributor to reduce her contribution, whereas at the same time the low provision level of the own group in comparison to the other group may ask her to increase the contribution. We show that the interplay of these opposed forces results in a mitigation of the overall decline of cooperation. The combination of insights from research in economics and social psychology allows us to hypothesize and show that if groups are not negatively interdependent, increased intra-group cooperation based on mere inter-group comparison may not only be in the interest of each group, but may also be in the collective interest. Therefore, social inter-group interactions by

means of mere inter-group comparisons are a powerful method to increase efficiency in human cooperation.

The Present Research

We conducted a laboratory experiment to investigate the effect of inter-group comparison on cooperation in a repeated linear public goods game. After each round, participants received feedback either about the average contributions of the own group members only (intra-group comparison only; *INTRA* treatment), or about the average contributions of the own *and* another group's members (intra- and inter-group comparison; *INTER* treatment). Following the proposed inter-group comparison – intra-group cooperation hypothesis, we expected larger intra-group cooperation (i.e., contribution to a public good) if actors are able to compare their own group with another group's level of cooperation (INTER) than when they may compare their personal with the own group's level of cooperation only (INTRA). Our design allows for testing the differential effects of advantageous/disadvantageous intra-group and inter-group comparisons on contributions.

Furthermore, we manipulated the provision environment. In one public goods environment (COOP+), provision took place in a smaller group ($n = 3$) with a higher individual return from cooperation (0.7). In the other environment (COOP–) the group was bigger ($n = 4$) and the individual return from cooperation was lower (0.4). In line with previous research [34], [35], we expected larger intra-group cooperation in COOP+ than in COOP–. Moreover, the variation of the provision environment allows for testing the robustness and generality of the proposed inter-group comparison – intra-group cooperation effect under different structural incentives for cooperation.

Methods

Ethics Statement

The experiment was conducted at a German University, where institutional review boards or committees are not mandatory (see guidelines of the German Psychological Society: http://www.bdp-verband.org/bdp/verband/ethic.shtml; particularly section C.II.4).

Participants and Design

Participants were 216 students (68 male, 148 female; $M_{age} = 22.24$, $SD = 3.24$) from various disciplines at the University of Erfurt, Germany. Treatment of participants was in agreement with the ethical guidelines of the German Research Foundation (Deutsche Forschungsgemeinschaft) and the German Psychological Society (DGP). All participants gave their written informed consent to participate voluntarily, assuring them that analyses and publication of experimental data would be without an association to their real identities. Decisions were incentivized; on average, participants earned 7.20 €. Moreover, random assignment to visually separated cubicles and private payment at the end of the experiment preserved the anonymity of participants. The experiment involved no deception of participants. As in other socio-economic experiments, there were no additional ethical concerns.

The experiment used a 2 (level of comparison: INTRA vs. INTER) × 2 (environment: COOP+ vs. COOP–) between-subjects design. There were nine experimental sessions, each consisting of 24 participants. Three sessions (72 participants) were randomly assigned to the INTER / COOP– condition, and two sessions (48 participants) were assigned to each of the other conditions (INTRA / COOP+, INTER / COOP+, and INTRA / COOP–). Due to the intra- and inter-group feedback provided in the INTER treatment, participants of the same intra-group as well as participants of matched groups were interdependent ($n = 6$ in COOP+ and $n = 8$ in COOP–), whereas in the INTRA treatments only participants of the same intra-group were interdependent ($n = 3$ in COOP+ and $n = 4$ in COOP-). Thus, the number of independent observations was 16 in INTRA / COOP+, 8 in INTER / COOP+, 12 in INTRA / COOP–, and 9 in INTER / COOP–.

Procedure

Participants were recruited to take part in a decision making experiment at the Erfurt laboratory for experimental economics (eLab). The experiment was computerized using the software z-Tree [36]. On arrival, participants drew an index card to determine their cubicle number. Participants received printed instructions, including some examples. The instructions were read aloud by the experimenter. To make sure that all participants understood the structure of the game, they had to correctly answer some control questions before the actual experiment started (see instructions and control questions in the Text S1). Each participant was randomly assigned to a group of three (COOP+) or four (COOP–) members (labeled the blue or the green group) that was matched with a group of the other color. Group assignment and matching remained constant over the entire experiment (*partner matching* protocol). The game consisted of 20 rounds. At the beginning of each round, participants received an endowment of 12 tokens and had to decide individually and independently how many (if any) of these tokens to contribute to a group project. For each token contributed to the group project, each member of the own group (including the contributor) received 0.7 (COOP+) or 0.4 (COOP–) experimental currency units (ECU; thus, it was multiplied by 2.1 or 1.6 and equally distributed among all three or four group members, respectively). For each token kept privately, the individual player only received 1 ECU. After each round, participants were informed about their individual earnings in the respective round and the average contributions of their own group (INTRA and INTER), as well as about the average contributions of the other group (INTER only). The experiment ended with a short post-experimental question-naire, assessing participants' demographics. Finally, the partici-pants were informed about their overall payoff and paid privately (exchange rate: 10 ECU = 0.25 €). The whole experiment took about 45 minutes.

Payoff Function

The individual payoff function is: $\pi_i = e_i - c_i + \alpha \sum_{i=1}^{n} c_i$ for player i in a group of n players, each endowed with e tokens, where c denotes the number of tokens contributed to the public good and α refers to the amount that each group member (including the contributor) receives for each contributed token.

Data Analysis

For non-parametric tests, we report two-tailed p-values of exact tests in all analyses to account for the relatively small number of independent observations. Additionally, we provide r-values as effect size approximations. Parametric feedback analyzes were computed using the statistical package *nlme* [37] in the R environment [38]. For analyzes of feedback in the INTRA treatment, intra-groups ($N = 28$) and participants ($N = 96$) were treated as random effects to control for their interdependent error terms (*random intercept model*; [39]). Similarly, for feedback analyzes in the INTER treatment, matched groups ($N = 17$), intra-groups ($N = 34$), and participants ($N = 120$) were treated as random effects.

Results

Our data provide clear evidence for the inter-group comparison – intra-group cooperation hypothesis (see Dataset S1). Overall, contributions in INTER were about 40% higher than in INTRA (Mann-Whitney U test: $z = -2.81$, $p = .004$, $r = -.42$). This result holds in each of the provision environments COOP+ and COOP– (Mann-Whitney U test: COOP+: $z = -2.51$, $p = .010$, $r = -.51$; COOP–: $z = -2.49$, $p = .012$, $r = -.54$). Figure 1 visualizes this result by depicting the mean individual contributions and its 95% confidence intervals per round and comparison treatment, separately for each provision environment treatment. Table 1 displays the mean values and standard deviations of all conditions at the appropriate level of comparison. As further expected, overall contributions in COOP+ were higher than in COOP– (Mann-Whitney U test: $z = -2.66$, $p = .007$, $r = -.40$).

Development of Contributions

As Table 1 shows, in the first round participants contributed about half of their endowment irrespective of the comparison treatment, $F(1, 215) = 2.91$, $p = .089$, and the environment treatment, $F(1, 215) = 2.09$, $p = .149$. Overall, we observe the usual decline in contributions. Contributions in the last 10 rounds were significantly lower than in the first 10 rounds, Wilcoxon signed-rank test: $z = -4.71$, $p < .001$, $r = -.70$. However, when evaluating the experimental conditions separately this result only holds in three of the four conditions. In INTER / COOP+ where participants could compare their own group's average contribu-tions to another group in an environment with high cooperative incentives, contributions exhibited no significant decline over the rounds, but remained rather constant, Wilcoxon signed-rank test: $z = -.14$, $p = .945$, $r = -.05$.

Reactions to Intra- and Inter-Group Feedback

To study the subjects' reactions to feedback, we investigated the contribution change (contribution in the actual round minus contribution in the previous round) in different mixed-effects models to account for observations' interdependence, while controlling for the provision environment (see Table 2). As predictors we used the deviation of the subject's contribution from the average of other members of the own group (own contribution minus average contributions of other group members) and additionally in INTER the contribution deviation of the own group's average from the other group's average (own group average contribution minus other group average contributions).

When only intra-group comparison was available (INTRA), the deviation from the own group's average significantly predicted the change in cooperation, $b = -.56$, $SE = .02$, $t_{1727} = -28.59$, $p < .0001$ (model 1 INTRA in Table 2). When the deviation was positive, i.e. the subject contributed more than the average of the other members of the own group, the negative estimate predicts a reduction in contribution, while the subjects increased the contribution when having contributed less than the other group members. This indicates conditional cooperation. When both intra- and inter-group comparison was available (INTER), both the intra- and the inter-group deviation significantly predicted changes of contributions, $b = -.55$, $SE = .02$, $t_{2158} = -30.96$, $p < .0001$ and $b = -.22$, $SE = .02$, $t_{2158} = -9.60$, $p < .0001$, respective-ly (model 1 INTER in Table 2). This shows that subjects not only conditionally cooperate within their own group, but also want their group to be ahead of the other group, yielding support for the inter-group comparison – intra-group cooperation hypothesis.

But how do subjects behave if these two motives – conditional cooperation within the own group and being ahead of the other

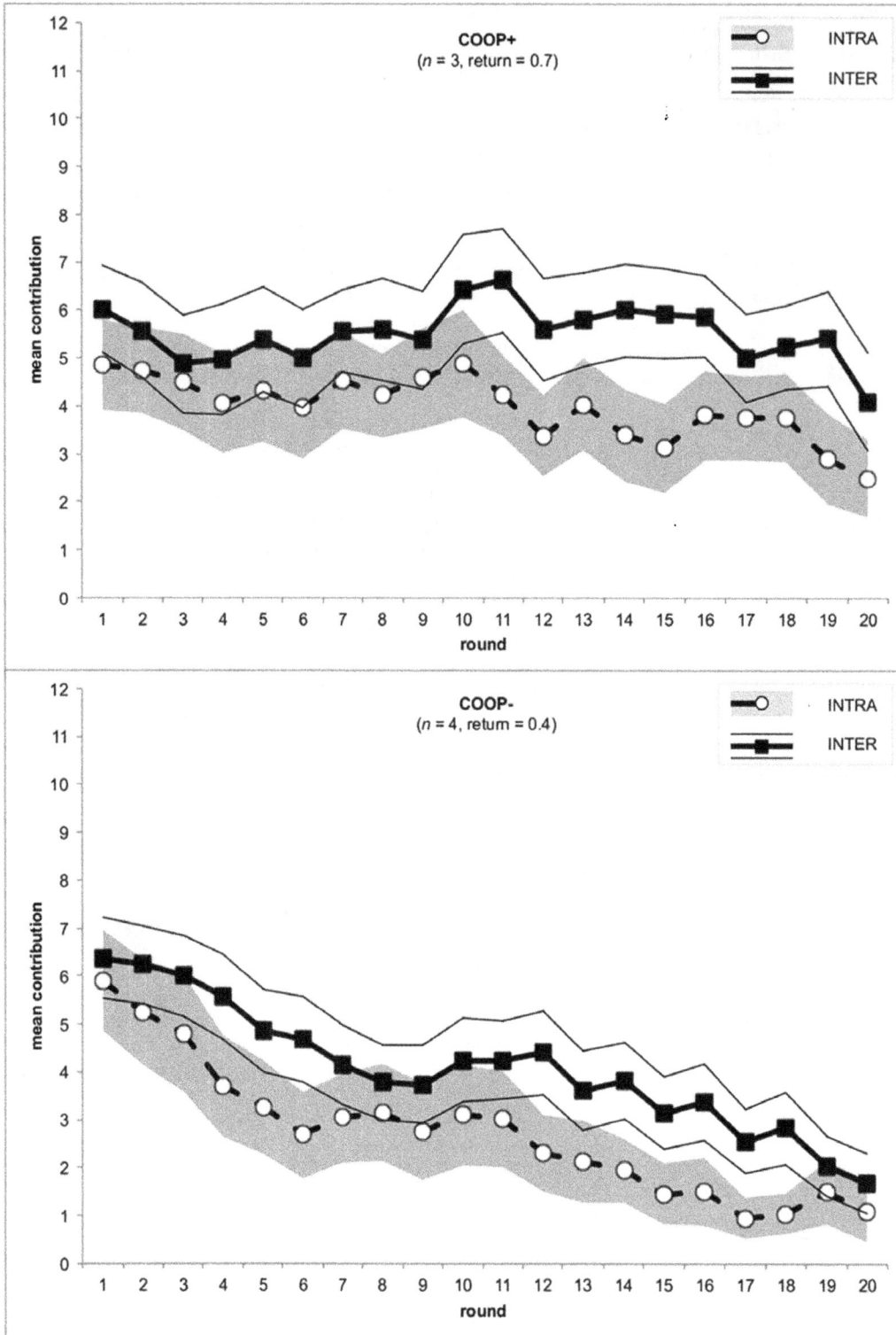

Figure 1. Mean contributions per round by comparison and environment treatments. Areas around mean values indicate 95% confidence intervals.

group – are in conflict? How does a subject react if conditional cooperation calls to reduce cooperation, but at the same time the desire to be ahead of the other group calls for increasing cooperation? To answer these questions we distinguish between positive and negative deviations in both the intra- and the inter-group comparison (model 2 in Table 2). In the case of intra-group comparison a positive deviation indicates a disadvantage (i.e. "I have contributed *more* than other group members."), while in the

Table 1. Mean values and standard deviations (in brackets) of the first round, the first 10 rounds, the last 10 rounds, and overall contributions by condition.

Contribution \\ Condition	INTRA		INTER	
	COOP+	COOP–	COOP+	COOP–
First round	4.85 (3.35)	5.90 (3.71)	6.02 (3.21)	6.37 (3.61)
Round 1–10	4.47 (1.71)	3.77 (2.20)	5.49 (1.30)	4.97 (1.40)
Round 11–20	3.49 (1.65)	1.70 (1.23)	5.56 (1.35)	3.17 (1.42)
Overall	3.98 (1.58)	2.73 (1.69)	5.52 (1.17)	4.07 (1.31)

INTRA: intra-group comparison only. INTER: intra- and inter-group comparison. COOP+: public good with $n = 3$ and individual return from cooperation = 0.7. COOP–: public good with $n = 3$ and individual return from cooperation = 0.4. Reported are mean values on the level of independent observations. Thus, values of first round contributions are mean values on the individual level. Values of contributions in round 1–10, round 11–20, and overall are mean values on the level of intra-groups (INTRA) or on the level of matched groups (INTER).

case of inter-group comparison it indicates an advantage (i.e. "My group has contributed *more* than the other group."). In contrast, a negative deviation in the case of intra-group comparison indicates an advantage (i.e. "I have contributed *less* than other group members."), while in the case of inter-group comparison it indicates a disadvantage (i.e. "My group has contributed *less* than the other group."). In the spirit of inequity aversion [21] the decrease in reaction to a disadvantageous intra-group comparison is stronger than the increase in reaction to an advantageous comparison, $b = -.67$, $SE = .03$, $t_{2156} = -21.13$, $p < .0001$ and $b = -$.42, $SE = .04$, $t_{2156} = -12.05$, $p < .0001$, respectively (see Figure 2: white and hatched bars on the left-hand side have smaller absolute values than on the right-hand side). Adding inter-group comparison has two remarkable effects. First, the increase in contribution triggered by a comparison group with higher contributions is much stronger than the decrease in contribution when being ahead of the other group, $b = -.28$, $SE = .04$, $t_{2156} = -6.50$, $p < .0001$ and $b = -.15$, $SE = .04$, $t_{2156} = -3.54$, $p = .0004$, respectively (see Figure 2: hatched bars have greater values than white bars). And second, the contribution adaptation due to the intra-group

Table 2. Parameter estimates of mixed-effects models predicting contribution change.

Predictor		Model 1 INTRA		Model 1 INTER		Model 2 INTER	
		Estimate	t-value	Estimate	t-value	Estimate	t-value
(Intercept)		−0.256	−1.99	−0.247	−1.85	−.056	−0.31
		(0.129)		(0.133)		(0.184)	
Environment		0.139	0.77	0.143	0.68	0.162	0.75
		(0.182)		(0.211)		(0.215)	
Intra-group comparison	Overall	−0.564	−28.59***	−0.549	−30.96***	–	–
		(0.020)		(0.018)			
	Positive (disadvantageous)	–	–	–	–	−0.666	−21.13***
						(0.032)	
	Negative (advantageous)	–	–	–	–	−0.421	−12.05***
						(0.035)	
Inter-group comparison	Overall	–	–	−0.219	−9.60***	–	–
				(0.023)			
	Positive (advantageous)	–	–	–	–	−0.149	−3.54**
						(0.042)	
	Negative (disadvantageous)	–	–	–	–	−0.276	−6.50***
						(0.042)	
Observations [subjects/intra-groups/matched groups]		1824 [96/28/−]		2280 [120/34/17]		2280 [120/34/17]	
REML model fit: AIC/BIC		9020/9053		11446/11492		11433/11491	

In model 1 INTRA subjects and intra-groups were treated as random effects, whereas in all other models subjects, intra-groups, and matched groups were treated as random effects. The presented models are superior regarding AIC/BIC to other model specifications (e.g. including interaction terms). REML = restricted maximum likelihood. AIC = Akaike information criterion. BIC = Bayesian information criterion. * $p < .01$, ** $p < .001$, *** $p < .0001$.

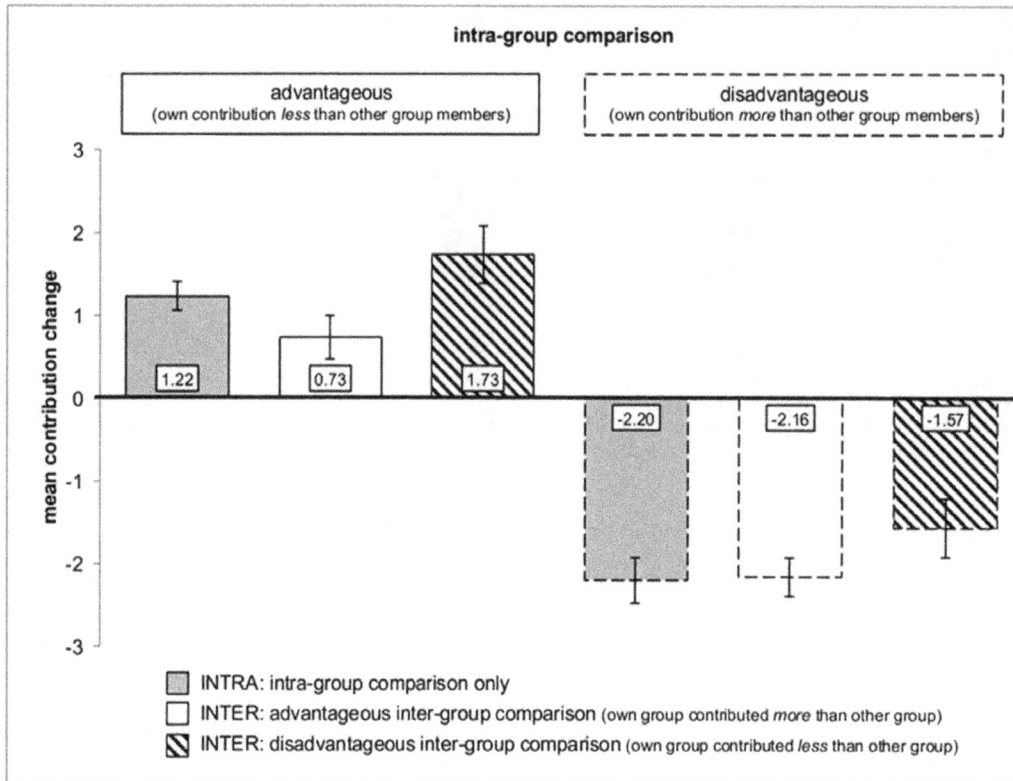

Figure 2. Mean contribution change by advantageous and disadvantageous intra- and inter-group comparison. Error bars indicate 95% confidence intervals.

comparison is always stronger than the one based on the inter-group comparison, resulting in a net effect in the direction of conditional cooperation (see Figure 2: the adaption according to advantageous intra-group comparison is always positive and the adaption according to disadvantageous intra-group comparison is always negative, irrespective of the inter-group comparison). Hence, the inter-group comparison may accelerate conditional cooperation among the low contributors of the group, but most importantly it dampens the downward trend in the high contributions. The desire to be ahead of the other group cushions the downward trend in the contributions of subjects contributing above average (see Figure 2: grey bar on the right-hand side has smaller value than hatched bar).

Discussion

The present research presents a new mechanism to increase cooperation and efficiency in public goods provision by integrating research on intra-group [20], [21] and inter-group social comparisons [25–29]. The mere inter-group comparison suffices to increase intra-group cooperation and overall efficiency. We found higher contributions to a public good in the presence of both intra-group and inter-group feedback of average group contributions than in the presence of intra-group feedback only. The effect appeared across two provision environments with differing individual incentives for intra-group cooperation, supporting its robustness and generality. Furthermore, analyzes provided evidence that the overall level of cooperation may be explained by the combination of intra-group conditional cooperation and a desire to be ahead of the other group.

The results also contribute to the empirical literature on the inter-group conflict – intra-group cooperation effect [12–16], that is, an increase of intra-group cooperation in the presence of a structural inter-group conflict. In contrast to this well-supported phenomenon, however, mere inter-group comparison does not increase cooperation at the cost of collective efficiency. Rather, it is in the interest of all individuals, irrespective of group membership, to maximize contributions to the (intra-group) public good. Therefore, the inter-group comparison – intra-group cooperation hypothesis proposes an efficiency-enhancing alternative to increase long-term human cooperation.

Practical Implications

Social comparisons are made frequently in everyday life [40]. It has been shown that inter-individual and intra-group social comparisons may increase individual performance, for instance in task performance [41], [42], academic performance [43], and sports performance [44], but also different kinds of prosocial behavior [45–48]. Our results suggest that social comparisons not only on the individual level but also on the group-level might further increase such effects. For instance, providing feedback about large-scale prosocial behavior (e.g., charity donations) of other groups (organizations, countries, etc.) might help to increase individual cooperation/prosocial behavior in the own group in order to receive a positive inter-group comparison. In a similar vein, if the level and success of environmental protection of different villages or towns may be quantified and made publically available (e.g. CO_2 emissions), this might motivate further activities on the individual and group level in order to improve one's own "group's" ranking. Also, because the success of a company requires coordination and cooperation among employees of its sub-units, the company could provide feedback not only on the individual level [46], [47] but also on the sub-units' level of cooperation. This might increase (intra-sub-unit) organizational collaboration without any additional costs for monetary incentives (in order to induce a resource conflict among the sub-units).

Acknowledgments

This paper is dedicated to our departed friend and colleague Gary Bornstein who inspired our research and supported the first steps of this project. We thank Cornelia Betsch and Stephen Benard for helpful comments on an earlier draft.

Author Contributions

Conceived and designed the experiments: RB BR. Performed the experiments: RB. Analyzed the data: RB. Contributed reagents/materials/analysis tools: RB BR. Wrote the paper: RB BR.

References

1. Hardin G (1968) The tragedy of the commons. Science 162: 1243–1248.
2. Chaudhuri A (2011) Sustaining cooperation in laboratory public goods experiments: A selective survey of the literature. Exp Econ 14: 47–83.
3. Ledyard JO (1995) A survey of experimental research. In: Kagel JH, Roth AE, editors.Handbook of Experimental Economics. Princeton, NJ: Princeton Univ Press.pp. 111–194.
4. Messick DM, Brewer MB (1983) Solving social dilemmas: A review. In: Wheeler L, Shaver P, editors. Review of Personality and Social Psychology.Beverly Hills, CA: Sage.pp. 11–14.
5. Fischbacher U, Gächter S (2010) Social preferences, beliefs, and the dynamics of free riding in public good experiments. Am Econ Rev 100: 541–556.
6. Fischbacher U, Gächter S, Fehr E (2001) Are people conditionally cooperative? Evidence from a public goods experiments. Econ Lett 71: 397–404.
7. Kelley HH, Stahelski AJ (1970) Social interaction basis of cooperators' and competitors' beliefs about others. J Pers Soc Psychol 16: 66–91.
8. Croson R (2007) Theories of commitment, altruism and reciprocity: Evidence from linear public goods games. Econ Inq 45: 199–216.
9. Neugebauer T, Perote J, Schmidt U, Loos M (2009) Selfish-biased conditional cooperation: On the decline of contributions in repeated public goods experiments. J Econ Psychol 30: 52–60.
10. Bowles S (2006) Group competition, reproductive leveling, and the evolution of human altruism. Science 314: 1569–1572.
11. Bowles S (2009) Did warfare among ancestral hunter-gatherers affect the evolution of human social behaviors? Science 324: 1293–1298.
12. Benard S, Doan L (2011) The conflict–cohesion hypothesis: Past, present, and possible futures. In: Thye SR, Lawler EJ, editors. Advances in Group Processes. Emerald. pp. 189–225.
13. Bornstein G (2003) Intergroup conflict: Individual, group and collective interests. Pers Soc Psychol Rev 7: 129–145.
14. Bernard S (2012) Cohesion from conflict: Does intergroup conflict motivate intragroup norm enforcement and support for centralized leadership? Soc Psychol Q 75: 107–130.
15. Sääksvuori L, Mappes T, Puurtinen M (2011) Costly punishment prevails in intergroup conflict. Proc R Soc B 2011: rspb.2011.0252v1–rspb20110252.
16. Maner JK, Mead NL (2010) The essential tension between leadership and power: When leaders sacrifice group goals for the sake of self-interest. J Pers Soc Psychol 99: 482–497.
17. Dawes RM (1980) Social dilemmas. Annu Rev Psychol 31: 169–193.
18. Gould RV (1999) Collective violence and group solidarity: Evidence from feuding society. Am Sociol Rev 64: 356–380.
19. Kreps D, Milgrom P, Roberts J, Wilson R (1982) Rational cooperation in the finitely repeated Prisoners' Dilemma. J Econ Theory 27: 245–252.
20. Bolton GE, Ockenfels A (2000) ERC: A theory of equity, reciprocity, and competition. Am Econ Rev 90: 166–193.
21. Fehr E, Schmidt KM (1999) A theory of fairness, competition, and cooperation. Q J Econ 114: 817–868.
22. Brewer MB, Weber JG (1994) Self-evaluation effects of interpersonal versus intergroup social comparison. J Pers Soc Psychol 66: 268–275.
23. Festinger L (1954) A theory of social comparison processes. Hum Relat 7: 117–140.
24. Suls J, Wheeler L (2000) Handbook of social comparison. New York, NY: Kluver.
25. Tajfel H, Turner JC (1979) An integrative theory of intergroup conflict. In: Austin WG, Worchel S, editors. The social psychology of intergroup relations. Monterey, CA: Brooks/Cole.pp. 33–47.

26. Tajfel H, Turner JC (1986) The social identity theory of intergroup conflict. In: Worchel S, Austin WG, editors. Psychology of intergroup relations. Chicago, IL: Nelson-Hall.pp. 7–24.
27. Turner JC, Hogg MA, Oakes PJ, Reicher SD, Wetherell MS (1987) Rediscovering the Social Group: A Self-categorization Theory. Cambridge, MA: Basil Blackwell.
28. Hogg MA (2000) Social identity and social comparison. In: Suls J, Wheeler L, editors. Handbook of social comparison: Theory and research. New York, NY: Kluver.pp. 401–421.
29. Corcoran K, Mussweiler T (2009) Comparative thinking styles in group and person perception: One mechanism – many effects. Soc Personal Psychol Compass 3: 244–259.
30. Campbell DT (1972) On the genetics of altruism and the counter-hedonic components in human culture. J Soc Issues 28: 21–27.
31. Sherif M (1966) In common predicament: Social psychology of intergroup conflict and cooperation. Boston, MA: Houghton Mifflin.
32. Turner JC (1975) Social comparison and social identity: Some prospects for intergroup behaviour. Eur J Soc Psychol 5: 5–14.
33. Hogg MA (2000) Subjective uncertainty reduction through self-categorization: A motivational theory of social identity processes. In Stroebe W, Hewstone M, editors. European review of social psychology. Oxford: John Wiley & Sons.
34. Brewer MB, Kramer RM (1986) Choice behavior in social dilemmas: Effects of social identity, group size, and decision framing. J Pers Soc Psychol 50: 543–549.
35. Isaac RM, Walker JM (1988) Group size effects in public goods provisions: The voluntary contribution mechanism. Q J Econ 103: 179–199.
36. Fischbacher U (2007) z-Tree: Zurich toolbox for ready-made economic experiments. Exp Econ 10: 171–178.
37. Pinheiro J, Bates D, Debroy S, Sarkar D, The R Core team (2007) nlme: Linear and nonlinear mixed effects models (Version 3.1-86). Available: http://cran.r-project.org/web/packages/nlme/index.html. Accessed 2009 January 12.
38. R Development Core Team (2008) A Language and Environment for Statistical Computing. Vienna: R Foundation for Statistical Computing. Available: http://www.R-project.org. Accessed 2009 April 13.
39. Pinheiro JC, Bates DM (2000) Mixed-effects models in S and S-PLUS. New York, NY:Springer.
40. Wheeler L, Miyake K (1992) Social comparisons in everyday life. J Pers Soc Psychol 62: 760–773.
41. Rijsman JB (1974) Factors in social comparison of performance influencing actual performance. Eur J Soc Psychol 4: 279–311.
42. Seta JJ (1982) The impact of comparison processes on coactors' task performance. J Pers Soc Psychol 42: 281–291.
43. Blanton H, Buunk BP, Gibbons FX, Kuyper H (1999) When better-than-others compare upward: Choice of comparison and comparative evaluation as independent predictors of academic performance. J Pers Soc Psychol 76: 420–430.
44. Triplett N (1897) The dynamogenic factors in pacemaking and competition. Am J Psychol 9: 507–553.
45. Bigoni M, Suetens S (2012) Feedback and dynamics in public good experiments. J Econ Behav Organ 82: 86–95.
46. Gächter S, Nosenzo D, Sefton M (2012) The impact of social comparisons on reciprocity. Scand J Econ 114: 1346–1367.
47. Kuhnen CM, Tymula A (2012) Feedback, self-esteem, and performance in organizations. Manage Sci 58: 94–113.
48. Shang J, Croson R (2009) A field experiment in charitable contribution: The impact of social information on the voluntary provision of public goods. Econ J 119: 1422–1439.

12

Short-Term Effects of Particulate Matter on Stroke Attack: Meta-Regression and Meta-Analyses

Xiao-Bo Yu[1]⑨, Jun-Wei Su[2]⑨, Xiu-Yang Li[3]*, Gao Chen[1]*

1 Department of Neurosurgery, the Second Affiliated Hospital of Zhejiang University School of Medicine, Hangzhou, P.R. China, **2** Key Laboratory of Infectious Diseases Ministry of Public Health of China, the First Affiliated Hospital of Zhejiang University School of Medicine, Hangzhou, P.R. China, **3** Department of Public Health, Zhejiang University, Hangzhou, P.R. China

Abstract

Background and Purpose: Currently there are more and more studies on the association between short-term effects of exposure to particulate matter (PM) and the morbidity of stroke attack, but few have focused on stroke subtypes. The objective of this study is to assess the relationship between PM and stroke subtypes attack, which is uncertain now.

Methods: Meta-analyses, meta-regression and subgroup analyses were conducted to investigate the association between short-term effects of exposure to PM and the morbidity of different stroke subtypes from a number of epidemiologic studies (from 1997 to 2012).

Results: Nineteen articles were identified. Odds ratio (OR) of stroke attack associated with particular matter ("thoracic particles" [PM_{10}]<10 μm in aerodynamic diameter, "fine particles" [$PM_{2.5}$]<2.5 μm in aerodynamic diameter) increment of 10 μg/m^3 was as effect size. PM_{10} exposure was related to an increase in risk of stroke attack (OR per 10 μg/m^3 = 1.004, 95%CI: 1.001~1.008) and $PM_{2.5}$ exposure was not significantly associated with stroke attack (OR per 10 μg/m^3 = 0.999, 95%CI: 0.994~1.003). But when focused on stroke subtypes, $PM_{2.5}$ (OR per 10 μg/m^3 = 1.025; 95%CI, 1.001~1.049) and PM_{10} (OR per 10 μg/m^3 = 1.013; 95%CI, 1.001~1.025) exposure were statistically significantly associated with an increased risk of ischemic stroke attack, while $PM_{2.5}$ (all the studies showed no significant association) and PM_{10} (OR per 10 μg/m^3 = 1.007; 95%CI, 0.992~1.022) exposure were not associated with an increased risk of hemorrhagic stroke attack. Meta-regression found study design and area were two effective covariates.

Conclusion: $PM_{2.5}$ and PM_{10} had different effects on different stroke subtypes. In the future, it's worthwhile to study the effects of PM to ischemic stroke and hemorrhagic stroke, respectively.

Editor: Yinping Zhang, Tsinghua University, China

Funding: This research was supported by a grant from Hangzhou Science and Technology Bureau (Grants 200513231344), the Fundamental Research Funds for the Central Universities (Grants 2010QNA7020), and a grant from Zhejiang University Undergraduate Zetetic Experiment Project of Public Health (2013). Additional support was provided by Zhejiang University Student Research Training Program (SRTP) (2011) and Zhejiang University Public Health Innovative Experiment Project (2011). The funders had no role in study design, data collection and analysis, decision to publish, or preparation of the manuscript.

Competing Interests: The authors have declared that no competing interests exist.

* E-mail: lixiuyang@zju.edu.cn (XL); d.chengao@163.com (GC)

⑨ These authors contributed equally to this work.

Introduction

Many studies regarded air pollution exposure as an important factor of hospitalization and mortality worldwide. PM, playing an important role in pollutants of major public health concern, had been confirmed that it could impair the respiratory and cardiovascular system through a series of changes in autonomic nervous system activity [1] and systemic inflammation [2], giving rise to alterations in oxidative stress [3,4], hematologic activation [5] and vascular endothelial dysfunction [6]. Most researches regarded PM_{10} and $PM_{2.5}$ as major harmful PMs.

Nevertheless, short-term effects of PM exposure on cerebral vessels were uncertain. Wordley et al. [7] and Tsai et al. [8] found that PM_{10} was associated with daily stroke attack positively. While, in the works of Chan et al. [9], Henrotin et al. [10] and Andersen et al. [11], no significant association was demonstrated between

PM_{10} and hemorrhagic stroke attack. Similarly, analyses on the relationship between $PM_{2.5}$ and stroke attack also appeared to divergent results. Villeneuve et al. [12] found $PM_{2.5}$ exposure wasn't related to an increased risk of ischemic stroke attack (OR per 10 μg/m^3 = 1.052, 95%CI: 0.996~1.160), while Wellenius et al. [13] found a positive association between $PM_{2.5}$ exposure and ischemic stroke attack (OR per 10 μg/m^3 = 1.278, 95%CI: 1.079~1.525).

Our previous research focused on the association between PM exposure and stroke attack in two study designs (time-series design and case-crossover design), and the result indicated that the effects of PM to stroke attack varied in different study designs [14]. However, in addition to study design, there were still many other covariates (e.g. age, gender, economic condition, area, lags times, historical disease and temperature) among studies, which could influence the results. Of special interest was that whether PM can act differently on different stroke subtypes. So in this article we determined to do meta-analyses, meta-regression and subgroup analyses of association between PM and different stroke attack.

Methods

1. Studies selection

We identified studies published in English and Chinese up to March 2013, by literature search using PubMed, Web of Science, MEDLINE, Google Scholar, China National Knowledge Infrastructure (CNKI) and reference lists of relevant articles. Search terms included "Air Pollution/Particulate Matter" plus "Cardiovascular disease/Stroke", besides, key terms "hospitalization/ Hospital Administration/Emergencies/Morbidity", "cardiovascular disease" were used to enlarge the searching range. We chose ICD9: 430–438 or ICD10: I60–I69 as the definition of "stroke" or "cerebrovascular disorders", ICD9: 430–432 or ICD10: I60–I62 for "hemorrhagic stroke", ICD9: 433–434 or ICD10: I63–I66 for "ischemic stroke".

Eligible studies were selected by two reviewers (X.L., J.S.) independently according to following inclusion criteria: (1) The outcome focused on the effect of PM to stroke or cerebrovascular disease (2) Published full-text articles (3) focused on PM_{10} and/or $PM_{2.5}$ (4) Studies with similar effects [e.g. risk ratios (RR), 95% CIs] that could approximate ORs. The exclusion criteria were: (1) Duplications (2) Reviews or Meta-analysis (3) Long-term effects articles (4) air pollution from industrial and occupational environment (5) Articles that did not provide calculable or reported ORs and 95% CIs. The two reviewer reached consensus on the eligibility of each study. When there was disagreement, a third reviewer (G.C.) took the final decision.

Using a standardized form, data from eligible studies were extracted by two reviewers (X.L., J.S.) independently, From each study, we collected name of first author, year of publication, number of participants, PM, country, age, gender, types of stroke, lags of air pollutants' concentrations, study design and ORs or RRs with 95% CI. In Meta-regression, covariates were gender, area, lag times, study design and research period. Gender was represented as female, male and the whole population. Areas included Asia, Europe, North America and Oceania. Lag times were the same day on stroke attack and the previous 1, 2, 3, 4, 5 day. Study designs were classified into time-series and case-crossover designs. Research period was divided into early stage (1992–1997), middle stage (1997–2004) and late stage (2004–2009). Given there were 5 covariates (gender, area, lag times of air pollutants' concentrations, study design and research period), we used Meta-regression model to detect any possible influence factors ($P<0.05$).

2. Statistical analysis

Pooled ORs of PM with 95%CI for stroke attack were calculated by using the fixed or random effects Meta-analysis of with Q and I^2 statistics given as the chosen measure of heterogeneity (The null hypothesis of this test is homogeneity). The Q and I^2 statistics were used to assess heterogeneity, where $P \le 0.05$ or $I^2 > 50\%$ were considered as significant heterogeneity. We presented random effects pooled estimates when heterogeneity was detected; otherwise, we used the fixed model. We also produced forest plots to show ORs from each of the individual studies included in the meta-analyses and the estimation of the pooled OR. The sizes of the markers of each OR in the plots represent the relative weight each study contributed to the pooled estimation. We assessed publication bias visually through funnel plots and a weighted Egger's test. We also performed sensitivity

analyses, whereby each article was omitted in turn, recalculating the pooled estimates under extreme conditions. Moreover, Meta-regression analysis was performed to figure out whether the association between PM and stroke attack was influenced by covariates. With a positive Meta-regression coefficient presented ($P \le 0.05$), we could recognize the influence of the given factor. All analyses were performed using software STATA version 12.0 (StataCorp LP, College Station, TX, USA).

Results

1. Study selection and data extraction

A total of 107 potentially relevant researches were identified by searching electronic databases and reference lists. 19 full-text articles were eligible for inclusion in this analysis and data were extracted from these studies [7–13,15–26]. The details of the selection process were presented in a flow chart in Figure 1. Publication years ranged from 1997 and 2012. In total, 9 countries were involved including Australia, Canada, China, Denmark, Finland, France, Italy, UK and US. 14 articles focused on PM_{10} and 9 were about $PM_{2.5}$. In our previous research [14], we only extracted adjusted maximum effective value in each study. However, in this article, we took advantage of all the effect values (33 for $PM_{2.5}$, 68 for PM_{10}, 29 for hemorrhagic stroke, 42 for ischemic stroke) that fulfilled our study aim. The basic overview of the 19 included articles is given in Table 1.

2. Meta-analysis of different PMs exposure to different stroke types

2.1 Effects of $PM_{2.5}$、PM_{10} exposure to stroke attack. There were nine articles (containing 33 studies) that referred to the association of $PM_{2.5}$ and stroke attack. Since heterogeneity existed among studies (Q = 67.09, $P<0.05$), the random effect model was conducted to calculate a pooled OR with 95%CI. The Meta-analysis results indicated that $PM_{2.5}$ exposure wasn't related to an increased risk of stroke attack (OR per 10 µg/ $m^3 = 0.999$, 95%CI: 0.994~1.003), Forest plot and Funnel plot were shown in Figure 2 and Figure 3(A), respectively. Egger's test ($t = 0.98$, $P = 0.336$) didn't find evidence of publication bias. Sensitivity analysis suggested that no individual study significantly affected the pooled effect size, indicating that the results for $PM_{2.5}$ and daily stroke attack were statistically robust (Table S1 in File S1).

Fourteen articles (containing 68 studies) were included. The heterogeneity was significant (Q = 223.25, $P<0.05$). With the random effect model, we pooled all 68 studies into the meta-analysis and found PM_{10} exposure was statistically significantly associated with an increased risk of stroke attack (OR per 10 µg/ $m^3 = 1.004$; 95%CI, 1.001~1.008). Forest plot was shown in Figure 4. Funnel plot was shown in Figure 3(B). Egger's test supported that publication bias was unlikely ($t = 0.80$, $P = 0.427$). Sensitivity analysis showed that results for PM_{10} and stroke attack were not robust to the inclusion of Tsai study [8] and Vidale study [25] (Table S1 in File S1).

2.2 Effects of $PM_{2.5}$、PM_{10} exposure to hemorrhagic stroke attack. Since the effect values of all studies included were not statistically significant, so it had no sense to do the Meta-analysis. As a result, we could not just simply determine the association between $PM_{2.5}$ exposure and hemorrhagic stroke attack.

We conducted meta-analysis for nineteen combinative effects of PM_{10} and hemorrhagic stroke attack. Heterogeneity was detected (Q = 39.82, $P = 0.002$) through heterogeneity test. With random effect model, we found no evidence for the association between PM_{10} exposure and hemorrhagic stroke attack (OR per 10 µg/

PRISMA 2009 Flow Diagram

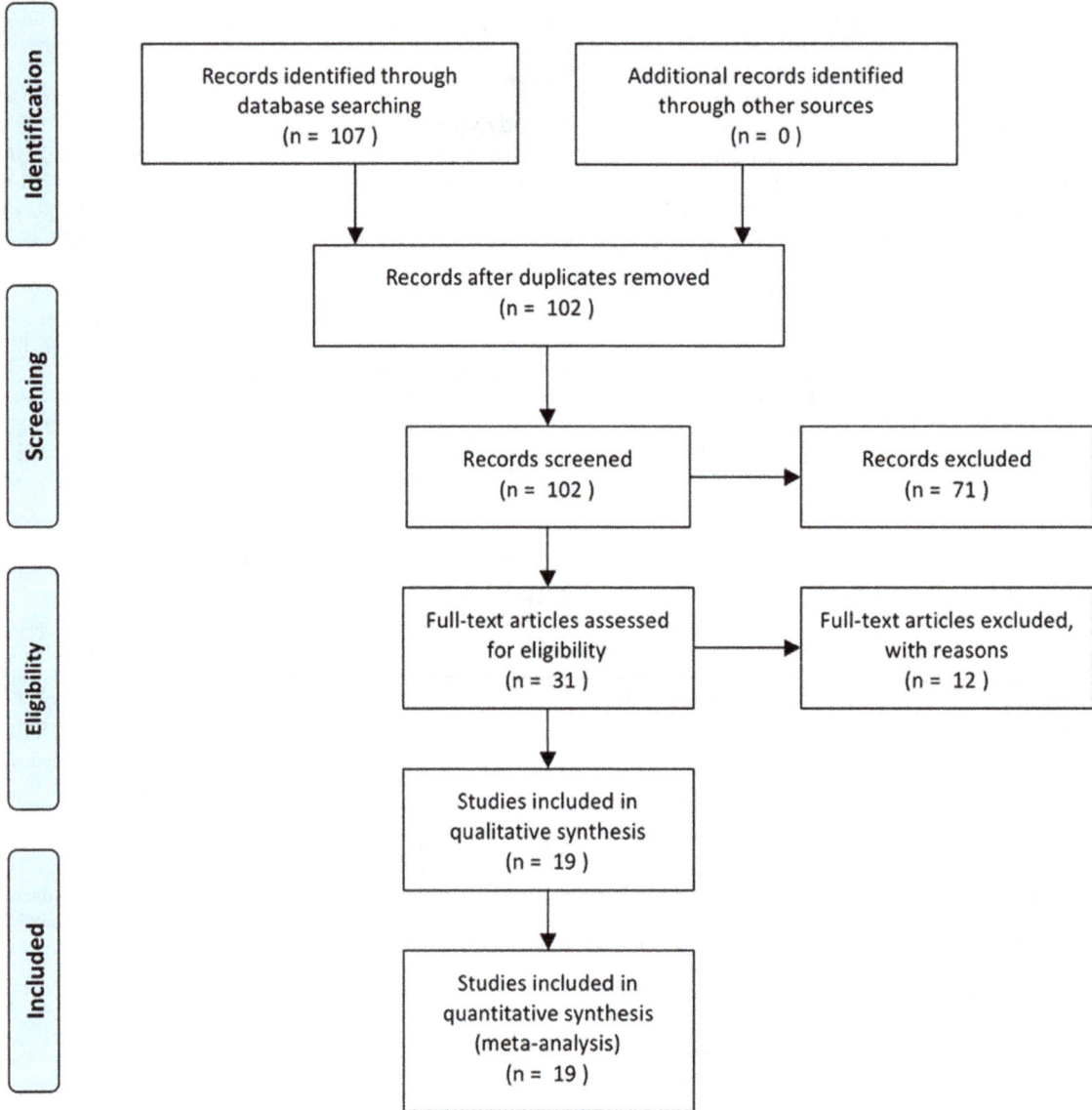

Figure 1. Flow chart of the selection process.

Table 1. The basic overview of the 19 included articles.

First author(year of publication)	Particulate matter	Case number	Country	Age	Gender	Study Design	Lag*	Research period	Types of stroke
Wordley(1997) [7]	PM$_{10}$	NM	UK	all	all	time-series	0	1992–1994	stroke
Wong(1999) [15]	PM$_{10}$	NM	China	all	all	time-series	2	1994–1995	stroke
Linn(2000) [16]	PM$_{10}$	108114	US	>29	all	time-series	0	1992–1995	stroke
Tsai(2003) [8]	PM$_{10}$	16067	China	all	all	case-crossover	0	1997–2000	HS and IS
Chan(2006) [9]	PM$_{2.5}$, PM$_{10}$	8582	China	all	all	time-series	0, 1, 2, 3	1997–2002	stroke, HS and IS
Dominici(2006) [17]	PM$_{2.5}$	11500000	US	≥65	all	time-series	0	1999–2002	stroke
Jalaludin(2006) [18]	PM$_{2.5}$, PM$_{10}$	20634	Australia	≥65	all	time-series	0, 1, 2, 3, 4	1997–2001	stroke
Villeneuve(2006) [19]	PM$_{2.5}$, PM$_{10}$	12034	Canada	≥65	all	case-crossover	0, 1, 3	1992–2002	HS and IS
Henrotin(2007) [10]	PM$_{10}$	1630	France	>40	F, M, all	case-crossover	0, 1, 2, 3	1994–2004	HS and IS
Bell(2008) [20]	PM$_{2.5}$, PM$_{10}$	11466	China	all	all	time-series	0, 1, 2, 3, 0–3	1995–2002	stroke
Guo(2008) [21]	PM$_{10}$	2990	China	all	all	case-crossover	0, 1, 2, 3	2004–2006	stroke
Lisabeth(2008) [22]	PM$_{2.5}$	3508	US	≥45	all	time-series	0, 1	2001–2005	IS
Halonen(2009) [23]	PM$_{2.5}$	10383	Finland	≥65	all	time-series	0, 1, 2, 3, 0–4	1998–2004	stroke
Ye(2009) [24]	PM$_{10}$	699	China	all	all	case-crossover	0	2002–2004	HS
Andersen(2010) [11]	PM$_{10}$	6369	Denmark	all	all	case-crossover	0, 1, 2, 3, 4, 0–4	2003–2006	HS and IS
Vidale(2010) [25]	PM$_{10}$	759	Italy	all	all	time-series	0, 1, 2, 3, 4, 5	2000–2003	IS
Villeneuve(2012) [12]	PM$_{2.5}$	5927	Canada	>18	all	case-crossover	0, 1, 3	2003–2009	stroke, HS and IS
Wellenius(2012) [13]	PM$_{2.5}$	1705	US	≥21	all	case-crossover	0–1	1999–2008	IS
Willocks(2012) [26]	PM$_{10}$	NM	UK	all	all	time-series	0, 1, 2, 3, 4, 5	2000–2006	stroke

NM indicates not mentioned; PM$_{10}$, particular matter with aerodynamic diameter ≤10 μm; PM$_{2.5}$, particular matter with aerodynamic diameter ≤2.5 μm; UK, the United Kingdom; US, the United States; F, female; M, male; HS, hemorrhagic stroke; IS, ischemic stroke.

*lag was the timing of the exposure; Lag0, the same day exposure; Lag1, exposure the day before; Lag2, exposure the 2 days before; Lag3, exposure the 3 days before; Lag4, exposure the 4 days before; Lag5, exposure the 5 days before; Lag0–3, mean exposure of previous 3 days and the same day; Lag0–4, mean exposure of previous 4 days and the same day.

Study ID		OR (95% CI)	% Weight
Chan(2006)		1.01 (0.99, 1.03)	3.83
Chan(2006)		1.01 (1.00, 1.02)	7.37
Chan(2006)		1.00 (1.00, 1.01)	8.10
Chan(2006)		1.00 (1.00, 1.01)	7.73
Dominici(2006)		1.01 (1.00, 1.01)	8.45
Jalaludin(2006)		0.99 (0.97, 1.00)	4.39
Jalaludin(2006)		0.99 (0.97, 1.00)	6.09
Jalaludin(2006)		0.99 (0.98, 1.00)	6.29
Jalaludin(2006)		0.99 (0.98, 1.01)	4.57
Jalaludin(2006)		0.99 (0.97, 1.00)	5.71
Villeneuve(2006)		1.11 (0.97, 1.27)	0.12
Villeneuve(2006)		1.08 (0.89, 1.32)	0.06
Villeneuve(2006)		1.00 (0.94, 1.06)	0.51
Villeneuve(2006)		1.02 (0.94, 1.10)	0.34
Villeneuve(2006)		0.98 (0.85, 1.13)	0.10
Villeneuve(2006)		1.00 (0.94, 1.08)	0.41
Bell(2007)		0.99 (0.98, 1.01)	5.05
Bell(2007)		1.01 (0.99, 1.02)	5.28
Bell(2007)		0.99 (0.98, 1.00)	5.03
Bell(2007)		1.00 (0.99, 1.01)	5.26
Bell(2007)		0.99 (0.97, 1.01)	3.21
Lisabeth(2008)		1.06 (1.00, 1.14)	0.47
Lisabeth(2008)		1.06 (0.98, 1.14)	0.36
Halonen(2009)		1.00 (0.97, 1.03)	2.04
Halonen(2009)		0.98 (0.96, 1.01)	1.98
Halonen(2009)		1.00 (0.97, 1.03)	2.14
Halonen(2009)		0.99 (0.95, 1.03)	1.31
Halonen(2009)		0.98 (0.95, 1.01)	2.13
Villeneuve(2012)		1.02 (0.97, 1.09)	0.58
Villeneuve(2012)		1.03 (0.95, 1.12)	0.30
Villeneuve(2012)		1.02 (0.97, 1.09)	0.58
Wellenius(2012)		1.18 (1.05, 1.33)	0.15
Wellenius(2012)		1.28 (1.08, 1.52)	0.07
Overall (I-squared = 52.3%, p = 0.000)		1.00 (0.99, 1.00)	100.00

NOTE: Weights are from random effects analysis

.656   1   1.52

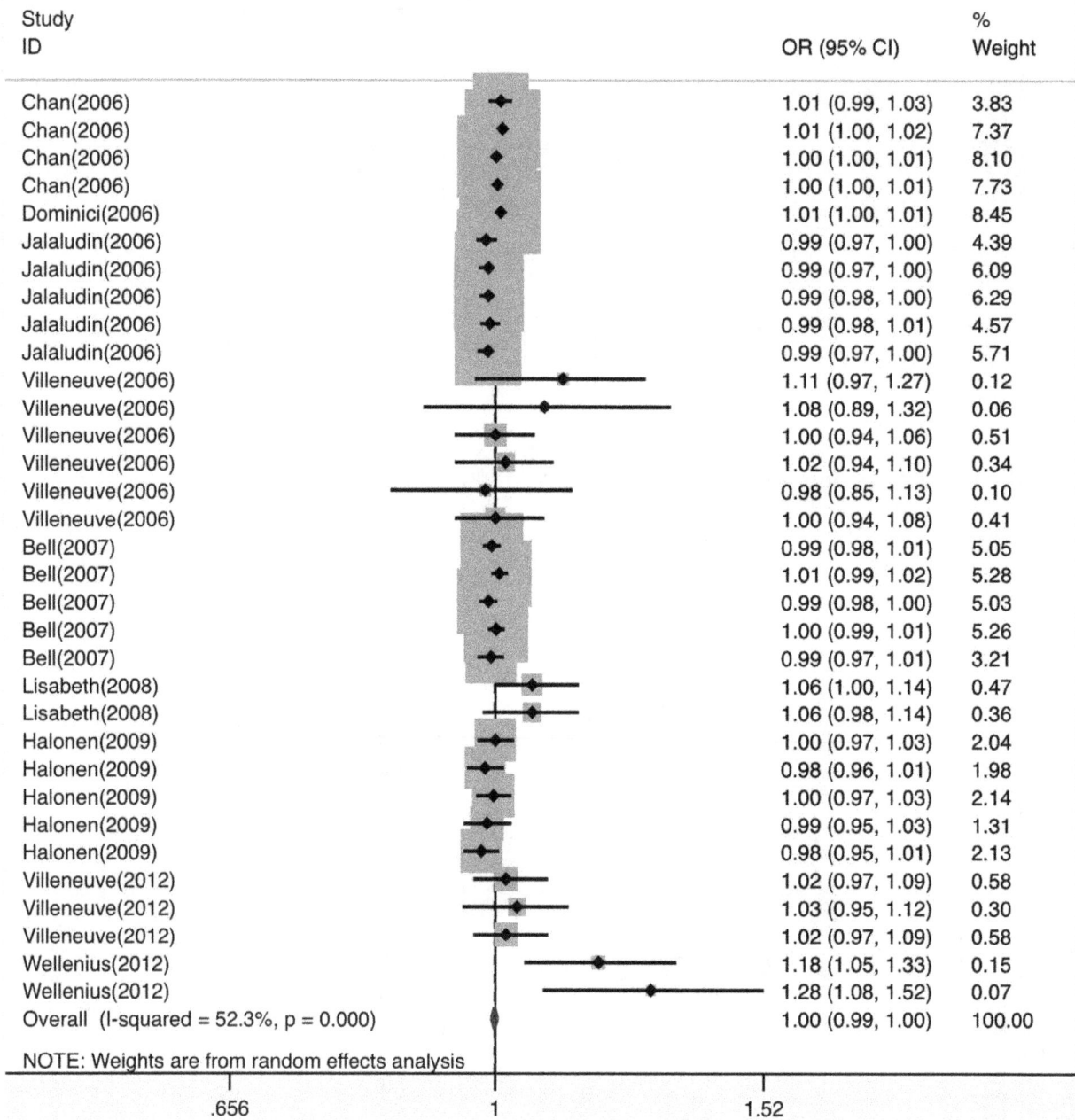

Figure 2. Forest plot of ORs for the association between PM$_{2.5}$ and stroke attack. OR indicates odds ratio; CI, confidence interval.

m^3 = 1.007; 95%CI, 0.992~1.022). Forest plot was shown in Figure 5, Funnel plot was shown in Figure 3(C). Egger's test showed that publication bias was unlikely in the meta-analysis on association between PM$_{10}$ and hemorrhagic stroke attack (t = 0.71, P = 0.487). Sensitivity analysis suggested the results for PM$_{10}$ and hemorrhagic stroke attack were statistically robust (Table S1 in File S1).

2.3 Effects of PM$_{2.5}$, PM$_{10}$ exposure to ischemic stroke attack. Heterogeneity was observed among five articles (Q = 24.00, P = 0.031). With random effect model, PM$_{2.5}$ exposure

was related to the risk of ischemic stroke attack (OR per 10 μg/m^3 = 1.025; 95%CI, 1.001~1.049). Forest plot was shown in Figure 6. Funnel plot was shown in Figure 3(D). Egger's test (t = 3.71, P = 0.003) showed publication bias existed among studies. Sensitivity analysis showed that results for PM$_{2.5}$ and ischemic stroke attack were not robust to the inclusion of Lisabeth study [22], Villeneuve study [12] and Wellenius study [13] (Table S1 in File S1).

Significant heterogeneity existed among six articles included (Q = 98.04, P = 0.000). Association was demonstrated between

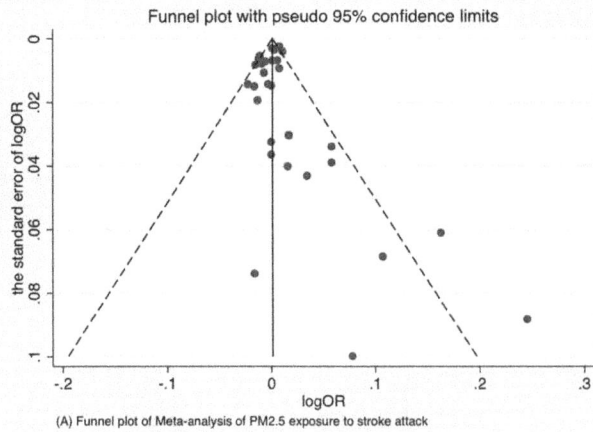

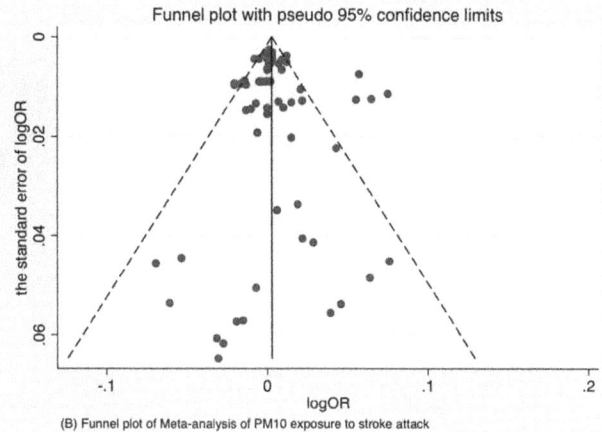

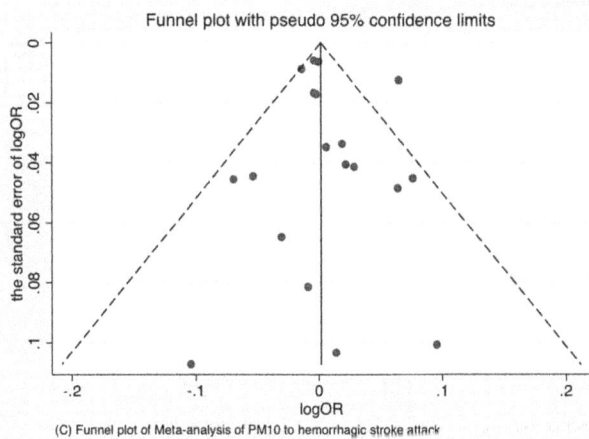

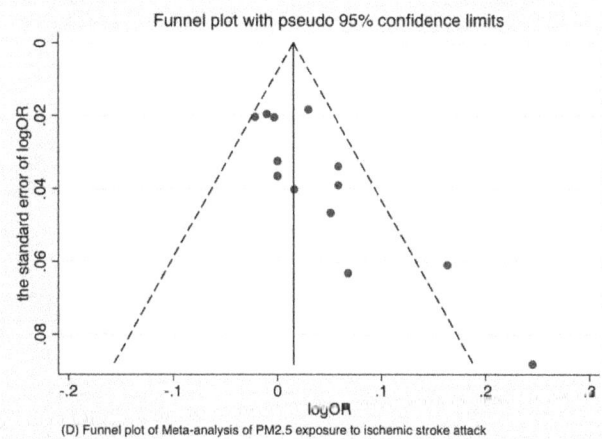

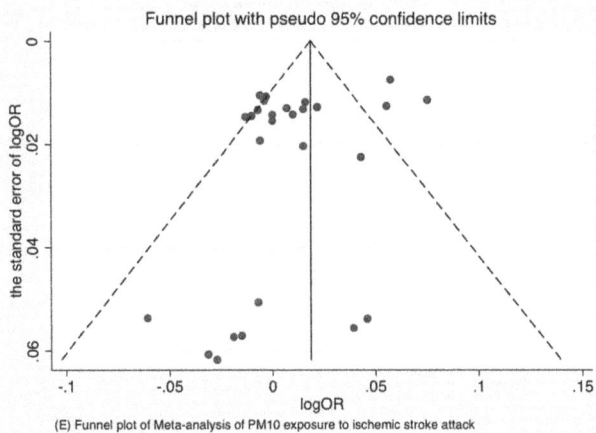

Figure 3. Funnel plots of Meta-analysis in different particular matters to different stroke types. OR indicates odds ratio. (A) Funnel plot of Meta-analysis of $PM_{2.5}$ exposure to stroke attack. (B) Funnel plot of Meta-analysis of PM_{10} exposure to stroke attack. (C) Funnel plot of Meta-analysis of PM_{10} exposure to hemorrhagic stroke attack. (D) Funnel plot of Meta-analysis of $PM_{2.5}$ exposure to ischemic stroke attack. (E) Funnel plot of Meta-analysis of PM_{10} exposure to ischemic stroke attack.

PM_{10} exposure and ischemic stroke attack (OR per 10 µg/m³ = 1.013; 95%CI, 1.001~1.025) using random effect model. Forest plot was shown in Figure 7. Funnel plot was shown in Figure 3(E). Egger's test ($t = -1.61$, $P = 0.120$) indicated no publication bias existed among studies of association between PM_{10} and ischemic stroke attack. Sensitivity analysis showed that results for PM_{10} and ischemic stroke attack were not robust to the inclusion of Andersen study [11] and Vidale study [25] (Table S1 in File S1).

Study ID	OR (95% CI)	% Weight
Wordley(1997)	1.02 (1.00, 1.04)	1.56
Linn(1999)	1.00 (1.00, 1.01)	2.97
Wong(1999)	1.00 (1.00, 1.01)	2.77
Tsai(2003)	1.06 (1.04, 1.07)	2.08
Tsai(2003)	1.07 (1.04, 1.09)	1.30
Chan(2006)	1.00 (0.99, 1.01)	2.29
Chan(2006)	1.01 (1.00, 1.02)	2.78
Chan(2006)	1.00 (0.99, 1.01)	2.68
Chan(2006)	1.01 (1.00, 1.02)	2.25
Jalaludin(2006)	0.98 (0.96, 1.00)	1.74
Jalaludin(2006)	0.98 (0.96, 1.00)	1.69
Jalaludin(2006)	0.99 (0.97, 1.01)	1.71
Jalaludin(2006)	0.98 (0.96, 1.00)	1.74
Jalaludin(2006)	0.98 (0.97, 1.00)	1.74
Villeneuve(2006)	0.99 (0.96, 1.03)	0.71
Villeneuve(2006)	0.99 (0.96, 1.02)	1.05
Villeneuve(2006)	1.02 (0.96, 1.09)	0.27
Villeneuve(2006)	1.08 (0.99, 1.18)	0.16
Villeneuve(2006)	1.01 (0.94, 1.07)	0.26
Villeneuve(2006)	1.00 (0.98, 1.03)	1.09
Bell(2007)	1.00 (0.99, 1.00)	2.68
Bell(2007)	0.99 (0.98, 1.00)	2.68
Bell(2007)	1.00 (0.99, 1.01)	2.24
Bell(2007)	1.00 (1.00, 1.01)	2.68
Bell(2007)	1.01 (1.00, 1.02)	2.64
Henrotin(2007)	0.98 (0.88, 1.10)	0.10
Henrotin(2007)	0.94 (0.85, 1.04)	0.11
Henrotin(2007)	0.99 (0.88, 1.10)	0.10
Henrotin(2007)	1.05 (0.94, 1.16)	0.11
Henrotin(2007)	1.04 (0.93, 1.16)	0.11
Henrotin(2007)	0.97 (0.86, 1.09)	0.09
Henrotin(2007)	0.97 (0.86, 1.10)	0.09
Henrotin(2007)	0.99 (0.90, 1.10)	0.13
Guo(2008)	1.00 (1.00, 1.01)	2.90
Guo(2008)	1.00 (1.00, 1.01)	2.90
Guo(2008)	1.00 (0.99, 1.01)	2.86
Guo(2008)	1.00 (1.00, 1.01)	2.90
Ye(2009)	0.99 (0.97, 1.00)	1.84
Andersen(2010)	1.02 (0.94, 1.10)	0.19
Andersen(2010)	1.03 (0.95, 1.12)	0.19
Andersen(2010)	0.95 (0.87, 1.04)	0.16
Andersen(2010)	1.07 (0.97, 1.17)	0.14
Andersen(2010)	1.02 (0.99, 1.04)	1.25
Andersen(2010)	1.02 (0.99, 1.04)	1.25
Andersen(2010)	1.01 (0.98, 1.06)	0.65
Andersen(2010)	0.93 (0.86, 1.02)	0.16
Andersen(2010)	0.97 (0.86, 1.10)	0.08
Andersen(2010)	1.01 (0.99, 1.04)	1.21
Andersen(2010)	1.01 (0.98, 1.03)	1.23
Andersen(2010)	0.99 (0.97, 1.02)	1.19
Vidale(2010)	1.08 (1.05, 1.10)	1.43
Vidale(2010)	1.06 (1.03, 1.08)	1.28
Vidale(2010)	1.00 (0.96, 1.02)	0.98
Vidale(2010)	1.01 (0.98, 1.04)	1.10
Vidale(2010)	0.99 (0.96, 1.02)	1.07
Vidale(2010)	1.04 (0.98, 1.07)	0.56
Willocks(2012)	1.00 (0.98, 1.02)	1.81
Willocks(2012)	1.00 (0.98, 1.01)	1.80
Willocks(2012)	1.00 (0.98, 1.02)	1.81
Willocks(2012)	1.00 (0.99, 1.01)	2.49
Willocks(2012)	1.00 (0.99, 1.01)	2.54
Willocks(2012)	1.00 (0.98, 1.01)	1.81
Willocks(2012)	1.01 (1.00, 1.02)	2.55
Willocks(2012)	0.98 (0.97, 1.00)	1.74
Willocks(2012)	1.00 (0.99, 1.01)	2.49
Willocks(2012)	1.00 (0.99, 1.01)	2.54
Willocks(2012)	1.01 (1.00, 1.02)	2.50
Willocks(2012)	1.00 (0.98, 1.01)	1.81
Overall (I-squared = 70.0%, p = 0.000)	1.00 (1.00, 1.01)	100.00

NOTE: Weights are from random effects analysis

.847   1   1.18

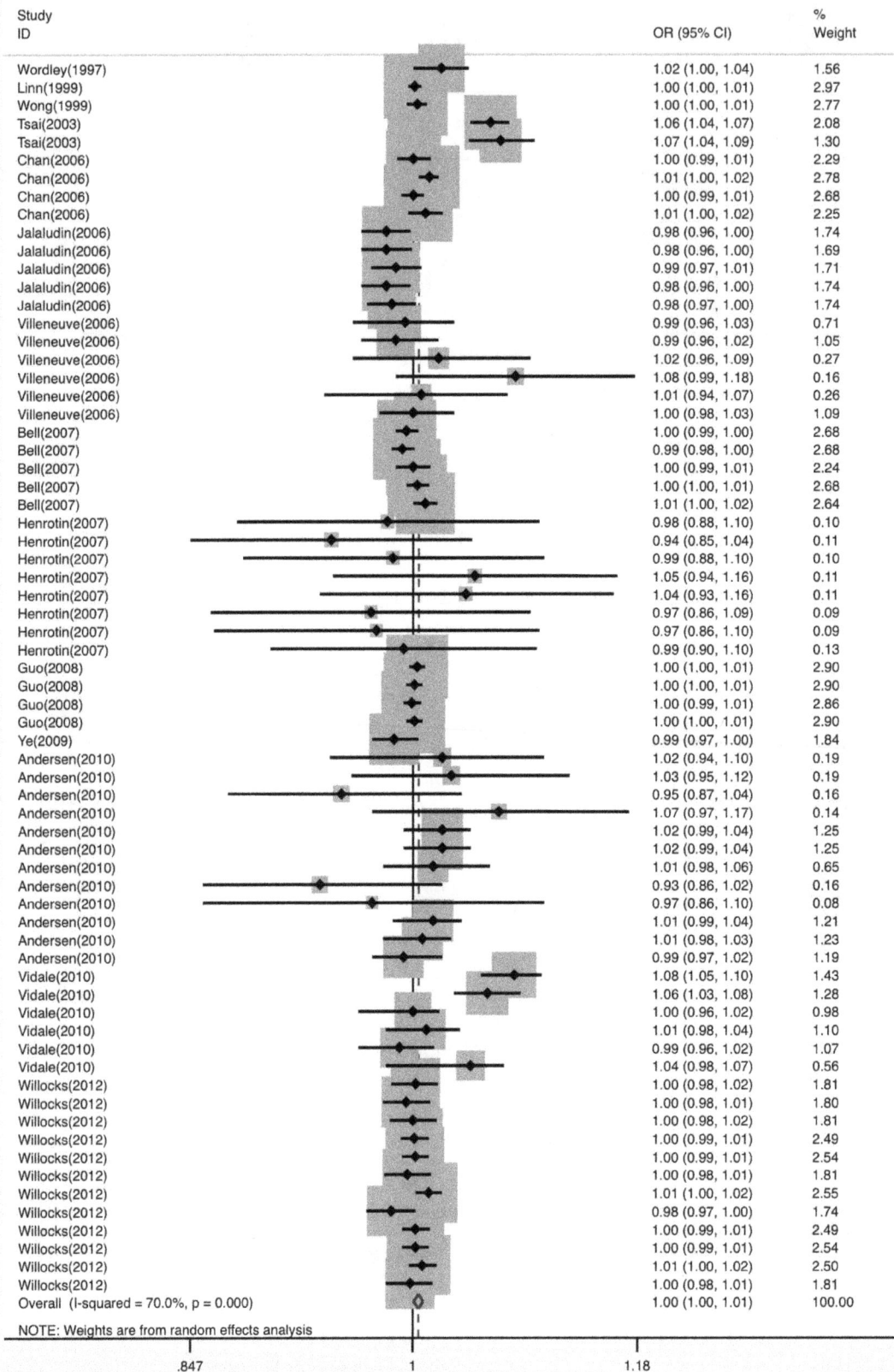

Figure 4. Forest plot of ORs for the association between PM$_{10}$ and stroke attack. OR indicates odds ratio; CI, confidence interval.

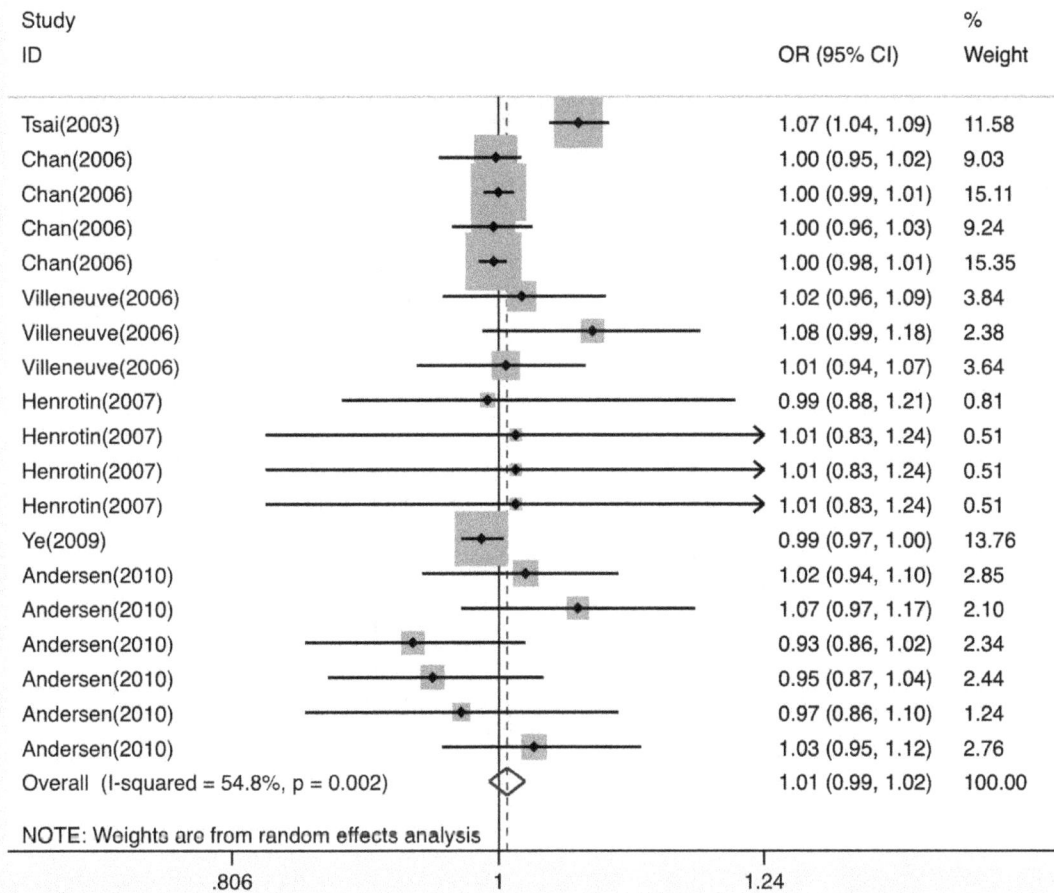

Figure 5. Forest plot of ORs for the association between PM$_{10}$ to hemorrhagic stroke attack. OR indicates odds ratio; CI, confidence interval.

3. Results of Meta-regression of different PMs exposure to different stroke types

We had detected heterogeneity among studies about PM$_{2.5}$ exposure to stroke attack, PM$_{10}$ exposure to stroke attack, PM$_{10}$ exposure to hemorrhagic stroke attack, PM$_{2.5}$ exposure to ischemic stroke and PM$_{10}$ exposure to ischemic stroke. Meta-regressions were done in these studies to detect the influence factors. (Table S2 in File S1).

When studying PM$_{2.5}$ exposure to stroke attack, we found study design (coefficient 0.032, 95% CI 0.002 to 0.062, P = 0.035) the influence factor. Among studies of PM$_{10}$ exposure to stroke attack, area (coefficient 0.007, 95% CI 0.002 to 0.012, P = 0.007) were found to be the possible influence factor. (Table S2 in File S1).

4. Subgroup analyses

Results above supported that design was the main covariate of studies about PM$_{2.5}$ exposure to stroke attack, so we divided these studies into subgroups (time-series group and case-crossover group) according to study design. Case-crossover group revealed the positive association between PM$_{2.5}$ exposure and stroke attack (OR per 10 µg/m^3 = 1.029; 95%CI, 1.003~1.055). And area was found to be the main covariate of studies about PM$_{10}$ exposure to

stroke attack, subgroups (Asia, Europe, North America and Oceania) analysis showed that in Asia and Europe, PM$_{10}$ was associated with stroke attack (Table 2).

Discussion

The pooled results of meta-analyses showed that PM$_{10}$ exposure was related to an increase in risk of stroke attack (OR per 10 µg/m^3 = 1.004, 95%CI: 1.001~1.008) and PM$_{2.5}$ exposure was not significantly associated with stroke attack (OR per 10 µg/m^3 = 0.999, 95%CI: 0.994~1.003). So it's meaningful to explore the effects of PM$_{10}$ and PM$_{2.5}$ to stroke attack, separately.

However, as we all know, ischemic stroke and hemorrhagic stroke are two different types of stroke. The mechanisms differ from each other. So the PM may affect different strokes in different ways. Thus, it is valuable to explore the association between PM and ischemic stroke or hemorrhagic stroke, respectively. This was the first Meta-analysis that focused on the association between particular matter and different stroke subtypes attack. Our research showed that PM$_{2.5}$ (OR per 10 µg/m^3 = 1.025; 95%CI, 1.001~1.049) and PM$_{10}$ (OR per 10 µg/m^3 = 1.013; 95%CI, 1.001~1.025) exposure were statistically significantly associated with an increased risk of ischemic stroke

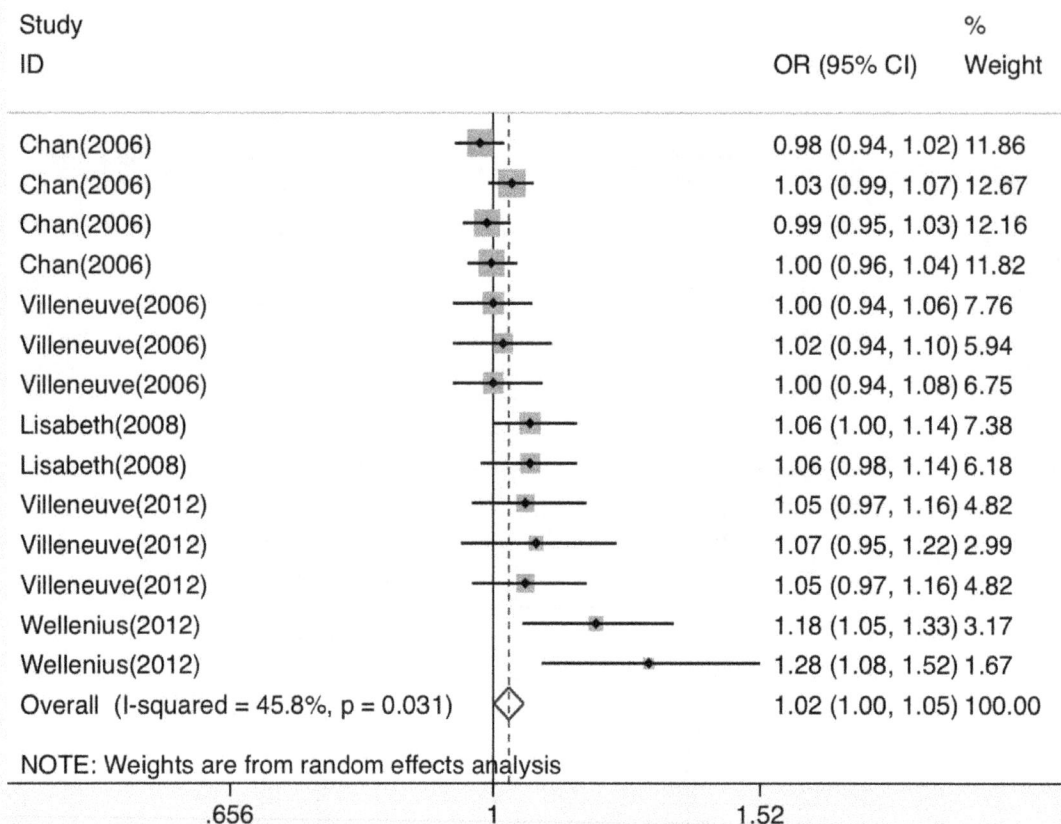

Figure 6. Forest plot of ORs for the association between PM$_{2.5}$ and ischemic stroke attack. OR indicates odds ratio; CI, confidence interval.

attack, while PM$_{2.5}$ (all the studies showed no significant association) and PM$_{10}$ (OR per 10 μg/m^3 = 1.007; 95%CI, 0.992~1.022) exposure were not associated with an increased risk of hemorrhagic stroke attack. We inferred that this phenomenon might be caused by the different formation mechanisms of different stroke types. For the biological mechanisms of PM to cerebrovascular disease are not clearly clarified at present, we can just supposed that PM's effect to ischemic stroke attack might be regulated by systemic inflammation and the activation of coagulation system, which leading to atherosclerosis, vasoconstriction, increase of fibrinogen and acceleration the formation of acute thrombus. And PM's effect to hemorrhagic stroke attack might be caused by systemic inflammation, giving rise to alterations in oxidative stress, vascular endothelial damage and rupture of plaque. According to our results, it might be easier for PM to activate coagulation system and constriction of blood vessels than simply destroy the vascular endothelial. But it was just a hypothesis that needed to be tested.

Heterogeneity in the studies we retrieved could come from inherent differences between study settings, as well as from differences in age, gender, area, lag times of air pollutants' concentrations, study design and so on. With respect to these influence factors, we used randomized effect model to minimize the heterogeneity and do Meta-regression to find the possible covariates. As a result, we only found design and area were the influence factors in the studies.

Sensitivity analysis revealed that results for PM and ischemic stroke were not robust, but results for PM and hemorrhagic were more robust than the former. Except for the studies of PM$_{2.5}$ and ischemic stroke, all the studies showed a symmetric inverted funnel shape, which indicated publication bias was unlikely. Besides, Egger's test, of which the results support funnel shape, was used to quantitatively assess the publication bias.

In this research, the following issues needed to be considered: (1) there were limitations for selecting studies. First, Meta-analysis articles [27–30] were not included in our research, because the methods authors used were different from ours, and they only provided pooled effects of PM exposure to stroke attack. Second, long-term effects articles [31–33] were not included, for there existed more covariates [e.g. body-mass index (BMI), smoking status, blood pressure, educational level, household income and historical disease] in long-term effect articles. Third, articles [34–37] proving quantitative relationship between PM and stroke attack but didn't provide calculable or reported ORs and 95% CIs were removed, too. Hence, removing these three kinds of articles might decrease bias but could lose some evidence of the association between PM and stroke attack. (2) Measurement of air pollution exposure varies within and between studies. The air pollution monitor apparatus itself had measurement error and was different from study to study. Besides, it was certainly known that personal pollution exposure levels were very different from those measured at a nearby fixed monitoring station [38]. Thus

Study ID		OR (95% CI)	% Weight
Tsai(2003)		1.06 (1.04, 1.07)	5.56
Chan(2006)		1.00 (0.97, 1.02)	4.97
Chan(2006)		1.00 (0.98, 1.02)	5.10
Chan(2006)		0.99 (0.97, 1.01)	5.13
Chan(2006)		1.02 (0.99, 1.04)	4.93
Villeneuve(2006)		0.99 (0.96, 1.03)	3.75
Villeneuve(2006)		1.00 (0.98, 1.03)	4.54
Villeneuve(2006)		0.99 (0.96, 1.02)	4.47
Henrotin(2007)		0.99 (0.90, 1.10)	1.15
Henrotin(2007)		0.94 (0.85, 1.04)	1.04
Henrotin(2007)		0.97 (0.86, 1.10)	0.82
Henrotin(2007)		1.05 (0.94, 1.16)	1.04
Henrotin(2007)		0.98 (0.88, 1.10)	0.93
Henrotin(2007)		1.04 (0.93, 1.16)	0.98
Henrotin(2007)		0.97 (0.86, 1.09)	0.85
Henrotin(2007)		0.99 (0.88, 1.10)	0.94
Andersen(2010)		1.01 (0.98, 1.06)	3.60
Andersen(2010)		1.02 (0.99, 1.04)	4.78
Andersen(2010)		1.01 (0.98, 1.03)	4.74
Andersen(2010)		0.99 (0.97, 1.02)	4.69
Andersen(2010)		1.02 (0.99, 1.04)	4.78
Andersen(2010)		1.01 (0.99, 1.04)	4.72
Vidale(2010)		1.00 (0.96, 1.02)	4.34
Vidale(2010)		1.04 (0.98, 1.07)	3.30
Vidale(2010)		1.01 (0.98, 1.04)	4.55
Vidale(2010)		0.99 (0.96, 1.02)	4.50
Vidale(2010)		1.08 (1.05, 1.10)	5.00
Vidale(2010)		1.06 (1.03, 1.08)	4.81
Overall (I-squared = 72.5%, p = 0.000)		1.01 (1.00, 1.03)	100.00

NOTE: Weights are from random effects analysis

.847   1   1.18

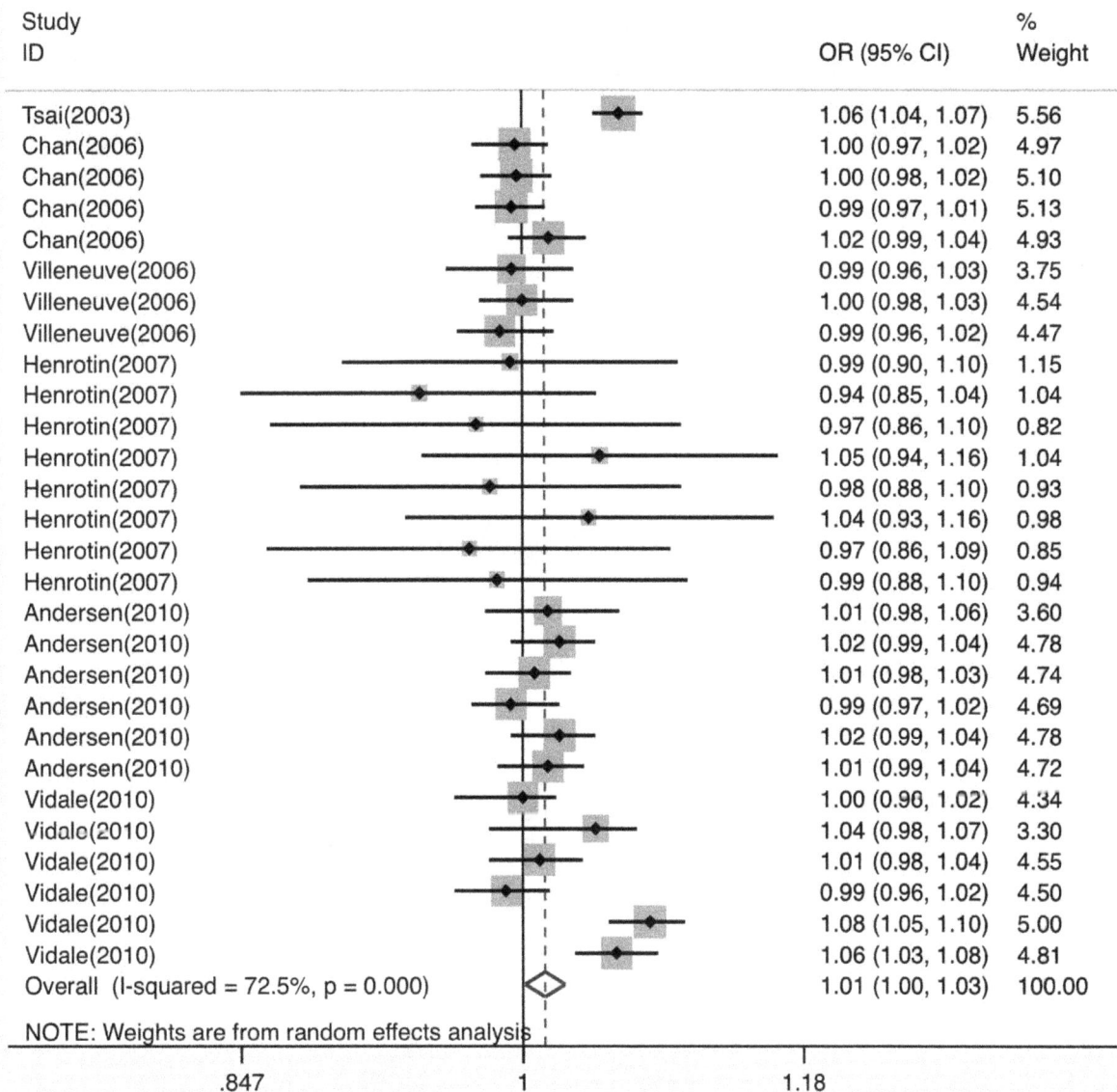

Figure 7. Forest plot of ORs for the association between PM_{10} and ischemic stroke attack. OR indicates odds ratio; CI, confidence interval.

standardized measurement should be established. (3) In our research, meteorological condition was not included as a covariate, because only some of retrieved studies provided mean or median temperature data. Besides, the association between PM_{10} and stroke attack was different from season to season [16]. Hence, it's not suitable to regard the mean or median temperature data as covariate in our research. Nevertheless, meteorological condition was an important covariate between studies. What's more, age is another important covariate to be considered, but current studies didn't provide sufficient data, they just provide the age range of study objects rather than the mean age. Future researches should provide more accurate data of age, temperature and pay more attention to the influence of age and temperature to stroke attack. (4) We only detect study design and area as important influence

factors among PM and stroke by Meta-regression and subgroup analysis, however, there might exist other covariates that we didn't focus on, such as age, historical disease and meteorological condition. Besides, adding more studies may bring us more valuable covariates. (5) Lots of studies showed that different compositions of PM could cause diverse health outcomes. However, till 2012, there were no such articles, except Halonen's [23] study, which could provide detail data concentrating on the short-term effects of some composition of PM on stroke attack. In some studies [10–13,15,18–20], authors just mentioned traffic exhaust emission was the main origin of PM without providing the accurate data about the association between some composition of PM and stroke attack. While in other studies [7,8,16,21,22,24–26], authors even didn't mention the effect of composition of PM to

Table 2. Meta-analysis of subgroup studies.

	Covariate	Level	Number of articles	Number of studies	Model	Pooled OR (95%CI)
The effect of $PM_{2.5}$ to stroke attack	study design	time-series study	6	22	random effect model	0.998 (0.993~1.002)
		case-crossover study	3	11	fixed effect model	1.029 (1.003~1.055)
The effect of PM_{10} to stroke attack	area	Asia	6	17	random effect model	1.006 (1.000~1.011)
		Europe	5	39	random effect model	1.008 (1.002~1.014)
		North America	2	7	fixed effect model	1.001 (0.996~1.006)
		Oceania	1	5	fixed effect model	0.982 (0.974~0.990)

$PM_{2.5}$ indicates particular matter with aerodynamic diameter ≤2.5 μm; PM_{10} indicates particular matter with aerodynamic diameter ≤10 μm; OR, odds ratio; CI, confidence interval.

stroke attack. As a result, we could only focus on the whole PM's effect on stroke attack. We urged future studies should pay more attention to the different composition of PM on health effect.

Conclusions

This Meta-regression and Meta-analysis raised significant issues that might help guide the future research in this area. $PM_{2.5}$ and PM_{10} exposure had different effect on different stroke subtypes. So it's worthwhile to study the effects of PM to ischemic stroke and hemorrhagic stroke, respectively. Standardizing of exposure measurement, bringing more covariates, exploring mechanisms and adding more studies in respective subtype are needed.

Supporting Information

File S1 Table S1, sensitivity analysis of Meta-analysis in different particular matter to different stroke types. PM10 indicates particular matter with aerodynamic diameter ≤10 μm; PM2.5, particular matter with aerodynamic diameter ≤2.5 μm; OR, odds ratio; CI, confidence interval. Table S2, meta-regression of

different particular matters exposure to different stroke types. PM2.5 indicates particular matter with aerodynamic diameter ≤ 2.5 μm; PM10, particular matter with aerodynamic diameter ≤ 10 μm; Coef., regression coefficient; Std. Err., standard error of logOR; CI, confidence interval. *Sex was not contained as a covariate for these studies were all about the whole population.

Acknowledgments

I would like to show my deepest gratitude to Dr. Li Xiuyang and Dr. Chen Gao, who have provided me with valuable guidance in every stage of doing this research. I'd like to thank my partner and all my friends for their encouragement and support.

Author Contributions

Conceived and designed the experiments: XL. Analyzed the data: XY JS. Wrote the paper: XY JS. Collected data: XL JS GC.

References

1. Liao D, Creason J, Shy C, Williams R, Watts R, et al. (1999) Daily variation of particulate air pollution and poor cardiac autonomic control in the elderly. Environ Health Perspect 107:521–525. PubMed: 1566669.
2. Ghio AJ, Kim C, Devlin RB (2000) Concentrated ambient air particles induce mild pulmonary inflammation in healthy human volunteers. Am J Respir Crit Care Med 162:981–988. doi: 10.1164/ajrccm.162.3.9911115. PubMed: 10988117.
3. Kelly FJ (2003) Oxidative stress: its role in air pollution and adverse health effects. OCCUPATIONAL AND ENVIRONMENTAL MEDICINE 60:612–616. doi:10.1136/oem.60.8.612. PubMed: 1740593.
4. Gurgueira SA, Lawrence J, Coull B, Murthy GG, Gonzalez-Flecha B (2002) Rapid increases in the steady-state concentration of reactive oxygen species in the lungs and heart after particulate air pollution inhalation. Environ Health Perspect 110:749–755. PubMed: 1240944.
5. Peters A, Doring A, Wichmann HE, Koenig W (1997) Increased plasma viscosity during an air pollution episode: a link to mortality? Lancet 349:1582–1587. doi: 10.1016/S0140-6736(97)01211-7. PubMed: 9174559.
6. Brook RD, Franklin B, Cascio W, Hong Y, Howard G, et al. (2004) Air pollution and cardiovascular disease: a statement for healthcare professionals from the expert panel on population and prevention science of the American heart association. Circulation 109:2655–2671. doi: 10.1161/01.CIR.0000128587.30041.C8. PubMed: 15173049.
7. Wordley J, Walters S, Ayres JG (1997) Short term variations in hospital admissions and mortality and particulate air pollution. Occup Environ Med 54:108–116. PubMed: 1128660.
8. Tsai SS, Goggins WB, Chiu HF, Yang CY (2003) Evidence for an association between air pollution and daily stroke admissions in kaohsiung, taiwan. Stroke 34:2612–2616. doi: 10.1161/01.STR.0000095564.33543.64. PubMed: 14551399.

9. Chan CC, Chuang KJ, Chien LC, Chen WJ, Chang WT (2006) Urban air pollution and emergency admissions for cerebrovascular diseases in taipei, taiwan. Eur Heart J 27:1238–1244. doi: 10.1093/eurheartj/ehi835. PubMed: 16537554.
10. Henrotin JB, Besancenot JP, Bejot Y, Giroud M (2007) Short-term effects of ozone air pollution on ischaemic stroke occurrence: a case-crossover analysis from a 10-year population-based study in dijon, france. Occup Environ Med 64:439–445. doi: 10.1136/oem.2006.029306. PubMed: 17409181.
11. Andersen ZJ, Olsen TS, Andersen KK, Loft S, Ketzel M, et al. (2010) Association between short-term exposure to ultrafine particles and hospital admissions for stroke in copenhagen, denmark. Eur Heart J 31:2034–2040. doi: 10.1093/eurheartj/ehq188. PubMed: 20538735.
12. Villeneuve PJ, Johnson JY, Pasichnyk D, Lowes J, Kirkland S, et al. (2012) Short-term effects of ambient air pollution on stroke: who is most vulnerable? Sci Total Environ 430:193–201. doi: 10.1016/j.scitotenv.2012.05.002. PubMed: 22647242.
13. Wellenius GA, Burger MR, Coull BA, Schwartz J, Suh HH, et al. (2012) Ambient air pollution and the risk of acute ischemic stroke. Arch Intern Med 172:229–234. doi: 10.1001/archinternmed.2011.732. PubMed: 3639313.
14. Li XY, Yu XB, Liang WW, Yu N, Wang L, et al. (2012) Meta-analysis of association between particulate matter and stroke attack. CNS Neurosci Ther 18:501–508. doi: 10.1111/j.1755-5949.2012.00325.x. PubMed: 22672304.
15. Wong TW, Lau TS, Yu TS, Neller A, Wong SL, et al. (1999) Air pollution and hospital admissions for respiratory and cardiovascular diseases in hong kong. OCCUPATIONAL AND ENVIRONMENTAL MEDICINE 56:679–683. PubMed: 1757671.
16. Linn WS, Szlachcic Y, Gong H, Kinney PL, Berhane KT (2000) Air pollution and daily hospital admissions in metropolitan los angeles. ENVIRONMENTAL

HEALTH PERSPECTIVES 108:427–434. doi: 10.2307/3454383. PubMed: 1638060.

17. Dominici F, Peng RD, Bell ML, Pham L, McDermott A, et al. (2006) Fine particulate air pollution and hospital admission for cardiovascular and respiratory diseases. JAMA 295:1127–1134. doi: 10.1001/jama.295.10.1127. PubMed: 3543154.

18. Jalaludin B, Morgan G, Lincoln D, Sheppeard V, Simpson R, et al. (2006) Associations between ambient air pollution and daily emergency department attendances for cardiovascular disease in the elderly (65+ years), sydney, australia. J Expo Sci Environ Epidemiol 16:225–237. doi: 10.1038/sj.jea.7500451. PubMed: 16118657.

19. Villeneuve PJ, Chen L, Stieb D, Rowe BH (2006) Associations between outdoor air pollution and emergency department visits for stroke in edmonton, canada. Eur J Epidemiol 21:689–700. doi: 10.1007/s10654-006-9050-9. PubMed: 17048082.

20. Bell ML, Levy JK, Lin Z (2008) The effect of sandstorms and air pollution on cause-specific hospital admissions in taipei, taiwan. Occup Environ Med 65:104–111. doi: 10.1136/oem.2006.031500. PubMed: 17626134.

21. Guo YM, Liu LQ, Chen JM, Yang MJ, Wich M, et al. (2008) Association between the concentration of particulate matters and the hospital emergency room visits for circulatory diseases: a case-crossover study. Zhonghua Liu Xing Bing Xue Za Zhi 29:1064–1068. PubMed: 19173924.

22. Lisabeth LD, Escobar JD, Dvonch JT, Sanchez BN, Majersik JJ, et al. (2008) Ambient air pollution and risk for ischemic stroke and transient ischemic attack. Ann Neurol 64:53–59. doi: 10.1002/ana.21403. PubMed: 2788298.

23. Halonen JI, Lanki T, Yli-Tuomi T, Tiittanen P, Kulmala M, et al. (2009) Particulate air pollution and acute cardiorespiratory hospital admissions and mortality among the elderly. Epidemiology 20:143–153. doi: 10.1097/EDE.0b013e31818c7237. PubMed: 19234403.

24. Ye Y, Li XY, Chen K, Liu QM, Xiang HQ (2009) A case-crossover study on the relationship between air pollution and acute onset of cerebral hemorrhage in hangzhou city. Zhonghua Liu Xing Bing Xue Za Zhi 30:816–819. PubMed: 20193205.

25. Vidale S, Bonanomi A, Guidotti M, Arnaboldi M, Sterzi R (2010) Air pollution positively correlates with daily stroke admission and in hospital mortality: a study in the urban area of como, italy. Neurol Sci 31:179–182. doi: 10.1007/s10072-009-0206-8. PubMed: 20119741.

26. Willocks LJ, Bhaskar A, Ramsay CN, Lee D, Brewster DH, et al. (2012) Cardiovascular disease and air pollution in scotland: no association or insufficient data and study design? BMC Public Health 12:227. doi: 10.1186/1471-2458-12-227. PubMed: 3476376.

27. Wellenius GA, Schwartz J, Mittleman MA (2005) Air pollution and hospital admissions for ischemic and hemorrhagic stroke among medicare beneficiaries. Stroke 36:2549–2553. doi: 10.1161/01.STR.0000189687.78760.47. PubMed: 16254223.

28. O'Donnell MJ, Fang J, Mittleman MA, Kapral MK, Wellenius GA (2011) Fine particulate air pollution (pm2.5) and the risk of acute ischemic stroke. Epidemiology 22:422–431. doi: 10.1097/EDE.0b013e3182126580. PubMed: 3102528.

29. Larrieu S, Jusot JF, Blanchard M, Prouvost H, Declercq C, et al. (2007) Short term effects of air pollution on hospitalizations for cardiovascular diseases in eight french cities: the psas program. Sci Total Environ 387:105–112. doi: 10.1016/j.scitotenv.2007.07.025. PubMed: 17727917.

30. Le Tertre A, Medina S, Samoli E, Forsberg B, Michelozzi P, et al. (2002) Short-term effects of particulate air pollution on cardiovascular diseases in eight european cities. J Epidemiol Community Health 56:773–779. PubMed: 1732027.

31. Miller KA, Siscovick DS, Sheppard L, Shepherd K, Sullivan JH, et al. (2007) Long-term exposure to air pollution and incidence of cardiovascular events in women. N Engl J Med 356:447–458. doi: 10.1056/NEJMoa054409. PubMed: 17267905.

32. Atkinson RW, Carey IM, Kent AJ, van Staa TP, Anderson HR, et al. (2013) Long-term exposure to outdoor air pollution and incidence of cardiovascular diseases. Epidemiology 24:44–53. doi: 10.1097/EDE.0b013e318276ccb8. PubMed: 23222514.

33. Maheswaran R, Pearson T, Smeeton NC, Beevers SD, Campbell MJ, et al. (2012) Outdoor air pollution and incidence of ischemic and hemorrhagic stroke: a small-area level ecological study. Stroke 43:22–27. doi: 10.1161/STROKEAHA.110.610238. PubMed: 22033998.

34. Corea F, Silvestrelli G, Baccarelli A, Giua A, Previdi P, et al. (2012) Airborne pollutants and lacunar stroke: a case cross-over analysis on stroke unit admissions. Neurol Int 4:e11. doi: 10.4081/ni.2012.e11. PubMed: 23139849.

35. Johnson JY, Rowe BH, Villeneuve PJ (2010) Ecological analysis of long-term exposure to ambient air pollution and the incidence of stroke in edmonton, alberta, canada. Stroke 41:1319–1325. doi: 10.1161/STROKEAHA.110.580571. PubMed: 20538697.

36. Oudin A, Stromberg U, Jakobsson K, Stroh E, Bjork J (2010) Estimation of short-term effects of air pollution on stroke hospital admissions in southern sweden. NEUROEPIDEMIOLOGY 34:131–142. doi: 10.1159/000274807. PubMed: 20068360.

37. Maheswaran R, Haining RP, Brindley P, Law J, Pearson T, et al. (2005) Outdoor air pollution and stroke in sheffield, united kingdom: a small-area level geographical study. Stroke 36:239–243. doi: 10.1161/01.STR.0000151363.71221.12. PubMed: 15604422.

38. Lebowitz MD (1995) Exposure assessment needs in studies of acute health effects. Sci Total Environ 168:109 117. PubMed: 7481728.

13

Short-Term Effect of Temperature on Daily Emergency Visits for Acute Myocardial Infarction with Threshold Temperatures

Suji Lee[1,4], **Eunil Lee**[1,2,4]*, **Man Sik Park**[3]*, **Bo Yeon Kwon**[1,4], **Hana Kim**[3], **Dea Ho Jung**[4], **Kyung Hee Jo**[4], **Myung Ho Jeong**[5], **Seung-Woon Rha**[6]

1 Department of Public Health, Graduate School, Korea University, Seoul, South Korea, 2 Department of Preventive Medicine, College of medicine, Korea University, Seoul, Korea, 3 Department of Statistics, Sungshin Women's University College of Natural sciences, Seoul, Korea, 4 Graduate School of Public Health, Graduate School, Korea University, Seoul, Korea, 5 Department of Cardiology, Chonnam National University Hospital, Gwangju, Korea, 6 Cardiovascular Center, Korea University Guro Hospital, Seoul, Korea

Abstract

Background: The relationship between temperature and myocardial infarction has not been fully explained. In this study, we identified the threshold temperature and examined the relationship between temperature and emergency admissions due to MI in Korea.

Methods: Poisson generalized additive model analyses were used to assess the short-term effects of temperature (mean, maximum, minimum, diurnal) on MI emergency visits, after controlling for meteorological variable and air pollution (PM10, NO$_2$). We defined the threshold temperature when the inflection point showed a statistically significant difference in the regression coefficients of the generalized additive models (GAMs) analysis. The analysis was performed on the following subgroups: geographical region, gender, age (<75 years or ≥75 years), and MI status (STEMI or non-STEMI).

Results: The threshold temperatures during heat exposure were for the maximum temperature as 25.5–31.5°C and for the mean temperature as 27.5–28.5°C. The threshold temperatures during cold exposure were for the minimum temperature as −2.5–1.5°C. Relative risks (RRs) of emergency visits above hot temperature thresholds ranged from 1.02 to 1.30 and those below cold temperature thresholds ranged from 1.01 to 1.05. We also observed increased RRs ranged from 1.02 to 1.65 of emergency visits when temperatures changes on a single day or on successive days.

Conclusions: We found a relationship between temperature and MI occurrence during both heat and cold exposure at the threshold temperature. Diurnal temperature or temperature change on successive days also increased MI risk.

Editor: Qinghua Sun, The Ohio State University, United States of America

Funding: This research was supported by a fund (2013E2100200) by Research of Korea Centers for Disease Control and Prevention. This work was funded by the Korea Meteorological Administration Research and Development Program under Grant CATER 2012-3110. The funders had no role in study design, data collection and analysis, decision to publish, or preparation of the manuscript.

Competing Interests: The authors have declared that no competing interests exist.

* E-mail: eunil@korea.ac.kr (EL); mansikpark@sungshin.ac.kr (MSP)

Introduction

Myocardial infarction (MI) is a major social and health issue, because acute MI remains a leading cause of morbidity and mortality worldwide [1]. A number of studies showed that cold temperature is associated with the increased occurrence of MI due to an increase in plasma viscosity and serum cholesterol levels, blood pressure, sympathetic nervous activities, and platelet aggregation [2–4]. Heat exposure is also reported to be associated with such physiological changes as increases in heart rate, blood viscosity, and coagulability [5], which could be risk factors for MI. However, only a few studies supporting heat exposure to MI have been published [6,7].

According to the Intergovernmental Panel on Climate Change, climate conditions have become more variable with more extreme heat episodes, unpredictable weather, including sudden cold, hot,

wet, or dry spells, and extreme weather events, including floods and droughts [8]. With climate change and a rapidly growing elderly population throughout the world, MI mortality from extreme heat and cold weather events is a significant public burden that may worsen in the future. In additions, an individual's susceptibility may be exacerbated by underlying chronic medical conditions and drug treatments that affect the body's capacity to adapt to temperature changes [9]. Therefore, temperature-associated episodes of MI may increase with aggravated climate conditions, especially in older people, those with underlying cardiovascular diseases, and those who are poor, uneducated, or isolated [10–12].

From a public health perspective, the identification of population subgroups vulnerable to heat and cold is important for effective prevention, and clinicians also should be aware that exposure to environmental heat and cold is a risk factor for MI and should consider this for risk prevention and management [13].

Public concerns about temperature-associated diseases began from thousands of heat-related deaths in Europe in 2003; one of the most important ways to prevent heat stroke is to establish a public warning system based on a threshold temperature above which heat stroke may increase rapidly [14]. A number of studies have examined the influence of meteorological factors and seasonal variations on MI morbidity and mortality with lag effect, However, no warning systems for temperature-related MI have been reported. Dilaveris et al. show that a minimum rate of MI occurred at a temperature of 23.3°C, with the rate of MI increasing both above and below this temperature [15]. Rossi et al. reported that high temperature(above 27°C) is also associated MI mortality on the same day [16]. The Myocardial Ischaemia National Audit Project (MINAP) registry study showed linearity only between cold temperature and MI without threshold temperatures [17]. And the threshold temperatures for diurnal temperature change(DTR) or successive daily temperature changes (SDTC) are also needed because these temperature variations are reported as the important risk factors for MI [14,18,19].

Therefore, we evaluated the effects of hot, cold, and DTR and SDTC on the number of emergency visits for MI with threshold temperatures according to geographical area, age, sex, and severity of MI by using the Korea Working Group of Myocardial Infarction (KorMI) data. KorMI was established in November 2005 as a Korean prospective multi-center on-line registry for investigating the risk factors of mortality in acute MI patients, and registry data are based on nationwide hospital emergency visits with the support of the Korean Circulation Society [20]. We also estimated risk ratios by a 1°C change above or below the thresholds using generalized additive models (GAMs). We demonstrated that threshold temperatures were different according to geographical locations with modifying seasonal effect. In addition, we also found that DTR in the spring, autumn, and winter, and SDTC in the spring increased MI risk.

Methods

Ethics Statement

This study was performed with the support of the Korean Circulation Society (KCS) to commemorate the 50th Anniversary of the KCS. The authors of this manuscript have certified that the information contained herein is true and correct as reflected in the records of the KUGH Medical Device Institutional Review Board.

Study Area

South Korea is located in the southern part of the Korean peninsula, including all its islands, lying between latitudes 33° and 39°N, and longitudes 124° and 130°E. Its total area is 100,032 square kilometers (38,622.57 square miles) [21]. Approximately 50 million residents lived in Korea in 2011. South Korea tends to have both a humid continental and subtropical climate, and is affected by the East Asian monsoon [22]. South Korea has four distinct seasons: spring, summer, autumn, and winter. Winter temperatures were higher along the southern coast and considerably lower in the mountainous interior. Summer can be uncomfortably hot and humid, with temperatures exceeding 30°C (86°F) in most parts of the country with heavy rainfall [23]. The weather of South Korea differs between the central and southern parts of the Korean peninsula, where the southern part is warmer than the central part.

Data

The KorMI registry covers a total of 62 general hospitals located in 16 major cities in Korea. We excluded data from the hospital on the Jeju island, which is located about 100 kilometers (60 miles) off the southern coast of the Korean peninsula, because the weather conditions on the island are quite different from those on the peninsula. We also excluded data from two hospitals in Gangwon Province, because the number of patients was not large enough to represent the area. We analyzed patient data from January 1, 2006 to December 31, 2010. The average number of MI patients who visited the emergency room was 4,564 per year. We defined the first medical contract time as the occurrence time of MI and used KorMI data for the statistical analysis, such as history of hypertension and diabetes, age, gender, and MI status, including ST-segment elevation myocardial infarction (STEMI) and non-STEMI.

We obtained weather data from the Korean Meteorological Administration. Data included daily mean, minimum, and maximum temperatures; DTR; SDTC; daily precipitation; humidity; dew point; sea level pressure; and wind speed for the study time period. Air pollution data, including ambient 24-h average concentrations of PM10, NO_2, SO_2, ozone, and CO, were provided by the National Institute of Environmental Research, Korea.

Statistical Analysis

MI-temperature Plotting. To calculate the average daily adjusted emergency visits (DAEVs) for MI according to daily

Table 1. General Characteristics of the Study Subjects.

	Central region	Southern region	All	P-value
Total MI	12,586	14,802	27,388	
Age, mean±SD	63.1±13.2	63.9±12.7	63.5±12.6	
Male, % (n)	72.1.7 (8,377)	69.3 (9,069)	70.5 (20,694)	<0.001
Female, % (n)	27.8 (3,239)	30.7 (4,022)	29.5 (8,643)	
STEMI, % (n)	60.1 (6,914)	55.0 (7,094)	57.0 (16,285)	<0.001
Non-STEMI, % (n)	40.2 (6,914)	45.0 (5,802)	42.2 (12,242)	
History of Hypertension % (n)	51.5 (5,934)	47.7 (6,218)	50.4 (14,407)	<0.001
History of Diabetes mellitus % (n)	27.6 (3,180)	27.2 (3,547)	27.8 (7,958)	0.2217

MI: Myocardial infarction.
STEMI: ST elevation myocardial infarction.
Non-STEMI: Non-ST elevation myocardial infarction.

Table 2. Summary Statistics for Temperature and other Meteorological Variables with the Level of Air pollutants in Study Areas.

Parameter	Central region				Southern region				Combined regions
	Mean (SD)	Range	Median	IQR	Mean (SD)	Range	Median	IQR	Mean (SD)
Temperature									
Mean (°C)	12.72 (10.18)	−13.2–30.7	14.1	17.8	14.32 (9.13)	−8.0–31.5	15.3	15.5	13.32 (9.83)
Minimum (°C)	8.74 (10.43)	−19.5–27.1	9.4	18.3	10.22 (9.57)	−15.3–28.2	10.6	16.7	9.3 (10.14)
Maximum (°C)	17.31 (10.32)	−10.7–36.2	19.2	17.9	19.3 (9.14)	−4.3–37.7	20.6	15.1	18.06 (9.94)
Diurnal temperature (°C)	8.57 (3.16)	1.0–24.0	8.4	4.1	9.09 (3.63)	0.8–25.7	8.7	4.8	8.77 (3.35)
Other meteorological variables									
Precipitation (mm)	4.03 (15.30)	0.0–272.5	0.0	0.5	3.5 (13.25)	0.0–310	0.0	0.5	3.83 (14.55)
Precipitation (log mm)	−0.004 (1.58)	−3.0–5.61	0.0	0.0	0.14 (1.39)	−3.0–5.74	0.0	0.0	0.05 (1.51)
Relative humidity (%)	63.72 (14.76)	19.9–99.0	64.6	21.7	63.68 (16.73)	16.1–99.0	65.5	24.4	63.7 (15.53)
Sea-level pressure (hPa)	1015.98 (8.04)	993.8–1038.2	1016.1	12.5	1015.89 (7.51)	993.8–1038.4	1016	11.4	1015.95 (7.84)
Level of air pollutants									
PM10 (µg/m³)	55.26 (37.76)	0.0–1153.25	48.09	36.21	52.29 (34.15)	0.0–974.0	45.95	33.0	54.14 (36.47)
NO$_2$ (ppm)	0.03 (0.01)	0.0–0.14	0.03	0.02	0.02 (0.01)	0.0–0.12	0.02	0.01	0.03 (0.01)
CO (ppm)	0.63 (0.32)	0.0–4.7	0.56	0.35	0.52 (0.30)	0.0–3.37	0.46	0.36	0.59 (0.31)
SO$_2$ (ppb)	5.59 (3.09)	0.0–33.87	4.96	3.74	5.81 (3.91)	0.0–45.78	4.71	4.08	5.67 (3.42)
O$_3$ (ppm)	0.02 (0.07)	0.0–0.05	0.02	0.02	0.02 (0.01)	0.0–0.09	0.02	0.02	0.02 (0.06)

SD: Standard deviation.
IQR: Interquartile range.

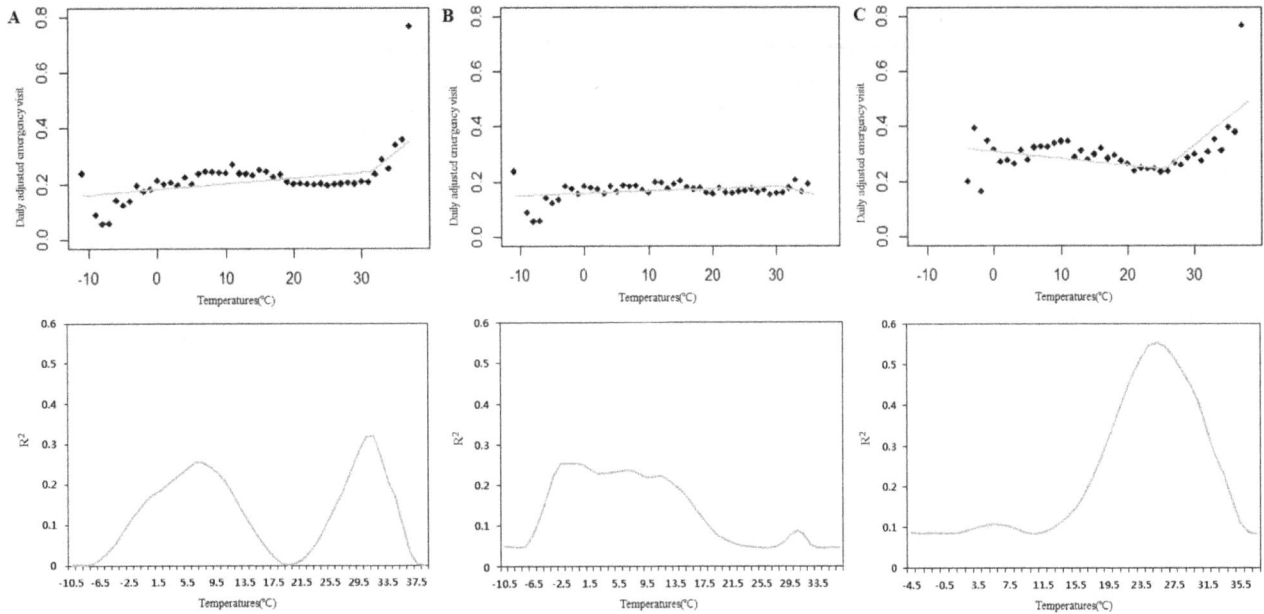

Figure 1. Daily adjusted emergency visit (DAEV) rate for MI according to maximum temperature by regions: A. Combined regions, B. Central region, C. Southern region. Lower figures showed change of R^2 values in each temperature by piecewise analysis and maximum R^2 value was chosen as the inflection point. The maximum R^2 value of the central region was 30.5°C; however, it did not show a threshold effect.

temperature or temperature change, we divided the total number of daily emergency visits that occurred on all days with a specific temperature (value was calculated for each 1°C temperature along the range from -X°C to X°C), the numerator, by the total number of days that temperature occurred over the X-day study period, the denominator. The DAEVs were plotted for each 1°C range, and piecewise regression (PR) analysis was applied to find the inflection points of the relationship between DAEVs and temperature with maximum R^2 values. The best fit was judged on the basis of the residual sum of squares and the value of the R^2 statistic by PR analysis.

Poisson GAM with Threshold Effect. We estimated the temperature- daily emergency visits relationship using generalized additive models (GAMs) with nonparametric smoothing functions (splines) to describe nonlinear relations [24]. Temperature variables were the mean, minimum, maximum temperature, DTR, SDTC. Moreover, to observe the independent effects of temperature on emergency visits for MI, we controlled for potential confounders, such as humidity, sea level pressure, and air pollutants (PM10, NO_2). To estimate the lag effect of the temperature- daily emergency visits relationship, temperature variables of the previous seven days were applied to the GAM using single-day lags from lag 0 (current day) to lag 7 (7 days before the event day).

We defined the threshold temperature when the inflection point showed a statistically significant difference in the regression coefficients of the GAM analysis between the temperature ranges above the inflection point and below the point. Each temperature range was treated as dummy variables. The relationship between the daily emergency visits for MI and temperature was also analyzed by subgroups in the following categories: geographical region, gender, age (under and over 75 years), and MI status (either STEMI or non-STEMI). Seasonal effects were analyzed especially for the relationships between daily emergency visits and DTR, or DAEV and SDTC.

Results

The total number of emergency visits for MI was 27,388 during the 5-year study period, and the general characteristics of subjects in the central region differed from those in the southern region (Table 1). The number of male MI patients was greater than female patients in both regions; however, the proportion of female patients was larger in the southern region (30.7%) than in the central region (27.8%). And the weather in the southern region was warmer than that of the central region (Table 2, File S1).

DAEVs for MI were plotted according to the daily maximum temperature for the central and southern regions combined (Figure 1A–1C). PR analysis revealed that the peak values indicated a significant change point. In the plot of combined regions, 31.5°C was identified for the inflection point for sudden increases in MI emergency visits. The southern region showed a prominent inflection point at 25.5°C with threshold effect. Meanwhile, the inflection point of the central region was 30.5°C, however, it did not show a threshold effect (Table 3). DAEVs according to the daily mean and minimum temperatures were plotted to find threshold temperatures (data not shown).

The threshold temperatures during heat exposure were for the maximum temperature as 25.5–31.5°C and for the mean temperature as 27.5–28.5°C. The threshold temperatures during cold exposure were −2.5–1.5°C.

A significant increased risk for MI with heat exposure was found above the threshold temperature both at the maximum and mean temperatures (Table 3). The RR of MI per a 1°C change in the maximum temperature was lowest in the southern region (RR = 1.02), and the RR in the old age group was greatest (RR = 1.12). Most subgroups showed a 4lag-day effects; however, male and non-STEMI subjects showed immediate daily temperature effects on MI with 0 lag-day effects. The RRs from mean temperatures were relatively higher than those from maximum temperatures. The RR of MI per a 1°C change in the mean temperature was lowest in the southern region (RR = 1.12), and

Table 3. Relative Risk of Myocardial Infarction per 1°C Change in Temperature above Threshold temperature by Subgroup.

Temperature(°C)	Threshold (°C)[f]	Lag (days)	RR (95% CI)
Effect of heat[a]			
Maximum[c]			
All	31.5*	4	1.07 (1.05–1.10)
Central	_[g]	-	-
Southern	25.5**	4	1.02 (1.00–1.03)
≥75 years	31.5*	4	1.12 (1.06–1.18)
≤75 years	31.5*	4	1.06 (1.03–1.09)
Male	30.5**	0	1.08 (1.04–1.12)
Female	-	-	-
STEMI	30.5**	4	1.06 (1.01–1.11)
Non-STEMI	31.5*	0	1.10 (1.02–1.19)
Mean[d]			
All	28.5*	0	1.26 (1.08–1.46)
Central	-	-	-
Southern	27.5*	0	1.12 (1.01–1.24)
≥75years	-	-	-
≤75years	28.5*	0	1.24 (1.04–1.47)
Male	28.5*	0	1.28 (1.07–1.53)
Female	-	-	-
STEMI	28.5*	0	1.23 (1.00–1.50)
Non-STEMI	28.5*	0	1.30 (1.03–1.64)
Effect of cold[b]			
Minimum[e]			
All	−1.5**	5	1.01 (1.00–1.02)
Central	−1.5**	1	1.04 (0.99–1.11)
Southern	−2.5*	5	1.05 (1.02–1.08)
≥75years	−1.5**	1	1.03 (1.00–1.06)
≤75 years	1.5**	5	1.01 (1.00–1.02)
Male	−1.5**	5	1.01 (1.00–1.03)
Female	−1.5**	5	1.02 (1.00–1.05)
STEMI	−1.5**	1	1.01 (0.99–1.02)
Non- STEMI	−1.5**	5	1.02 (1.00–1.04)

Model adjusted for precipitation, humidity, sea level pressure, and air pollutants (PM10, NO_2) using a spline function.
RR = Relative risk.
STEMI: ST elevation myocardial infarction.
Non-STEMI: Non-ST elevation myocardial infarction.
Spring: March–May, Summer: June–August, Autumn: September–November, Winter: December–February.
*$P < 0.05$;
** $P < 0.001$.
[a]For heat exposure, temperature increase of 1°C above threshold.
[b]For cold exposure, temperature decrease of 1°C below threshold.
[c]Maximum temperature.
[d]Mean temperature.
[e]Minimum temperature.
[f]Threshold temperature.
[g]No threshold effect was identified.

the RR in the non-STEMI group was highest (RR = 1.30). Lag effects were shown on the current day. Increased risk of MI visits was also found below the threshold temperature from the minimum temperature. The RRs associated with the minimum temperature ranged from 1.01 to 1.05. In contrast to exposure at higher temperatures, the RR of MI per a 1°C change in the minimum threshold temperature was greatest in the southern region (RR = 1.05).

Both the range of temperatures on a single day or on successive days showed increased MI risk (Tables 4 and 5). The DTR above 7.5 or 8.5°C in the spring and autumn showed threshold effects for increased MI visits. Non-STEMI patients in the spring showed an increased risk for MI above 6.5°C of the DTR. The threshold

Table 4. Relative Risk of Myocardial Infarction per 1°C Change in Diurnal Temperature Range (DTR) above the Threshold Temperature in All Regions by Season.

DTR (°C)	Spring			Autumn			Winter		
	Threshold(°C)[a]	Lag (days)	RR (95% CI)	Threshold (°C)[a]	Lag (days)	RR (95% CI)	Threshold (°C)[a]	Lag (days)	RR (95% CI)
All	7.5**	1	1.03 (1.02–1.04)	-	-	-	6.5**	1	1.02 (1.01–1.03)
Central region	8.5**	1	1.03 (1.01–1.05)	7.5**	2	1.02 (1.01–1.04)	6.5**	1	1.02 (1.00–1.04)
Southern region	-[b]	-	-	-	-	-	-	-	-
Male	7.5**	1	1.03 (1.02–1.04)	-	-	-	6.5**	7	1.02 (1.01–1.03)
Female	7.5**	1	1.04 (1.03–1.06)	8.5**	2	1.04 (1.01–1.06)	6.5**	1	1.03 (1.01–1.05)
≥75 years	7.5**	1	1.03 (1.01–1.05)	7.5**	2	1.03 (1.01–1.06)	6.5**	4	1.02 (1.00–1.05)
≤75 years	7.5**	1	1.03 (1.02–1.05)	-	-	-	5.5**	1	1.02 (1.01–1.04)
STEMI	8.5**	1	1.03 (1.01–1.04)	7.5**	2	1.02 (1.01–1.04)	7.5**	1	1.03 (1.01–1.04)
Non-STEMI	6.5**	1	1.03 (1.02–1.05)	-	-	-	4.5**	1	1.02 (1.00–1.03)

Model adjusted for precipitation, humidity, sea level pressure, and air pollutants ($PM10$, NO_2) using a spline function.

RR = Relative risk.

STEMI: ST elevation myocardial infarction.

Non-STEMI: Non-ST elevation myocardial infarction.

Spring: March–May, Summer: June–August, Autumn: September–November, Winter: December–February.

* $P < 0.05$;

** $P < 0.001$.

[a]Threshold temperature.

[b]No threshold effect was identified.

Table 5. Relative Risk of Myocardial Infarction per 1°C Change in Successive Daily Temperature Changes by Subgroup.

Successive daily temperature changes (°C)	Spring			
	Threshold (°C)[a]	RR (95% CI)	Threshold (°C)[b]	RR (95% CI)
All	4.5**	1.20 (1.01–1.43)*	−4.5**	1.10 (1.02–1.18)*
Central region	4.5**	1.65 (1.01–2.70)*	−7.0**	1.30 (1.04–1.62)*
South region	-[c]	-	−6.5**	1.49 (1.23–1.81)*
Male	4.5**	1.26 (1.03–1.55)*	−4.5**	1.09 (1.00–1.20)
Female	-	-	−4.5**	1.12 (0.98–1.28)
≥75 years	-	-	-	-
≤75 years	4.5**	1.31 (1.08–1.60)*	−4.5**	1.09 (1.00–1.18)*
STEMI	4.5**	1.30 (1.05–1.60)*	−4.5**	1.15 (1.04–1.26)*
Non-STEMI	-	-	-	-

Model adjusted for precipitation, humidity, sea level pressure, and air pollutants (PM10, NO_2) using a spline function.
RR = Relative risk.
STEMI: ST elevation myocardial infarction.
Non-STEMI: Non-ST elevation myocardial infarction.
Spring: March–May, Summer: June–August, Autumn: September–November, Winter: December–February.
*$P < 0.05$;
** $P < 0.001$.
[a]Temperature rise between consecutive days.
[b]Temperature fall between consecutive days.
[c]No threshold effect was identified.

temperature of DTR in the winter was 4.5 to 6.5°C, which was lower than in the spring or autumn. Lag-day effects for DTR were 1 or 2 days; however, delayed lag effects were evident in males at 7 days and in the old age group at 4 days in winter (Table 4).

Increases and decreases in successive daily mean temperatures showed significant effects only in the spring (Table 5). Significant increases in MI risk with a 4.5°C increase in temperature was evident in several subgroups, including the central region, males, the young age group, and STEMI patients. A decrease in temperature over 4.5°C showed significant effects among all subgroups except for old age and non-STEMI patients.

Discussion

Many reports showed various threshold temperatures using different estimation methods; however, they did not estimate the actual threshold temperatures that would show an increase in the risk of MI. Most reports used a Poisson regression model with a natural spline function to estimate the relationship between temperature and risk of MI [18,25,26]. But the spline function in GAM analysis cannot provide an exact inflection point. Therefore, PR analysis after plotting temperature and DEAV would be a better approach for identifying the threshold temperature. We calculated the threshold temperature using regression coefficients from dummy variables (both below and above the inflection point) of the GAM analysis after finding the inflection point from the PR analysis. We estimated threshold temperatures in all subgroups. Several subgroups showed no threshold temperature, but most subgroups showed similar threshold temperatures for maximum temperature (25.5 to 31.5°C) and mean temperature (27.5 to 28.5°C). In the southern region, the threshold temperature was relatively lower than that for most other groups (25.5°C), and the rate of MI increased rapidly over 32°C(Figure 1C). These findings suggest that patients in the southern region are vulnerable to higher temperatures and temperature changes above the threshold temperature.

Many reports showed various threshold temperatures each using different estimation methods, or showed no threshold temperature. The Myocardial Ischemia National Audit Project registry study in England and Wales showed linearity between cold temperature and MI, but not with threshold temperature [17]. Dilaveris et al. [15] showed that the rate of MI events increased smoothly both above and below the minimum event rate at 23.3°C based on the U-shape of the MI mortality and temperature association, which showed a similar pattern to our results in the southern region. Rossi et al. [16] reported an increase in MI mortality above 27°C compared to MI mortality when 14°C was the reference temperature. Gasparrini et al. [27] used the 93rd percentile of year-round maximum temperature as the threshold temperature, including 20.9 to 24.7°C, based on the region.

The highest RRs by mean temperature were found for males, the young age group, and non-STEMI patients who may participate in many outdoor activities when compared with other subgroups. These findings suggest that outdoor daily activity is strongly related to the effects of temperature on MI risk. Consistent with our findings, Na et al. [28] found that heat-related illnesses largely influence the age group from 20 to 64 years in Korea. Goggins et al. and Morabito et al. [29,30] also found a higher RR in males than in females in Italia City and Taiwan, respectively. However, Bhaskaran et al. [31] reported that only STEMI patients showed a significant RR above 20°C. This inconsistency between our results and Bhaskaran's report may be explained by the difference in average temperatures in the summer and the outdoor activity of the study population.

Many studies reported a cold effect on MI when temperatures were above the freezing temperature. Wang et al. [32] reported that daily MI events occurred more frequently below 10°C compared with above 20°C in Hiroshima, Japan, and MI admission increased in Hong Kong per 1°C drop below a mean temperature of 24°C [29]. MI mortality increased in the USA under a maximum temperature of 17°C [33]. However, several studies showed that a threshold temperature of −1.5°C or −2.5°C for MI risk were similar to our study just below 0°C [34,35]. The differences in cold threshold temperature were related with the diverse weather in each country.

Additionally, the RR of the cold effect ranged from 1% (95% CI: 0.2%–2.4%) to 5% (95% CI: 2.0%–7.9%) below the threshold temperature in our study, which is larger than that seen in other studies [17]. A multi-country study using World Health Organization data found increases in age-standardized MI rates between 0.1% and 2.3% per a 1°C drop across 24 locations [36]. This report showed general cold effects without using threshold temperature.

We found increased risks of MI with DTR in spring, autumn, and winter, but not in summer. DTR changes over 6.5–8.5°C increased MI risk in the spring and those over 4.5–7.5°C increased MI risk in the winter, suggesting an increased risk of MI due to relatively small changes in DTR in the winter, especially for non-STEMI patients. These findings may explain the higher risk of MI in the winter, similar to the results of other studies [37,38]. Ebi et al. [14] reported that changes in the daily maximum and minimum temperatures resulted in increased hospitalizations of elderly people for MI by 6–13% in several American cities.

We also found that the difference in mean temperature between successive days either rise or fall, related to increased MI risk. Messner et al. [18] reported similarly that increases in temperature between consecutive days are associated with increases in MI hospitalization. We also found an increased risk of MI only in the spring. This risk was especially apparent for STEMI patients and in the younger age group (<75 years). No other reports found an increased risk of MI in the spring because of SDTC.

Our study showed a U-shaped association between temperature and MI risk, including both hot and cold effects, which is consistent with several previously published studies [15,25,33]. In addition, we found that diurnal temperature or temperature change on successive days also increased MI risk. We could not adjust for other confounding factors, including indoor temperature, outdoor daily activity, smoking and other behavioral factors, socioeconomic status, air conditioning of the house or working site, and effects of other pre-existing diseases. Despite these limitations, our study provides useful information about actual threshold temperatures with regard to RR of MI and could be used to establish a warning system for MI in hot and cold temperatures.

In conclusion, climate change, including extreme weather or increases in average temperatures, may increase the risk of MI in susceptible populations. Our findings provide useful information for identifying the risk of MI in vulnerable groups for establishing climate change adaptation strategies.

Supporting Information

File S1 Summary of Monthly Average Temperature and Number of Emergency visits for Myocardial Infarction in the Central and Southern regions: Table S1. Summary Statistics of Central region for Temperature and other Meteorological Variables with the Level of Air pollutants by Season: Table S2. Summary Statistics of Southern region for Temperature and other Meteorological Variables with the Level of Air pollutants by Season: Table S3.

Author Contributions

Conceived and designed the experiments: SL EL. Performed the experiments: SL. Analyzed the data: SL HK. Contributed reagents/materials/analysis tools: SL MSP BYK DHJ KHJ. Wrote the paper: SL EL. Clinically reviewed and collected data : MHJ SWR.

References

1. White HD, Chew DP (2008) Acute myocardial infarction. The Lancet 372: 570–584.
2. Keatinge WR, Coleshaw SR, Cotter F, Mattock M, Murphy M, et al. (1984) Increases in platelet and red cell counts, blood viscosity, and arterial pressure during mild surface cooling: factors in mortality from coronary and cerebral thrombosis in winter. Br Med J (Clin Res Ed) 289: 1405–1408.
3. Ockene IS, Chiriboga DE, Stanek EJ 3rd, Harmatz MG, Nicolosi R, et al. (2004) Seasonal variation in serum cholesterol levels: treatment implications and possible mechanisms. Arch Intern Med 164: 863–870.
4. Wilkinson P, Pattenden S, Armstrong B, Fletcher A, Kovats RS, et al. (2004) Vulnerability to winter mortality in elderly people in Britain: population based study. BMJ 329: 647.
5. Keatinge WR, Coleshaw SR, Easton JC, Cotter F, Mattock MB, et al. (1986) Increased platelet and red cell counts, blood viscosity, and plasma cholesterol levels during heat stress, and mortality from coronary and cerebral thrombosis. Am J Med 81: 795–800.
6. Bhaskaran K, Hajat S, Haines A, Herrett E, Wilkinson P, et al. (2009) Effects of ambient temperature on the incidence of myocardial infarction. Heart 95: 1760–1769.
7. Turner LR, Barnett AG, Connell D, Tong S (2012) Ambient temperature and cardiorespiratory morbidity: a systematic review and meta-analysis. Epidemiology 23: 594–606.
8. McMichael T, Montgomery H, Costello A (2012) Health risks, present and future, from global climate change. BMJ 344: e1359.
9. Bouchama A, Knochel JP (2002) Heat stroke. N Engl J Med 346: 1978–1988.
10. Costello A, Abbas M, Allen A, Ball S, Bell S, et al. (2009) Managing the health effects of climate change: Lancet and University College London Institute for Global Health Commission. Lancet 373: 1693–1733.
11. Hajat S, Armstrong B, Baccini M, Biggeri A, Bisanti L, et al. (2006) Impact of high temperatures on mortality: is there an added heat wave effect? Epidemiology 17: 632–638.
12. McMichael AJ, Lindgren E (2011) Climate change: present and future risks to health, and necessary responses. J Intern Med 270: 401–413.
13. Hajat S, Kovats RS, Atkinson RW, Haines A (2002) Impact of hot temperatures on death in London: a time series approach. J Epidemiol Community Health 56: 367–372.
14. Ebi KL, Exuzides KA, Lau E, Kelsh M, Barnston A (2004) Weather changes associated with hospitalizations for cardiovascular diseases and stroke in California, 1983–1998. Int J Biometeorol 49: 48–58.
15. Dilaveris P, Synetos A, Giannopoulos G, Gialafos E, Pantazis A, et al. (2006) CLimate Impacts on Myocardial infarction deaths in the Athens TErritory: the CLIMATE study. Heart 92: 1747–1751.
16. Rossi G, Vigotti MA, Zanobetti A, Repetto F, Gianelle V, et al. (1999) Air pollution and cause-specific mortality in Milan, Italy, 1980–1989. Arch Environ Health 54: 158–164.
17. Bhaskaran K, Hajat S, Haines A, Herrett E, Wilkinson P, et al. (2010) Short term effects of temperature on risk of myocardial infarction in England and Wales: time series regression analysis of the Myocardial Ischaemia National Audit Project (MINAP) registry. BMJ 341: c3823.
18. Messner T, Lundberg V, Wikström B (2002) A temperature rise is associated with an increase in the number of acute myocardial infarctions in the subarctic area. International Journal of Circumpolar Health 61: 201–207.
19. Tam WW, Wong TW, Chair SY, Wong AH (2009) Diurnal temperature range and daily cardiovascular mortalities among the elderly in Hong Kong. Arch Environ Occup Health 64: 202–206.
20. Lee SR, Jeong MH, Ahn YK, Chae SC, Hur SH, et al. (2008) Clinical safety of drug-eluting stents in the Korea acute myocardial infarction registry. Circ J 72: 392–398.
21. AsianInfo Korea's Geography. Available: http://www.asianinfo.org/asianinfo/korea/geography.htm#TERRITORY. Accessed 2010 July 13.
22. Climate-Zone Average temperature, rainfall and snowfall information in South Korea. Available: http://www.climate-zone.com/climate/south-korea/. Accessed 2006 March March 27.
23. Wikipedia South Korea definition. Available: http://en.wikipedia.org/wiki/South_Korea. Accessed 2013 July 1.
24. Hastie T, Tibshirani R (1995) Generalized additive models for medical research. Stat Methods Med Res 4: 187–196.
25. Sharovsky R, César LA, Ramires JA (2004) Temperature, air pollution, and mortality from myocardial infarction in Sao Paulo, Brazil. Braz J Med Biol Res 37: 1651–1657.
26. Danet S, Richard F, Montaye M, Beauchant S, Lemaire B, et al. (1999) Unhealthy Effects of Atmospheric Temperature and Pressure on the Occurrence of Myocardial Infarction and Coronary Deaths : A 10-Year Survey: The Lille-World Health Organization MONICA Project (Monitoring Trends and Determinants in Cardiovascular Disease). Circulation 100: e1–e7.
27. Gasparrini A, Armstrong B, Kovats S, Wilkinson P (2012) The effect of high temperatures on cause-specific mortality in England and Wales. Occup Environ Med 69: 56–61.

28. Na W, Jang JY, Lee KE, Kim H, Jun B, et al. (2013) The effects of temperature on heat-related illness according to the characteristics of patients during the summer of 2012 in the Republic of Korea. J Prev Med Public Health 46: 19–27.

29. Goggins WB, Chan EY, Yang CY (2012) Weather, pollution, and acute myocardial infarction in Hong Kong and Taiwan. Int J Cardiol 168: 243–249.

30. Morabito M, Modesti PA, Cecchi L, Crisci A, Orlandini S, et al. (2005) Relationships between weather and myocardial infarction: A biometeorological approach. International Journal of Cardiology 105: 288–293.

31. Bhaskaran K, Armstrong B, Hajat S, Haines A, Wilkinson P, et al. (2012) Heat and risk of myocardial infarction: hourly level case-crossover analysis of MINAP database. BMJ 345: e8050.

32. Wang H, Matsumura M, Kakehashi M, Eboshida A (2006) Effects of atmospheric temperature and pressure on the occurrence of acute myocardial infarction in Hiroshima City, Japan. Hiroshima J Med Sci 55: 45–51.

33. Medina-Ramón M, Schwartz J (2007) Temperature, temperature extremes, and mortality: a study of acclimatisation and effect modification in 50 US cities. Occup Environ Med 64: 827–833.

34. Gerber Y, Jacobsen SJ, Killian JM, Weston SA, Roger VL (2006) Seasonality and daily weather conditions in relation to myocardial infarction and sudden cardiac death in Olmsted County, Minnesota, 1979 to 2002. J Am Coll Cardiol 48: 287–292.

35. Ohlson CG, Bodin L, Bryngelsson IL, Helsing M, Malmberg L (1991) Winter weather conditions and myocardial infarctions. Scand J Soc Med 19: 20–25.

36. Barnett AG, Dobson AJ, McElduff P, Salomaa V, Kuulasmaa K, et al. (2005) Cold periods and coronary events: an analysis of populations worldwide. J Epidemiol Community Health 59: 551–557.

37. Ornato JP, Peberdy MA, Chandra NC, Bush DE (1996) Seasonal Pattern of Acute Myocardial Infarction in the National Registry of Myocardial Infarction. Journal of the American College of Cardiology 28: 1684–1688.

38. Spencer FA, Goldberg RJ, Becker RC, Gore JM (1998) Seasonal distribution of acute myocardial infarction in the second National Registry of Myocardial Infarction. J Am Coll Cardiol 31: 1226–1233.

14

Molecular and Neurodevelopmental Benefits to Children of Closure of a Coal Burning Power Plant in China

Deliang Tang[1]*, Joan Lee[1], Loren Muirhead[1], Ting Yu Li[2], Lirong Qu[1], Jie Yu[1], Frederica Perera[1]

1 Department of Environmental Health Sciences, Columbia Center for Children's Environmental Health, Mailman School of Public Health, Columbia University, New York, New York, United States of America, **2** Department of Pediatrics, Chongqing Medical University, Chongqing, China

Abstract

Polycyclic aromatic hydrocarbons (PAH) are major toxic air pollutants released during incomplete combustion of coal. PAH emissions are especially problematic in China because of their reliance on coal-powered energy. The prenatal period is a window of susceptibility to neurotoxicants. To determine the health benefits of reducing air pollution related to coal-burning, we compared molecular biomarkers of exposure and preclinical effects in umbilical cord blood to neurodevelopmental outcomes from two successive birth cohorts enrolled before and after a highly polluting, coal-fired power plant in Tongliang County, China had ceased operation. Women and their newborns in the two successive cohorts were enrolled at the time of delivery. We measured PAH-DNA adducts, a biomarker of PAH-exposure and DNA damage, and brain-derived neurotrophic factor (BDNF), a protein involved in neuronal growth, in umbilical cord blood. At age two, children were tested using the Gesell Developmental Schedules (GDS). The two cohorts were compared with respect to levels of both biomarkers in cord blood as well as developmental quotient (DQ) scores across 5 domains. Lower levels of PAH-DNA adducts, higher concentrations of the mature BDNF protein (mBDNF) and higher DQ scores were seen in the 2005 cohort enrolled after closure of the power plant. In the two cohorts combined, PAH-DNA adducts were inversely associated with mBDNF as well as scores for motor (p = 0.05), adaptive (p = 0.022), and average (p = 0.014) DQ. BDNF levels were positively associated with motor (p = 0.018), social (p = 0.001), and average (p = 0.017) DQ scores. The findings indicate that the closure of a coal-burning plant resulted in the reduction of PAH-DNA adducts in newborns and increased mBDNF levels that in turn, were positively associated with neurocognitive development. They provide further evidence of the direct benefits to children's health as a result of the coal plant shut down, supporting clean energy and environmental policies in China and elsewhere.

Editor: Silvana Allodi, Federal University of Rio de Janeiro, Brazil

Funding: This study was supported by the Energy Foundation, the Rockefeller Brothers Fund and the Schmidt Foundation. The funders had no role in study design, data collection and analysis, decision to publish, or preparation of the manuscript.

Competing Interests: The authors have declared that no competing interests exist.

* E-mail: dt14@columbia.edu

Introduction

Polycyclic aromatic hydrocarbons (PAH) are a group of compounds formed from the incomplete combustion of coal and other organic material. PAH are ubiquitous in the environment, present in outdoor and indoor air from coal combustion, diesel and other motor vehicle emissions, tobacco smoking and cooking of food [1,2]. In China, rapid economic expansion coupled with escalating fossil fuel-based energy production has resulted in massive amounts of air pollution. Specifically, coal-fired power plants produce over 70 percent of China's electricity, and new power plants being designed are also run on coal, which contributes to unsafe levels of toxic contaminants, including PAH [3,4,5].

Inhalation of airborne PAH leads to formation of DNA adducts, which are considered a valid marker of the biologically effective dose of PAH, reflecting individual variation in metabolism of PAH and DNA repair [6]. PAH are carcinogenic and neurotoxic. Previous studies by the Columbia Center for Children's Environmental Health (CCCEH) in New York City, Krakow and the present Tongliang cohort have shown that the developing fetus is more susceptible than the adult to PAH-DNA adduct formation.

They have also shown that prenatal PAH exposure is associated with adverse impacts on child neurodevelopment, including developmental delay, reduced IQ, behavioral problems in childhood [7,8,9,10,11,12] and reduced intelligence scores later in childhood [10,12].

In Tongliang County, the high pollutant concentrations in ambient air prompted the government to shut down the local power plant in May of 2004 to significantly improve community health [4,13]. This action, announced in advance, provided a unique opportunity to compare air monitoring, biomarker and health outcome data in two successive cohorts of children with and without prenatal exposure to emissions from the coal-fired power plant. In partnership with Chongqing Children's Hospital, the CCCEH carried out two prospective cohort studies between 2002 and 2005, pre-plant shutdown and post-plant shutdown, respectively. We previously reported evidence from the 2002 cohort that prenatal exposure to PAH adversely affected the neurodevelopment of children on the GDS at age 2 [11] and that the later cohort had significantly lower levels of PAH-DNA adducts in cord blood [9]. Additionally, the significant associations observed in the 2002 cohort between PAH-DNA adducts and adverse outcomes on the GDS at age 2 were no longer seen in the 2005 cohort [9].

Here we have analyzed levels of BDNF and their relationship to both adducts and developmental outcomes in the two cohorts. BDNF is critical for neurological survival and cognitive development of the central nervous system (CNS) [14,15] and is the most widely distributed neurotrophin [16]. BDNF is first released as the proBDNF precursor before being cleaved into mBDNF [14]. Regulation of BDNF depends on its site of release: postsynaptic release is regulated by a Ca^{2+} influx through ionotropic glutamate receptors and voltage gated Ca^{2+} channels [17]; release from presynaptic sites is also dependent on mobilizing Ca^{2+} influx from intracellular stores [17,18]. Following release, BDNF binds to two different transmembrane proteins: tropomyosin- related TrkB receptor and neurotrophin receptor p75, with higher affinity for the TrkB receptor [17]. Specifically, the binding of TrkB receptor triggers three signaling cascades that ultimately phosphorylate and activate the cAMP responsive element binding protein (CREB) transcription factor to encourage genes transcription essential for neuronal development and synaptic plasticity [17,19].

Because BDNF and TrkB are expressed in the hippocampus to reinforce and stabilize synaptic connections, BDNF has been widely recognized as a key regulator in long-term potentiation (LTP), one of several phenomena underlying synaptic plasticity, learning and memory [16,20]. LTP has been divided into early LTP and late LTP, where early LTP (characterized as short term potentiation) is shorter lasting and depends on the modification and translocation of proteins [21], and late LTP lasts several days and depends on de novo protein synthesis [20]. BDNF is an active mediator of neuronal processes in both the developing and mature brain, promoting differentiation, growth, and survival of neurons during development [22]. For late LTP, studies have shown that higher endogenous levels of BDNF are required to activate multiple signaling cascades that may act concertedly to regulate downstream cellular effects for memory formation and maintenance [23]. Therefore, in the mature brain, memory formation and development are associated with increased levels of BDNF and TrkB activation [24].

The goal of this paper is to determine whether mBDNF can provide a potential risk marker in assessing the neurodevelopment effects of exposure to PAH during fetal development, and in documenting the benefits of a regulatory intervention. Because previous studies have shown adverse effects of lower BDNF levels on learning and memory processes [24], we hypothesized that increased levels of PAH-DNA adducts would be associated with decreased levels of mBDNF that, in turn, would mediate the effects on neurodevelopmental outcomes determined by the DQ.

Methods

The overall approach used in this investigation and the comparison of the two cohorts with respect to adducts and DQ scores have been presented elsewhere [9].

Ethics Statement

This study was approved by the Institutional Review Board of Columbia University. All subjects gave informed written consent by completing a form approved by both the Columbia University IRB and by Chongqing Medical University.

Population

The city of Tongliang has a population of approximately 810,000 and is situated in a small basin approximately 3 km in diameter [3]. A coal-fired power plant located south of the town center operated during the dry season from 1 December to 31 May each year prior to 2004 in order to compensate for insufficient hydroelectric power during that time period. This plant was the principal source of local air pollution because in 1995 nearly all domestic heating and cooking units were converted to natural gas, motor vehicles were not a major pollution source, and there were no major coal-burning sources within 20 km of the city [4,13]. In May 2004, the power plant was closed and replaced by the national grid system of electrical energy.

Study subjects

In the 2002 cohort, as previously reported [9], the subjects were 150 nonsmoking mothers and their newborns enrolled between 4 March 2002 and 19 June 2002 at four hospitals in Tongliang: the Tongliang County Hospital, the Traditional Chinese Medicine Hospital, the Tongliang Maternal Children's Health Hospital, and the Bachuan Hospital. In the 2005 cohort, the subjects were 158 children born at the same hospitals from 2 March 2005 to 23 May 2005 and recruited and studied using the same methods. The women were selected using a screening questionnaire when they checked in for delivery. Eligibility criteria included current nonsmoking status, ≥ 20 years of age, and residence within 2.5 km of the Tongliang power plant. All but one eligible woman agreed to enter the study. The demographic characteristics of the two cohorts are presented in Table 1.

Personal interview

A 45-min questionnaire was administered by a trained interviewer after delivery. The questionnaire elicited demographic information, lifetime residential history (location of birth and duration of residence), history of active and passive smoking (including number of household members who smoke), occupational exposure, medication use, alcohol consumption during each trimester of pregnancy, and consumption of PAH-containing meat (frequency of eating fried, broiled, or barbecued meat during the last 2 weeks). Socioeconomic information related to income and education was also collected.

Biological sample collection and biomarker analysis

40–60 mL of umbilical cord blood was collected at delivery and 10 mL of maternal blood (10 mL) was collected within 1 day postpartum. Samples were transported to the field laboratory at the Tongliang County Hospital immediately after collection for processing. Blood samples, the buffy coat, packed red blood cells, and plasma were separated and stored at $-70°C$. B[a]P is widely used as a representative of PAH because concentrations of individual PAH in the urban setting are highly intercorrelated. Therefore B[a]P –DNA adducts were analyzed as a proxy for PAH-DNA adducts [25]. Details of laboratory methods for analyzing B[a]P-DNA adducts have been described[11]. Immunoassays for plasma levels of BDNF were performed using the BDNF E_{max} ImmunoAssay System (Promega) according to the manufacturer's instruction.

Measurement of child neurodevelopment using GDS

Two-year-old children in the cohort were administered the Chinese version of the standardized GDS for 0- to 3-year-old children adapted to the Chinese population by the Chinese Pediatric Society from the department of Pediatric Psychiatry in Xinghua Hospital in Shanghai, China. Each child is assigned a DQ in each of four areas: motor, adaptive, language, and social. The standardized mean ($\pm$SD) of the DQ is 100 ± 15; a score <85 indicates developmental delay [26]. Testing was conducted by physicians in the same group who were certified in the GDS to

Table 1. Demographic and Exposure Characteristics of the Cohorts.

		2002 Cohort	2005 Cohort
Mother's Age, years*		25.18±3.152, N=110	27.909±4.585, N=107
Mother's Education (%)	<High School	48 (43.64%)	59(55.14%)
	≥High School	62 (56.36%)	48(44.86%)
Gestational Age, days		277.35±11.27, N=110	276.692±9.188, N=107
Sex of Newborn, % female		56 (50.91%), N=110	48(44.86%), N=107
Birth Head Circumference, cm*		33.77±1.151, N=110	34.130±1.283, N=106
Infant's Birth Weight, g		3346.64±398.03, N=110	3378.505±399.653, N=107
Infant's Birth Length, cm		50.38±1.690, N=110	50.212±1.635, N=104
Heavy Metals	Cord Lead (mg/dL)	3.60±1.59	3.74±1.50
	Cord Mercury (ppb)	6.67±4.43	6.61±2.77
PAH-DNA Adducts, adducts/10^8 nucleotides*		0.324±0.139, N=110	.204±.081, N=107
ETS(hours/day)		0.293±0.586, N=110	0.297±0.539, N=107
Income**	Total N	77	107
	<10000	2	22
	10001 to 20000	13	52
	20001 to 30000	30	17
	30001 to 40000	22	6
	40001 to 50000	7	5
	50001 to 60000	3	3
	60001 to 70000	0	0
	>70000	0	2
BDNF, µg/dL*		752.87±463.71, N=108	1266.57±619.77, N=107
Gesell Scores	Average	99.42±10.74, N=110	100.3±7.157, N=107
	Motor	97.53±11.47, N=110	97.83±7.821, N=107
	Adaptive	98.71±14.90, N=110	101.18±10.96, N=107
	Language*	102.1±12.83, N=110	100.47±9.777, N=107
	Social	99.40±11.79, N=110	101.83±6.808, N=107

*significant difference between the two cohorts at alpha =0.05 level, using t-test.
** significant difference between the two cohorts at alpha =0.05 level, using chi-square test.

maximize reliable assessment and valid interpretation, minimizing both inter-examiner and intra-examiner variability.

Statistical Analysis

We compared characteristics of the two cohorts using the t-test or chi-square test. Multiple linear regression and logistic regression were used to test the relationships among cord adducts, BDNF, and DQ. Covariates in the regression models were selected based on previous findings and included cord blood lead, cord mercury, environmental tobacco smoke (ETS), mother's education, mother's age, gestational age, and gender. Cord adducts were calculated on the original scale, while cord lead was dichotomized at the median of detectable lead in the multiple linear regression models. BDNF values were log transformed to normalize data and remove outliers. DQ scores were handled on the original scale. Cord blood BDNF levels of newborns in the two cohorts were compared using the Mann-Whitney test. We explored whether BDNF was a mediator of the relationship between PAH-DNA adducts and DQ scores by comparing the change in the beta coefficient of cord adducts in two regression models, with and without BDNF as a variable.

Results

The demographics, exposure, and developmental characteristics of each cohort are provided and compared in Table 1. There were no significant differences between the cohorts with respect to these characteristics except in mother's age and income level. The mean level of mBDNF was significantly higher in the 2005 cohort as compared to the 2002 cohort (1266.56 pg/ml vs. 752.87 pg/ml, P<0.05). The mean birth head circumference of the 2005 infants was significantly greater than that of the 2002 cohort (33.766 cm vs. 34.130 cm, P<0.05). The mean PAH-DNA cord adduct level was 0.324 adducts/10^8 nucleotides in the 2002 cohort and 0.204 adducts/10^8 nucleotides in the 2005 cohort, also a significant difference (P<0.05). BDNF levels were also significantly higher in the 2005 cohort, with a mean of 752.871 µg/dL in the 2002 cohort compared to 1266.568 µg/dL in the 2005 cohort (P<0.05). Although the differences were not statistically significant in all areas, DQ scores were consistently higher in the 2005 cohort than the 2002 cohort for motor, adaptive, social, and average DQ (Table 1).

In cord blood, PAH-DNA adducts were modestly but significantly inversely correlated with BDNF (r = −0.233, p<0.01,

N = 269). The results of multiple regression analysis of the association between biomarkers and DQ scores are shown in Table 2 for levels of PAH-DNA adducts and BDNF protein separately. The regression coefficients (betas) represent the change in DQ score per unit increase in cord adducts or per log unit BDNF. Combining the two cohorts, a unit increase in cord adducts was negatively associated with average ($\beta = -12.113$, p = 0.014), motor ($\beta = -10.699$, p = 0.050), and adaptive ($\beta = -16.472$, p = 0.022) DQ scores after adjusting for cord blood lead, cord mercury, ETS, mother's education, mother's age, gestational age, and gender. A log unit increase in log transformed BDNF level was positively associated with average ($\beta = 2.496$, p = 0.017), motor area ($\beta = 2.117$, p = 0.018), and social area DQ ($\beta = 3.222$, p = 0.001) scores, after adjusting for covariates including cord PAH-DNA adducts, prenatal ETS, cord lead, cord mercury, gender, gestational age, mother's education, mother's age and income. BDNF was deemed to be a potential mediator variable as the betas displayed a change of over 10% after including BDNF as a covariate in the model with adducts as the independent variable and DQs as the dependent variable: the change in average DQ was 21.5%; in motor DQ 33.6%; in adaptive DQ 18.2%; and in social DQ 37.3%.

Discussion

As hypothesized, comparison of the two cohorts in Tongliang, China, before and after closure of the local power plant, has provided additional evidence of significant measurable benefits on children's development. As previously reported, we found a significant reduction in mean PAH-DNA adduct levels (0.204 adducts/10^8 nucleotides in the 2005 cohort compared to 0.324 adducts/10^8 nucleotides in the 2002 cohort). Also, as reported previously, PAH-DNA adducts in cord blood were significantly associated with decrements in motor and language areas, along with overall DQ score in the 2002 cohort [11], consistent with previous evidence that prenatal exposure to B[a]P produces neurodevelopmental effects in the offspring of study animals [27] as well as in children at age 3 years in a New York City cohort [28].

Further, the 2005 cohort had a significantly higher average mBDNF level compared to the 2002 cohort (1266.56 pg/ml vs. 752.87 pg/ml, P<0.05). As PAH-DNA cord adducts were inversely correlated with BDNF levels (r = −0.233, p<0.01), we hypothesized that reduction in BDNF levels as a result of prenatal PAH exposure may have contributed to the adverse neurocognitive effects in the 2002 cohort. Overall, DQ scores were positively associated with BDNF levels. Although the associations were not statistically significant, motor, adaptive, social area, and the average scores were consistently higher in the 2005 cohort compared to the 2002 cohort. Language DQ scores were significantly higher in the 2002 cohort; however language is the

least reliable measure for cognitive development because environmental enrichment, such as more parental interaction time, has been shown to reverse potential language deficits [29]. Moreover, our exploratory analyses suggest that BDNF may mediate in part the association between cord adducts and neurocognitive outcomes (average, motor, adaptive and social DQ scores).

The adverse developmental effects observed in the 2002 cohort are consistent with PAH being a maternal contaminant that crosses the placenta to affect neurodevelopment and cognitive function [30]. Furthermore, we have reported that prenatal PAH exposure adversely affects cognitive development [10,12,28]. This paradigm of prenatal PAH-induced neurotoxic effects is consistent with results from animal studies, which explained the relationship between downregulation of the gene encoding the receptor tyrosine kinase, MET, a target of B[a]P, which mediates neuronal process growth and synapse formation [31]. In another study, in utero exposure to aerosolized B[a]P altered the expression profile of genes involved in synaptogenesis such as Sp4 transcription factor, as well as glutamate homeostasis and metabolism of reactive oxygen intermediates [32].

These results are consistent with findings of Toledo-Rodriguez et al. that prenatal exposure to ETS (which contains PAH) was associated with increased methylation of the BDNF exon and down-regulation of BDNF[33], and reports of dysfunction in the BDNF system and compromised cognitive function in adult life from fetal exposure to unfavorable intrauterine conditions [34]. This paper is the first to assess BDNF and cognitive development with respect to prenatal exposure to PAH. The mechanisms by which PAH-related reduction of BDNF may cause child developmental problems have not been determined but recent studies have pointed to several potential mechanistic pathways: alteration in the conversion of proBDNF to mBDNF which influences levels of signaling proteins involved in LTP; disruption of activation of TrkB; and induction of methylation changes in BDNF.

First, urban air pollution has been shown to increase levels of plasminogen activator (PA) inhibitor [35], a serine protease necessary to cleave plasminogen into its active form of plasmin [36,37]. Tissue PA (tPA) is involved in cleaving proBDNF to mBDNF. Defects in tPA-dependent cleavage of proBDNF to mBDNF may downregulate mBDNF, resulting in impaired LTP [36,37]. Second, the effects of BDNF are believed to be mediated by TrkB, which follows receptor tyrosine kinase activation to initiate three major signaling pathways: phospholipase Cγ (PLCγ), phosphatidylinositol 3-kinase (PI3K), and cascade governed by extracellular signal-regulated kinases (ERK) [38]. Experimentally, prenatal stress decreases hippocampal mBDNF expression and subsequent TrkB- regulated signaling cascades for LTP [39]. The absence of TrkB forces BDNF to selectively bind to p75, resulting in death of the hippocampal neurons [40]. Similarly, higher levels of endogenous proBDNF and low levels of mBDNF were observed

Table 2. Regression Analysis for DQ and DNA-Adducts and BDNF.

	Average	Motor	Adaptive	Language	Social
Adducts[1]	−12.113, p = .014	−10.699, p = .050	−16.472, p = .022	−11.679, p = .057	−9.544, p = .078
(N = 215)	(−21.786, −2.440)	(−21.409, .010)	(−30.545, −2.398)	(−23.720, .363)	(−20.180, 1.093)
BDNF[2]	2.496, p = .017	2.117, p = .018	1.844, p = .086	.368, p = .518	3.222, p = .001
(N = 207)	(.454, 4.539)	(.467, 4.965)	(−.377, 5.604)	(−1.73, 3.416)	(1.694, 6.068)

[1]Adjusting for cord lead (Ln), cord mercury (Ln), ETS (hours/day), mother's education, mother's age, gestational age and gender.
[2]Adjusting for income, cord lead (Ln), cord mercury (Ln), ETS (hours/day), mother's education, mother's age, and gestational age.

in prenatally stressed rats, which influenced synaptic plasticity later in life [37]. Finally, BDNF expression is also regulated by the dynamic methyl CpG binding protein 2 (MeCP2), which may function as a molecular linker between DNA methylation, chromatin remodeling, and ultimately gene repression by recruiting and interacting with histone deacetylases (HDACs) and transcription repressor mSin3A [41,42]. Experimentally MeCP2 has been shown to bind to methylated CpG sequences in the BDNF promoter III [43] and in BDNF exon IV promoter to induce chromatin compaction and repress gene expression [41]. Prenatal exposure to maternal smoking was associated with enhanced DNA methylation in BDNF exon IV that may lead to long-term deregulation of BDNF [33]. Thus, mBDNF expression is regulated by changes in DNA methylation patterns that have previously been associated with levels of ambient air pollution and prenatal tobacco smoke exposure.

Strengths of the present study are the prospective cohort design, the ability to control for confounders, and the measurement of PAH-DNA adducts as a biological dosimeter for environmental exposure. Additionally, mBDNF is a mechanistically relevant measure of preclinical response/potential risk. A limitation is that we were not able to adjust for postnatal exposure to PAH. However, in this population the major source of PAH was the coal-burning plant emissions prior to its closure.

This study provides insight into the relationship between PAH, PAH-DNA adducts, BDNF, and developmental outcomes. Further studies are needed to determine whether specific enrichment environments can mitigate the damage caused by prenatal environmental exposure to PAH. In conclusion, this provides further evidence that the power plant shutdown in Tongliang, China in May of 2004 directly benefited the health and development of children.

Author Contributions

Conceived and designed the experiments: DT TYL FP. Performed the experiments: LRQ JY. Analyzed the data: DT JL LM TYL FP. Contributed reagents/materials/analysis tools: DT TYL FP. Wrote the paper: DT JL LM FP.

References

1. Lijinsky W (1991) The formation and occurrence of polynuclear aromatic hydrocarbons associated with food. Mutat Res 259: 251–261.
2. Bostrom CE, Gerde P, Hanberg A, Jernstrom B, Johansson C, et al. (2002) Cancer risk assessment, indicators, and guidelines for polycyclic aromatic hydrocarbons in the ambient air. Environ Health Perspect 110 Suppl 3: 451–488.
3. Millman A, Tang D, Perera FP (2008) Air pollution threatens the health of children in China. Pediatrics 122: 620–628.
4. (2004) The WADE economic model: China, in World Survey of Decentralized Energy. (WADE) World Alliance for Decentralized Energy.
5. Zhang J, Mauzerall DL, Zhu T, Liang S, Ezzati M, et al. (2010) Environmental health in China: progress towards clean air and safe water. Lancet 375: 1110–1119.
6. Godschalk RW, Van Schooten FJ, Bartsch H (2003) A critical evaluation of DNA adducts as biological markers for human exposure to polycyclic aromatic compounds. J Biochem Mol Biol 36: 1–11.
7. Jedrychowski W, Whyatt RM, Camann DE, Bawle UV, Peki K, et al. (2003) Effect of prenatal PAH exposure on birth outcomes and neurocognitive development in a cohort of newborns in Poland. Study design and preliminary ambient data. Int J Occup Med Environ Health 16: 21–29.
8. Perera F, Li TY, Lin C, Tang D (2012) Effects of prenatal polycyclic aromatic hydrocarbon exposure and environmental tobacco smoke on child IQ in a Chinese cohort. Environ Res 114: 40–46.
9. Perera F, Li TY, Zhou ZJ, Yuan T, Chen YH, et al. (2008) Benefits of reducing prenatal exposure to coal-burning pollutants to children's neurodevelopment in China. Environ Health Perspect 116: 1396–1400.
10. Edwards SC, Jedrychowski W, Butscher M, Camann D, Kieltyka A, et al. (2010) Prenatal exposure to airborne polycyclic aromatic hydrocarbons and children's intelligence at 5 years of age in a prospective cohort study in Poland. Environ Health Perspect 118: 1326–1331.
11. Tang D, Li TY, Liu JJ, Zhou ZJ, Yuan T, et al. (2008) Effects of prenatal exposure to coal-burning pollutants on children's development in China. Environ Health Perspect 116: 674–679.
12. Perera FP, Li Z, Whyatt R, Hoepner L, Wang S, et al. (2009) Prenatal airborne polycyclic aromatic hydrocarbon exposure and child IQ at age 5 years. Pediatrics 124: e195–202.
13. Chow JC, Watson JG, Chen LW, Ho SS, Koracin D, et al. (2006) Exposure to PM2.5 and PAHs from the Tong Liang, China epidemiological study. J Environ Sci Health A Tox Hazard Subst Environ Eng 41: 517–542.
14. Numakawa T, Suzuki S, Kumamaru E, Adachi N, Richards M, et al. (2010) BDNF function and intracellular signaling in neurons. Histol Histopathol 25: 237–258.
15. Cohen-Cory S, Kidane AH, Shirkey NJ, Marshak S (2010) Brain-derived neurotrophic factor and the development of structural neuronal connectivity. Dev Neurobiol 70: 271–288.
16. Lu Y, Christian K, Lu B (2008) BDNF: a key regulator for protein synthesis-dependent LTP and long-term memory? Neurobiol Learn Mem 89: 312–323.
17. Cunha C, Brambilla R, Thomas KL (2010) A simple role for BDNF in learning and memory? Front Mol Neurosci 3: 1.
18. Balkowiec A, Katz DM (2002) Cellular mechanisms regulating activity-dependent release of native brain-derived neurotrophic factor from hippocampal neurons. J Neurosci 22: 10399–10407.
19. Minichiello L (2009) TrkB signalling pathways in LTP and learning. Nat Rev Neurosci 10: 850–860.
20. Kandel ER (2004) The molecular biology of memory storage: a dialog between genes and synapses. Biosci Rep 24: 475–522.
21. Malenka RC, Bear MF (2004) LTP and LTD: an embarrassment of riches. Neuron 44: 5–21.
22. Huang EJ, Reichardt LF (2003) Trk receptors: roles in neuronal signal transduction. Annu Rev Biochem 72: 609–642.
23. Haapasalo A, Sipola I, Larsson K, Akerman KE, Stoilov P, et al. (2002) Regulation of TRKB surface expression by brain-derived neurotrophic factor and truncated TRKB isoforms. J Biol Chem 277: 43160–43167.
24. Yamada K, Nabeshima T (2003) Brain-derived neurotrophic factor/TrkB signaling in memory processes. J Pharmacol Sci 91: 267–270.
25. Perera FP, Tang D, Tu YH, Cruz LA, Borjas M, et al. (2004) Biomarkers in maternal and newborn blood indicate heightened fetal susceptibility to procarcinogenic DNA damage. Environ Health Perspect 112: 1133–1136.
26. Hudon L, Moise KJ, Jr., Hegemier SE, Hill RM, Moise AA, et al. (1998) Long-term neurodevelopmental outcome after intrauterine transfusion for the treatment of fetal hemolytic disease. Am J Obstet Gynecol 179: 858–863.
27. Wormley DD, Chirwa S, Nayyar T, Wu J, Johnson S, et al. (2004) Inhaled benzo(a)pyrene alters cognitive functions in the F1 generation rat dentate gyrus. Cell Mol Biol (Noisy-le-grand) 50: 715–721.
28. Perera FP, Rauh V, Whyatt RM, Tsai WY, Tang D, et al. (2006) Effect of prenatal exposure to airborne polycyclic aromatic hydrocarbons on neurodevelopment in the first 3 years of life among inner-city children. Environ Health Perspect 114: 1287–1292.
29. Sarsour K, Sheridan M, Jutte D, Nuru-Jeter A, Hinshaw S, et al. (2011) Family socioeconomic status and child executive functions: the roles of language, home environment, and single parenthood. J Int Neuropsychol Soc 17: 120–132.
30. Wormley DD, Ramesh A, Hood DB (2004) Environmental contaminant-mixture effects on CNS development, plasticity, and behavior. Toxicol Appl Pharmacol 197: 49–65.
31. Sheng L, Ding X, Ferguson M, McCallister M, Rhoades R, et al. (2010) Prenatal polycyclic aromatic hydrocarbon exposure leads to behavioral deficits and downregulation of receptor tyrosine kinase, MET. Toxicol Sci 118: 625–634.
32. Li Z, Chadalapaka G, Ramesh A, Khoshbouei H, Maguire M, et al. (2012) PAH particles perturb prenatal processes and phenotypes: protection from deficits in object discrimination afforded by dampening of brain oxidoreductase following in utero exposure to inhaled benzo(a)pyrene. Toxicol Sci 125: 233–247.
33. Toledo-Rodriguez M, Lotfipour S, Leonard G, Perron M, Richer L, et al. (2010) Maternal smoking during pregnancy is associated with epigenetic modifications of the brain-derived neurotrophic factor-6 exon in adolescent offspring. Am J Med Genet B Neuropsychiatr Genet 153B: 1350–1354.
34. Gomez-Pinilla F, Vaynman S (2005) A "deficient environment" in prenatal life may compromise systems important for cognitive function by affecting BDNF in the hippocampus. Exp Neurol 192: 235–243.
35. Su TC, Chan CC, Liau CS, Lin LY, Kao HL, et al. (2006) Urban air pollution increases plasma fibrinogen and plasminogen activator inhibitor-1 levels in susceptible patients. Eur J Cardiovasc Prev Rehabil 13: 849–852.
36. Greenberg ME, Xu B, Lu B, Hempstead BL (2009) New insights in the biology of BDNF synthesis and release: implications in CNS function. J Neurosci 29: 12764–12767.
37. Yeh CM, Huang CC, Hsu KS (2012) Prenatal stress alters hippocampal synaptic plasticity in young rat offspring through preventing the proteolytic conversion of pro-brain-derived neurotrophic factor (BDNF) to mature BDNF. J Physiol 590: 991–1010.

38. Cunha C, Brambilla R, Thomas KL (2010) A simple role for BDNF in learning and memory? Frontiers in Molecular Neuroscience 3.

39. Neeley EW, Berger R, Koenig JI, Leonard S (2011) Prenatal stress differentially alters brain-derived neurotrophic factor expression and signaling across rat strains. Neuroscience 187: 24–35.

40. Friedman WJ (2000) Neurotrophins induce death of hippocampal neurons via the p75 receptor. J Neurosci 20: 6340–6346.

41. Martinowich K, Hattori D, Wu H, Fouse S, He F, et al. (2003) DNA methylation-related chromatin remodeling in activity-dependent BDNF gene regulation. Science 302: 890–893.

42. Li H, Zhong X, Chau KF, Williams EC, Chang Q (2011) Loss of activity-induced phosphorylation of MeCP2 enhances synaptogenesis, LTP and spatial memory. Nat Neurosci 14: 1001–1008.

43. Chen WG, Chang Q, Lin Y, Meissner A, West AE, et al. (2003) Derepression of BDNF transcription involves calcium-dependent phosphorylation of MeCP2. Science 302: 885–889.

15

National Patterns in Environmental Injustice and Inequality: Outdoor NO$_2$ Air Pollution in the United States

Lara P. Clark[1], Dylan B. Millet[1,2], Julian D. Marshall[1]*

1 Department of Civil Engineering, University of Minnesota, Minneapolis, Minnesota, United States of America, **2** Department of Soil, Water and Climate, University of Minnesota, Minneapolis, Minnesota, United States of America

Abstract

We describe spatial patterns in environmental injustice and inequality for residential outdoor nitrogen dioxide (NO$_2$) concentrations in the contiguous United States. Our approach employs Census demographic data and a recently published high-resolution dataset of outdoor NO$_2$ concentrations. Nationally, population-weighted mean NO$_2$ concentrations are 4.6 ppb (38%, $p<0.01$) higher for nonwhites than for whites. The environmental health implications of that concentration disparity are compelling. For example, we estimate that reducing nonwhites' NO$_2$ concentrations to levels experienced by whites would reduce Ischemic Heart Disease (IHD) mortality by ~7,000 deaths per year, which is equivalent to 16 million people increasing their physical activity level from inactive (0 hours/week of physical activity) to sufficiently active (>2.5 hours/week of physical activity). Inequality for NO$_2$ concentration is greater than inequality for income (Atkinson Index: 0.11 versus 0.08). Low-income nonwhite young children and elderly people are disproportionately exposed to residential outdoor NO$_2$. Our findings establish a national context for previous work that has documented air pollution environmental injustice and inequality within individual US metropolitan areas and regions. Results given here can aid policy-makers in identifying locations with high environmental injustice and inequality. For example, states with both high injustice and high inequality (top quintile) for outdoor residential NO$_2$ include New York, Michigan, and Wisconsin.

Editor: Yinping Zhang, Tsinghua University, China

Funding: This material is based upon work partially supported by University of Minnesota and by the National Science Foundation under Grant No. 0853467. Any opinions, findings, and conclusions or recommendations expressed in this material are those of the author(s) and do not necessarily reflect the views of the funders. The funders had no role in study design, data collection and analysis, decision to publish, or preparation of the manuscript.

Competing Interests: The authors have declared that no competing interests exist.

* E-mail: julian@umn.edu

Introduction

Environmental injustice often places disproportionate health risks on people who are already the most vulnerable or susceptible to those risks. Since the earliest US environmental justice studies [1–6] in the 1960s–1980s, disparities in exposures to environmental risks (e.g., landfills, hazardous waste sites, polluting industries, vehicle traffic) by socioeconomic status (SES) have been widely documented [7–9]. Air pollution is a priority environmental risk in the United States (US): urban outdoor air pollution is one of the top ten causes of death in high-income nations [10]. Low-SES communities are often disproportionately exposed to air pollution [11] and also may be more susceptible to air pollution owing to other underlying disparities in, for example, access to health care [12].

Although relationships between air pollution exposure and SES have been documented in certain US cities, little is known about the broader patterns in ambient air pollution environmental justice within and across US geographies (cities, regions, states, urban versus rural areas). This previous lack of understanding is largely because of the limited coverage and spatial resolution of ambient air pollution data. Recent work exploring air pollution environmental justice in US cities or regions has been based on industrial emissions-based air pollution concentration estimates [13–16], or has focused on people living near regulatory monitor locations

[17–19]. Those multi-city and national studies reported differences in environmental injustice by US region [18], metropolitan area [13] and urban form characteristics of metropolitan areas [15–17].

Here, we employ a recently developed ambient air pollution dataset [20] to explore patterns in environmental justice within and across US geographies, including rural and urban populations. The work applies a national land use regression with high spatial resolution (~0.1 km) to examine residential outdoor nitrogen dioxide (NO$_2$) air pollution in the US. NO$_2$, which is one of the six US Environmental Protection Agency criteria pollutants, in the US is mainly emitted (as NO$_x$) from combustion in vehicles and power plants [21]; it is a marker for traffic emissions [22] and has high within-urban variability [23,24]. NO$_2$ and other traffic emissions are linked to asthma [25] and decreased lung function [26] in children, low birth-weights [27], and cardiovascular and respiratory mortality (e.g., ischemic heart disease mortality) [28,29]. Previous work in specific US cities suggests that ambient NO$_2$ (and/or NO$_x$) concentrations tend to be higher in low- than in high-SES communities [30–33].

This paper applies a national-scale analysis to quantify US-wide NO$_2$ concentration patterns by SES characteristics. It provides quantitative information for understanding how environmental equality and justice for air pollution vary among communities and regions across the US. A goal of this study is to identify US locations with highest priority environmental justice and equality concerns attributable to NO_2 and co-emitted air pollutants.

Methods

1. Data

Our analysis covers the year-2000 population of the contiguous US (280 million people). The spatial unit of analysis is the Census Block Group (BG), which is the smallest Census geography with demographic data (race-ethnicity, household income, poverty status, education status, and age) reported in the 2000 Census. Of all BGs ($n = 207,492$), 64% are urban, 14% are rural, and 21% are mixed urban-rural (i.e., contain both urban and rural Census Blocks). The mean BG sizes are 1.1 km^2 (urban), 185 km^2 (rural), and 45 km^2 (mixed); the mean (standard deviation) BG population is 1,350 (890) people.

Air pollution data are year-2006 annual average ground-level NO_2 concentration estimates from a recently published national land use regression (LUR) [20]. This LUR predicts NO_2 concentrations at the Census Block level for the contiguous US based on satellite- and ground-based measurements of NO_2, combined with land use data (e.g., road locations, elevation, tree cover, impervious-surface coverage, population density). To match the Census BG level demographic data, we calculate the mean concentration among all Blocks in each BG. Nationally, the mean NO_2 concentration for all BGs is 11.4 ppb.

2. Statistical Analyses

We calculate population-weighted mean NO_2 concentrations by race-ethnicity, poverty status, household income, education status, and age, using annual mean BG concentrations (from year-2006 LUR data) and population estimates (from year-2000 Census data). For example, the national population-weighted mean NO_2 concentration for nonwhites is the mean of BG mean concentrations weighted by the population of nonwhites in each BG. We then calculate environmental injustice and inequality metrics by US region, state, county, and Urban Area (UA), and rural versus urban location.

Our primary comparison metric for environmental injustice is the difference (ppb) in population-weighted mean NO_2 concentration between lower-income nonwhites (LIN; nonwhites in the lowest annual household income quintile [<$20,000]) and higher-income whites (HIW; whites in the highest annual household income quintile [>$75,000]). Our primary comparison metric for environmental inequality is the Atkinson Index ($\varepsilon = 0.75$ [34–38]), which measures the extent to which NO_2 concentrations are evenly distributed across the population: Atkinson Index = 0 indicates perfect equality (i.e., concentrations are equal for all people); higher values indicate greater inequality (maximum = 1). The US Census information about race covers 100% of the population, whereas combined race-income categories (e.g., whites with income >$75,000) are only available for 38% of the population (one person per household; "householders"). Our injustice metric includes 10% of the total Census population (26% of householders): lower-income nonwhite householders are 2.9% of the total Census population; higher-income white householders are 7.0%. In contrast, the inequality metric and straightforward white/nonwhite comparisons include 100% of the total Census population. See Supporting Information (**Figures S1–S2** and **Table S1** in **File S1**) for sensitivity analyses regarding metric selection.

Results and Discussion

Our results reveal significant disparities in NO_2 concentrations for specific socioeconomic groups (**Table 1; Table 2**). For example, average NO_2 concentrations are 4.6 ppb (38%, $p<0.01$) higher for nonwhites than for whites, 1.2 ppb (10%, $p<0.01$) higher for people below versus above poverty level, and 3.4 ppb (27%, $p<0.01$) higher for lower-income nonwhites than for higher-income whites. Likewise, NO_2 concentrations are higher for residents with less than a high school education compared to those with a high school education or above (difference: 0.9 ppb [8%], $p<0.01$). Among urban residents, NO_2 concentrations for Black Hispanics (the most exposed race-ethnicity group) are 6.1 ppb (38%, $p<0.01$) higher than for American Indians (the least exposed race-ethnicity group) and 4.7 ppb (28%, $p<0.01$) higher than for the total urban population. Urban-rural differences abound: in urban areas, NO_2 concentrations are higher for nonwhites than for whites, and higher for low- than for high-income groups; in contrast, NO_2 concentrations in rural areas are similar for nonwhites and for whites but are slightly lower for low-than for high-income groups. Urban areas exhibit more low- than high-income communities in NO_2-polluted areas (e.g., adjacent to busy roadways), whereas the same trend does not emerge in rural areas. Among race-ethnicity groups, American Indians have the lowest NO_2 exposures in urban areas, but the second highest NO_2 exposures (after Hispanics) in rural areas. Overall, for seven of the eight nonwhite race-ethnicity groups considered (upper portion of **Table 1**), NO_2 concentrations are higher for that group than for whites.

Young children and the elderly are especially vulnerable to air pollution. We find that NO_2 concentrations for these groups correlate with SES. Population-weighted mean NO_2 concentrations are similar (within 3% [0.3 ppb]) for those two subpopulations (elderly: greater than 65 years; young: less than 5 years) as for other age groups (5 to 65 years). However, for below-poverty level nonwhite individuals, NO_2 concentrations are notably higher for young children (3.0 ppb; 23%, $p<0.01$) and elderly people (3.1 ppb; 24%, $p<0.01$) than for the rest of the population (age 5 to 65 years, including whites and nonwhites).

An important issue is whether the NO_2 disparities described above are relevant to public health. To investigate that question, we consider here one illustrative example: ischemic heart disease (IHD) annual deaths associated with NO_2 concentration disparities between nonwhites and whites. Assuming a 6.6% change in IHD mortality rate per 4.1 ppb NO_2 [39] and US-average IHD annual mortality rates (109 deaths per 100,000 people [40]), reducing NO_2 concentrations to levels experienced by whites (a 4.6 ppb [38%] reduction) for all nonwhites (87 million people) would be associated with a decrease of ~7,000 IHD deaths per year. For comparison, interventions with a similar benefit (a decrease in ~7,000 IHD deaths per year) include: 16 million people increasing physical activity level from inactive (0 h/wk) to sufficiently active (>2.5 h/wk)[41]; 25 million people increasing physical activity level from insufficiently active (<2.5 h/wk) to sufficiently active (>2.5 h/wk); or, 3.2 million fewer adults (age 30–44) beginning smoking [42]. Calculations in this paragraph (details in **Table S2** in **File S1**) may underestimate true health impacts because we ignore here differences in vulnerability and susceptibility to air pollution and differences in underlying IHD mortality rates; also, the analysis above considers only one health outcome (IHD mortality) and one pollutant (outdoor NO_2).

Within individual urban areas, even after controlling for urban area size and household income group, nonwhites are generally more exposed to residential outdoor NO_2 air pollution than

Table 1. Population-weighted mean NO_2 concentration in ppb (percent of total population[1]).

	Total	Urban	Mixed	Rural
Total	11.3 (100%)	14.2 (63%)	7.3 (25%)	4.4 (12%)
Race-ethnicity[2]				
White	9.9 (69%)	12.9 (38%)	7.1 (20%)	4.4 (11%)
Nonwhite	14.5 (31%)	16.4 (24%)	8.1 (4.6%)	4.5 (1.6%)
Hispanic	15.6 (13%)	17.2 (10%)	8.6 (1.8%)	5.8 (0.4%)
Black	13.3 (12%)	15.3 (9.4%)	7.4 (1.9%)	3.7 (0.8%)
Asian	16.5 (3.4%)	17.5 (3.0%)	9.7 (0.4%)	4.8 (0.03%)
Two or more races	13.1 (1.6%)	15.3 (1.2%)	7.9 (0.3%)	4.5 (0.1%)
Amer. Indian/Alaska Native	8.8 (0.7%)	12.8 (0.3%)	7.2 (0.2%)	5.4 (0.2%)
Black Hispanic	17.4 (0.3%)	18.9 (0.2%)	9.0 (0.03%)	4.2 (0.01%)
Other race	15.0 (0.2%)	16.9 (0.1%)	8.3 (0.03%)	4.7 (0.01%)
Nat. Hawaiian/Pacific Islander	14.2 (0.1%)	15.7 (0.1%)	8.4 (0.01%)	4.7 (0.003%)
Poverty status				
Below poverty level	12.4 (12%)	15.3 (8.2%)	7.3 (2.3%)	4.3 (1.5%)
Above poverty level	11.2 (85%)	14.1 (53%)	7.3 (22%)	4.5 (10%)
Household income quintile				
<$20,000	11.4 (8.3%)	14.4 (5.3%)	7.3 (1.8%)	4.3 (1.2%)
$20,000–$35,000	11.0 (7.3%)	13.9 (4.6%)	7.2 (1.7%)	4.4 (1.0%)
$35,000–$50,000	10.9 (6.2%)	13.9 (3.8%)	7.2 (1.5%)	4.4 (0.8%)
$50,000–$75,000	11.0 (7.3%)	13.9 (4.5%)	7.3 (1.9%)	4.5 (0.9%)
>$75,000	11.7 (8.4%)	14.2 (5.5%)	7.7 (2.3%)	4.6 (0.6%)
Education level for population >25 years old				
Less than high school degree	12.0 (13%)	15.5 (8.0%)	7.2 (2.8%)	4.3 (1.9%)
High school degree	10.5 (19%)	13.9 (10%)	7.1 (5.0%)	4.4 (3.1%)
Some post-secondary	11.0 (18%)	13.8 (11%)	7.3 (4.6%)	4.5 (2.0%)
Bachelor's degree	11.7 (10%)	14.0 (6.8%)	7.6 (2.5%)	4.5 (0.7%)
Graduate degree	12.1 (5.7%)	14.3 (4.0%)	7.7 (1.4%)	4.5 (0.4%)
Age				
<5 years	11.6 (6.8%)	14.4 (4.4%)	7.4 (1.7%)	4.5 (0.8%)
5 to 18 years	11.2 (19%)	14.2 (12%)	7.2 (4.8%)	4.5 (2.4%)
18 to 40 years	11.8 (32%)	14.5 (21%)	7.4 (7.4%)	4.4 (3.3%)
40 to 65 years	11.0 (30%)	14.1 (18%)	7.2 (7.9%)	4.4 (4.0%)
>65 years	11.0 (12%)	13.9 (7.7%)	7.3 (3.1%)	4.4 (1.7%)
Children (<5 years) below poverty level				
White	9.1 (0.4%)	12.5 (0.2%)	6.9 (0.1%)	4.3 (0.1%)
Nonwhite	14.3 (0.8%)	16.1 (0.6%)	7.9 (0.1%)	4.7 (0.1%)
Elderly (>65 years) below poverty level				
White	9.9 (0.8%)	13.5 (0.4%)	7.1 (0.2%)	4.2 (0.2%)
Nonwhite	14.5 (0.2%)	16.9 (0.2%)	7.7 (0.03%)	4.3 (0.02%)

[1]Population totals may be less than 100% because of rounding, nonresponses in Census data, and category definitions (e.g., population >25 years old is 66% of total population).
[2]Each race-ethnicity category in **Table 1** includes people who reported a single race category and non-Hispanic ethnicity (i.e., "White" category is "White alone; non-Hispanic"), except for the "Hispanic" category, which includes people who reported any race(s) and Hispanic ethnicity, and the "Black Hispanic" category, which includes people who reported Black race alone and Hispanic ethnicity.

whites. **Figure 1** presents regression models predicting population-weighted mean NO_2 concentration as a function of household income for all 16 Census-defined household income categories and for the 4 largest race-ethnicity groups (Whites, Hispanics, Blacks, Asians) by urban area size (small; medium; large; defined by urban population tertiles). Each within-urban model reveals an inverse relationship between population-weighted NO_2 concentration and household income with high statistical significance ($R^2 > 0.86$; model p-value<0.01; **Tables S3–S18** in **File S1**). Across household income groups, urban NO_2 concentrations are often highest for Asians or Hispanics and lowest for Whites.

Table 2. Comparisons between population-weighted mean NO_2 concentrations for specific populations.

Group 1 (concentration in ppb)	Group 2 (concentration in ppb)	Difference[1] (ppb)	Relative Difference (%)
National comparisons			
Nonwhites (14.5)	Whites (9.9)	4.6	38
Below poverty (12.4)	At or above poverty (11.2)	1.2	10
Low-income nonwhites (14.4)	High-income whites (11.0)	3.4	27
Less than high school degree (12.0)	High school degree or above (11.1)	0.9	8
Children<5 years (11.6)	Age 5 to 65 years (11.3)	0.2	2
Nonwhite children below poverty level (14.3) poverty	Age 5 to 65 years (11.3)	3.0	23
Elderly>65 years (11.0)	Age 5 to 65 years (11.3)	−0.3	−3
Nonwhite elderly below poverty level (14.5)	Age 5 to 65 years (11.3)	3.1	24
Urban comparisons			
Black Hispanics (18.9)	American Indians (12.8)	6.1	38
Black Hispanics (18.9)	Total (14.2)	4.7	28

[1]Difference in population-weighted mean concentration [Group 1 - Group 2]. For all rows, differences are statistically significant with $p<0.001$.

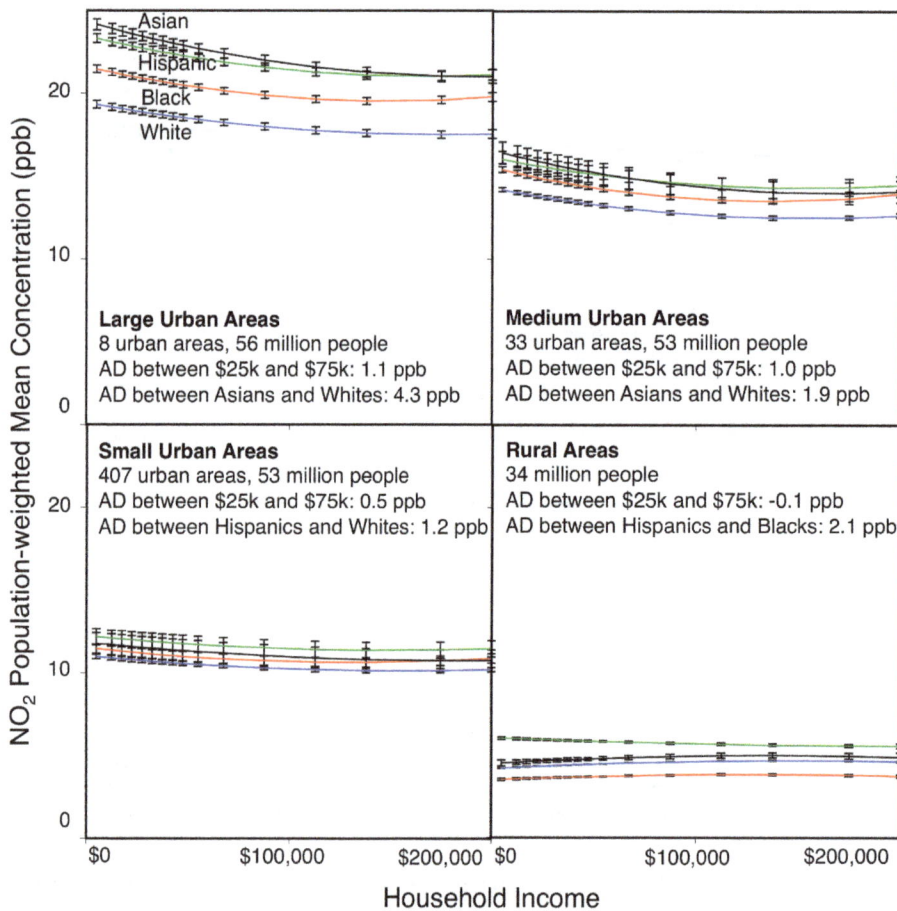

Figure 1. Within-urban and within-rural population-weighted mean NO_2 concentrations (105 million householders) by Census household income category, race, and urban category (large UA population tertile, medium UA population tertile, small UA population tertile, or rural). Concentrations shown are modeled by UA population tertile (linear regressions: $R^2>0.98$ [large UAs], >0.96 [medium UAs], >0.86 [small UAs], >0.47 [rural]; all models are statistically significant at $p<0.01$; see **Tables S3–S18** in **File S1**). For visual display, plots use the population-weighted mean UA-specific dummy variable for each UA population tertile. Error bars show the 95% confidence intervals on linear regression model predictions. AD = average difference, UA = Urban Area. AD values shown are for interquartile range incomes ($25k, $75k) and for race-ethnicity groups with highest and lowest concentrations for that panel.

Within individual urban areas, on average, NO_2 concentration disparities by race (after controlling for income) are more than 2 times greater than NO_2 concentration disparities by income (after controlling for race). The relative importance of race versus income for environmental injustice increases with urban area size. For each urban area size category, we compared average differences in NO_2 concentrations between the race group (of the 4 largest race groups) with the highest versus the lowest NO_2 concentrations (controlling for household income group) to the average differences in NO_2 concentrations between the $25,000 versus $75,000 income groups (approximate income interquartile range; controlling for race group; **Figure 1**). In large urban areas, disparities by race are ~4 times greater than by income. In medium and small urban areas, disparities by race are ~2 times greater than by income. For rural residents, differences by race are ~20 times greater than by income (despite significantly lower average concentrations for rural versus urban residents: 4.4 ppb [rural population-weighted mean] versus 14.2 ppb [urban population-weighted mean]). For rural areas, differences by income are small (0.1 ppb) and in the opposite direction as for the US as a whole (i.e., in rural areas, concentrations are higher for higher- than for lower-income groups).

As an alternative analysis, we developed NO_2 regression models for which each observation is a Block Group concentration rather than population-weighted concentration (by location, income and race category; **Tables S19–S30** in **File S1**). Results for the Block Group and population-weighted analyses cannot be compared directly. Block Group analyses indicate a more varied relationship with race and with income, but in general suggest that NO_2 concentrations are higher for nonwhites than for whites and are higher for lower-income than for higher-income communities; and, on average, disparities are greater by race (percent white) than by income.

Inequality metrics are presented in **Table 3**. On a national scale, we find that inequality levels are higher for NO_2 (Atkinson Index = 0.11) than for income (Atkinson Index = 0.08), despite the fact that the US has a high degree of income inequality compared to most developed nations [43].

Figure 2. shows national spatial patterns in environmental injustice and inequality in outdoor NO_2 air pollution. States with high levels (top quintile) of both injustice and inequality include New York, Michigan, and Wisconsin. Given previous work documenting inequality and injustice in NO_2 concentrations (among other environmental hazards) it is not surprising that we observe injustice and inequality in NO_2 concentrations on a national basis. What is unexpected, however, are the spatial patterns in **Figure 2**. Environmental injustice and inequality do not exhibit clear spatial coherence with respect to regional race or income characteristics. For example, among urban areas, environmental inequality (Atkinson Index) has a low correlation with race (percent nonwhite) and average income [Pearson's $r<0.2$]. Understanding the processes driving these spatial distributions of environmental injustice and inequality is thus a priority need for future research.

Inequality and injustice metrics vary by location. NO_2 inequality (Atkinson Index) is slightly higher among rural residents than among urban residents, but environmental injustice may be higher for urban residents: NO_2 concentration differences between lower-income nonwhites and higher-income whites are an order of magnitude higher and in the opposite direction for urban residents as for rural residents (2.8 ppb versus -0.3 ppb; see **Table 1**). Across the 448 urban areas in the US, there is variation in injustice (difference range [ppb]: -1.1 to 6.0) and inequality (Atkinson Index range: 0.00008 to 0.04) for NO_2 air pollution, consistent with a previous multi-city study [13]. In 426 of 448 urban areas (accounting for 99% of the total US urban population), NO_2 concentrations are higher for the lower-income nonwhite group than for the higher-income white group, with injustice and inequality tending to be higher in large urban areas. Supporting Information (**File S2**) provides environmental injustice and inequality rankings by urban area, county, and state.

A contribution of this work is that it covers the entire contiguous US population, including both urban and rural populations, with

Table 3. Environmental injustice and inequality metric mean (population-weighted mean) [range].

	Environmental Injustice	Environmental Inequality
	Difference[1] between low-income nonwhites and high-income whites (ppb)	Atkinson Index[2]
National	3.4	0.11
Urban	2.8	0.059
Mixed	0.4	0.062
Rural	−0.3	0.080
Regions (n = 10)	3.6 (3.7) [1.1 to 7.1]	0.083 (0.083) [0.064 to 0.12]
States (n = 49)	2.5 (3.5) [−0.6 to 7.2]	0.068 (0.073) [0.006 to 0.14]
Counties[3] (n = 3,109)	0.8 (1.9) [−2.6 to 7.0]	0.031 (0.027) [0.000006 to 0.17]
Urban Areas (n = 448)	1.3 (2.8) [−1.1 to 6.0]	0.009 (0.016) [0.00008 to 0.040]
Large Urban Areas (n = 8)	3.6 (4.0) [0.8 to 6.0]	0.018 (0.020) [0.009 to 0.031]
Medium Urban Areas (n = 33)	2.6 (2.7) [1.1 to 5.0]	0.015 (0.015) [0.005 to 0.039]
Small Urban Areas (n = 407)	1.1 (1.7) [−1.1 to 4.7]	0.009 (0.012) [0.0001 to 0.040]

[1]Larger positive differences indicate greater injustice (concentrations are higher for low-income nonwhites than for high-income whites). A negative value denotes concentrations being lower for low-income nonwhites than for high-income whites.
[2]Larger Atkinson Indices indicate greater inequality. Inequality aversion coefficient: $\varepsilon = 0.75$.
[3]This analysis excludes counties that consist of 1 Block Group (n = 29; total population = 21,500 people) or contain 0 low-income nonwhites and/or 0 high-income whites (n = 16; total population = 65,800 people).

Environmental Injustice
Difference Between LIN and HIW

Environmental Inequality
Atkinson Index

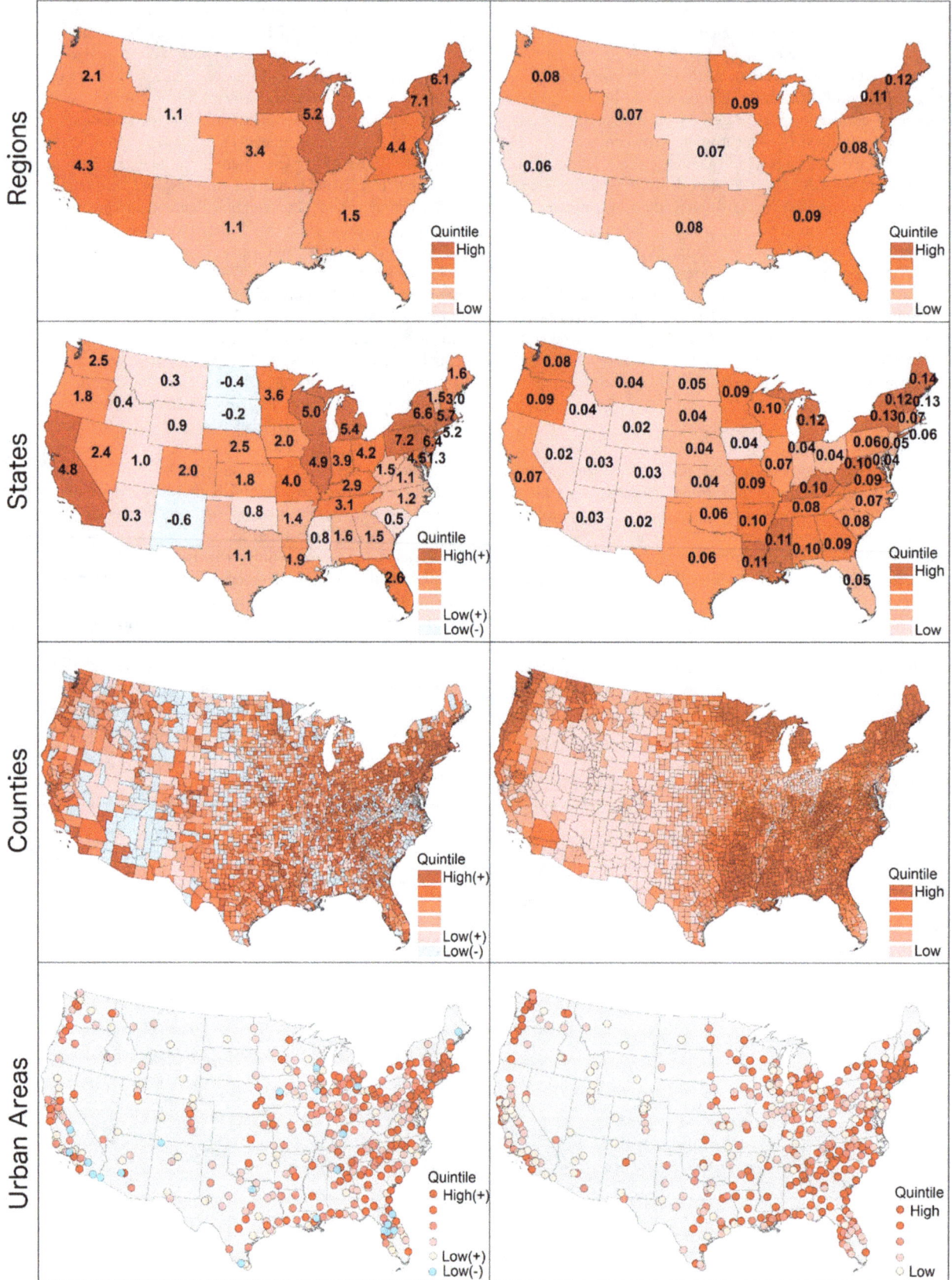

Figure 2. Environmental injustice and inequality in residential outdoor NO$_2$ concentrations for US regions, states, counties and urban areas. The left column shows differences in population-weighted mean NO$_2$ concentrations between low-income nonwhites (LIN) and high-income whites (HIW), with larger positive differences (red colors) indicating higher injustice (larger concentration difference between LIN and HIW). The right column shows the Atkinson Index, with higher values indicating greater inequality.

higher spatial precision in urban areas (urban BG-scale: ~1-km; LUR scale: ~0.1-km) relative to previous regional or multi-city air quality environmental equality and/or justice studies (typical air quality model-scale: ~12-km grid or coarser). Although the spatial resolution is higher than in previous work, resolution is still a limitation: because we are using Census demographic data, we are unable to study within-BG variations. As a second limitation, we measure inequality for one pollutant (NO$_2$); inequality may differ for other pollutants (e.g., ozone [44]) or for multi-pollutant cumulative exposure [32]. As a third limitation, we study only ambient pollution; disparities may also exist for indoor NO$_2$ emissions (e.g., owing to indoor sources such as natural gas combustion), for indoor-outdoor pollution relationships (e.g., because low-income households may live in comparatively older, leakier buildings), and for occupational and commute exposures. As a fourth limitation, there is a temporal mismatch between the year-2000 Census data and year-2006 air pollution data. We expect demographic changes during that time to be small compared to the cross-sectional differences explored here.

We investigated environmental injustice and inequality in residential outdoor NO$_2$ air pollution for the contiguous US population. Nationally, inequality in average NO$_2$ concentration is greater than inequality in average income. Nonwhites experience 4.6 ppb (38%) higher residential outdoor NO$_2$ concentrations than whites – an exposure gap that has potentially large impacts to public health. Within individual urban areas, after controlling for income, nonwhites are on average exposed to higher outdoor residential NO$_2$ concentrations than whites; and, after controlling for race, lower-income populations are exposed to higher outdoor

residential average NO$_2$ concentrations than higher-income populations. The spatial patterns observed for inequality and injustice nationally (**Figure 2**) are not predicted by region, race, or income. Our results highlight a need for future work exploring the reasons behind these spatial distributions of environmental injustice and inequality. Results given here provide strong US-wide evidence of ambient NO$_2$ air pollution injustice and inequality, establish a national context for studies of individual metropolitan areas and regions, and enable comprehensive tracking over time. Hopefully results given here will usefully allow policy-makers to identify counties and urban areas with highest priority NO$_2$ air pollution environmental justice and equality concerns.

Acknowledgments

Matthew Bechle calculated Block Group mean NO$_2$ concentrations. The Minnesota Supercomputing Institute provided computational resources.

Author Contributions

Conceived and designed the experiments: LPC DBM JDM. Analyzed the data: LPC DBM JDM. Wrote the paper: LPC DBM JDM.

References

1. Anderson SJ, Gardner BW, Moll BJ, Tribble GL, Webster TF, et al. (1978) Correlation between air pollution and socio-economic factors in Los Angeles County. Atmos Environ 12: 1531–1535.

2. Council on Environmental Quality (1971) *Environmental quality: the second annual report of the council on environmental quality* (US Government Printing Office, Washington, DC).

3. General Accounting Office (1983) *Siting of hazardous waste landfills and their correlation with racial and economic status of surrounding communities* (Rep. GAO/RCED-83-168, General Accounting Office, Washington, DC).

4. United Church of Christ Commission for Racial Justice (1987) *Toxic wastes and race in the United States: A national report on the racial and socio-economic characteristics of communities surrounding hazardous waste sites* (UCCRJ: New York, NY, USA).

5. Van Arsdol MD (1966) Metropolitan growth and environmental hazards: an illustrative case. Ekistics 21: 48–50.

6. Van Arsdol MD, Sabagh G, Alexander F (1964) Reality and the perception of environmental hazards. J Health Hum Behav 5: 144–153.

7. Brown P (1995) Race, class, and environmental health: a review and systematization of the literature. Environ Res 69: 15–30.

8. Chakraborty J, Maantay JA, Brender JD (2011) Disproportionate proximity to environmental health hazards: methods, models, and measurement. Amer J Pub Health 101: S27–S36.

9. Mohai PM, Pellow D, Roberts TJ (2009) Environmental justice. Annu Rev Env Resour 34: 405–430.

10. World Health Organization. Global health risks: mortality and burden of disease attributable to selected major risks. Available: www.who.int/healthinfo/global_burden_disease/GlobalHealthRisks_report_full.pdf. Accessed 2013 April 1.

11. Schweitzer L, Valenzuela A (2004) Environmental injustice and transportation: the claims and the evidence. J Plan Lit 18: 383–398.

12. O'Neill MS, Jerrett M, Kawachi I, Levy JI, Cohen AJ, et al. (2003) Health, wealth, and air pollution: advancing theory and methods. Environ Health Persp 111: 1861–1870.

13. Downey L, Dubois S, Hawkins B, Walker M (2008) Environmental inequality in metropolitan America. Organ Environ 21: 270–294.

14. Lopez R (2002) Segregation and black/white differences in exposure to air toxics in 1990. Environ Health Persp 110(S2): 289–295.

15. Morello-Frosch R, Jesdale BM (2006) Separate and unequal: residential segregation and estimated cancer risks associated with ambient air toxics in US metropolitan areas. Environ Health Persp 114: 386–393.

16. Brooks N, Sethi R (1997) The distribution of pollution: community characteristics and exposure to air toxics. J Environ Econ Manag 32: 233–250.

17. Schweitzer L, Zhou J (2010) Neighborhood air quality, respiratory health, and vulnerable populations in compact and sprawled regions. J Am Plann Assoc 76: 363–371.

18. Miranda ML, Edwards SE, Keating MH, Paul CJ (2011) Making the environmental justice grade: the relative burden of air pollution exposure in the United States. Int J Environ Res Public Health 8: 1755–1771.

19. Bell ML, Ebisu K (2012) Environmental inequality in exposures to airborne particulate matter components in the United States. Environ Health Persp 120: 1699–1704.

20. Novotny EV, Bechle MJ, Millet DB, Marshall JD (2011) National satellite-based land use regression: NO$_2$ in the United States. Environ Sci Technol 45: 4407–4414.

21. U. S. Environmental Protection Agency. Our nation's air: status and trends through 2010. Available: www.epa.gov/airtrends/2011/report/fullreport.pdf. Accessed 2013 April 1.

22. Beckerman B, Jerrett M, Brook JR, Verma DK, Arain MA, et al. (2008) Correlation of nitrogen dioxide with other traffic pollutants near a major expressway. Atmos Environ 42: 275–290.

23. Hewitt CN (1991) Spatial variations in nitrogen dioxide concentrations in an urban area. Atmos Environ B 25: 429–434.

24. Jerrett M, Arain MA, Kanaroglou P, Beckerman B, Crouse D, et al. (2007) Modeling the intraurban variability of ambient traffic pollution in Toronto, Canada. J Toxicol Env Heal A 70: 200–212.

25. Brauer M, Hoek G, Van Vliet P, Meliefste K, Fischer PH, et al. (2002) Air pollution from traffic and the development of respiratory infections and asthmatic and allergic symptoms in children. Am J Respir Crit Care Med 166: 1092–1098.

26. Gauderman WJ, Avol E, Gilliland F, Vora H, Thomas D, et al. (2004) The effect of air pollution on lung development from 10 to 18 years of age. N Engl J Med 351: 1057–1067.

27. Brauer M, Lencar C, Tamburic L, Koehoorn M, Demers P, et al. (2008) A cohort study of traffic-related air pollution impacts on birth outcomes. Environ Health Persp 116: 680–686.
28. Chiusolo M, Cadum E, Stafoggia M, Galassi C, Berti G, et al. (2011) Short-term effects of nitrogen dioxide on mortality and susceptibility factors in ten Italian cities: the EpiAir Study. Environ Health Persp 119: 1233–1238.
29. Filluel L, Rondeau V, Vandentorren S, Le Moual N, Cantagrel A, et al. (2005) Twenty five year mortality and air pollution: results from the French PAARC survey. Occup Environ Med 62: 453–460.
30. Grineski S, Bolin B, Boone C (2007) Criteria air pollution and marginalized populations: environmental inequity in metropolitan Phoenix, Arizona. Soc Sci Quart 88: 535–554.
31. Stuart AL, Zeager M (2011) An inequality study of ambient nitrogen dioxide and traffic levels near elementary schools in the Tampa area. J Environ Manag 92: 1923–1930.
32. Su JG, Jerrett M, Morello-Frosch R, Jesdale BM, Kyle AD (2012) Inequalities in cumulative environmental burdens among three urbanized counties in California. Environ Int 40: 79–87.
33. Yanosky JD, Schwartz J, Suh HH (2008) Associations between measures of socioeconomic position and chronic nitrogen dioxide exposure in Worcester, Massachusetts. J Toxicol Env Heal A 71: 1593–1602.
34. Levy JI, Chemerynski SM, Tuchman JL (2006) Incorporating concepts of inequality and inequity into health benefits analysis. Int J Equity Health 5: 10.1186/1475-9276-5-2.
35. Levy JI, Wilson AM, Zwack LM (2007) Quantifying the efficiency and equity implications of power plant and air pollution control strategies in the United States. Environ Health Persp 115: 743–750.
36. Levy JI, Greco SL, Melly SJ, Mukhi N (2009) Evaluating efficiency-equality tradeoffs for mobile source control strategies in an urban area. Risk Anal 29: 34–47.
37. Fann N, Roman HA, Fulcher CM, Gentile MA, Hubbell BJ, et al. (2011) Maximizing health benefits and minimizing inequality: incorporating local-scale data in the design and evaluation of air quality policies. Risk Anal 36: 1–15.
38. Post ES, Belova A, Huang J (2011) Distributional benefit of a national air quality rule. Int J Environ Res Public Health 8: 1872–1892.
39. Jerrett M, Burnett RT, Beckerman BS, Turner MC, Krewski D, et al. (2013) Spatial analysis of air pollution and mortality in California. Am J Resp Crit Care 188: 593–599.
40. U. S. Centers for Disease Control. National Vital Statistics Reports: Deaths, Preliminary Data for 2011. Available: http://www.cdc.gov/nchs/data/nvsr/nvsr61/nvsr61_06.pdf. Accessed 2013 October 1.
41. World Health Organization. Comparative Quantification of Health Risks: Global and Regional Burden of Disease Attribution to Selected Major Risk Factors. Available: http://www.who.int/publications/cra/chapters/volume1/0729-0882.pdf. Accessed 2013 October 1.
42. Danaei G, Ding EL, Mozaffarian D, Taylor B, Rehm J, et al. (2009) The preventable causes of death in the United States: comparative risk assessment of dietary, lifestyle, and metabolic risk factors. PLOS Med 6: 10.1371/journal.pmed.1000058.
43. U. S. Central Intelligence Agency. World Fact Book 2013–2014: Distribution of Family Income: Gini Index. Available: https://www.cia.gov/library/publications/the-world-factbook/index.html. Accessed 2014 March 10.
44. Marshall JD (2008) Environmental inequality: air pollution exposures in California's south coast air basin. Atmos Environ 42: 5499–5503.

16

Contribution of Climate and Air Pollution to Variation in Coronary Heart Disease Mortality Rates in England

Peter Scarborough[1]*, Steven Allender[1], Mike Rayner[1], Michael Goldacre[2]

1 British Heart Foundation Health Promotion Research Group, Department of Public Health, University of Oxford, Headington, Oxford, United Kingdom, 2 Unit of Health Care Epidemiology, Department of Public Health, University of Oxford, Headington, Oxford, United Kingdom

Abstract

There are substantial geographic variations in coronary heart disease (CHD) mortality rates in England that may in part be due to differences in climate and air pollution. An ecological cross-sectional multi-level analysis of male and female CHD mortality rates in all wards in England (1999–2004) was conducted to estimate the relative strength of the association between CHD mortality rates and three aspects of the physical environment - temperature, hours of sunshine and air quality. Models were adjusted for deprivation, an index measuring the healthiness of the lifestyle of populations, and urbanicity. In the fully adjusted model, air quality was not significantly associated with CHD mortality rates, but temperature and sunshine were both significantly negatively associated ($p < 0.05$), suggesting that CHD mortality rates were higher in areas with lower average temperature and hours of sunshine. After adjustment for the unhealthy lifestyle of populations and deprivation, the climate variables explained at least 15% of large scale variation in CHD mortality rates. The results suggest that the climate has a small but significant independent association with CHD mortality rates in England.

Editor: Colin Simpson, The University of Edinburgh, United Kingdom

Funding: PS, SA and MR were supported by the British Heart Foundation. The funders had no role in study design, data collection and analysis, decision to publish, or preparation of the manuscript.

Competing Interests: The authors have declared that no competing interests exist.

* E-mail: peter.scarborough@dph.ox.ac.uk

Introduction

Geographical inequalities in coronary heart disease (CHD) mortality rates in England are substantial and persistent. Since the late 1970s, male CHD mortality rates have been at least 30% higher in the North of England than in the South East, and the differences between North and South for female rates have been even larger [1]. Small scale geographic variations also exist, with female mortality rates for CHD in local authorities in the South East of England more than double those of the lowest in the same region, and neighbouring wards within local authorities experiencing CHD mortality rates that are considerably different [2]. If all local authorities shared the same CHD mortality rate as Kensington & Chelsea then there would be over 32,000 fewer deaths from CHD in England every year [1]. The fact that low mortality rates are attained in some areas implies that they are an achievable target with modern standards of prevention and treatment.

It is unclear how much of the geographic variation in CHD in England is a result of differences in the physical environmental. This paper explores the impact of climate and air pollution on geographic variation in CHD mortality rates. Plausible mechanisms for the effect of these factors on CHD have been suggested. Cold weather increases blood pressure, blood cholesterol, blood viscosity (thereby increasing the risk of thrombosis), and could induce a mild inflammatory response thereby increasing blood coagulability [3]. Low exposure to sunlight could increase blood cholesterol levels, since laboratory studies have shown that sunlight is a catalyst for the synthesis of a precursor for cholesterol (squalene) into vitamin D [4]. Exposure to air pollution can provoke an inflammatory response, which increases blood

coagulability (and hence risk of thrombosis), the association between air pollution and lung disease could also affect CHD via hypoxia, and air pollution may possibly affect the autonomic nervous system leading to heart rate variability [5]. The temporal influence of climate and air pollution on CHD rates has previously been demonstrated either in time-series analyses [6–8] or in seasonal mortality patterns [9], and geographic variation in cardiovascular disease mortality rates in Sheffield [10] and in the US is associated with air pollution [11]. Previous studies that have addressed geographical variations in CHD have either used data on individuals collected from different sites but have been under-powered at the area-level to consider more than one environmental variable simultaneously [12,13], or have used area-level data and have been unable to adjust analyses adequately for behavioural risk factors for CHD [14,15]. This paper addresses these gaps in the literature by reporting an analysis of the association between climate and air pollution and CHD mortality rates in a large dataset of small areas, using an area-level measure of the prevalence of behavioural risk factors introduced, that we have previously used to investigate the role of deprivation and unhealthy lifestyle on geographic variations in CHD [16]. The aim of this is to estimate the amount of geographic variation in CHD mortality rates in England that is a result of climate and air pollution after adjustment for the behavioural risk factor profile of populations, deprivation and urbanicity.

Methods

The analyses reported in this paper utilise ecological regression models, with all standard table wards as the unit of analysis.

Standard table wards are a statistical set of boundaries based on the electoral ward boundaries as of 1st January 2003. Henceforth these areas are referred to simply as 'wards'. There are 7,929 wards in England, which can be grouped into 355 local authorities (LAs). Mortality data were provided by the Office for National Statistics for the years 1999 to 2004 (inclusive) stratified by sex, ward and five year age group. The mortality data included all deaths in England where CHD was recorded as the primary cause of death (for 1999 and 2000, ICD codes 410–414; for 2001–2004, ICD codes I20–25). Change in ICD coding over the data collection period is thought to have had little impact on reporting of CHD mortalities [17]. Rates were constructed using mid-2001 population data stratified by sex, ward and five year age group, collected for the 2001 UK census, and were directly standardised to the European Standard Population.

Data on the physical environment

Data on mean maximum temperature and total hours of sunshine were provided by the Meteorological Office for 37 English weather stations for every month between 2000 and 2002. The data were used to generate model-based ward-level estimates of mean maximum temperature and total hours of sunshine for each month between 2000 and 2002 using second order trend surface modelling [18], where the climate estimates from the weather stations were used as the dependent variables in a regression model with grid references of the weather stations as the independent variables. The resulting models were used to estimate mean maximum temperature and total hours of sunshine for all wards in England, using the central grid reference for each ward. The modelled monthly estimates were then combined to produce aggregated estimates for the period 2000–2002.

Air pollution data were collected in 2001 for the development of the physical environment domain of the Index of Multiple Deprivation 2004 [19]. The data were drawn from the National Atmospheric Emissions Inventory which estimated annual mean concentrations of benzene, nitrogen dioxide, sulphur dioxide and particulates for all 1 km grid scores within the United Kingdom, using data on location of roads, housing, agriculture and point sources of emissions (e.g. power stations) [20]. These data were used to model estimated annual mean concentrations for each super output area in England. In addition, a single measure – the air quality index – was constructed that is a standardised index of levels of the four pollutants with comparison to recognised safe levels [21]. The air quality index was used in the analyses reported here, after aggregation to ward level by producing averages of the super output area estimates, weighted by population.

Data on unhealthy lifestyle

An index of unhealthy lifestyle was used as a measure of the behavioural risk factor profile of populations. This index was derived from a principal components analysis of five sets of ward-level synthetic estimates of the prevalence of cardiovascular risk factors, specifically consumption of less than five portions of fruit and vegetables per day [22], body mass index $>= 30$ kg/m^2 [22], blood pressure $>= 160/95$ mmHg [22], blood cholesterol $>= 6.5$ mmol/l [22], and current smoking [23]. The development of the index of unhealthy lifestyle is described elsewhere [16], and an assessment of the validity of the included synthetic estimates is described elsewhere [24].

Data on deprivation and urbanicity

Deprivation and urbanicity are other potential confounders of the relationship between climate, air pollution and CHD mortality rates. Deprivation in England is higher in the North than in the South (following a similar gradient to mean temperature and hours of sunshine), and air pollution is higher in more urban areas. The deprivation index used in these analyses was the ward-level Carstairs index [25], generated using data from the 2001 census at ward level [26]. The index is a sum of the z scores of census variables regarding unemployment, overcrowding, car ownership and low social class. The urbanicity variable was a categorisation of all wards into one of three groups: coastal and countryside, urban, and metropolitan. This categorisation was based on the Office for National Statistics area classification variable, which categorises all wards in the United Kingdom into nine supergroups, 17 groups and 27 subgroups, based on a cluster analysis on demographic structure, household composition, housing, socioeconomic status, employment, and industry [27]. The categorisation of English wards into the nine supergroups, and then into the urbanicity variable used in this paper, is displayed in table 1.

Statistical techniques

Initially exploratory data analysis techniques were used to investigate correlations between the exposure variables and assess the distribution of the outcome variables. Then baseline multi-level regression models (wards nested in local authorities (LAs)) were built with male and female CHD mortality rates as outcome variables, in order to get a baseline measurement of residual variance at ward-level and LA-level. Then univariate and multivariate multi-level models were built with the physical environment, unhealthy lifestyle index and deprivation variables as exposure variables. Inclusion of variance at ward-level and LA-level is important as climate and air pollution vary on different spatial scales. Finally, equivalent spatial error regression models were built with the same exposure and outcome variables. These were built to assess whether the associations derived in the multi-level models were adversely affected by spatial autocorrelation bias. Results from the multi-level models were the primary outcomes, as they allow for an assessment of how much variance is explained by the exposure variables both at ward-level and LA-level. These results are used as proxies for explanation of 'small-scale' variation (e.g. variation in CHD mortality rates within a city) and 'large-scale' variation (e.g. variation in CHD mortality rates between regions of England, such as the North and South). The estimation technique used for the multi-level modelling was iterative generalised least squares (IGLS), and the spatial error

Table 1. Categorisation of English wards (n = 7,932) by the Office for National Statistics (ONS) area classification variable and the urbanicity variable used for this paper.

Urbanicity variable	ONS area classification	Wards (%)	Population (%)
Coastal and countryside	Coastal and countryside	1,838 (23)	8.14M (16)
	Accessible countryside	899 (11)	2.79M (6)
Urban	Industrial hinterlands	1,211 (15)	9.46M (19)
	Traditional manufacturing	524 (7)	4.69M (9)
	Built up areas	163 (2)	0.95M (2)
	Student communities	306 (4)	2.64M (5)
	Suburbs and small towns	2,504 (32)	14.90M (30)
Metropolitan	Prospering metropolitan	169 (2)	1.86M (4)
	Multicultural metropolitan	318 (4)	4.01M (8)

modelling used maximum likelihood techniques, ensuring that the results of the multi-level models and the spatial error models are comparable.

Results

Both male and female ward-level age-standardised CHD mortality rates were reasonably normally distributed, and hence suitable for regression analyses. Table 2 shows descriptive statistics for the dependent and independent variables. Eight wards featured zero female CHD deaths in the six year data collection period – these wards were retained in the data analysis as they had little impact on the distribution of the outcome variables. Both maximum temperature and hours of sunshine showed little variance (a range of only 3.2°C and 400 hours of sunshine annually). These two variables were also correlated (r = 0.63), and were negatively correlated with both the unhealthy lifestyle and deprivation indices. Air pollution was significantly higher in more urban areas.

The ward-level and LA-level variance in the baseline models is shown in table 3. For men, 73% of the total geographic variation in CHD mortality rates was at ward-level, and 74% of the total geographic variation in female CHD mortality rates was at ward-level, with the remainder at local authority-level. This suggests that the average variance in CHD rates for wards within a local authority was three times higher than the variance between local authorities within England, and hence that small scale geographic variations in CHD rates are larger than large scale geographic variations. Univariate analyses (models A, B and C, table 4) showed that each of the exposure variables were strongly associated with both male and female CHD mortality rates. In both male and female multivariate models, the beta coefficient for sunshine was strongly attenuated when included alongside temperature and air pollution (model D, table 4). The multivariate models containing only climate and air pollution variables explained a considerable amount of variance at LA-level (56% and 60% in male and female mortality models, respectively) but very little of the ward-level variance.

Table 4 also shows the results for the multi-level model that includes all of the exposure and confounding variables (model F). Nearly 80% of LA-level variance in both male and female CHD mortality rates was explained, and around 20% of the ward-level variance. Beta coefficients for the climate and air pollution variables were heavily attenuated after inclusion of the confounding variables. The maximum temperature variable showed a significant negative association with both male and female CHD rates after adjustment for deprivation, urbanicity and unhealthy lifestyle, and sunshine was also independently (though weakly) associated with CHD rates. The air quality index variable showed only a small association with CHD mortality rates after adjustment for confounding variables (this association was non-significant for men).

The physical environment variables contribute little to the explanation of ward-level variation. However, they clearly contribute to the explanation of LA-level variance in mortality, even after adjustment for urbanicity, the unhealthy lifestyle and deprivation indices: the models containing only the confounding variables (model E) explained around 65% of the LA-level variance, whereas this increased to nearly 80% in the final model (model F).

The spatial error univariate and multivariate models showed good agreement with the multi-level models, suggesting that spatial autocorrelation bias has not substantially affected these findings. The parameter estimates in the spatial error models tended to be closer to zero than in the multi-level models, demonstrating that spatial autocorrelation (when unaccounted for) tends to result in a bias away from the null hypothesis. The difference in the parameter estimates between the multi-level and spatial error models was generally in the region of around 10% to 20% (results not shown).

Discussion

Statement of principal findings

Two local climate measures (mean daily maximum temperature and total hours of sunshine) and a measure of air pollution were found to explain - without accounting for other factors - nearly 60% of large scale geographic variation in CHD mortality rates but did little to explain small scale geographic variations in CHD rates. The strength of the relationships was strongly attenuated when deprivation, urbanicity and behavioural risk factor profiles of populations were added as explanatory variables. A substantial amount of large scale geographic variation in CHD rates is explained by physical environment variables even after adjustment for deprivation, urbanicity and behavioural risk factor profiles of populations – at least 10% of large scale variation in mortality rates. These models suggest that the climate has a small but independent association with CHD mortality rates in England – a ward with the lowest observed temperature had 40 more male deaths per 100,000 and 25 more female deaths per 100,000 than a ward with highest observed temperature, all else being equal. In comparison, applying excess winter mortality from CHD for England in 2004/05 [9] to temporal differences in temperature in England [28] suggests an increase in CHD mortality of approximately 3 deaths per 100,000 for men and 2 deaths per 100,000 for women. This suggests that the association between climate and CHD mortality rates shown in these analyses may be due to residual confounding, but it should be noted that temporal variations and geographic variations in CHD mortality rates due to temperature are not directly comparable. If environmental exposures contribute more to long-term cumulative risk rather than short-term risk, then it's plausible that geographic variation is indeed a much larger contributor than seasonal variation. The fully adjusted analyses suggest that air pollution has a small association with geographic variation in CHD mortality rates, however this finding may be due to over-adjustment - one of the mechanisms of the impact of urbanicity on health is via air pollution levels. However, without adjusting for urbanicity (such as model D), the association between air pollution and CHD mortality rates may be confounded by other mechanisms for the urban-health relationship, such as access to healthcare. Since the air quality index is a more direct measure of air pollution than the urbanicity variable, the limited association of air pollution with CHD shown in the fully adjusted model seems the most plausible interpretation of these results.

Strengths and weaknesses of the study

This is the first instance of a study of geographic variation in small area CHD rates that accounts for behavioural risk factor profiles of populations, deprivation, and a number of measures of the physical environment within the same set of analyses. The multi-level design of the analyses allowed for the explanation of large scale and small scale geographic variation in CHD rates simultaneously, which allowed for disentanglement of the influence of variables that are effective at the different scales. The spatial error models allowed for an assessment of whether the multi-level models were prone to spatial autocorrelation bias, which was shown not to be the case. The systematic approach to

Table 2. Summary statistics, correlation co-efficient matrix of the continuous exposure variables, and mean of exposure variables by urbanicity category (wards, n = 7,929).

Variable	Range	Interquartile range	Standard deviation	Mean	Median
CHD mortality rate per 100,000, men	24.4 to 525.3	142.5 to 212.1	53.6	179.9	174.9
CHD mortality rate per 100,000, women	0.0 to 336.2	63.0 to 100.6	29.7	83.6	80.5
Mean max. temp (°C)	11.2 to 14.4	13.5 to 14.4	0.6	13.9	14.1
Sunshine (000s hrs/yr)	1.3 to 1.7	1.4 to 1.6	0.1	1.5	1.5
Air quality index (SDs)	0.4 to 2.2	0.9 to 1.3	0.3	1.1	1.1
Unhealthy lifestyle, men (SDs)	−6.7 to 5.3	−1.2 to 1.2	1.8	0.0	−0.1
Unhealthy lifestyle, women (SDs)	−6.2 to 5.6	−1.3 to 1.3	1.8	0.0	−0.1
Deprivation (SDs)	−5.7 to 16.5	−5.4 to 15.1	3.5	−0.1	−1.0

	Coastal & countryside (n, %)			Urban (n, %)	Metropolitan (n, %)
Urbanicity	2737, 35%			4708, 59%	484, 6%

Correlation co-efficient matrix

	Mean max. temp	Sunshine	Air quality index	Unhealthy lifestyle, men	Unhealthy lifestyle, women	Deprivation
Mean max. temp	1.00					
Sunshine	0.63	1.00				
Air quality index	0.21	−0.07	1.00			
Unhealthy lifestyle, men	−0.43	−0.40	−0.07	1.00		
Unhealthy lifestyle, women	−0.44	−0.39	−0.12	0.99	1.00	
Deprivation	−0.19	−0.17	0.42	0.57	0.51	1.00

Mean of continuous exposure variables by urbanicity category

Variable	Coastal & countryside	Urban	Metropolitan	p for trend
Mean max. temp (°C)	13.8	14.0	14.7	<0.001
Sunshine (000s hrs/yr)	1.5	1.5	1.6	<0.001
Air quality index (SDs)	0.9	1.2	1.6	<0.001
Unhealthy lifestyle, men (SDs)	−0.1	0.2	−1.4	0.001
Unhealthy lifestyle, women (SDs)	−0.1	0.3	−2.1	<0.001
Deprivation (SDs)	−1.8	0.3	6.0	<0.001

SDs = Standard Deviations.

building models that was utilised here allowed for a comprehensive assessment of the impact of confounding, and for some disentanglement of the amount of geographic variation that is explained by the climate and air pollution variables.

Table 3. Residual variance at ward-level (n = 7,929) and local authority-level (n = 354) for baseline (no exposure variables) and final models (MODEL L).

		BASELINE		FINAL	
		Variance	Standard Error	Variance	Standard Error
MEN	Ward-level	2,096.4	34.1	1,580.2	25.7
	LA-level	779.7	66.3	166.1	18.1
WOMEN	Ward-level	660.8	10.7	547.6	8.9
	LA-level	226.8	19.5	53.5	6.0

The results presented in this paper are derived from ecological cross-sectional analyses. Because of the cross-sectional nature of the studies, the results cannot confirm causal relationships. The relationship between climate and CHD rates presented here may be a result of residual confounding. Economic deprivation, unhealthy lifestyles and the climate generally follow the same North-South gradient in England, and the associations shown in the analyses may be a result of errors in the measurement of economic deprivation and unhealthy lifestyles, or could be due to unmeasured and potentially confounding factors such as utilisation and quality of health care. A previous study of women in 23 towns in Great Britain suggested that controlling for aspirin and statin use (as a proxy for health service utilisation) removed the residual variance in adjusted cardiovascular prevalence rates in England (but not in Scotland) [13], suggesting that this residual confounding could explain the associations with climate found here. However, the longitudinal impact of climate on CHD mortality rates is well established, so a potential impact of climate on CHD mortality rates is plausible. The ecological nature of the study design implies that the results cannot provide any information about how the explanatory variables affect individuals [29]. For example, the results imply that the average temperature of an area

Table 4. Beta coefficients for multi-level regression models for physical environment exposure variables in univariate (MODELS A–C) and multivariate (MODEL D) analyses, and after further adjustement for confounding variables (MODELS E–F).

	MODEL A	MODEL B	MODEL C	MODEL D	MODEL E	MODEL F
Beta coefficients in models for male CHD mortality rates						
Mean max. temp (°C)	−27.7**			−32.7**		−12.5**
Sunshine (000s hrs/yr)		−162.6**		−18.2		−27.3*
Air quality index (SDs)			54.4**	56.6**		5.8
Urban[†]					2.0	1.9
Metropolitan[†]					−5.9	−8.0*
Unhealthy lifestyles (SDs)					6.5**	5.0**
Deprivation (SDs)					7.2**	7.2**
Ward-level variance explained	0%	0%	3%	3%	25%	25%
LA-level variance explained	43%	34%	−8%	56%	68%	79%
Beta coefficients in models for female CHD mortality rates						
Mean max. temp (°C)	−15.5**			−17.2**		−7.9**
Sunshine (000s hrs/yr)		−90.2**		−14.2		−14.3*
Air quality index (SDs)			23.7**	26.5**		5.7**
Urban[†]					0.5	0.4
Metropolitan[†]					2.9	1.2
Unhealthy lifestyles (SDs)					4.1**	3.3**
Deprivation (SDs)					3.0**	3.0**
Ward-level variance explained	0%	0%	2%	2%	17%	17%
LA-level variance explained	45%	35%	−2%	60%	62%	76%

SDs – Standard Deviations;
[†]in comparison to coastal and countryside wards;
*significant at p<0.05;
**significant at p<0.01.

has an impact on CHD rates within that area, but they do not tell us anything directly about how the temperature of an area affects the individuals living in the area, or whether certain individuals within the area are more at risk than others. Risk factors for CHD tend to accrue over the life course [30], and so examining a cross-sectional relationship with the physical environment and unhealthy lifestyle will tend to under estimate the impact of these variables. This is particularly problematic as the analyses did not take account of migration between wards in England. The exposure and the confounding variables used in these analyses are derived from a number of different data sources and using different techniques. It is therefore difficult to assess the degree of uncertainty in the results that is due to measurement error, but this is likely to have had some impact on the results.

Comparison with other studies

The results presented here are in general agreement with the UK literature on geographic variation in CHD rates, in that not all of the variation in CHD rates can be explained by lifestyle factors alone. The British Regional Heart Study (BRHS) provides the most comparable results for the impact of climate on geographic variation in heart disease in England, despite the widely differing methodology employed in the study compared with the analyses reported here. Analysis of phase one of the BRHS (which utilised ecological analyses of CHD mortality rates in 253 towns) suggested that in 1969–73 climate variables had a modest effect on variation in local CHD mortality rates after adjustment for deprivation [14], which is a similar result to those reported here. Phase two of the

study (a cohort study of 7735 men in 24 British towns, followed up for fifteen years) showed that temperature explains around 30% of the between-towns variance in CHD incidence rates that remained after adjustment for social class and individual-level risk factors [12]. Again, this is in broad agreement with the results reported here – that the climate has a modest effect on CHD rates after adjustment for differences in the behavioural risk factor profile of populations and socio-economic status.

The results of this paper extend the results of phase one of the BRHS in the following ways: all wards in England were included in the analysis; a measure of the behavioural risk factor profile of populations of areas was included; an exploration of both small scale and large scale geographic variation in CHD rates was conducted; including wards from rural areas allowed for urbanicity to be included as a potential explanatory variable; more sophisticated estimates of air pollution and climate were used, which allowed for modelled estimates of these measures to be applied to all wards in England. The results of this paper complement the results of phase two of the BRHS, but extend the interpretations to women and to men of all ages. In addition, the analyses reported here were sufficiently powered at the area-level to allow for inclusion of several environmental variables in the models simultaneously.

The results presented here suggest that air pollution has a small positive association with CHD mortality rates in small areas. A similar finding was shown by Maheswaran et al. in an analysis of census enumeration districts in Sheffield [10], where nitrogen oxide levels were significantly associated with increased CHD mortality rates (smaller, non-significant associations were also shown for carbon monoxide and particulates). Interestingly, the Sheffield analysis showed no association between air pollution and CHD hospital admissions, suggesting that air pollution may increase the risk of sudden death from CHD (although residual confounding could not be ruled out). Secondary analysis of a cohort study restricted to US metropolitan areas with estimates of particulate air pollution [11] also showed small but significant increases in cardiovascular deaths for residents in areas with increased air pollution (for both current and former smokers).

Implications and further research

The analyses reported here suggest that, on top of excess winter mortality, CHD mortality rates in the coldest parts of England are generally higher compared to the warmest parts (although this association may be due to residual confounding). Whilst this difference is small compared to differences in the lifestyle of populations, if the relationship is shown to be causal then it is an area which could be targeted in order to reduce geographic inequalities in CHD. Analyses of excess winter mortality in different regions of Europe have shown that the excess mortality is generally greater in countries with milder climates and this has led researchers to suggest that the impact of a cold climate on cardiovascular health can be substantially reduced if the population were better prepared for cold weather by improving household heating and insulation and wearing more appropriate clothing during cold periods of the year [31,32]. Interventions such as these would be beneficial for reasons other than improving

cardiovascular health. Cold weather has been implicated in the development of a number of conditions such as respiratory disease, particularly in elder people. Improvements in home heating have the potential to improve quality of life, and increased insulation of homes would reduce fuel use thereby saving household finances and reducing greenhouse gas emissions. Further research should be conducted to determine cost-effective interventions to reduce the impact of climate on coronary heart disease mortality. Such interventions have the potential to reduce geographic inequalities in health in England. With regard to air pollution, the results of this study are inconclusive as to whether raised levels of air pollution in urban areas lead to increased levels of CHD in comparison to rural areas. This needs further investigation, using more refined small area data of air pollution (preferably directly measured), CHD incidence and confounding variables (e.g. small area prescription rates for aspirins/statins, access to health care etc.), and including small areas drawn from rural and urban areas.

Prior publication

The work reported in this manuscript has not previously been published elsewhere, or submitted for publication elsewhere.

Author Contributions

Conceived and designed the experiments: PS SA MR MG. Performed the experiments: PS. Analyzed the data: PS. Contributed reagents/materials/analysis tools: PS. Wrote the paper: PS SA MR MG.

References

1. Scarborough P, Allender S, Peto V, Rayner M (2008) Regional and social differences in coronary heart disease. London: British Heart Foundation.
2. Allender S, Scarborough P, Keegan T, Rayner M (2011) Relative deprivation between neighbouring wards is predictive of coronary heart disease mortality after adjustment for absolute deprivation of wards. J Epidemiol Comm Health, In press.
3. Toledano M, Shaddick G, Elliott P (2005) Seasonal variations in all-cause and cardiovascular mortality and the role of temperature. In: Marmot M, Elliott P, eds. Coronary heart disease epidemiology. From aetiology to public health. 2nd edition. Oxford: Oxford Medical Publications.
4. Grimes D, Hindle E, Dyer T (1996) Sunlight, cholesterol and coronary heart disease. Q J Med 89: 579–589.
5. Pope C (2005) Air pollution. In: Marmot M, Elliott P, eds. Coronary heart disease epidemiology. From aetiology to public health. 2nd edition. Oxford: Oxford Medical Publications.
6. Bhaskaran K, Hajat S, Haines A, Herrett E, Wilkinson P, et al. (2010) Short term effects of temperature on risk of myocardial infarction in England and Wales: time series regression analysis of the Myocardial Ischaemia National Audit Project (MINAP) registry. BMJ 341: c3823.
7. Bell M, Peng R, Dominici F, Samet J (2009) Emergency hospital admissions for cardiovascular diseases and ambient levels of carbon monoxide: results for 126 United States urban counties, 1999–2005. Circulation 120: 949–955.
8. Stieb D, Szyszkowicz M, Rowe B, Leech J (2009) Air pollution and emergency department visits for cardiac and respiratory conditions: a multi-city time-series analysis. Environmental Health 8: 25.
9. Allender S, Peto V, Scarborough P, Kaur A, Rayner M (2008) Coronary heart disease statistics 2008. London: British Heart Foundation.
10. Maheswaran R, Haining R, Brindley P, Law J, Pearson T, et al. (2005) Outdoor air pollution, mortality, and hospital admissions from coronary heart disease in Sheffield, UK: a small-area level ecological study. European Heart Journal 26: 2543–2549.
11. Pope C, Burnett R, Thurston G, Thun M, Calle E, et al. (2004) Cardiovascular mortality and long-term exposure to particulate air pollution: epidemiological evidence and general pathophysiological pathways to disease. Circulation 109: 71–77.
12. Morris R, Whincup P, Lampe F, Walker M, Wannamethee S, et al. (2001) Geographic variation in incidence of coronary heart disease in Britain: the contribution of established risk factors. Heart 86: 277–283.
13. Lawlor D, Bedford C, Taylor M, Ebrahim S (2003) Geographical variation in cardiovascular disease, risk factors, and their control in older women: British Women's Heart and Health Study. J Epidemiol Comm Health 57: 134–140.
14. Pocock S, Shaper A, Cook D, Packham R, Lacey R, et al. (1980) British Regional Heart Study: geographic variations in cardiovascular mortality, and the role of water quality. BMJ 280: 1243–1249.

15. Maheswaran R, Morris S, Falconer S, Grossinho A, Perry I, et al. (1999) Magnesium in drinking water supplies and mortality from acute myocardial infarction in north west England. Heart 82: 455–460.
16. Scarborough P, Allender S, Rayner M, Goldacre M (2011) An index of unhealthy lifestyle is associated with coronary heart disease mortality rates for small areas in England after adjustment for deprivation. Health & Place 17: 691–695.
17. Griffiths C, Brock A, Rooney C (2004) The impact of introducing ICD-10 on trends in mortality from circulatory diseases in England and Wales. Health Stat Q 22: 14–20.
18. Cressie N (2000) Geostatistical methods for mapping environmental exposures. In: Elliott P, Wakefield J, Best N, Briggs D, eds. Spatial epidemiology. Oxford: Oxford University Press.
19. Office of the Deputy Prime Minister (ODPM) (2004) The English indices of deprivation 2004 (revised). London: ODPM.
20. Bush T, Tsagatakis I, King K, Passant N (2008) NAEI UK emission mapping methodology 2006. London: DEFRA.
21. Neighbourhood statistics. Combined air quality indicator 2001. www.neighbourhood.statistics.gov.uk. Accessed 2011 July.
22. Dibben C, Sims A, Watson J, Barnes H, Smith T, et al. (2004) The Health Poverty Index. Oxford: South East Public Health Observatory, http://www.hpi.org.uk/index.php Accessed 2011 July.
23. Twigg L, Moon G, Walker S (2004) The smoking epidemic in England. London: Health Development Agency.
24. Scarborough P, Allender S, Rayner M, Goldacre M (2009) Validation of model-based estimates (synthetic estimates) of the prevalence of risk factors for coronary heart disease for wards in England. Health & Place 15: 596–605.
25. Carstairs V, Morris R (1990) Deprivation and health in Scotland. Health Bull 48: 162–175.
26. Morgan O, Baker A (2006) Measuring deprivation in England and Wales using 2001 Carstairs scores. Health Stat Q 31: 28–33.
27. Office for National Statistics (ONS) (2001) Area classification for statistical wards. London: ONS.
28. The Meteorological Office (2011) Rainfall, sunshine and temperature time-series. http://www.metoffice.gov.uk/climate/uk/actualmonthly/. Accessed 2011 November.
29. Subramanian S, Jones K, Kaddour A, Krieger N (2009) Revisiting Robinson: the perils of individualistic and ecologic fallacy. Int J Epidemiol 38: 342–360.
30. Lynch J, Smith GD (2005) A life course approach to chronic disease epidemiology. Annual Review of Public Health 26: 1–35.
31. Keatinge W, Donaldson G, Bucher K, Jendritsky G, Cordioli E, et al. (1997) Cold exposure and winter mortality from ischaemic heart disease, cerebrovascular disease, respiratory disease, and all causes in warm and cold regions of Europe. Lancet 349: 1341–1346.
32. Mercer J (2003) Cold – an underrated risk factor for health. Environmental Research 92: 8–13.

17

Fetal Window of Vulnerability to Airborne Polycyclic Aromatic Hydrocarbons on Proportional Intrauterine Growth Restriction

Hyunok Choi[1]*, **Lu Wang**[2], **Xihong Lin**[3], **John D. Spengler**[1], **Frederica P. Perera**[4]

1 Department of Environmental Health, Harvard School of Public Health, Boston, Massachusetts, United States of America, **2** Department of Biostatistics, The University of Michigan, Ann Arbor, Michigan, United States of America, **3** Department of Biostatistics, Harvard School of Public Health, Boston, Massachusetts, United States of America, **4** Columbia Center for Children's Environmental Health, Mailman School of Public Health, New York, New York, United States of America

Abstract

Background: Although the entire duration of fetal development is generally considered a highly susceptible period, it is of public health interest to determine a narrower window of heightened vulnerability to polycyclic aromatic hydrocarbons (PAHs) in humans. We posited that exposure to PAHs during the first trimester impairs fetal growth more severely than a similar level of exposure during the subsequent trimesters.

Methods: In a group of healthy, non-smoking pregnant women with no known risks of adverse birth outcomes, personal exposure to eight airborne PAHs was monitored once during the second trimester for the entire cohort (n = 344), and once each trimester within a subset (n = 77). Both air monitoring and self-reported PAH exposure data were used in order to statistically estimate PAH exposure during the entire gestational period for each individual newborn.

Results: One natural-log unit increase in prenatal exposure to the eight summed PAHs during the first trimester was associated with the largest decrement in the Fetal Growth Ratio (FGR) (−3%, 95% Confidence Interval (CI), −5 to −0%), birthweight (−105 g, 95% CI, −188 to −22 g), and birth length (−0.78 cm, 95% CI, −1.30 to −0.26 cm), compared to the unit effects of PAHs during the subsequent trimesters, after accounting for confounders. Furthermore, a unit exposure during the first trimester was associated with the largest elevation in Cephalization Index (head to weight ratio) (3 μm/g, 95% CI, 1 to 5 μm/g). PAH exposure was not associated with evidence of asymmetric growth restriction in this cohort.

Conclusion: PAH exposure appears to exert the greatest adverse effect on fetal growth during the first trimester. The present data support the need for the protection of pregnant women and the embryo/fetus, particularly during the earliest stage of pregnancy.

Editor: Alex R. Cook, National University of Singapore, Singapore

Funding: This work was supported by The National Institute of Environmental Health Sciences (NIEHS) (grant numbers 5 P01 ES009600, R01ES014939, 5 R01 ES008977, 5 R01ES11158, 5 R01 ES012468, 5 R01ES10165), the US Environmental Protection Agency (grant numbers R827027, 82860901, RD-832141), the US National Research Service Award (T32 ES 07069), Irving General Clinical Research Center (grant number RR00645), and The Gladys & Roland Harriman Foundation. None of the authors has any actual or potential competing financial interests. E. Evans, R. Whyatt, H. Andrews, L. Hoepner, W. Jedrychowski, R. Jacek, E. Mroz, A. Pac, E. Flak, W. Tsai, D. Tang, and J. Yu contributed to the study. The funders had no role in study design, data collection and analysis, decision to publish, or preparation of the manuscript.

Competing Interests: The authors have declared that no competing interests exist.

* E-mail: hchoi@albany.edu

Introduction

Polycyclic aromatic hydrocarbons (PAHs) are multiphasic fused aromatic rings of carbon compounds. Ubiquitous human dependence on combustion of carbon-containing materials—primarily fossil fuel—has contributed to PAH air pollution as a global issue [1,2,3]. Some PAHs are potent mutagens, genotoxins and known human carcinogens [4].

Transplacental mutagenicity and genotoxicity of some PAHs are well established in several experimental animal species [5,6]. For example, *in utero* exposure of mice to benzo[a]pyrene (B[a]P) and dibenzo[a,l]pyrene could induce the formation of DNA-adducts in thymocytes and splenocytes of the offspring [7]. Such adducts are vital precursors to PAH-mediated carcinogenesis [7].

In utero PAH exposure could also induce lung and liver tumors, as well as lymphoma in mice offspring [8,9,10].

In humans, maternal exposure to certain carcinogenic PAHs (c-PAHs) during pregnancy could induce DNA damage, histone modification, and chromosome abnormalities in the fetus [11] at an environmentally relevant exposure range. PAH-DNA adducts have been detected in human fetal umbilical cord blood DNA, as well as in maternal blood after exposure to ambient airborne PAHs [12]. Prenatal PAH exposure quantified using personal air monitoring significantly predicted dose-responsive elevation in chromosomal aberrations in cord blood [13]. Accordingly, prenatal PAH exposure may increase cancer risk in humans.

Furthermore, prenatal exposure to PAHs through maternal inhalation is associated with a wide range of non-carcinogenic fetotoxic effects, including intrauterine growth restriction [14], small-for-gestational age [15], and preterm delivery [15]. When the prenatally monitored group of newborns was followed to school age, the prenatal PAH exposure furthermore impaired neurodevelopmental performances [4], and increased the likelihood of asthma-related symptoms [11].

However, to date, the precise window of fetal susceptibility to the airborne PAHs on adverse birth outcomes has never been directly examined in human populations. The importance of this question stems from the observed variability in transplacental fetotoxic effects according to the gestational window of exposure in several animal models [8,16,17]. Thus, not only the concentration and the components of the PAH mixture in air, but also the fetal age, might influence the type and the severity of adverse clinical health outcomes in humans.

The goal of the present analysis is to address this critical gap in knowledge. Several lines of experimental and clinical evidence suggest that the embryo/fetus is most vulnerable to a number of PAHs during the first trimester, or the period of organogenesis. Two modes of PAH effects have been identified during the earliest gestational weeks, the first being the interference with placental development [18,19,20], and the second being direct injury of the embryo. In the first mechanism, PAHs, in particular B[a]P, disrupt early trophoblast endovascular proliferation and their ability to infiltrate into the fetal envelope [21]. Such proliferative failure results in an altered vascular labyrinthine structure of the placenta, a lowered fetoplacental vascular surface area, and altered apoptosis in fetal endothelia and syncytiotrophoblast cells [10,22]. This critically impairs the development of several vital fetal organ systems, including the central nervous systems and the heart, as they undergo terminal cell lineage commitment during this period [10,22]. In the second mode of PAH fetotoxicity, the direct impairment of embryonic growth differs with the gestational age of exposure as well as the target organs [17]. For example, B[a]P administration in the patas monkey model has shown that the fetal brain is most vulnerable to B[a]P DNA adduct formation during the first trimester, while the fetal liver is most vulnerable to the same adducts during the second trimester [17].

A small body of epidemiological evidence suggests that ambient PAH concentrations pose the greatest risk during the first trimester. In Teplice, Czech Republic, an area noted for high environmental pollution, exposure of pregnant women to high ambient concentration of c-PAHs and PM_{10} during the first gestational month was associated with significantly increased risk of intrauterine growth restriction [14,23].

Considering both biological and epidemiological evidence, we tested the hypothesis that an embryo/fetus is most vulnerable during the first trimester per unit PAH exposure, compared to a comparable unit of exposure during the second or the third trimester in a cohort of non-smoking, healthy pregnant women. We examined this by 1) using a newly developed model to estimate personal exposure to airborne PAHs during the entire gestational period for each newborn; 2) comparing the birth outcome effect sizes per unit PAH exposure during given window of interest (i.e. trimester or gestational month) to those during other periods, including the sixth gestational month as the reference period; 3) examining whether the gestational period of IUGR onset is associated with an IUGR subtype as well as its severity.

Methods

Site Characterization

In a prospective birth cohort study, pregnant women were recruited from prenatal care clinics during their first trimester in Krakow, Poland. In the city of Krakow, coal combustion for domestic heating represents the major air pollution source [24]. In contrast, automobile traffic emissions and coal-combustion for industrial activities are relatively minor contributors [24]. During typical winter days in 2005, ambient PM_{10} concentrations in Krakow have been shown to peak at 400 $\mu g/m^3$ [24]. During such episodes, ambient B[a]P and other PAHs were spatially homogeneous in their concentrations over the city, suggesting that the city's population could have been exposed to a narrow range of concentrations [24].

Study Subjects

Details regarding subject enrollment and methods are discussed elsewhere [25]. We targeted Caucasian pregnant women of ethnic Polish background during the 8th to 13th weeks of gestation. To reduce confounding, only young (age, 18–35) and healthy women with no known risks for adverse birth outcomes were eligible. Those who met all the eligibility criteria were simultaneously monitored for their personal (n = 344), home indoor (n = 76), and outdoor (n = 70) levels of PAHs and $PM_{2.5}$ during the second trimester of pregnancy between November 2000 and January 2003 [25]. The women also answered a questionnaire on health, lifestyle and exposure history. In the subset of women (n = 77), they were monitored for their personal exposure, in addition to the indoor and outdoor monitoring of PAHs. In the subset, personal monitoring was repeatedly taken once during each trimester (see Table S1). For personal monitoring, each woman carried her a personal air monitor which operated for a consecutive 48-hour period. The split flow inlet, placed near the woman's breathing zone, drew in the particulate or semi-volatile vapor PAHs and particles ≤2.5 μm ($PM_{2.5}$) on a pre-cleaned quartz microfiber filter and polyurethane foam backup. The filters were analyzed for pyrene and eight PAHs known to be carcinogenic as well as having other toxicities: benz(a)anthracene, chrysene/isochrysene, benzo(b)-fluoranthene, benzo(k)fluoranthene, benzo(a)pyrene, indeno(1,2,3-cd)pyrene, dibenz(a,h)anthracene and benzo(g,h,i)perylene.

Full enrollment criteria required the completion of a prenatal interview, compliance with air monitoring with minimal problems in data quality, and the donation of a maternal and/or newborn cord blood sample at delivery. The cotinine concentration (0.319±0.882 ng/ml serum, mean ± S.D.), which was available for 228 newborns, indicated no secondhand smoke exposure (≥15 ng/mL) in women up to 48-hours prior to delivery. The study was reviewed and approved by the institutional review boards of Jagiellonian University in Krakow and Columbia Presbyterian Medical Center, New York City, US. Written informed consent was obtained from all study participants.

Outcomes

In addition to abstracting birthweight (g), birth length (cm) and birth head circumference (cm) from the medical record, we calculated fetal growth ratio (FGR, %), Cephalization Index (cm/g), and Ponderal Index (g/cm³) for each newborn. In order to test our main hypothesis, we chose fetal growth restriction (FGR) as a marker for symmetric IUGR because of its clinical predictiveness of neonatal morbidities [26], advantage in linear regression analysis, and its consistency with small for gestational age (SGA) outcome in our New York City birth cohort [15]. FGR indicates percent underweight relative to the population mean [26]. Moderate to severe (i.e. <80%) FGR has been associated with a greater risk of delayed neurodevelopment [27,28], shorter stature, cardiovascular disease, insulin resistance and diabetes during adulthood [29]. In order to meet our third specific aim of detecting severe IUGR subtypes, we chose two indices of asymmetric fetal growth, the Cephalization Index and the Ponderal Index. The Cephalization Index has been validated as a marker for a severe IUGR subtype for an impaired fetal brain development [26]. Ten-year old children born with larger head size relative to their body weight (i.e. "brain spared") scored significantly lower in neurodevelopmental tests, intelligence quotient, and school performance [30]. In a murine model, fetal cranium and neural tissues are most exquisitely sensitive to PAHs during the period of organogenesis [31]. Accordingly, we speculated that onset of PAH-mediated impairment in fetal neural tissues occurs in late first to early second trimester. On the other hand, the Ponderal Index was chosen because earlier epidemiologic studies observed that the SGA newborns who were particularly "thin" were at a greater risk of perinatal asphyxia and extended hospitalization, compared with the proportional SGA cases [32]. Low Ponderal Index furthermore predicted shorter stature, cardiovascular disease, insulin resistance, and diabetes during adulthood [29]. As fetal weight gain is highest during the third trimester, such impairment is expected to result in low Ponderal Index [33]. Accordingly, we posited that the Ponderal Index captures an acute IUGR phenomenon with the third trimester as the time point of onset. The Ponderal Index was calculated as [birthweight (g)/(birthlength)^3 (cm)^3×100] [32]. The Cephalization Index was calculated as the ratio of the [birth head circumference (μm)/birthweight (g)] [30]. FGR was calculated as observed birthweight/mean birthweight at a given gestational age and gender based on 1994–1996 Polish birthweight distribution [34]. Cephalization index has been validated as a marker for a severe IUGR subtype [26].

Statistical Analysis

As in prior analyses, the main exposure variable was calculated as a summed concentration of the eight carcinogenic PAHs (Σ8 c-PAHs) [35]. The exposure unit of interest was a one natural-log (ln) unit increase in concentration of Σ8 c-PAHs.

Personal Exposure Concentration Prediction. We considered 48-hour personal exposure monitored data in the overall cohort and the subset (see Table S1) in order to predict chronic prenatal exposure of the same newborns during the unmonitored periods. Each individual newborn's exposure to Σ8 c-PAHs over the entire gestational period was estimated using two approaches – a random effects model and a semi-parametric mixed effects model.

First, based on our earlier analysis [25], the individual newborn's chronic prenatal exposure was estimated using a random effects model as shown below in [1].

$$\ln(\sum 8 \text{ c-PAHs})_{ij} = b_{i1} + (b_{i2} \times t_{ij}) + X_{ij}^T \beta + \varepsilon_{ij}. \quad (1)$$

Personal exposure to $\ln(\sum 8 \text{ c-PAHs})_{ij}$ for the ith subject at the jth measurement was modeled as a function of the random intercept (b_{i1}), the product of random slope (b_{i2}) and calender time t_{ij}, and a vector of other covariates, which are time-independent (e.g., living in City center, secondhand smoke exposure), denoted as $X_{ij}^T \beta$. The measurement error is noted as ε_{ij}. The superscript ^{T} represents the vector of each covariate, and β the associated regression coefficient. Here, b_{i1}, b_{i2} and ε_{ij} are expected to be normally distributed with a mean of 0. By including both random intercept and random slope, the predicted individual PAH exposure trajectory not only shift from the population mean curve by a subject specific amount, b_{i1}, but also has a subject specific slope, b_{i2}. To reduce within-person collinearity in predicted personal exposure, mean over three months period of personal

exposure to $\ln(\sum 8 \text{ c-PAHs})_{ij}$ was taken for each person during each trimester.

Alternatively, we estimated individual's chronic gestational exposure using a semi-parametric mixed model [25] as shown in [2].

$$\ln(\sum 8 \text{ c-PAHs})_{ij} = b_{i1} + (b_{i2} \times t_{ij}) + X_{ij}^T \beta + f(t_{ij}) + \varepsilon_{ij}. \quad (2)$$

Personal exposure to $\ln(\sum 8 \text{ c-PAHs})_{ij}$ for the ith subject at the jth measurement was modeled as a function of the random intercept (b_{i1}), the product of random slope (b_{i2}) and calender time t_{ij}, and a vector of other time-independent covariates (i.e., living in city center, secondhand smoke, parity, maternal pre-pregnancy body mass index, gestational age, newborn gender, and c-section delivery for birth head circumference and Cephalization Index only) ($X_{ij}^T \beta$), where ^{T} stands for a transpose of a vector of the covariates, and β denotes the corresponding coefficients. The nonlinear trend in ambient PAH levels was captured by a fully nonparametric function, $f(t_{ij})$. Term, e_{ij}, denotes a measurement error. The distributions of b_{i1}, b_{i2} and ε_{ij} were assumed to be normal with a mean of 0. Overall, the semi-parametric model yielded more precise predicted concentrations of personal exposure than the estimates using the linear mixed effects model during each gestational month (see Figure S1). Considering the observed cohort's mean personal exposure throughout the monitoring period as the gold—standard, Pearson's coefficients between predicted and observed personal exposure levels were 0.91, 0.98 and 0.96 for the third, sixth and the eighth gestational month using semi-parametric mixed model (Figure S1). In comparison, the linear mixed effects model, Pearson's coefficient between predicted and observed concentration was 0.76 during the sixth gestational month.

Parametric Analysis of the Birth Outcomes. Functional linear models of the ith subject's birth outcome is shown as,

$$\text{Birth Outcome}_i = \int_0^{T_{delivery_i}} \ln\left(\sum 8 \, c-PAHs\right)_i (s) \, \alpha \, (s) + \tilde{x}_i^T \tilde{\beta} + \varepsilon_i, \quad (3)$$

where the fetal growth is a function of the integrated chronic exposure between conception and delivery, $\int_0^{T_{delivery_i}} \ln (\sum 8 \, c-PAHs)_i(s) \, \alpha \, (s)$, and other time-independent, potential confounders (i.e., gestational age, newborn gender, parity, and the mother's pre-pregnancy body mass index) $\left(\tilde{X}_i^T \tilde{\beta}\right)$, and measurement error (ε_i) [36]. Here, $(s) \, \alpha \, (s)$ denote instantaneous time unit and the coefficient curve, respectively. The measurement error (ε_i) is assumed to be distributed normally with mean 0.

Non-Parametric Analysis of the Birth Outcomes. Indicator variables for the season during each given gestational month were used as a proxy of maternal exposure to PAHs. Our prior observations demonstrated that the season of PAH monitoring is a precise and accurate indicator of personal exposure ($R^2 = 0.74$; all regression coefficients >0.96) [25]. The calendar months between conception and delivery dates were calculated based on last menstrual date- or sonogram-based gestational age [35]. To preclude multicollinearity in the birth outcome models, the indicator variables were calculated for each newborn's first, third and the last gestational month. The season during given gestational month was coded as winter if it corresponded to the months December–February. The remaining calendar months were set as the reference.

$Birth\ Outcome_i =$

$$Z_i^T \bar{\delta} + \ln\left(\sum 8\ c - PAHs\right)_i^{sixth\ gestational\ month} + \tilde{X}_i^T \tilde{\beta} + \varepsilon_{i.} \tag{4}$$

Here, Z represents a vector of the winter indicator variable during the first, third, and the ninth gestational months, and $\bar{\delta}$ represents the respective regression coefficients. Since personal air monitoring was conducted in all women in the cohort during sixth gestational month, we accounted for this with the term, $\ln(\sum 8\ c - PAHs)_i^{sixth\ gestational\ month}$.

Cross-Validation of the predicted PAH exposure. The prediction ability of our final model for a future observation is evaluated using "leaving-one-out" strategy of cross validation. Specifically, we fit the model with one observation deleted at each time and use that observation as the test sample to calculate the prediction error, defined as the difference square between observed and predicted. We do this for each observation and average the prediction error over. That is, we calculate

$$CV = n^{-1} \sum_{i=1}^{n} \{Y_i - \hat{Y}_i^{(-i)}\}^2 \tag{5}$$

where Y_i denote the observed outcome of the ith observation and $\hat{Y}_i^{(-i)}$ denote the predicted outcome of that observation with regression applied to the data with (X_i, Y_i) deleted. The statistical analysis was conducted in R, version 2.5.1 [37], SAS version 9.1.3 (SAS Institute Inc., Cary, NC, USA), and PASW Statistics version 14.0 (SPSS Inc., 2009, Chicago, IL, USA). Figures were generated in R 2.5.1 and PASW Statistics version 14.0.

Results

The demographic characteristics of the mother-newborn pairs are shown in Table 1. The majority of the pregnant women did not have any risk factors for adverse birth outcomes (e.g., secondhand cigarette smoke, low educational attainment, or absence of spouse during the present pregnancy) or other sources of PAH exposure (i.e. routine consumption of grilled/barbecued food).

The prevalence of preterm delivery (4.7%, n = 16) and low birthweight (0.6%, n = 2) was low. Overall, mean FGR (mean ± S.D., 104±12%, range, 70–149%) and mean Ponderal Index (mean ± S.D., 2.11±0.21, range, 1.60–3.12) suggest that the birthweight distribution of the present cohort were comparable to the Polish population means. In this cohort, FGR demonstrates a high internal validity as a summary index of IUGR, and is consistent with trends in birthweight, birth length, birth head circumference, and Cephalization Index (Table 2). The severity of FGR was inversely correlated with the Cephalization Index (Pearson's coefficient = −0.68, p-value<0.001). At the same time, severe FGR newborns (<75%) were not significantly thinner (i.e. lower Ponderal Index) than non-FGR newborns (Table 2).

I. Estimated Personal Chronic PAH Exposure

Consistent with our earlier analysis [25], the three illustrative subjects shown in Figure 1 demonstrate that their chronic individual-level of exposure is predominantly influenced by ambient PAHs. The same subjects show that within-person variability in the estimated PAH exposure across the seasons was markedly larger than the between-person variability in the estimated PAH exposure during a same season (Figure 1). Prediction error between the predicted vs. observed exposure to $\Sigma 8$ c-PAHs was 0.43 for the 344 newborns (see Figure S1).

II. Mean PAH Effect Sizes by Trimester on Symmetric Fetal Growth Restriction

a. Random Effects Model-Estimated PAH Exposure and Mean Effect By Trimester. Using random effects model, one ln-unit exposure to $\Sigma 8$ c-PAHs during the first trimester was associated with the largest mean reduction in FGR (−3%, 95% CI, −5 to −0%), compared to the mean reduction during the second or the third trimester (see Figure 2).

In addition, the same unit increase in exposure was associated with the largest mean reduction in birthweight during the first trimester (−105 g, 95% CI, −188 to −22 g), compared to the second (−36 g, 95% CI, −76 to 4 g), or the third trimester (−44 g, 95% CI, −123 to 36 g). For the birth length, a one ln-unit increase in exposure during the first trimester was associated with the largest mean reduction (−0.78 cm, 95% CI, −1.30 to −0.26 cm) compared to the same unit of exposure during the second trimester (−0.24 cm, 95% CI, −0.49 to 0.01 cm) or during the third trimester (−0.57 cm, 95% CI, −1.07 to −0.07 cm). However, one ln-unit of exposure was associated with a similar reduction in birth head circumference during the first (−0.11 cm, 95% CI, −0.39 to 0.17 cm) and the second trimesters (−0.13 cm, 95% CI, −0.26 to 0.00 cm). All models controlled for the mother's pre-pregnancy body mass index, gestational age, gender and parity.

b. Semi-Parametric Mixed Effects Model-Based Exposure Estimation. Figure 3 shows the point-wise PAH effect on FGR throughout the gestational period. Within this cohort, the newborns that experienced their highest exposure during the earliest gestational weeks were associated with the largest point-wise FGR decrement, compared to other newborns that experienced lower exposure during the same period. Among the full-term newborns, a high ambient PAH concentration during the earliest gestational weeks is positively correlated with an elevated exposure during the last gestational months due to seasonal trend. Thus, a unit of PAH exposure during the first four (or, the last four) gestational weeks is associated with a largest reduction in FGR. Using the semi-parametric exposure estimation approach, a unit of PAH exposure during the first gestational month was associated with a mean reduction in FGR of 0.244% (95% CI, −0.574, 0.086%), compared to the effect size during the reference period (6th month). Similarly, a unit PAH exposure during the 9th gestational month was associated with a 0.178% (95% CI, −0.558, 0.201%) reduction in FGR after adjusting for the effects during other gestational months.

c. Non-parametric Comparison of ambient PAH Levels and Birth Outcomes. Compared to the newborns who experienced winter during their 6th gestational month, those that experienced winter (December–February period) during their 1st gestational month had a significantly larger reduction in birthweight (−191 g, 95% CI, −316 to −67 g), birth length (−1.14 cm, 95% CI, −1.93 to −0.35 cm) and FGR (−5%, 95% CI, 9 to −2%) respectively. The models accounted for the same set of potential confounders (Table 3).

III. Prenatal PAH Exposure and Asymmetric Fetal Growth Restriction

One ln-unit $\sum 8$ PAHs during the first trimester (using random effects model for prediction of chronic exposure) was associated with the largest mean elevation in Cephalization Index (3 $\mu m/g$, 95% CI, 1 to 5 $\mu m/g$), adjusting for the same set of confounders (Figure 4). Based on the semi-parametric exposure estimation method, the first month's exposure was associated with a highest point-wise elevation in Cephalization Index (0.22 $\mu m/g$, 95% CI, −0.04, 0.48 $\mu m/g$), compared to the ratio during the 6th month

Table 1. Demographic and exposure characteristics of mother-newborn pairs.

Characteristics	N (%)	Mean ± S.D.
Mother's characteristics		
Age [years]	344 (100%)	28±4
Annual household income[a]		
Low <37,024 PLN	232 (67.4%)	
Medium 37,024–74,048 PLN	16 (4.7%)	
High >74,048 PLN	1 (0.3%)	
Refused/Don't know	95 (27.6%)	
Education		
<high school	37 (10.8%)	
high school graduate	91 (26.5%)	
Attained>high school	216 (62.8%)	
Parity [yes]	126 (36.6%)	
Currently married [yes]	320 (93.0%)	
Routinely consumed high level of PAH-laden food items[b]	49 (14.2%)	
Pregnant women with daily alcohol intake[c]	4 (1.2%)	
Season of delivery		
Winter (Dec–Feb)	85 (24.7%)	
Spring (March–May)	94 (27.3%)	
Summer (June–Aug)	90 (26.2%)	
Fall (Sep–Nov)	75 (21.8%)	
Secondhand cigarette smoke		
Non-smoker home	286 (83.1%)	
Exposed to ≤4 hrs/day	46 (13.4%)	
Exposed to 5+ hrs/day	12 (3.5%)	
Newborn characteristics		
Gender [female]	175 (50.9%)	
Newborn cotinine (ng/ml)	228 (66.3%)	0.319±0.882
Birthweight (g)		3430±491
Birth length (cm)		55±3
Birth head circumference (cm)		33.93±1.48

[a]For 2003, 1 US dollar was equal to 4.07 Poland Zlotych (PLN).
[b]Reported taking at least one smoked, grilled or barbequed food >twice/wk.
[c]Drank at least one glass of wine, beer, or liquor per day during pregnancy.

(0.01 μm/g, 95% CI, −0.12, 0.14 μm/g, Figure 5). As the exposure during the earliest and the last gestational months are correlated, the same unit exposure was also associated with an elevation in Cephalization Index (e.g., 0.05 μm/g) during the 9th month. Alternatively, based on consideration of non-parametric indicators, the newborns who experienced winter (December February) during their 1st gestational month, on average, had the largest elevation in Cephalization Index (5 μm/g, 95% CI, 1 to 8 μm/g), compared to the newborns who experienced winter during their 6th gestational month (Table 3). On the other hand, prenatal PAH exposure was not associated with a significant reduction in Ponderal Index during any gestational period (Figure 6).

Discussion

The identification of a "window of critical vulnerability" to ubiquitous air pollutants such as PAHs is a particularly important, yet challenging, question. Critical hurdle with answering this question regards the dose-response relationship of the xenotox-

icant during a given age, which is inherently related to the host's susceptibility as well as the host's adaptiveness. Such exquisite sensitivity of the fetus and newborn to xenotoxicants is thought to be related to the immaturity of the developing immune systems; the rapid development of fetal organs; epigenetic mediation; and the fact that exposure per body weight is much higher than for adult exposure [38,39,40]. Thus, the age-specific measurement of PAH exposure is critical for the clarification of the severity and the type of IUGR [10,22].

The present analysis supports the hypothesis that PAH exposure during the first trimester imparts the largest reduction in the markers of symmetric fetal growth restriction. Furthermore, we originally posited that elevated Cephalization Index represents a marker of asymmetric growth. However, our results suggest that it is, in fact, correlated with a symmetric fetal growth restriction. Prenatal PAH exposure during the earliest gestational months was associated with a consistent effect on Cephalization Index and FGR. Also, we saw no evidence that PAH exposure increases the

Table 2. Correlation between fetal growth ratio with other anthropometric indicators[a].

fetal growth ratio (%)N		Birthweight [g]	Birth Length [cm]	Birth Head Circumference [cm]	Ponderal Index [g/cm³×100]	Cephalization Index [μm/g]
		Mean ± SD	Mean ± SD	Mean ± SD	Mean ± SD	Mean ± SD
		(min – max)	(min – max)	(min – max)	(min – max)	(min – max)
Non-case	326	3482±436	55±3	34.07±1.34	2.12±0.21	99±11
(fetal growth ratio ≥85%)		(2150–4700)	(46–64)	(31.00–39.00)	(1.60–3.12)	(74–153)
Mild	11	2657±231	51±2	31.91±0.70	1.96±0.20	121±12
(80–84.99%)		(2210–2850)	(50–55)	(31.00–33.00)	(1.68–2.27)	(109–145)
Moderate	2	2225±460	48±3	31.50±0.71	2.00±0.06	144±27
(75–79.99%)		(1900–2550)	(46–50)	(31.00–32.00)	(1.95–2.04)	(125–163)
Severe	3	2190±295**	48±3**	30.67±0.58**	2.03±0.20	142±21**
(<75%)		(1870–2450)	(44–50)	(30.00–31.00)	(1.80–2.20)	(127–166)
Overall mean[b]	342	3436±477	55±3	33.96±1.42	2.11±0.21	100±13
		(1870–4700)	(44–64)	(30.00–39.00)	(1.60–3.12)	(75–166)

[a]Severity was defined as non-case (≥85%), mild (80–84.99%), moderate (75–79.99%), and severe (<75%) [46].
[b]Sample size is reduced from 344 to 342 because two newborns had gestational age of 43 and 29 weeks, which fell outside the plausible range [34].
**Test of linear trend with increasing severity of FGR, p<0.001.

likelihood of acute fetal growth restriction, as indicated by the Ponderal Index.

The strong inverse correlation between FGR and Cephalization Index in our results is also supported by similar results in the murine model [16]. B[a]P administration during organogenesis interfered with preliminary synapse formation during the earliest weeks of gestation and induced the largest disproportionate increase in Cephalization Index [16]. Furthermore, gestational B[a]P exposure postnatally inhibited the cortical region for learning and memory [16].

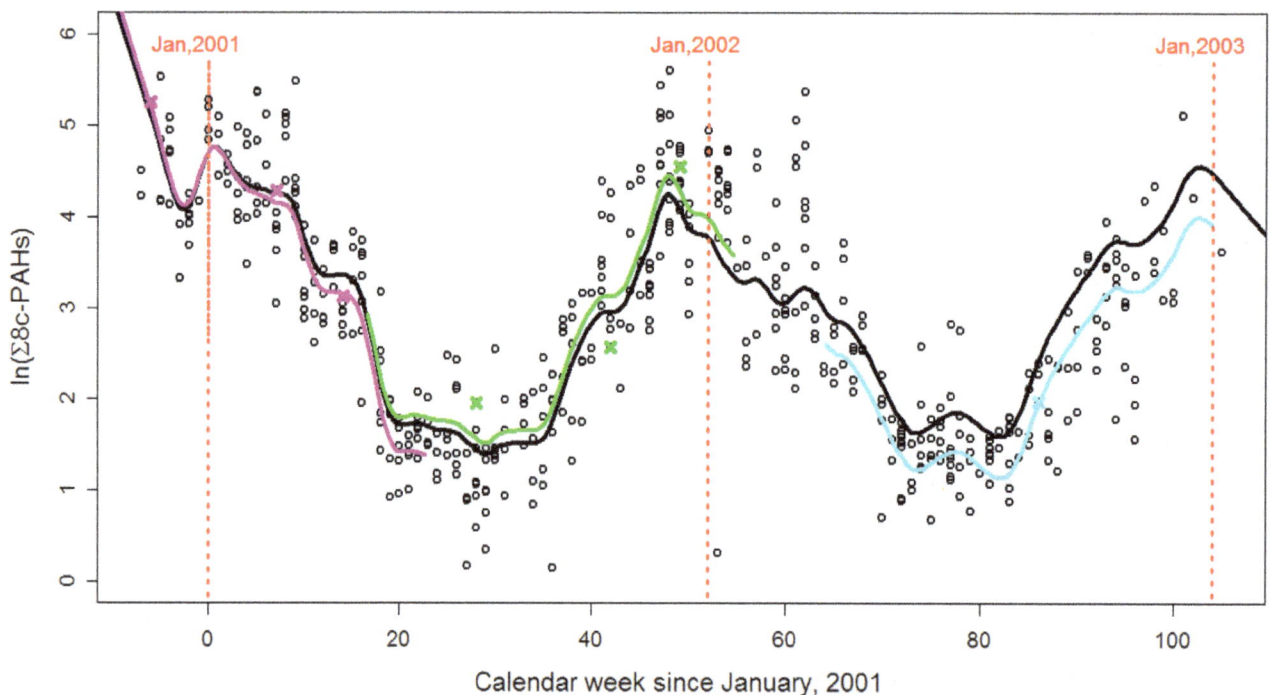

Figure 1. The observed and predicted individual gestational Σ8 c-PAH exposure using semi-parametric mixed effects model[a]. a. Black, open circle represents the observed concentration of Σ8 c-PAHs. Black line represents the pooled cohort mean during the entire monitoring period. Three persons were randomly selected to demonstrate estimated personal Σ8 c-PAHs exposure during her entire pregnancy period (in color). Based on the semi-parametric mixed effects model, Pearson's correlation coefficients between observed vs. predicted prenatal exposure were 0.91, 0.98 and 0.96 for the third, sixth and the eighth month.

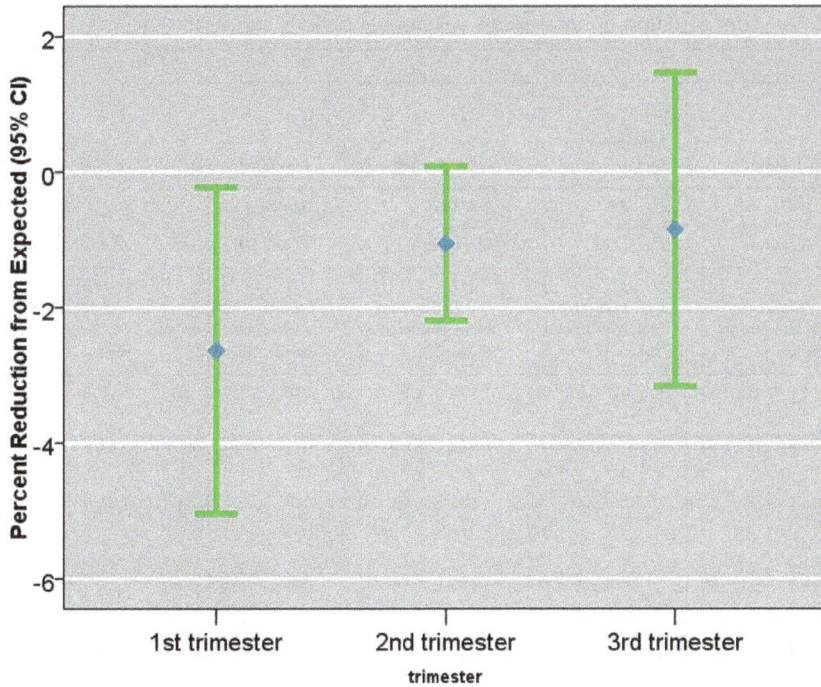

Figure 2. Mixed Effects Model Estimated ln-unit Σ8 C-PAH Exposure and Their Effects on Fetal Growth Ratio. The ln Σ8 c-PAHs effects on FGR was estimated as a mean effect per trimester-wise exposure and 95% confidence interval.

Our present observation of the largest unit effect during the first trimester suggests that PAHs might influence the rate of fetal growth. Other epidemiologic investigations independently observed that fetal growth rate is programmed during the earliest gestational period, resulting in a progressively larger deficit as gestation matures [18,19,20].

An adverse intrauterine environment, particularly during the early pregnancy period, is hypothesized to switch on the survival mechanism of the fetus by protecting vital organs such as brain and heart while suppressing the development of other systems [41]. Recent clinical examinations in various populations have demonstrated that a significantly slower growth rate begins in the

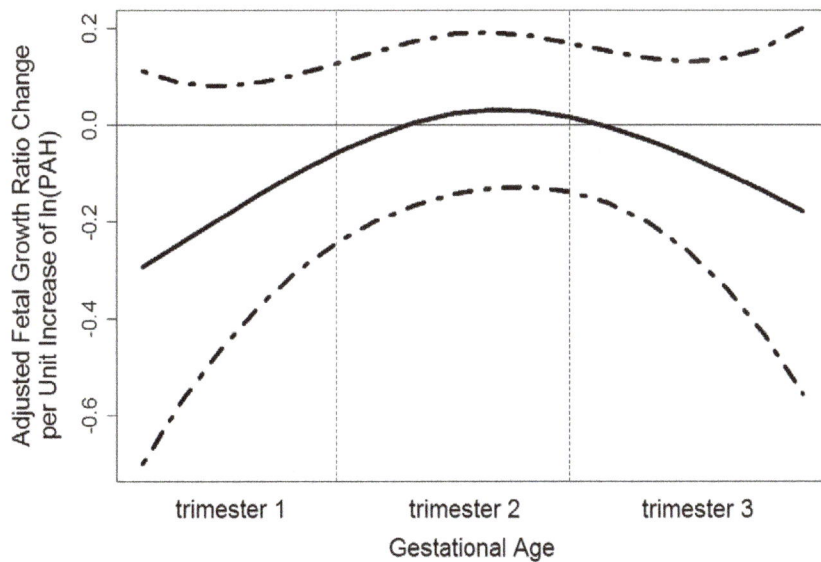

Figure 3. Semi-Parametric Mixed Model Estimated ln-Unit Σ8 C-PAH Exposure and Their Point-Wise Effects on Fetal Growth Ratio. Exposure was estimated using semi-parametric mixed effects model. Figure shows continuous, point-wise effect throughout the trimester based on functional linear model. The bold line shows regression coefficient per natural-log (ln) unit exposure to airborne PAHs. The dotted lines show point-wise 95% confidence interval.

Table 3. Non-parametric markers of ambient PAH concentration during various gestational periods and birth outcomes[a].

Nonparametric markers of ambient PAH concentration during various gestational periods	Birthweight	Birth Length	Birth head circumference	Fetal Growth Ratio	Ponderal Index	Cephalization Index
	(n = 344)	(n = 344)	(n = 344)	(n = 340)	(n = 344)	(n = 344)
	$R^2 = 0.461$	$R^2 = 0.348$	$R^2 = 0.275$	$R^2 = 0.093$	$R^2 = 0.018$	$R^2 = 0.505$
	[g] reduction from the mean (95% CI)	[cm] reduction from the mean (95% CI)	[cm] reduction from the mean (95% CI)	[%] reduction from the mean (95% CI)	[g/cm³×100] reduction from the mean (95% CI)	[μm/g] reduction from the mean (95% CI)
Season during 1st month (winter vs. other)[b]	−191 (−316, −67)	−1.14 (−1.93, −0.35)	−0.38 (−0.80, 0.03)	−5.37 (−8.92, −1.82)	0.02 (−0.05, 0.08)	5 (1, 8)
Season during 3rd month (winter vs. other)[b]	−122 (−226, −17)	−0.62 (−1.29, 0.05)	−0.37 (−0.72, −0.02)	−3.04 (−6.06, −0.03)	0.00 (−0.06, 0.06)	3 (0, 6)
Directly monitored personal c-PAH exposure, second trimester[c]	−67 (−110, −23)	−0.48 (−0.76, −0.20)	−0.20 (−0.34, −0.05)	−1.85 (−3.09, −0.60)	0.01 (−0.01, 0.04)	1 (0, 3)
Season during 9th month (winter vs. other)[b]	−40 (−148, 68)	−0.36 (−1.05, 0.33)	−0.05 (−0.41, 0.32)	−0.69 (−3.78, 2.39)	0.02 (−0.04, 0.08)	1 (−2, 4)

[a]Model controls for gestational age (centered at mean and square term of the centered at mean), newborn gender, parity, and the mother's pre-pregnancy body mass index. C-section delivery is included only for the head circumference.
[b]To reduce multicollinearity in the models, the indicator variable was coded as the winter vs. other.
[c]Direct measurement of personal exposure was determined using a personal monitor for 48-hour period. The exposure variable was coded in natural-log scale.

earliest gestational weeks for growth restricted newborns [18,19,20]. Among a group of singleton newborns who were longitudinally followed from the 12th week of gestation to delivery by ultrasound, a significant difference in inter-individual rate of fetal growth was evident in terms of fetal abdominal and head circumference, as well as femur diaphysis length [20]. Starting around the 13th week, significantly slower fetal growth velocity was evident for those who were born in the lowest third percentile of birth weight. This difference in fetal growth velocity remained constant throughout gestation after maternal anthropomorphic characteristics (including pre-pregnancy weight, height and weight gain) were accounted for [20].

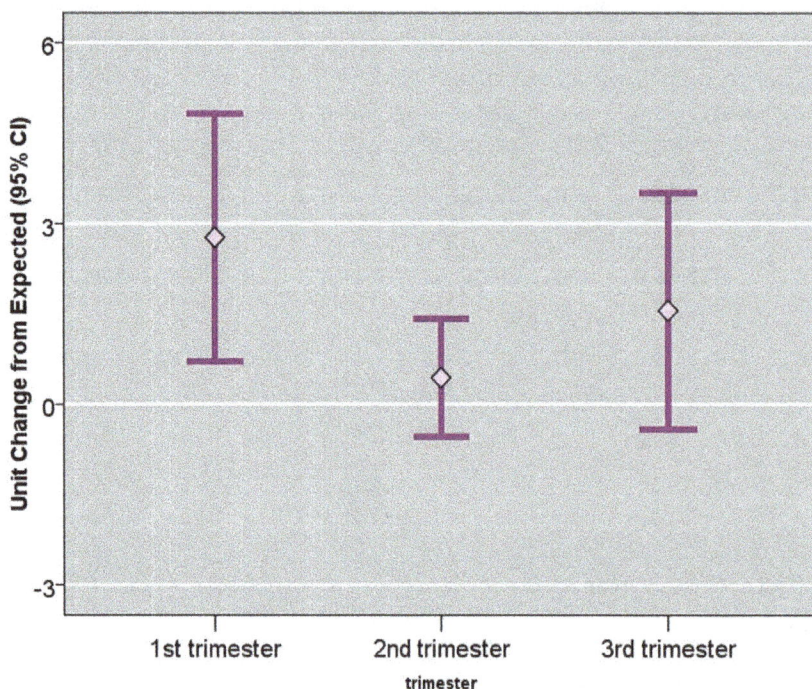

Figure 4. Mixed Effects Model Estimated ln-unit Σ8 C-PAH Exposure and Their Effects on Cephalization Index. The ln Σ8 c-PAHs effects on the outcome was estimated as a mean effect per trimester-wise exposure and 95% confidence interval.

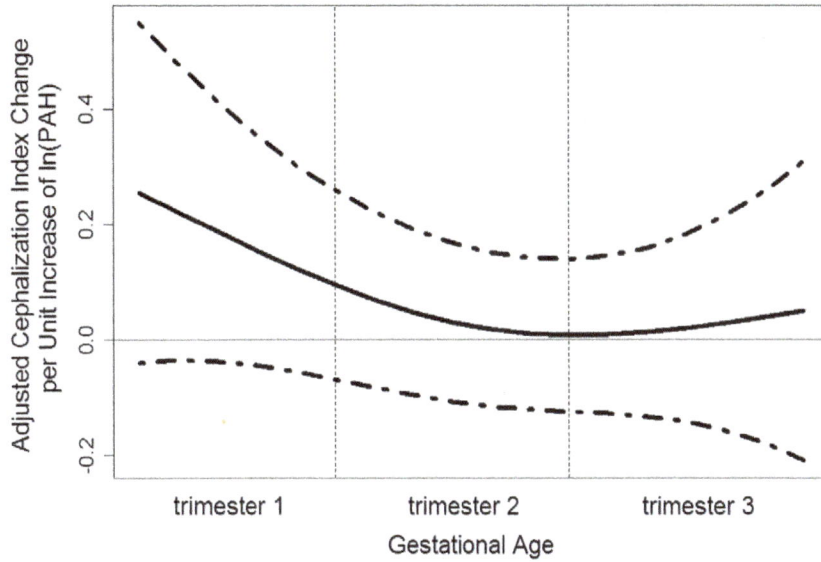

Figure 5. Semi-Parametric Mixed Model Estimated ln-Unit Σ8 C-PAH Exposure and Their Point-Wise Effects on Cephalization Index. Exposure was estimated using semi-parametric mixed effects model. Figure shows continuous, point-wise effect throughout the trimester based on functional linear model. The bold line shows regression coefficient per natural-log (ln) unit exposure to airborne PAHs. The dotted lines show point-wise 95% confidence interval.

In some populations, fetuses that are born small-for-gestational age or with low birthweight, are at greater risk of neurodevelopmental delays [20], impaired lung function [42], asthma symptoms [43] throughout childhood, as well as cardiopulmonary diseases [41] during adulthood, including hypertension, and atherosclerosis, as well as diabetes [44]. Our personal, indoor, and outdoor air monitoring of 344 women represents one of the largest and most comprehensive PAH exposure assessment campaigns to date. The airborne PAH exposure in Krakow, Poland, represents a typical exposure scenario in countries dependent on coal-burning for heat and power generation [24]. Ambient PAH concentrations differ by more than two orders of magnitude between summer and winter with overall spatial homogeneity in concentration during given season [24]. The extreme seasonal fluctuation of ambient PAH

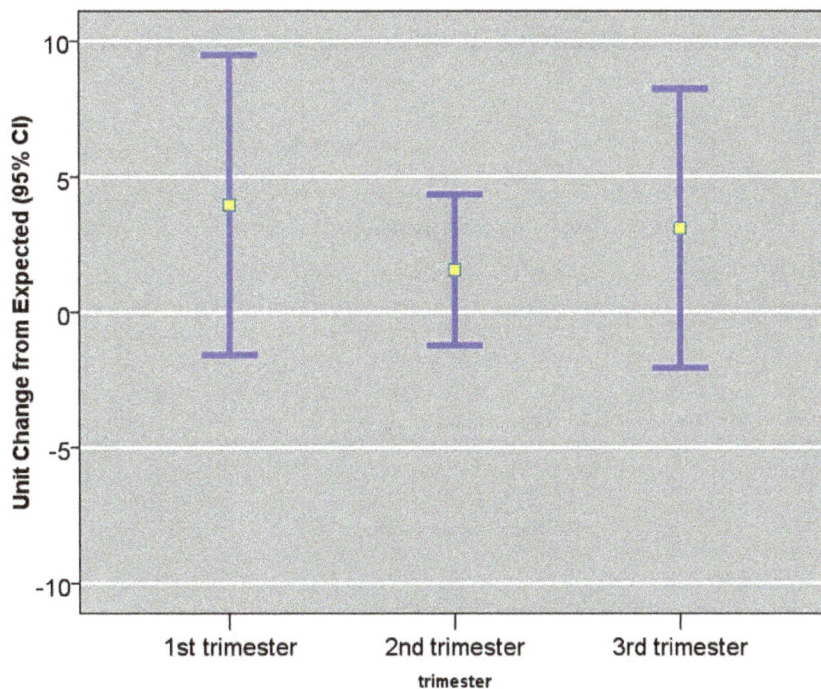

Figure 6. Mixed Effects Model Estimated ln-unit Σ8 C-PAH Exposure and Their Effects on Ponderal Index. The ln Σ8 c-PAHs effects on the outcome was estimated as a mean effect per trimester-wise exposure and 95% confidence interval.

levels was instrumental in obtaining valid and precise predicted personal PAH exposure concentrations during the unmonitored months. Spearman's coefficients for concurrent personal, indoor and outdoor measurements ranged between 0.96–0.98 [25]. Thus, the spatial variability in personal, indoor and outdoor PAH concentration (within a given household) was very small during a typical air pollution episode (Figure 1). For example, the maximum values for personal B[a]P exposure concentration (42.23 ng/m^3) in our cohort and ambient B[a]P level (200 ng/m^3), as reported by Junninen et al. (2009), represent two of the highest values of the compound that have ever been documented. Accordingly, the cohort's mean exposure over the monitoring period was weighed heavily in estimating the personal exposure of other newborns during their unmonitored months. Pearson's correlation coefficients between observed vs. predicted personal exposure concentration were 0.91, 0.98, and 0.96 at the 3rd, 6th, and 8th gestational month, respectively.

Another key strength of the study is the application of stringent enrollment criteria. Only young Polish women (18–35 years of age), non-smoking, healthy (i.e., free of diabetes, hypertension, or known HIV, nonusers of other tobacco products or illicit drugs) were targeted. In addition, only those who received adequate prenatal care (i.e. enrolled in the clinic between 8th to 13th week of pregnancy) were included. Thus, several important confounders have been precluded. At the same time, several study limitations warrant consideration. As our strict enrollment criteria precluded notable sources of confounding, our mothers and newborns cohort is not representative of the general population. Furthermore, only those fetuses who survived beyond the 8–13th gestational weeks were recruited into the study. Thus, PAH effects in the general population might be different than in the present cohort.

It is currently unknown whether other constituents of coal combustion by-products, including Cadmium (Cd), Nickel (Ni), Arsenic (As), and Lead (Pb) affect birth outcomes through a mechanism common with PAHs. Thus, our results cannot rule out confounding effects from other airborne correlates of PAHs, particularly during the winter. Another limitation of the study is the fact that we did not consider the genetic polymorphisms of the developing fetus and/or the mothers for the risk of IUGR. In our earlier analyses, both maternal and newborn haplotypes of the cytochrome 450 genes CYP1A1 significantly augmented PAH effects on children's neurocognitive development once they reached the age of 1, 2, and 3 respectively [45]. It is plausible that the genes involved in PAH metabolic activation or detoxification may also modify PAH exposure risks on IUGR.

Cyclic fluctuation in ambient PAH concentration influenced the mean personal exposure during the first gestational month so that it was positively correlated with exposure during the ninth gestational month. While we statistically adjusted for this correlation in exposure, future research should quantify the fetal growth rate in real-time during gestation, rather than determine the absolute size decrement at birth. Such a measurement would also be useful in early detection of IUGR cases.

Conclusion

The identification of a "window of critical vulnerability" to ubiquitous air pollutants such as PAHs is a particularly important, yet challenging, question. The challenge surrounding this question stems, at least partly, from the fact that the dose-response relationship of the xenotoxicant during a given age is inherently related to the host's susceptibility as well as the host's adaptiveness. Furthermore, exposure duration is chronic, yet variable. Our results based on several alternate exposure estimation approaches suggest that one ln-unit PAH exposure during the first trimester, and the first gestational month in particular, increases the risk of FGR reduction and Cephalization Index elevation, respectively. On the other hand, no gestational period was associated with a marked reduction in Ponderal index. Reduction in birthweight, birth length and FGR, as well as an elevated CI predict mortality and morbidity risks in newborns and compromised cognitive development in children. In addition, ambient PAH concentrations in Krakow are typical of regions dependent on coal-burning for heat and power generation [24]. The present data support the need for a multinational coal-combustion abatement strategy for the protection of pregnant women and the embryo/fetus, particularly during the earliest stage of pregnancy.

Acknowledgments

E. Evans, R. Whyatt, H. Andrews, L. Hoepner, W. Jedrychowski, R. Jacek, E. Mroz, A. Pac, E. Flak, W. Tsai, D. Tang, J. Yu contributed to the study.

Author Contributions

Conceived and designed the experiments: JDS FPP. Performed the experiments: HC LW XL. Analyzed the data: HC LW XL. Wrote the paper: HC LW.

References

1. ATSDR (1995) Toxicological Profile for Polycyclic Aromatic Hydrocarbons (PAHs). Atlanta, GA.: Agency for Toxic Substances and Registry.
2. Bostrom CE, Gerde P, Hanberg A, Jernstrom B, Johansson C, et al. (2002) Cancer risk assessment, indicators, and guidelines for polycyclic aromatic hydrocarbons in the ambient air. Environ Health Perspect 110 Suppl 3: 451–488.
3. Finlayson-Pitts BJ, Pitts JN, Jr. (1997) Tropospheric air pollution: ozone, airborne toxics, polycyclic aromatic hydrocarbons, and particles. Science 276: 1045–1052.
4. Perera FP, Li Z, Whyatt R, Hoepner L, Wang S, et al. (2009) Prenatal Airborne Polycyclic Aromatic Hydrocarbon Exposure and Child IQ at Age 5 Years. Pediatrics 124: e195–202.
5. IARC (1983) Polynuclear aromatic compounds. Part 1. Chemical, environmental and experimental data. (IARC Monographs on the evaluation of the carcinogenic risk of chemicals to humans. Vol 32). Lyon: International Agency for Research Cancer.
6. WHO (2000) Air quality guidelines for Europe. WHO, editor. Copenhagen.
7. Rodriguez JW, Kohan MJ, King LC, Kirlin WG (2002) Detection of DNA adducts in developing CD4+ CD8+ thymocytes and splenocytes following in utero exposure to benzo[a]pyrene. Immunopharmacol Immunotoxicol 24: 365–381.
8. Castro DJ, Lohr CV, Fischer KA, Pereira CB, Williams DE (2008) Lymphoma and lung cancer in offspring born to pregnant mice dosed with dibenzo[a,l]pyrene: the importance of in utero vs. lactational exposure. Toxicol Appl Pharmacol 233: 454–458.
9. WHO (1998) International programme on chemical safety. Environmental Health Criteria 202. Selected non-heterocyclic policyclic aromatic hydrocarbons.
10. Yu Z, Loehr CV, Fischer KA, Louderback MA, Krueger SK, et al. (2006) In utero exposure of mice to dibenzo[a,l]pyrene produces lymphoma in the offspring: role of the aryl hydrocarbon receptor. Cancer Res 66: 755–762.
11. Perera F, Tang WY, Herbstman J, Tang D, Levin L, et al. (2009) Relation of DNA methylation of 5'-CpG island of ACSL3 to transplacental exposure to airborne polycyclic aromatic hydrocarbons and childhood asthma. PLoS One 4: e4488.

12. Perera F, Tang D, Whyatt R, Lederman SA, Jedrychowski W (2005) DNA damage from polycyclic aromatic hydrocarbons measured by benzo[a]pyrene-DNA adducts in mothers and newborns from Northern Manhattan, the World Trade Center Area, Poland, and China. Cancer Epidemiol Biomarkers Prev 14: 709–714.

13. Bocskay KA, Tang D, Orjuela MA, Liu X, Warburton DP, et al. (2005) Chromosomal Aberrations in Cord Blood Are Associated with Prenatal Exposure to Carcinogenic Polycyclic Aromatic Hydrocarbons. Cancer Epidemiol Biomarkers Prev 14: 506–511.

14. Dejmek J, Solansky I, Benes I, Lenicek J, Sram RJ (2000) The impact of polycyclic aromatic hydrocarbons and fine particles on pregnancy outcome. Environmental Health Perspectives 108: 1159–1164.

15. Choi H, Rauh V, Garfinkel R, Tu Y, Perera FP (2008) Prenatal exposure to airborne polycyclic aromatic hydrocarbons and risk of intrauterine growth restriction. Environ Health Perspect 116: 658–665.

16. Brown LNA, Khousbouei H, Goodwin JS, Irvin-Wilson CV, Ramesh A, et al. (2007) Down-regulation of early ionotrophic glutamate receptor subunit developmental expression as a mechanism for observed plasticity deficits following gestational exposure to benzo(a)pyrene. NeuroToxicology 28: 965–978.

17. Lu L-JW, Anderson LM, Jones AB, Moskal TJ, Salazar JJ, et al. (1993) Persistence, gestation stage-dependent formation and interrelationship of benzo[a]pyrene-induced DNA adducts in mothers, placentae and fetuses of Erythrocebus patas monkeys. Carcinogenesis 14: 1805–1813.

18. Smith GC (2004) First trimester origins of fetal growth impairment. Semin Perinatol 28: 41–50.

19. Neufeld L, Pelletier DL, Haas J (1999) The timing hypothesis and body proportionality of the intra-terine growth retarded infants. American Journal of Human Biology 11: 638–646.

20. Milani S, Bossi A, Bertino E, Battista ED, Coscia A, et al. (2005) Differences in size at birth are determined by differences in growth velocity during early prenatal life. Pediatric Research 57: 205–210.

21. Chaddha V, Viero S, Huppertz B, Kingdom J (2004) Developmental biology of the placenta and the origin of placental insufficiency. Seminars in Fetal and Neonatal Medicine 9: 357–369.

22. Detmar J, Rennie MY, Whiteley KJ, Qu D, Taniuchi Y, et al. (2008) Fetal growth restriction triggered by polycyclic aromatic hydrocarbons is associated with altered placental vasculature and AhR-dependent changes in cell death. Am J Physiol Endocrinol Metab 295: E519–530.

23. Sram RJ, Binkova B, Rossner P, Rubes J, Topinka J, et al. (1999) Adverse reproductive outcomes from exposure to environmental mutagens. Mutation Research 428: 203–215.

24. Junninen H, Mønster J, Rey M, Cancelinha J, Douglas K, et al. (2009) Quantifying the Impact of Residential Heating on the Urban Air Quality in a Typical European Coal Combustion Region. Environmental Science & Technology 43: 7964–7970.

25. Choi H, Perera F, Pac A, Wang L, Flak E, et al. (2008) Estimating individual-level exposure to airborne polycyclic aromatic hydrocarbons throughout the gestational period based on personal, indoor, and outdoor monitoring. Environ Health Perspect 116: 1509–1518.

26. Kramer MS, Olivier M, McLean FH, Willis DM, Usher RH (1990) Impact of Intrauterine Growth Retardation and Body Proportionality on Fetal and Neonatal Outcome. Pediatrics 86: 707.

27. Sizonenko SV, Borradori-Tolsa C, Bauthay DM, Lodygensky G, Lazeyras F, et al. (2006) Impact of intrauterine growth restriction and glucocorticoids on brain development: insights using advanced magnetic resonance imaging. Mol Cell Endocrinol 254–255: 163–171.

28. Van Wassenaer A (2005) Neurodevelopmental consequences of being born SGA. Pediatric endocrinology reviews 2: 372–377.

29. Xu Y, Williams SJ, O'Brien D, Davidge ST (2006) Hypoxia or nutrient restriction during pregnancy in rats leads to progressive cardiac remodeling and impairs postischemic recovery in adult male offspring. FASEB J 20: 1251–1253.

30. Leitner Y, Fattal-Valevski A, Geva R, Eshel R, Toledano-Alhadef H, et al. (2007) Neurodevelopmental outcome of children with intrauterine growth retardation: a longitudinal, 10-year prospective study. J Child Neurol 22: 580–587.

31. Sanyal MK, Li YL (2007) Deleterious effects of polynuclear aromatic hydrocarbon on blood vascular system of the rat fetus. Birth Defects Res B Dev Reprod Toxicol 80: 367–373.

32. Villar J, de Onis M, Kestler E, Bolanos F, Cerezo R, et al. (1990) The differential neonatal morbidity of the intrauterine growth retardation syndrome. Am J Obstet Gynecol 163: 151–157.

33. Hemachandra AH, Klebanoff MA (2006) Use of serial ultrasound to identify periods of fetal growth restriction in relation to neonatal anthropometry. 18: 791–797.

34. Malinowski A, Chlebny-Sokol D (1998) Lodzki child -The Methods of Examine and the Norms of Biological Development. Lodz. pp 70–89.

35. Choi H, Jedrychowski W, Spengler J, Camann DE, Whyatt RM, et al. (2006) International studies of prenatal exposure to polycyclic aromatic hydrocarbons and fetal growth. Environ Health Perspect 114: 1744–1750.

36. Ramsay JO, Silverman BW (1997) Functional Data Analysis: Springer.

37. R Development Core Team (2008) R: A language and environment for statistical computing. 2.5.1 ed. Vienna, Austria: R Foundation for Statistical Computing.

38. Guyda HJ, Mathieu L, Lai W, Manchester D, Wang SL, et al. (1990) Benzo(a)pyrene inhibits epidermal growth factor binding and receptor autophosphorylation in human placental cell cultures. Molecular Pharmacology 37: 137–143.

39. Sanyal MK, Li YL, Biggers WJ, Satish J, Barnea ER (1993) Augmentation of polynuclear aromatic hydrocarbon metabolism of human placental tissues of first-trimester pregnancy by cigarette smoke exposure. American Journal of Obstetrics & Gynecology 168: 1587–1597.

40. Zhang L, Shiverick KT (1997) Benzo(a)pyrene, but not 2,3,7,8-tetrachlorodibenzo-p-dioxin, alters cell proliferation and c-myc and growth factor expression in human placental choriocarcinoma JEG-3 cells. Biochem Biophys Res Commun 231: 117–120.

41. Barker DJ (2006) Adult consequences of fetal growth restriction. Clin Obstet Gynecol 49: 270–283.

42. Lipsett J, Tamblyn M, Madigan K, Roberts P, Cool JC, et al. (2006) Restricted fetal growth and lung development: a morphometric analysis of pulmonary structure. Pediatr Pulmonol 41: 1138–1145.

43. Nepomnyaschy L, Reichman NE (2006) Low birthweight and asthma among young urban children. Am J Public Health 96: 1604–1610.

44. Martin-Gronert MS, Ozanne SE (2007) Experimental IUGR and later diabetes. Journal of Internal Medicine 261: 437–452.

45. Wang S, Chanock S, Tang D, Li Z, Edwards S, et al. (2010) Effect of gene-environment Interactions on mental development in African American, Dominican, and Caucasian mothers and newborns. Ann Hum Genet 74: 46–56.

46. Kramer MS, McLean FH, Olivier M, Willis DM, Usher RH (1989) Body Proportionality and Head and Length 'Sparing' in Growth-Retarded Neonates: A Critical Reappraisal. Pediatrics 84: 717.

18

Longitudinal Relationship between Personal CO and Personal PM$_{2.5}$ among Women Cooking with Woodfired Cookstoves in Guatemala

John P. McCracken[1], Joel Schwartz[1], Anaite Diaz[2], Nigel Bruce[3], Kirk R. Smith[4]*

1 Department of Environmental Health, Harvard School of Public Health, Boston, Massachusetts, United States of America, **2** Center for Health Studies, Universidad del Valle de Guatemala, Guatemala City, Guatemala, **3** Department of Public Health and Policy, University of Liverpool, Liverpool, United Kingdom, **4** School of Public Health, University of California, Berkeley, California, United States of America

Abstract

Household air pollution (HAP) due to solid fuel use is a major public health threat in low-income countries. Most health effects are thought to be related to exposure to the fine particulate matter (PM) component of HAP, but it is currently impractical to measure personal exposure to PM in large studies. Carbon monoxide (CO) has been shown in cross-sectional analyses to be a reliable surrogate for particles<2.5 μm in diameter (PM$_{2.5}$) in kitchens where wood-burning cookfires are a dominant source, but it is unknown whether a similar PM$_{2.5}$-CO relationship exists for personal exposures longitudinally. We repeatedly measured (216 measures, 116 women) 24-hour personal PM$_{2.5}$ (median [IQR] = 0.11 [0.05, 0.21] mg/m^3) and CO (median [IQR] = 1.18 [0.50, 2.37] mg/m^3) among women cooking over open woodfires or chimney woodstoves in Guatemala. Pollution measures were natural-log transformed for analyses. In linear mixed effects models with random subject intercepts, we found that personal CO explained 78% of between-subject variance in personal PM$_{2.5}$. We did not see a difference in slope by stove type. This work provides evidence that in settings where there is a dominant source of biomass combustion, repeated measures of personal CO can be used as a reliable surrogate for an individual's PM$_{2.5}$ exposure. This finding has important implications for the feasibility of reliably estimating long-term (months to years) PM$_{2.5}$ exposure in large-scale epidemiological and intervention studies of HAP.

Editor: Mehrdad Arjomandi, University of California San Francisco, United States of America

Funding: This work was supported by the National Institutes of Health (grants R01ES010178, P01-ES09825, T-32 ES07069-25, ES-0002, and ES015172), the World Health Organization, and the U.S. Environmental Protection Agency (grants R827353 and R832416). The funders had no role in study design, data collection and analysis, decision to publish, or preparation of the manuscript.

Competing Interests: The authors have declared that no competing interests exist.

* E-mail: krksmith@berkeley.edu

Introduction

Household air pollution (HAP) from use of solid fuels is estimated to be a major risk factor for diseases, including acute respiratory, chronic respiratory, cancer, and cardiovascular outcomes [1,2]. Most of the epidemiological evidence for these relationships comes from studies using categorical exposure assignments based on stove and fuel types, which does not allow exposure-response analyses and limits comparability between studies in different settings. An ideal study design would include personal measures of exposure to the component of HAP that is causally related to the health effects being investigated. Fine particulate matter (PM$_{2.5}$) is often considered the best pollutant to measure for studies of health effects from combustion-generated pollutant mixtures, including HAP, secondhand tobacco smoke, and ambient air pollution [3,4]. Because of the size and weight of the monitoring equipment that has been available, personal PM measurements are generally burdensome and for infants infeasible, a particularly important limitation given the importance of quantifying the exposure-response relationship between HAP and pneumonia during infancy [5].

To overcome this problem, some HAP epidemiological studies have used area measurements of pollutant concentrations as surrogates for personal exposures. Kitchen area measures have been found to be poor surrogates of personal exposures to HAP [6–8], which may be largely attributable to differences in people's time-location patterns and the wide variability across small distances within the household and over short time periods [9]. Indirect exposure assessment, using time-activity patterns combined with area measurements [9,10], may improve exposure assessment, but one study with simultaneous personal exposure measures indicated that this method has low validity [8].

An alternative approach to HAP exposure assessment is personal measurement of a surrogate pollutant for PM, such as carbon monoxide (CO), which is relatively easy and inexpensive to measure, for example with very small passive dosimeter tubes that can be attached to an infant's clothing. Both pollutants are products of incomplete combustion and are major components of biomass smoke [11]. Strong correlation has been found between CO and fine PM levels in kitchens where biomass fuels are used for cooking [12–14]. It has been unknown, however, whether the relationship between these pollutants in a fixed location can be extrapolated to personal exposures. Additionally, the aim of most HAP epidemiological studies is to investigate effects of long-term (several months to years) exposures, whereas the relationships between CO and PM have previously been evaluated only in cross-sectional designs [12,14] or analyses [13].

The RESPIRE (Randomized Exposure Study of Pollution Indoors and Respiratory Effects) trial in Guatemala, the first randomized trial of an HAP exposure-reduction intervention, a chimney woodstove [5], for the prevention of pneumonia, included personal exposure measurements among a subset of women living in the study households. This short note presents a longitudinal analysis of the relationship between personal CO and PM among these women.

Methods

The study population and exposure assessment methodology have been described previously [15,16]. Briefly, women ≥ 38 years of age living in households participating in RESPIRE were recruited for a cardiovascular study. The study villages are located in the San Marcos department at approximately 2600 meters elevation above sea level. Smoking is uncommon, automobile traffic is low, and study households used only biomass fuels for cooking. The exposure assessment included a gravimetric (pump flow rate at 1.5 liters/minute, BGI Inc. sharp-cut cyclone inlet, 37 mm Teflon filter weighed before and after) measure of 24-hour personal exposure to particles with median aerodynamic diameter<2.5 μm ($PM_{2.5}$). Simultaneously, continuous measurement of personal CO was performed with the span-gas calibrated Hobo (Onset Inc.) passive electrochemical datalogger, with conversion of CO ppm values to mass concentration for comparison with the PM mass concentrations [17]:

$$mg/m^3 = (ppmv) \times ((12.187) \times (MW)/(273.15 + C)) \times (0.9877^A) \quad (1)$$

where the molecular weight (MW) of CO is 28.01,

C, the mean temperature at the site, is 12 deg celsius

A is the elevation of each house in 100 meters (range 2250–2960 m)

We analyze measures (up to three per subject) taken during the trial period, when the intervention group used the chimney stove and the control group used the open fire for cooking.

Pollution measures were right-skewed, so we applied a natural log transformation to the data before assessing the relationship between personal CO and personal $PM_{2.5}$ by scatterplot, correlation coefficients, and regression models. We used linear mixed effects models with personal $PM_{2.5}$ as the dependent variable and random subject intercepts to account for correlation among repeated measures within subjects and to estimate the within- and between-subjects variance components. The model residuals were consistent with being derived from a normal distribution. We compared the variance of the random subject intercept between models to measure the extent to which between-subjects differences in typical personal $PM_{2.5}$ are explained by covariates ($R^2_{between}$). For example, we estimated the $R^2_{between}$ for a model with CO as the independent variable by calculating the proportional reduction in the variance of the random subject intercept compared to the null model (no independent variable). The fixed effects in these models can be used to estimate personal $PM_{2.5}$ based on covariates (stove, personal CO). To test for differences in the slope of $PM_{2.5}$ on CO by stove type, we added a stove-by-CO interaction term. We tested for nonlinearity using a penalized spline for CO in a generalized additive mixed model (R software, GAMM function).

Protocols were approved by the Comité de Ética de la Universidad del Valle de Guatemala and the Harvard School of Public Health, Office of Human Research Administration. Written consents were obtained from all participants.

Results

We obtained 216 simultaneous 24-hour measures of CO and $PM_{2.5}$ among 116 women, 40 on one occasion, 52 on two occasions, and 24 on three occasions. The median (interquartile range) personal $PM_{2.5}$ was 0.20 mg/m^3 (0.11, 0.32) in the open fire group (67 women, n = 104) and 0.07 mg/m^3 (0.04, 0.12) in the chimney stove group (49 women, n = 112), and personal CO was 2.02 mg/m^3 (1.20, 3.35) in the open fire group and 0.63 mg/m^3 (0.33, 1.22) in the chimney stove group. Figure 1 shows a direct relationship between the natural log-transformed values of personal CO and $PM_{2.5}$ exposures. The Spearman rank correlation coefficient was 0.70 (p-value<0.001) between these two pollutant exposures (see Table 1).

In linear mixed effects models, the variance of the random intercept decreased from 0.31 to 0.07 when CO was added as an independent variable, equivalent to an $R^2_{between}$ = 0.78. A further reduction in random between-subject variability to 0.04 was achieved when stove type (chimney stove versus open fire) was added to the model ($R^2_{between}$ = 0.85).

The estimated population-mean personal $PM_{2.5}$ based on personal CO alone can be calculated with the following equation:

$$PM_{2.5} = e^{(-2.13 + 0.61 * \ln(CO) - 0.36 * chimney)}, \quad (2)$$

where chimney = 1 for the chimney stove and chimney = 0 for open fire.

We did not find evidence of a difference in the slope by stove type (interaction p-value = 0.986), and we also did not find evidence of nonlinearity in these log-transformed data using generalized cross validation, which chose a spline with one degree of freedom (Figure 1).

Discussion

Absent or minimal assessment of exposure to combustion-generated PM has been a major weakness of most epidemiological

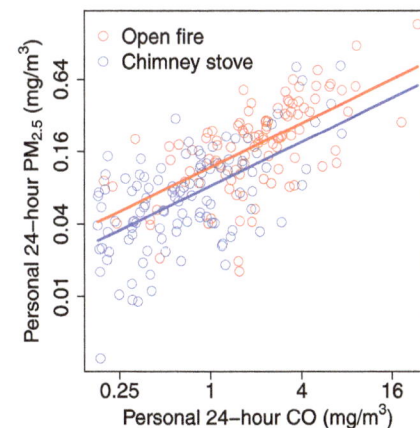

Figure 1. Scatter plot of simultaneous 24-hour personal fine particles (PM2.5) and personal carbon monoxide (CO). Lines for each stove type (red for open fire, blue for chimney stove) using equation $PM_{2.5} = e^{(-2.13 + 0.61 * \ln(CO) - 0.36 * chimney)}$ estimated from linear mixed effects regression model with natural log-transformed exposures (216 measurements among 116 women).

Table 1. Effect estimates (95% confidence intervals) and variance components from linear mixed effects models to predict natural log personal $PM_{2.5}$ (216 24-hour exposure measures among 116 subjects).

Independent variables	Chimney stove Effect	CO slope (per log-unit)	Between-subject variance	Within-subject variance	$R^2_{between}$
Null			0.31	0.76	
CO		0.69	0.07	0.48	0.78
		(0.59, 0.79)			
Stove type	−1.00		0.07	0.75	0.77
	(−1.25, −0.74)				
Stove type and CO	−0.36	0.61	0.05	0.48	0.85
	(−0.59, −0.12)	(0.50, 0.72)			
Plus stove by CO	−0.36	0.61	0.05	0.48	0.85
interaction*	(−0.60, −0.12)	(0.45, 0.77)			

*Stove by CO interaction effect = −0.00 (−0.22, 0.22).

studies of HAP in developing countries, particularly those with the additional challenges presented by assessing these exposures among infants. Previous studies have shown that CO is strongly correlated with $PM_{2.5}$ in kitchens where there is a single major source of smoke, but it was unclear whether this relationship could be extrapolated to personal exposures. We performed a longitudinal analysis of personal exposures among women from households in the RESPIRE trial in Guatemala, and found a moderately, strong correlation between personal CO and personal $PM_{2.5}$. Repeated personal CO levels explain 78% of the between-subject variability in personal $PM_{2.5}$. The estimated slope for the relationship between log-transformed measures is a 61% increase in personal $PM_{2.5}$ per 100% increase in personal CO.

Our results contrast with those from a study conducted among children <5 years of age in the Gambia by Dionisio et al [18], who did not find evidence of correlation between personal CO and personal $PM_{2.5}$ (r = −0.04), but there are a number of potential explanations for the weak correlations. In that population, there were a mixture of wood and charcoal stoves in use, which have substantially different ratios of CO to $PM_{2.5}$ in their emissions [19], which would reduce the local correlation between personal $PM_{2.5}$ and personal CO [9]. Similarly, the peri-urban children in the Gambian study may have been exposed to high levels of traffic emissions, with an even greater difference in CO:$PM_{2.5}$ in its emissions [20]. Moreover, the durations of personal $PM_{2.5}$ (48 hours) and personal CO (72 hours) measures differed in The Gambian study, which is expected to lower the correlation because of day-to-day variability in exposure levels. Finally, it is possible that instrument measurement error may have led to underestimation of the true correlation. Duplicate measures of each pollutant in a subset of participants should be collected in future studies to account for measurement error. Alternatively, longitudinal analyses can be used to separate within-subject variation (both true changes over time and measurement error) from between-subjects variation, as herein presented.

The form of the relationship between personal CO and personal $PM_{2.5}$ in our models suggests a smaller increment of $PM_{2.5}$ per unit of CO at higher exposure levels. This contrasts with the constant slope on the linear scale reported previously for kitchen concentrations in Guatemala [12–14], and there are several reasons why personal measurements may exhibit this relationship. Since combustion-generated $PM_{2.5}$ is an irritant and the $PM_{2.5}$-CO emissions ratio varies throughout the solid fuel burn cycle [11], it is possible that avoidance of the discomfort of PM and associated irritating compounds in the smoke by the householders may decrease the $PM_{2.5}$-CO slope at higher exposure levels. In addition, particles tend to adhere to surfaces over time whereas CO does not. If people tend to be in the kitchen more after emissions have been exposed to surfaces around the household, this would also decrease the $PM_{2.5}$-CO slope. Finally, the $PM_{2.5}$-CO relationship may be different in microenvironments where people spend time other than the kitchen and the relative contribution of each microenvironment may vary by total exposure level.

Conclusions

Our findings demonstrate that personal CO, which is relatively inexpensive and easy to measure can be a reliable surrogate for personal $PM_{2.5}$ in some settings. We emphasize that the association was observed among women living in Guatemalan villages with a single dominant source of combustion, but may be modified by time-activity patterns associated with demographic characteristics, and is unlikely to be generalizable to settings with mixtures of pollution source types.

Acknowledgments

We appreciate insights gained by discussions with Steven Chillrud, Alan Hubbard, Mark Nicas, and Charles Weschler. We thank the Randomized Exposure Study of Pollution Indoors and Respiratory Effects (RESPIRE) fieldworkers for their dedication to the project, and the Guatemalan Ministry of Health for their collaboration.

Author Contributions

Conceived and designed the experiments: JPM JS KRS NB. Performed the experiments: JPM AD. Analyzed the data: JPM KRS. Wrote the paper: JPM NB KRS.

References

1. Smith KR, Mehta S (2003) The burden of disease from indoor air pollution in developing countries: comparison of estimates. International journal of hygiene and environmental health 206: 279–289.
2. Balakrishnan K, Ramaswamy P, Sambandam S, Thangavel G, Ghosh S, et al. (2011) Air pollution from household solid fuel combustion in India: an overview of exposure and health related information to inform health research priorities. Global health action 4.
3. Pope CA, 3rd, Burnett RT, Turner MC, Cohen A, Krewski D, et al. (2011) Lung cancer and cardiovascular disease mortality associated with ambient air pollution and cigarette smoke: shape of the exposure-response relationships. Environmental health perspectives 119: 1616–1621.
4. Smith KR, Peel JL (2010) Mind the gap. Environmental health perspectives 118: 1643–1645.
5. Smith KR, McCracken JP, Weber MW, Hubbard A, Jenny A, et al. (2011) Effect of reduction in household air pollution on childhood pneumonia in Guatemala (RESPIRE): a randomised controlled trial. Lancet 378: 1717–1726.
6. Bruce N, McCracken J, Albalak R, Schei MA, Smith KR, et al. (2004) Impact of improved stoves, house construction and child location on levels of indoor air pollution exposure in young Guatemalan children. Journal of exposure analysis and environmental epidemiology 14 Suppl 1: S26–33.
7. Baumgartner J, Schauer JJ, Ezzati M, Lu L, Cheng C, et al. (2011) Patterns and predictors of personal exposure to indoor air pollution from biomass combustion among women and children in rural China. Indoor air 21: 479–488.
8. Cynthia AA, Edwards RD, Johnson M, Zuk M, Rojas L, et al. (2008) Reduction in personal exposures to particulate matter and carbon monoxide as a result of the installation of a Patsari improved cook stove in Michoacan Mexico. Indoor air 18: 93–105.
9. Ezzati M, Saleh H, Kammen DM (2000) The contributions of emissions and spatial microenvironments to exposure to indoor air pollution from biomass combustion in Kenya. Environmental health perspectives 108: 833–839.
10. Zuk M, Rojas L, Blanco S, Serrano P, Cruz J, et al. (2007) The impact of improved wood-burning stoves on fine particulate matter concentrations in rural Mexican homes. Journal of exposure science & environmental epidemiology 17: 224–232.
11. Smith K (1987) Biofuels, air pollution, and health: a global review. New York: Plenum.
12. Naeher LP, Smith KR, Leaderer BP, Neufeld L, Mage DT (2001) Carbon monoxide as a tracer for assessing exposures to particulate matter in wood and gas cookstove households of highland Guatemala. Environmental science & technology 35: 575–581.
13. Northcross A, Chowdhury Z, McCracken J, Canuz E, Smith KR (2010) Estimating personal PM2.5 exposures using CO measurements in Guatemalan households cooking with wood fuel. Journal of environmental monitoring : JEM 12: 873–878.
14. Naeher LP, Leaderer BP, Smith KR (2000) Particulate matter and carbon monoxide in highland Guatemala: indoor and outdoor levels from traditional and improved wood stoves and gas stoves. Indoor air 10: 200–205.
15. McCracken JP, Smith KR, Diaz A, Mittleman MA, Schwartz J (2007) Chimney stove intervention to reduce long-term wood smoke exposure lowers blood pressure among Guatemalan women. Environmental health perspectives 115: 996–1001.
16. McCracken J, Smith KR, Stone P, Diaz A, Arana B, et al. (2011) Intervention to lower household wood smoke exposure in Guatemala reduces ST-segment depression on electrocardiograms. Environmental health perspectives 119: 1562–1568.
17. Seinfeld JH, Pandis SN (1998) Atmospheric Chemistry and Physics. New York: John Wiley & Sons. 1326 p.
18. Dionisio KL, Howie SR, Dominici F, Fornace KM, Spengler JD, et al. (2012) Household concentrations and exposure of children to particulate matter from biomass fuels in The Gambia. Environmental science & technology 46: 3519–3527.
19. Smith KR, Uma R, Kishore VVN, Zhang J, Joshi V, et al. (2000) Greenhouse implications of household stoves: an analysis for India. Annual Review Energy Environment 25: 741–763.
20. Wallington TJ, Sullivan JL, Hurley MD (2008) Emissions of CO2,CO, NOx, HC, PM, HFC-134a, N2O and CH4 from the global light duty vehicle fleet. Meteorologische Zeitschrift 17: 109–116.

19

In Utero Exposure to Diesel Exhaust Air Pollution Promotes Adverse Intrauterine Conditions, Resulting in Weight Gain, Altered Blood Pressure, and Increased Susceptibility to Heart Failure in Adult Mice

Chad S. Weldy[1,2], Yonggang Liu[1], H. Denny Liggitt[3], Michael T. Chin[1,2]*

1 Division of Cardiology, Department of Medicine, University of Washington School of Medicine, Seattle, Washington, United States of America, **2** Department of Pathology, University of Washington School of Medicine, Seattle, Washington, United States of America, **3** Department of Comparative Medicine, University of Washington School of Medicine, Seattle, Washington, United States of America

Abstract

Exposure to fine particulate air pollution ($PM_{2.5}$) is strongly associated with cardiovascular morbidity and mortality. Exposure to $PM_{2.5}$ during pregnancy promotes reduced birthweight, and the associated adverse intrauterine conditions may also promote adult risk of cardiovascular disease. Here, we investigated the potential for *in utero* exposure to diesel exhaust (DE) air pollution, a major source of urban $PM_{2.5}$, to promote adverse intrauterine conditions and influence adult susceptibility to disease. We exposed pregnant female C57Bl/6J mice to DE (≈ 300 µg/m^3 $PM_{2.5}$, 6 hrs/day, 5 days/week) from embryonic day (E) 0.5 to 17.5. At E17.5 embryos were collected for gravimetric analysis and assessed for evidence of resorption. Placental tissues underwent pathological examination to assess the extent of injury, inflammatory cell infiltration, and oxidative stress. In addition, some dams that were exposed to DE were allowed to give birth to pups and raise offspring in filtered air (FA) conditions. At 10-weeks of age, body weight and blood pressure were measured. At 12-weeks of age, cardiac function was assessed by echocardiography. Susceptibility to pressure overload-induced heart failure was then determined after transverse aortic constriction surgery. We found that *in utero* exposure to DE increases embryo resorption, and promotes placental hemorrhage, focal necrosis, compaction of labyrinth vascular spaces, inflammatory cell infiltration and oxidative stress. In addition, we observed that *in utero* DE exposure increased body weight, but counterintuitively reduced blood pressure without any changes in baseline cardiac function in adult male mice. Importantly, we observed these mice to have increased susceptibility to pressure-overload induced heart failure, suggesting this *in utero* exposure to DE 'reprograms' the heart to a heightened susceptibility to failure. These observations provide important data to suggest that developmental exposure to air pollution may strongly influence adult susceptibility to cardiovascular disease.

Editor: Michelle L. Block, Virginia Commonwealth University, United States of America

Funding: YL and MTC were supported by NIEHS DISCOVER Center grant P50 ES015915 and CSW was supported by the Experimental Pathology of Cardiovascular Disease Training Grant (T32HL007312). Additional support was provided by the NIEHS Center for Ecogenetics and Environmental Health though a pilot project award funded by grant P30ES007033. The funders had no role in study design, data collection and analysis, decision to publish, or preparation of the manuscript.

Competing Interests: The authors have declared that no competing interests exist.

* E-mail: mtchin@u.washington.edu

Introduction

The inhalation of fine ambient particulate matter ($PM_{2.5}$; PM < 2.5 µm in diameter) has long been associated with increased risk of cardiovascular morbidity and mortality [1,2,3]. Ambient $PM_{2.5}$ air pollution was recently listed as the ninth-ranked cause of disease worldwide, and the fourth-ranked cause of disease in East Asia [4]. To understand the pathological mechanisms that may explain these population-wide observations, investigators utilizing controlled exposure facilities and animal models have begun to elucidate potential biological mechanisms by which $PM_{2.5}$ adversely affects systemic cardiovascular function. Studies investigating the acute as well as the chronic effects of $PM_{2.5}$ inhalation have demonstrated that $PM_{2.5}$ inhalation is capable of promoting multiple deleterious effects on the cardiovascular system through a variety of mechanisms, including: 1) inciting pulmonary inflammation, resulting in the spillover of reactive products into the circulation, including proinflammatory cytokines, 2) dysregulating the autonomic nervous system, and 3) directly translocating particles from the lung into the circulation, influencing vascular inflammation and function [5]. All of these pathways have been suggested to play a role in the observed acceleration of atherosclerosis [6,7,8,9], impaired vascular function and loss of bioavailable nitric oxide (NO) [10,11], and induction of cardiac remodeling and increased susceptibility to heart failure [12,13] in rodent $PM_{2.5}$ exposure models.

As the inhalation of $PM_{2.5}$ can have adverse effects on systemic vascular function, there has been an increased interest to understand the potential effect of *in utero* $PM_{2.5}$ exposure on the developing fetus. During development, the highly vascularized placenta is the primary component of the fetomaternal interface, regulating oxygen tension, nutrient levels, certain immune functions, and intrauterine growth [14,15,16]. In a recent multicenter meta-analysis investigating the effect of PM_{10} and

$PM_{2.5}$ exposure during pregnancy and birthweight, Dadvand and colleagues (2013) reported PM exposure to be associated with reduced birthweight, as both a term low-birthweight outcome variable and as a continuous variable [17]. These data provide evidence that PM exposure can adversely affect intrauterine growth, likely in a linear manner. As much of traffic related $PM_{2.5}$ is derived from diesel exhaust (DE) [18], controlled DE exposures using mouse models have demonstrated *in utero* DE exposure to promote placental and fetal inflammation in late gestation [19,20]. These effects of DE on *in utero* inflammation are associated with adult susceptibility to weight gain and obesity [20] as well as airway hyperreactivity following subsequent ozone exposure [19].

Barker and colleagues have previously revealed fetal development to be a critical window that strongly influences adult susceptibility to disease, leading to the concept of 'fetal origins of adult disease' [21,22,23,24,25,26,27]. Importantly, reduced birthweight has been observed to be associated with increased adult risk of cardiovascular mortality [21] as well as increased ventricular hypertrophy [27] and hypertension [26], suggesting that adverse intrauterine conditions that promote reduced birthweight, may also promote lifelong susceptibility to cardiovascular diseases that increase overall risk of cardiovascular mortality. In addition, Eriksson and colleagues (2000) reported that reduced placental weight was associated with an increased risk of hypertension in patients with diabetes [22], suggesting that placental insufficiency may exert a profound influence on adult cardiovascular physiology.

In our previous work, we compared the effects of 'adult' versus 'developmental' exposure to DE on adult susceptibility to heart failure in mice [28]. Interestingly, we observed that offspring exposed to DE *in utero* and early in life, where exposure began maternally 3-weeks prior to pregnancy, throughout pregnancy, and until offspring were 3 weeks of age, had increased sensitivity to pressure overload-induced heart failure as adults. This effect was not observed in mice exposed to DE for two months post weaning, suggesting that developmental exposure to DE produces long lasting effects on adult myocardial susceptibility to injury. Although our previous work demonstrates the importance of developmental exposures to air pollution on adult susceptibility to disease, it did not determine the potential for *in utero* exposure to air pollution alone to influence adult health. Understanding the potential for *in utero* air pollution exposure to influence adult cardiovascular risk will likely have strong public health implications.

From these prior observations, we hypothesized that *in utero* exposure to DE air pollution would promote placental oxidative stress and inflammation, resulting in adverse intrauterine conditions leading to increased embryo resorption as well as decreased placental and fetal weights. We also hypothesized that *in utero* exposure would promote weight gain, increased blood pressure, and increased susceptibility to pressure overload induced heart failure in adult mice. To test these hypotheses, we exposed pregnant female C57BL6/J mice to DE (300 $\mu g/m^3$ $PM_{2.5}$, 6 hrs/day, 5 days/week) or filtered air (FA) from embryonic day 0.5 (E0.5) through E17.5, at which point we assessed embryo resorption, placental and fetal weights, placental oxidative stress as measured by 3-nitrotyrosine protein modification, placental histology by hematoxylin and eosin (H&E) staining, and inflammatory cell infiltration by α-CD45 immunohistochemistry. In addition, following birth, offspring were raised in FA conditions to adulthood for subsequent measurement of adult body weight, blood pressure, cardiac function and susceptibility to pressure overload-induced heart failure. In this study, we report that *in utero* DE exposure promotes an adverse intrauterine environment that

results in adult weight gain, altered blood pressure and increased susceptibility to pressure overload-induced heart failure.

Materials and Methods

Ethics Statement

All animal work was conducted according to relevant national and international guidelines. This study was carried out in strict accordance with the recommendations in the Guide for the Care and Use of Laboratory Animals of the National Institutes of Health. All animal experiments were approved by the University of Washington Institutional Animal Care and Use Committee (PHS Animal Assurance Welfare # A3464-01).

Diesel Exhaust Exposure and Mice

Male and female C57Bl/6J mice were purchased from The Jackson Laboratory (Bar Harbor, Maine, USA). All mice were housed in specific pathogen free (SPF) conditions on a 12/12-light/dark cycle. Female and male mice between the ages of 12 to 14 weeks were transferred to our Northlake Diesel Exposure Facility located near the University of Washington (UW) and housed under SPF conditions in Allentown caging systems (Allentown, NJ, USA) as previously described [29,30,31]. Diesel exhaust (DE) was generated from a single cylinder Yanmar diesel engine (Model YDG5500EV-6EI) operating on 82% load. A detailed analysis of DE particulate components in this system has been previously reported [29]. DE exposures were conducted for 6 hours per day (9 am –3 pm) five days a week (Monday – Friday) and DE concentrations were regulated to 300 $\mu g/m^3$ of $PM_{2.5}$. A 300 $\mu g/m^3$ concentration of $PM_{2.5}$ six hours/day, five days/week equates to a time weighted hourly average of 53 $\mu g/m^3$.

Exposure characteristics detailing gas, particle-bound polycyclic aromatic hydrocarbons (PAH), and particle diameter were recently measured and reported [28,32]. Briefly, oxides of nitrogen concentrations were 1800 ppb NO_x and 60 ppb NO_2, carbon monoxide was 2 ppm, and carbon dioxide was 1000 ppm. The mass fraction of particle-bound PAH was 20 ng/μg $PM_{2.5}$ and the ratio of the organic carbon to elemental carbon mass concentration was 0.10. The mass median aerodynamic diameter of particles was 85 nm (GSD 1.2) and the count median thermodynamic equivalent diameter was 87 nm (GSD 3.0).

E17.5 Fetal and Placental Collection

For our assessment of the embryonic effects of *in utero* DE exposure, timed matings were initiated on Monday afternoon and subsequently checked for a visible vaginal plug the following morning. If plugs were visible, mice were immediately transferred to either FA or DE. Noon of the day vaginal plug was detected was considered embryonic day 0.5 (E0.5), and exposure ended at E17.5, at which point pregnant mice were euthanized by overdose with tribromoethanol (intraperitoneal injection, i.p., 650 mg/kg). Seven pregnant dams were used for this E17.5 study, 3 FA and 4 DE. The uterus of each dam was collected and a visual inspection was conducted to detect evidence of embryo resorption. Following removal of the uterus, a gravimetric analysis was performed on the following tissues: uterus, individual whole embryos (including placenta, amnion, fetus, and membranes), individual placentas, and individual fetuses. Three placentas and fetuses from each dam were fixed in 10% neutral buffered formalin and processed with paraffin for further histological analysis.

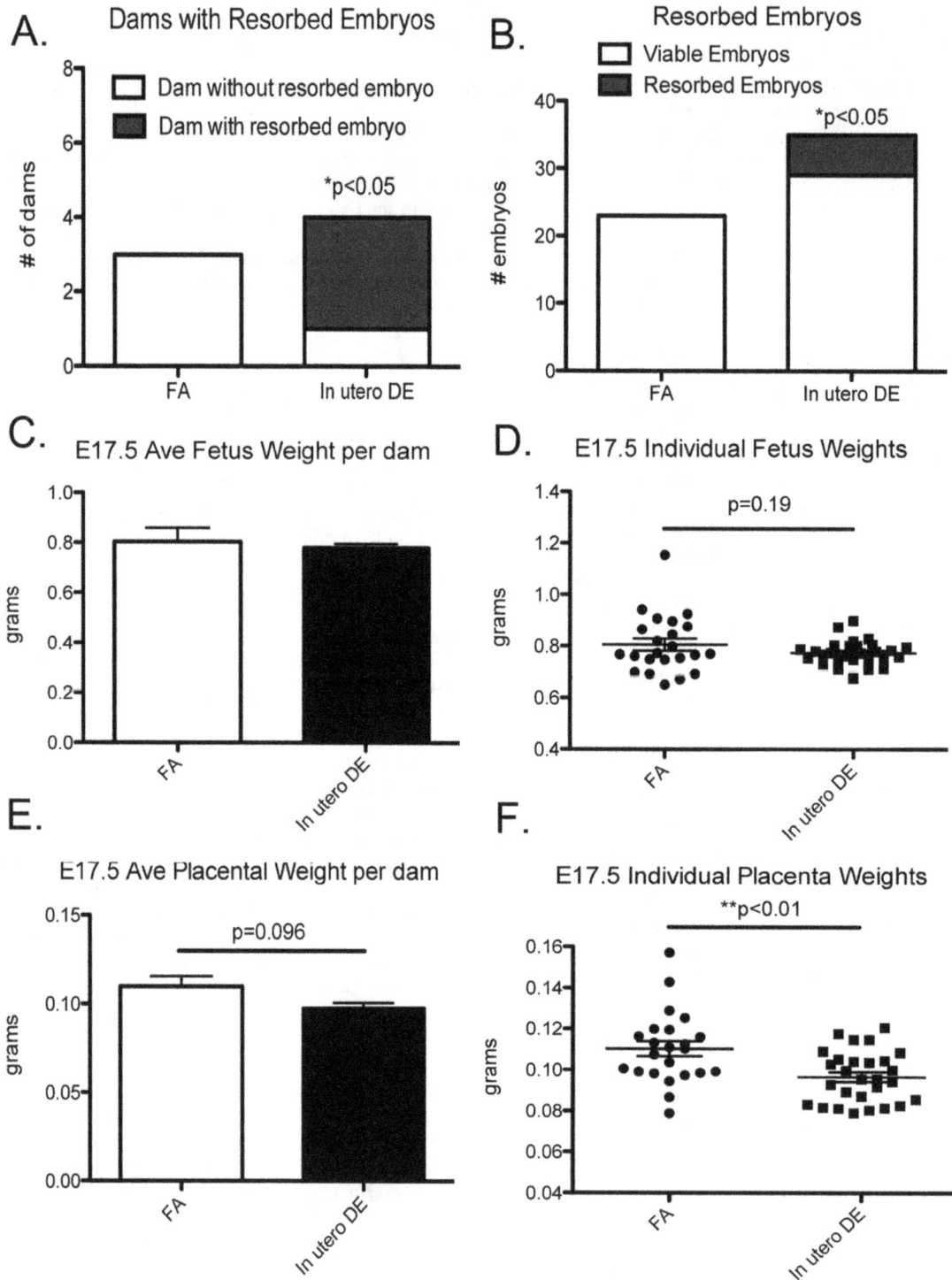

Figure 1. Embryonic day 17.5 (E17.5) embryo collection in FA and DE exposed dams. (A) Number of dams with resorbed embryos and (B) number of embryos resorbed vs number of viable embryos in FA and DE exposed dams. (C) Average fetus weight per dam and (D) individual fetus weights. (E) Average placental weight per dam and (F) individual placental weights. FA dams (n = 3), DE dams (n = 4).

Tissue Histology, Placental 3-nitrotyrosine Immunofluorescence, and α-CD45 Immunohistochemistry

To assess the placental pathology, tissues were collected, fixed in 10% neutral buffered formalin and processed following standard protocols. Seven-micron thick sections from both sagittal and *en face* orientations of the placenta were de-paraffinized in xylene, and hydrated using 100%, 95%, and 70% ethanol incubation steps prior to incubation in distilled H2O and subsequent staining with hematoxylin and eosin (H&E). Following staining, sections were mounted with Permount Mounting Medium (Fisher Scientific,

A. FA - Decidua

B. In utero DE - Decidua

C. FA - Spongiotrophoblast

D. In utero DE - Spongiotrophoblast

E. FA - Labyrinth

F. In utero DE - Labyrinth

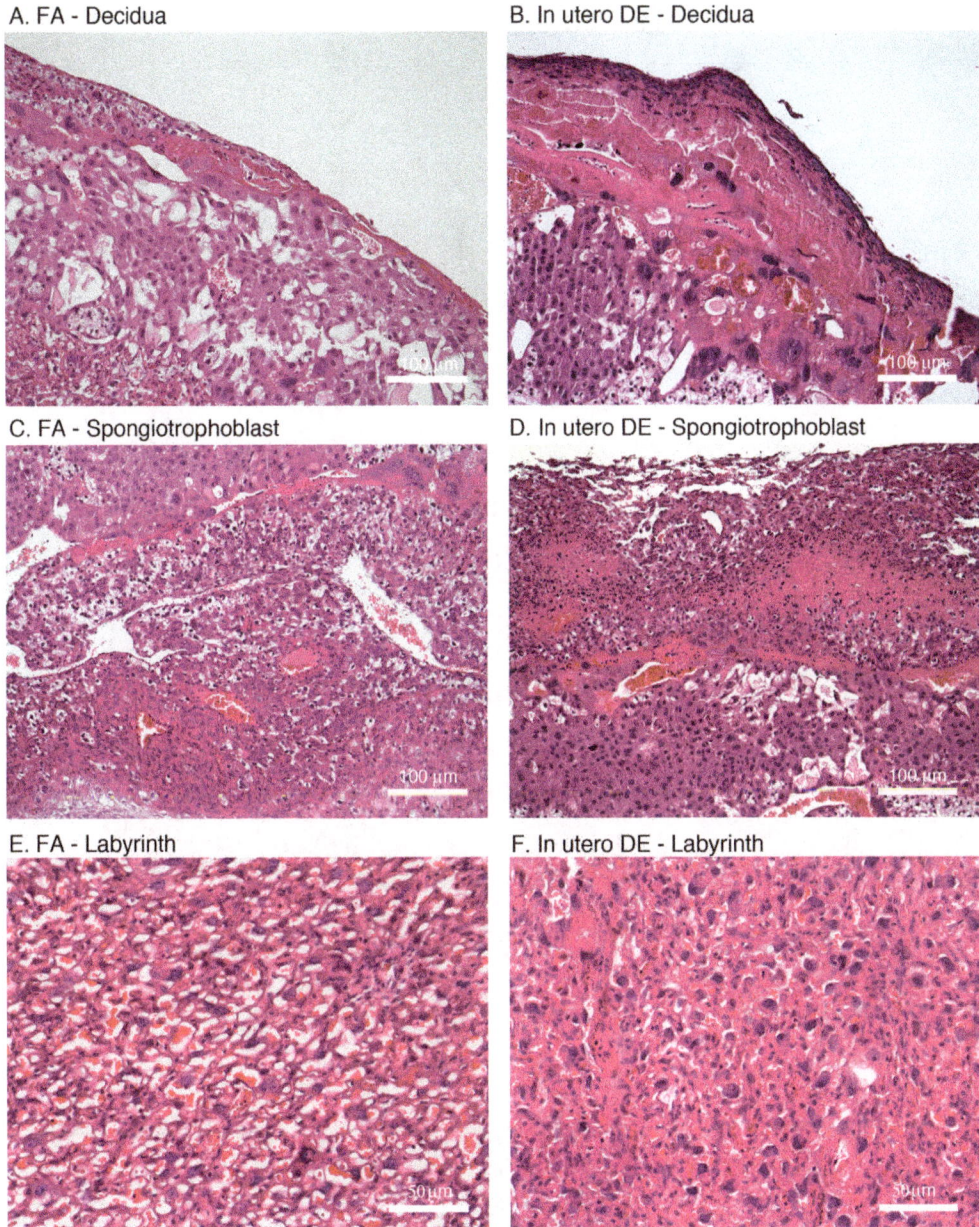

Figure 2. Images of placental cross sections stained with hematoxylin and eosin. (A) FA – Decidua layer, (B) DE – Decidua layer, excessive densely packed fibrin and enmeshed red cells and nuclear debris deep in the spongiotrophoblast layer, (C) FA – spongiotrophoblast layer, (D) DE – spongiotrophoblast layer, focal-extensive area of necrosis, congestion and hemorrhage at the labyrinth/spongiotrophoblast interface, (E) FA – labyrinth layer, (F) DE – labyrinth layer, increased stromal density and compaction of the vascular spaces. Scale bars = panels A–D, 100 μm; panels E and F, 50 μm.

Hampton, NH, USA) and evaluated by a veterinary pathologist unaware of treatment status.

3-nitrotyrosine (3-NT) immunofluorescence was performed as previously described, with modifications [33]. Seven-micron thick sagittal sections of the placenta were processed as described above. Sections were blocked in 10% goat serum in PBS at room temperature for 2 hrs, and then incubated at 4°C in 10% goat serum with primary antibody directed against 3-nitrotyrosine (3-NT)(1:200 dilution; Millipore 06–284, rabbit IgG). Following overnight 3-NT antibody incubation, slides were rinsed with PBS, and then incubated in the dark for 1 hour with goat anti rabbit Alexa 546 secondary antibody (1:400 dilution). Slides were

subsequently rinsed with PBS, and then fixed with 10% neutral buffered formalin for 5 minutes to preserve secondary stain. Slides were rinsed with PBS, and then mounted using Vectashield, mounting medium for fluorescence with DAPI (Vector Laboratories, Inc., Burlingame, CA, USA). Fluorescence images were captured with a Nikon Eclipse E400 microscope equipped with a TRITC filter (Nikon, Tokyo, Japan), a Mercury-100W lamp (Chiu Technical Corp., Kings Park, NY, USA) and a Nikon Digital Sight (Nikon, Tokyo, Japan) camera. Two sections from each placenta were prepared following this procedure, and 3 pictures under a 20X objective were taken from each section for a total of 6 pictures per placenta. Following background subtraction, number of pixels

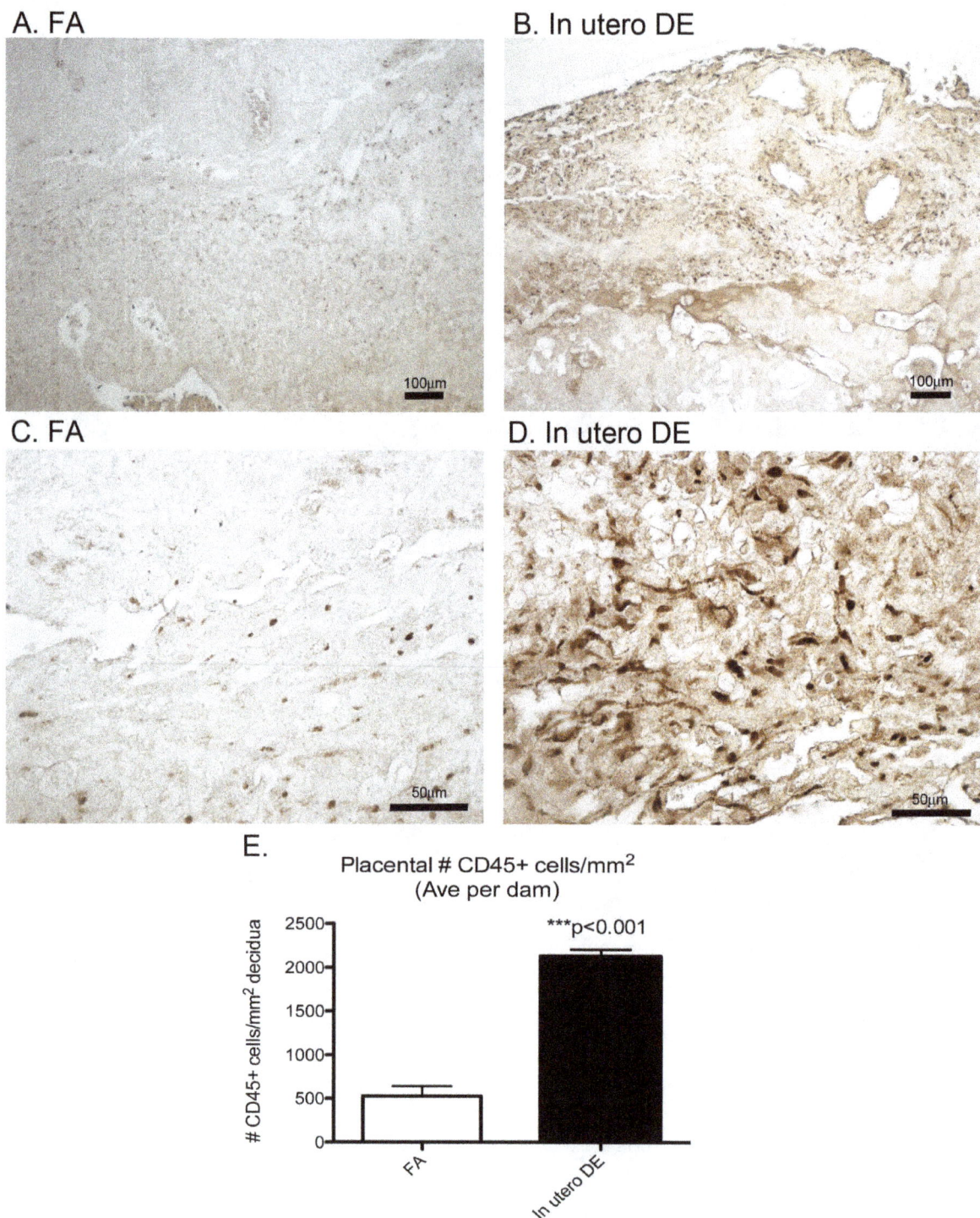

Figure 3. Images of placental cross-sections with immunohistochemistry staining against CD45 in FA (A and C) and *in utero* DE (B and D). Number of CD45+ cells normalized to decidua cross sectional area is quantified (E). FA dams (n = 3), FA placentas (n = 3 per dam), DE dams (n = 4), DE placentas (n = 3 per dam). Scale bars = 100 μm (panels A and B) and 50 μm (panels C and D).

that are 3-NT positive were quantified by Image J (NIH, Bethesda, MD, USA), and average positive 3-NT staining from 3 placentas for each dam were compared.

To assess inflammatory cell infiltration into the placenta, α-CD45 immunohistochemistry was conducted on seven micron sagittal sections using a Diaminobenzidine (DAB) histochemistry kit with streptavidin-HRP (Cat. # D22187, Life Technologies, Carlsbad, CA, USA) following the manufacturer's protocol. Rat α-mouse CD45 IgG (1:50 dilution) (Mat. # 553076, BD Pharmingen, Franklin Lakes, NJ, USA) was used as a primary, and biotinylated α-rat IgG raised in goat (1:100) was used as a secondary (BA-9400, Vector Labs, Burlingame, CA, USA).

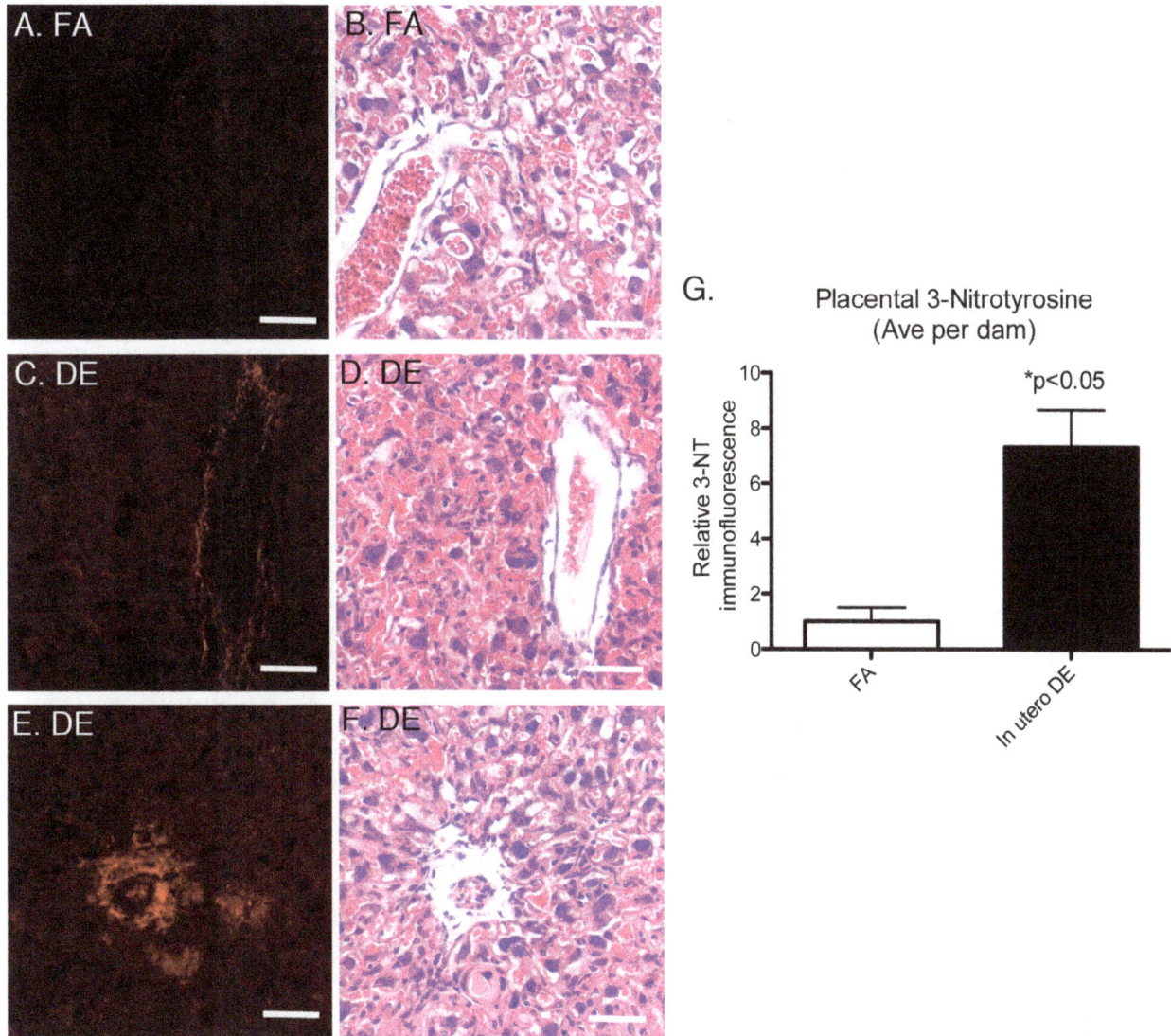

Figure 4. Placental 3-nitrotyrosine (3-NT) staining by immunofluorescence and adjacent H&E stained sections. Representative images from sagittal cross-section from (A and B) FA exposed dam, (C and D) DE exposed dam showing perivascular 3-NT staining, (E and F) DE exposed dam showing perivascular 3-NT staining, and (G) quantification of relative 3-NT fluorescence. FA dams (n = 3), FA placentas (n = 3 per dam), DE dams (n = 4), DE placentas (n = 3 per dam). Scale bars = 50 μm.

Following de-paraffinization and hydration steps, placental sections underwent a 30-minute antigen revealing step with 0.1% Triton X-100 in 0.1% sodium citrate. Following DAB treatment and completion of IHC staining, sections were dehydrated with serial ethanol rinses, incubated in xylene, and mounted with permount mounting medium.

Assessment of Adult Endpoints Following in Utero Exposure to DE

To examine the effects of *in utero* DE exposure on adult endpoints, a total of 9 female mice became pregnant, 4 were exposed to FA and 5 were exposed to DE. Female mice exposed to DE throughout pregnancy were transferred to FA at the end of exposure on E17.5. Birth was observed to occur between E18.5 and E19.5. All offspring from these 9 litters were raised in FA conditions and were weaned at postnatal day (PND) 21. A total of 26 male offspring were used for this study, 12 FA and 14 DE. At

10 weeks of age, male offspring were transferred from the Northlake facility to the UW Medicine South Lake Union (SLU) SPF vivarium, where mice underwent blood pressure, body weight, echocardiographic assessment, and surgery. Male mice were chosen for this study to eliminate the effect of cyclic hormonal variation.

Blood Pressure Measurement by Tail Cuff

Blood pressure was measured in male mice by use of the CODA 6 tail-cuff blood pressure system (Kent Scientific, Torrington, CT, USA). Prior to blood pressure measurement, mice were acclimated to restraint by following a 7-day training protocol. Each day, male mice were placed into a plastic restraining tube used for tail-cuff and blood pressure measurement for increasing amounts of time, beginning at 5 min/day. At the end of the 7-day period, mice were accustomed to spending 45 min restrained in individual restraining tubes without overt discomfort. Upon blood pressure

A.

10 week body weight

B.

10 week body weight/tibia length

C.

Tibia length

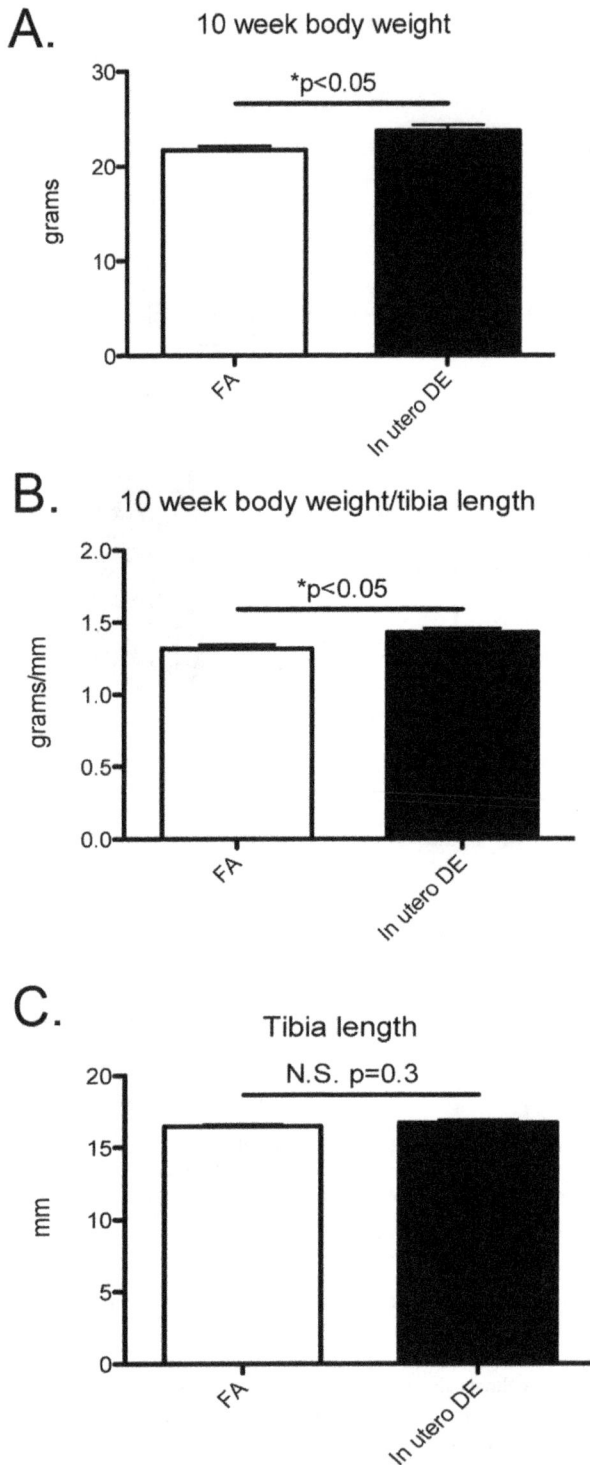

Figure 5. Effects of *in utero* DE exposure on adult body weight. (A) Male 10-week body weight, (B) body weight normalized to tibia length, and (C) tibia length. FA males (n = 10), DE males (n = 9).

measurement, mice were restrained and warm water-bags, monitored at a temperature between 35–37°C, were placed on top and around mice to keep temperature stable. Tails of mice were fitted with CODA O-Cuff and VPR-Cuffs according to the manufacturer's protocol to ensure proper BP measurements.

Settings of the tail-cuff were set to measure BP 10 times without recording data, followed by 10 times while recording data. This procedure was practiced 1X without data collection, and a second time the following day with complete data collection. The average of the second 10X data measurements on the second day were used for BP measurements.

Echocardiography and Transverse Aortic Constriction Surgeries

Upon transfer of male offspring to the SLU SPF vivarium, mice underwent baseline echocardiographic assessment at 11–12 weeks of age. Under 0.5% isoflurane anesthesia, cardiac size and function were assessed using a Visual Sonics (Toronto, Canada) VEVO 770 system equipped with a 707B scan head as previously described [34,35]. When the heart rate of the mouse had returned to normal following anesthesia (>520 bpm), parasternal short axis views were obtained under M-mode. Cine loops collected from M-mode views were analyzed for anterior and posterior LV wall thickness as well as LV internal diameter at diastole and systole. Percentage fractional shortening (%FS) was calculated from Visual Sonics Standard Measurements and Calculations. One week after baseline echocardiographic assessment was completed, male mice were randomly assigned to either transverse aortic constriction (TAC) or sham surgeries. For surgeries, male mice between 12 and 14 weeks of age were anesthetized using ketamine (130 mg/kg i.p.) and xylazine (8.8 mg/kg i.p.) and subjected to transverse aortic constriction using a 27-gauge needle as described [28,34,35]. To measure cardiac response to surgery, echocardiographic assessments were completed for each mouse at 3 weeks post surgery. All echocardiographic measurements were performed by a blinded observer. Consistent with our previous report [28], within 17 mice subjected to TAC surgery, we observed surgery associated death, as defined by death within 72 hours of surgery, in 4/17 (23.5%) of mice. The rate of this mortality was 1/8 (12.5%) in FA mice, and 3/9 (33%) in *in utero* DE exposed mice. Although this mortality rate appeared to be elevated in *in utero* DE exposed mice, this trend did not reach statistical significance by χ^2 test.

Necropsy, Gravimetric Tissue Analysis, and Myocardial Histology for Fibrosis

Three weeks after surgery, mice were sacrificed by overdose with tribromoethanol (i.p. 650 mg/kg) followed by exsanguination. Following sacrifice, tissue weights (ventricle weight, lung weight, liver weight, kidney weight, spleen weight, kidney weights, brain weight) and right tibia length were recorded. Tissues were either stored for histology or snap frozen in liquid nitrogen for subsequent biochemical analysis. Ventricles were cut in half along the sagittal axis to expose a two-chamber view. The first half of the ventricle was placed into formalin for fixation, while the second half was frozen in liquid nitrogen. Fixed ventricles were processed and embedded in paraffin as described above. Seven-micron thick sections were made of the heart, and Masson's Trichrome staining was performed using standard techniques. Extent of myocardial fibrosis was determined by Masson's Trichrome stain where percentage of blue stain was quantified over total tissue area from 4X images of the left ventricle free wall using NIH Image J (Bethesda, MD, USA) as previously done [28,34]. Using NIH Image J, low power images of the LV free wall were assessed in a blinded fashion. The border of the myocardium was traced manually to get a baseline area, then the area that would be considered 'highly fibrotic', where there is clear evidence of fibroblast collagen deposition and blue staining, is manually traced to get a subsequent area. The area of the myocardium that would

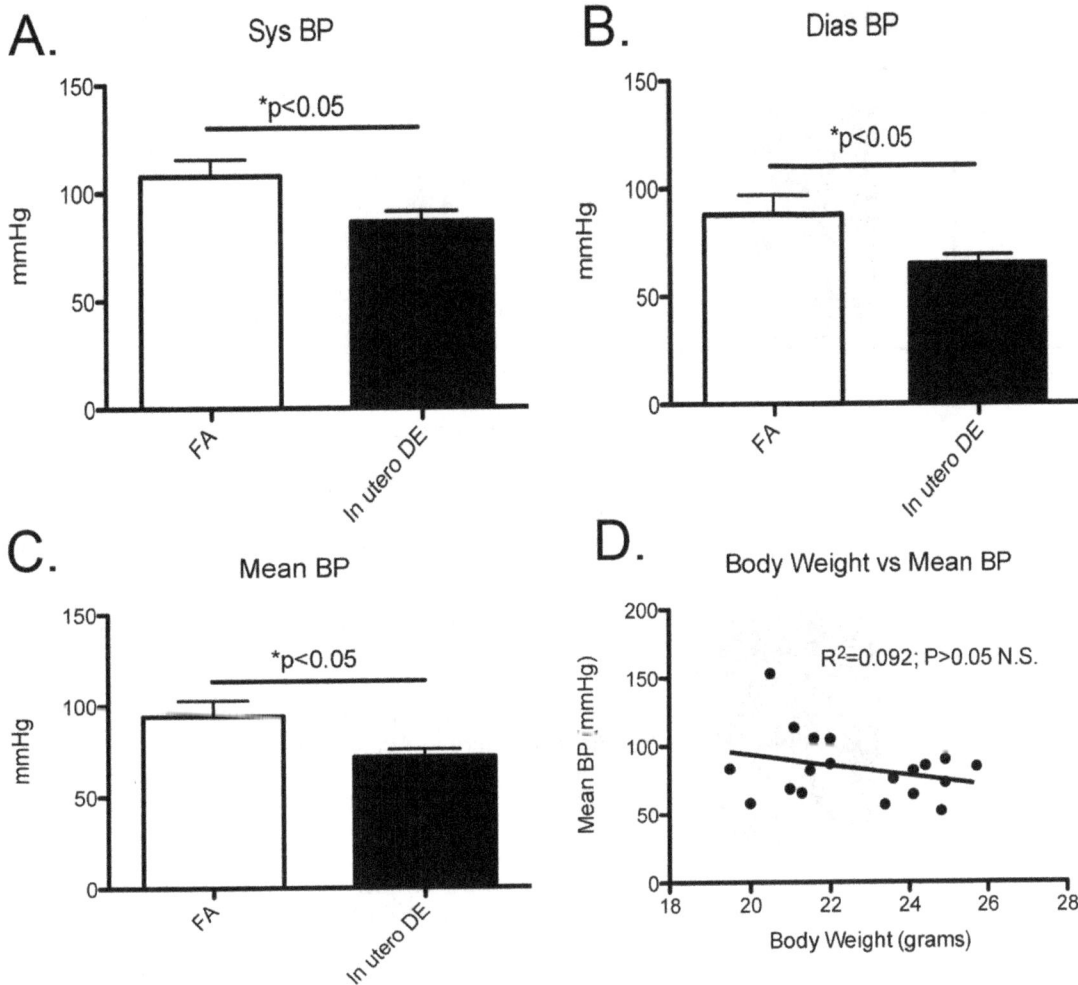

Figure 6. Systolic (A), diastolic (B), and mean (C) arterial pressure as measured by tail-cuff in 10-week old male mice. (D) Linear regression between body weight and mean BP. FA males (n = 10), DE males (n = 9).

be 'highly fibrotic' is expressed as a 'percentage highly fibrotic area'. Images of representative fibrotic regions were selected for presentation. To assess individual cardiomyocyte area, three 40X images of the LV free wall were taken from non-fibrotic areas of each heart. The individual cross-sectional cell area was quantified using NIH Image J, where the areas of 100 cardiomyocytes were manually traced (~33 randomly selected cells per 40X image) and averaged for each section as described previously [34]. Of 100 cardiomyocytes within the LV, area is averaged to get a single value that represents average myocyte area per heart. We then averaged cardiomyocyte area within exposure and surgery groups to assess individual cardiomyocyte hypertrophy.

Statistical Analysis

Statistical analyses were performed using GraphPad Prism 5 for Microsoft OS X (GraphPad Software, Inc.; San Diego, CA, USA). Differences between two groups were determined by Student's T-Test using an α-value of 0.05. To test the effect of heart failure stimulation and exposure, a two-way ANOVA was used followed by a Bonferroni post hoc comparison to test differences between means. All error bars in figures represent mean ± standard error of the mean; *, **, *** represent significant differences with $p < 0.05$, $p < 0.01$, and $p < 0.001$ respectively.

Results

In Utero DE Exposure Increases Embryo Resorption, Placental Inflammation, and Placental Oxidative Stress

To investigate the effect of *in utero* DE exposure on embryonic development, we exposed C57BL6/J mice to DE beginning on E0.5, 6 hrs/day, 5 days a week (M-F) until E17.5. At E17.5 dams were sacrificed. Uterine tissues were examined for evidence of embryo resorption, and weights were collected for each individual fetus and placenta. Interestingly, we observed the presence of resorbed embryos to be significantly increased within the dams exposed to DE compared to dams exposed to FA (Figure 1A and B). In a further assessment of fetal and placental weight from viable embryos, we did not observe DE to affect fetal weight (Figure 1C and D), but we did find evidence that DE seemed to promote a reduction in placental weight (Figure 1E and F). Although statistical significance is not reached when the average placental weight per dam is compared between FA and DE (n = 3 vs n = 4, respectively, p = 0.096), when statistical analysis is compared from each individual placenta (n = 23 vs n = 28), a strong statistical difference is achieved (**$p < 0.01$).

We next assessed the general placental pathology from FA and DE dams by H&E staining. Relative to FA control placentas, those

A. LVAW;d baseline

B. LVID;d baseline

C. %FS baseline

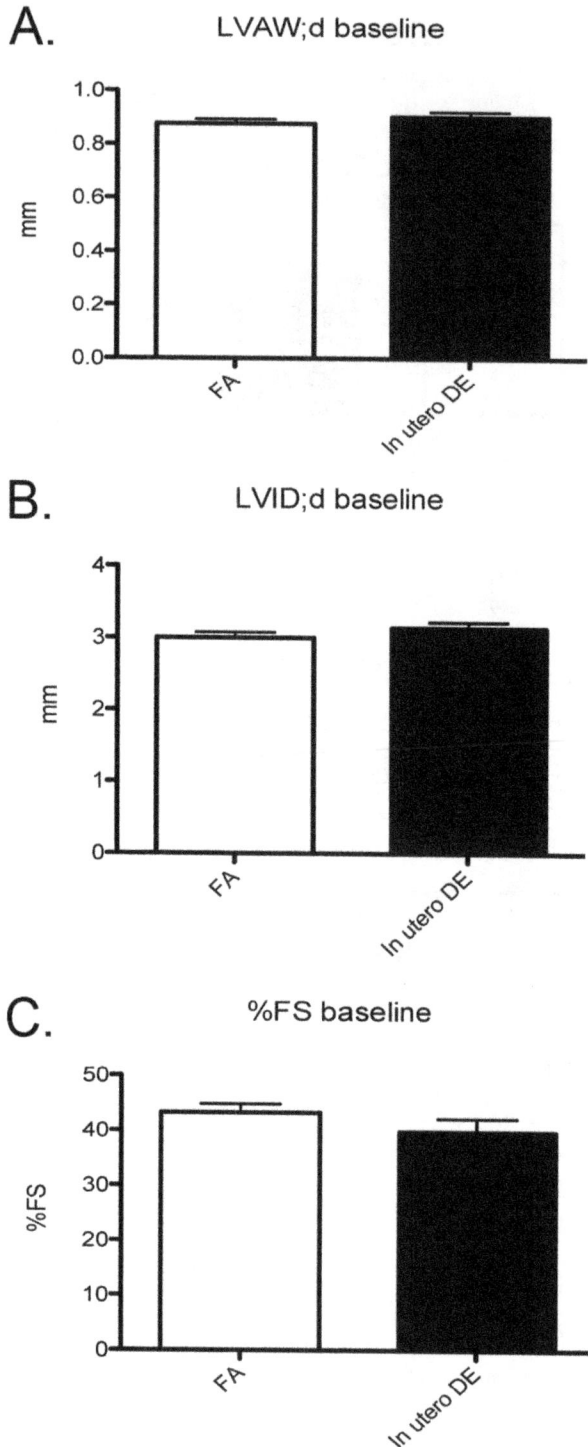

Figure 7. Baseline left ventricular wall thickness and contractile function in 12 week old male mice as measured by echocardiography. (A) Left ventricular anterior wall thickness at diastole, (B) left ventricular internal diameter at diastole, (C) percentage fractional shortening. FA males (n = 10), DE males (n = 9).

from DE dams showed evidence of increased necrosis, focal congestion, and hemorrhage at the labyrinth/spongiotrophoblast interface (Figure 2A–D). In addition, in utero DE exposure was associated with diffuse to focally-extensive compaction and/or

increased stromal density of the highly vascularized labyrinth, potentially diminishing vascular space available for blood flow (Figure 2E and F).

To investigate whether the observed placental pathologies associated with DE exposure are associated with inflammatory cell infiltration, immunohistochemistry against CD45 was performed. CD45+ cells can be observed within the decidua layer in placentas from both FA and DE dams (Figure 3), but the number of cells was increased in placentas exposed to DE (Figure 3B and D). Quantification of the number of CD45+ cells within the decidua layer reveals a significant increase in CD45+ cellular infiltration in the in utero DE exposed placentas (Figure 3E).

To investigate if there is any evidence of DE-induced placental oxidative stress, we measured 3-nitrotyrosine (3-NT) protein modification, a robust marker of oxidative/nitrosative injury, by immunofluorescence in sagittal cross-sections of the placenta. We did not observe any clear evidence of overt 3-NT protein modification in the placental cross-sections from FA exposed dams (Figure 4A). In contrast, 3-NT immunofluorescence from placental cross-sections of DE exposed dams revealed clear evidence of elevated 3-NT protein modification, predominantly within perivascular regions (Figure 4C–F) in the fetal labyrinth layer. This spatial distribution of 3-NT immunofluorescence suggests that in utero DE promotes vascular oxidative stress.

In Utero DE Exposure Increases Body Weight and Alters Blood Pressure in Adult Male Mice

Since in utero DE exposure is associated with significant placental pathology, we investigated whether in utero exposure to DE promotes long-term effects on adult body weight and blood pressure. We observed 10-week body weight to be elevated in male mice exposed to in utero DE (Figure 5A). In addition, to control for effects on body habitus rather than animal size, 10-week body weight was normalized to tibia length, a measure collected at necropsy at the end of the study. Body weight normalized to tibia length remained elevated in male mice exposed to DE in utero (Figure 5B), whereas no significant differences were observed when comparing tibia lengths (Figure 5C), indicating that effects on body weight are mediated through effects on body habitus.

As adverse intrauterine conditions have been associated with elevations in adult blood pressure [21], we measured blood pressure by tail-cuff in the same cohorts of male mice that demonstrated an increase in body weight. Interestingly, in contrast to our hypothesis, we observed systolic (SBP), diastolic (DBP), as well as mean arterial blood pressure (MAP) to be significantly decreased in 10-week old male mice exposed to DE in utero (Figure 6A–C). To test if these observed decreases in BP were associated with our observed increases in body weight, we performed a linear regression between body weight and MAP (Figure 6D). We did not observe any significant linear relationship between MAP and body weight.

In Utero DE Exposure does not Alter Adult Baseline Cardiac Function but Increases Susceptibility to Pressure Overload-induced Heart Failure

At 12-weeks of age, male mice underwent a baseline echocardiographic assessment. We did not observe in utero exposure to DE to have any effect on baseline ventricular wall thickness, chamber dimension, and contractile function (Figure 7A–C).

Between 13 and 14 weeks of age, male mice were randomly assigned to undergo sham or transverse aortic constriction (TAC) surgery to promote pressure overload-induced heart failure. At 3 weeks post TAC surgery, an echocardiographic assessment was

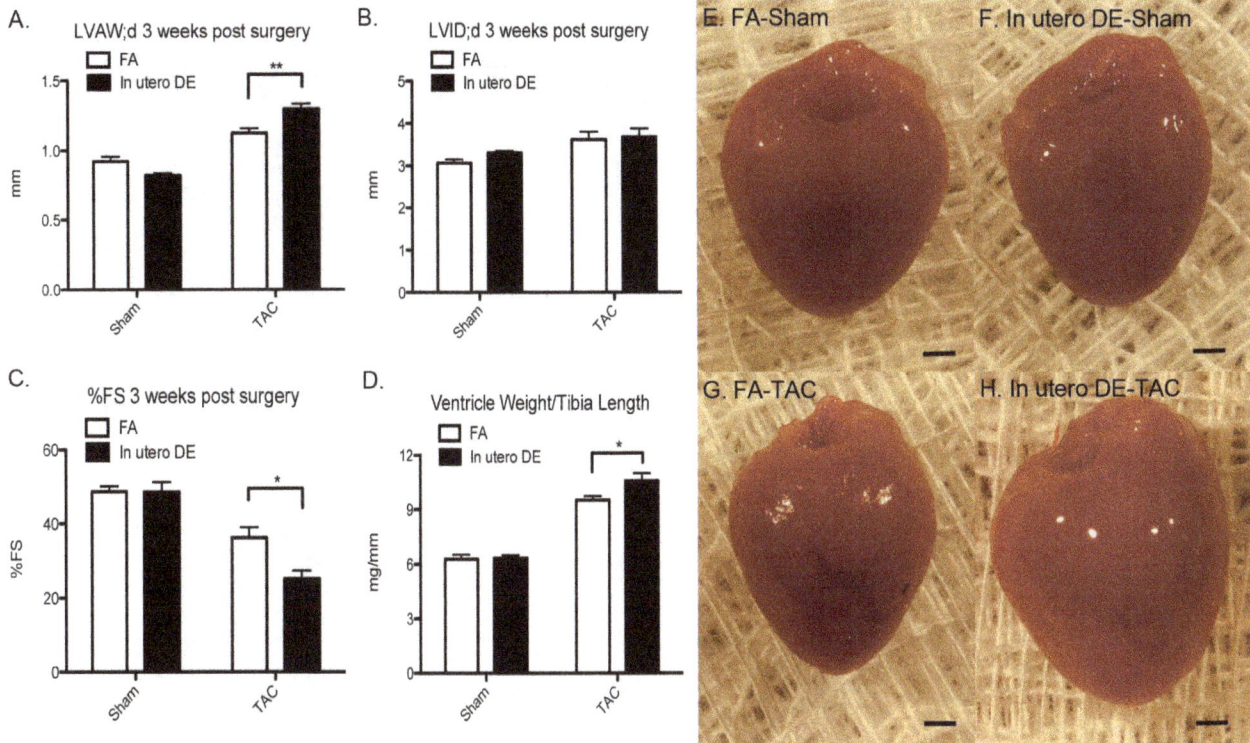

Figure 8. Three weeks following sham or transverse aortic constriction (TAC) surgeries to stimulate pressure overload-induced heart failure, (A) left ventricular anterior wall thickness at diastole, (B) left ventricular interior wall thickness at diastole, (C) percentage fractional shortening, and (D) ventricle weight normalized to tibia length at necropsy. (E–H) Representative images of ventricles at necropsy. FA sham (n = 4), FA TAC (n = 7), DE sham (n = 5), DE TAC (n = 6). Scale bars = 1 mm.

performed to determine the effect of TAC surgery on cardiac size, dimension, and function, followed by euthanasia and gravimetric analysis (Figure 8). TAC surgery induced cardiac hypertrophy and systolic dysfunction in both FA and DE mice, but mice exposed to DE *in utero* have significantly increased ventricular wall thickness (Figure 8A) in the absence of any change in LV internal diameter (Figure 8B), decreased fractional shortening (Figure 8C), and increased ventricle weight normalized to tibia length (Figure 8D–H).

In Utero DE Exposure Promotes Myocardial Fibrosis but not Cardiac Myocyte Hypertrophy Following Pressure Overload

To assess whether *in utero* exposure to DE promotes myocardial fibrosis following sham or TAC surgeries, we performed Masson's Trichrome staining on sagittal ventricle cross sections and assessed area of highly fibrotic regions as previously described [28,34]. We did not observe any evidence of fibrosis in the hearts from FA-Sham mice (Figure 9A), and only mild perivascular fibrosis in FA-TAC and in utero DE-Sham hearts (Figure 9B and C). In contrast, the hearts from *in utero* DE-TAC mice showed extensive perivascular and interstitial fibrosis distributed diffusely across the left ventricular free wall. Quantification of fibrotic area demonstrated that fibrosis within the *in utero* DE-TAC mice was significantly greater than FA-TAC mice (FA-TAC n = 7, *in utero* DE-TAC n = 6)(**p < 0.01)(Figure 9E).

To assess whether the observed effect of *in utero* DE exposure on heart weight is mediated at the myocyte level, we assessed the extent of individual cardiomyocyte hypertrophy by measuring individual myocyte cross-sectional area in exposed sham and TAC mice. We observed TAC to significantly increase individual cardiomyocyte cross-sectional area in both FA and *in utero* DE exposed mice (Figure 10) but did not observe an additive effect of *in utero* exposure to DE (Figure 10E).

Discussion

In this report, we provide evidence that *in utero* exposure to diesel exhaust air pollution (≈ 300 µg/m^3 PM$_{2.5}$, 6 hrs/day, 5 days/week, a 53 µg/m^3/hr PM$_{2.5}$ time weighted average) promotes significant placental injury, manifested by hemorrhage, vascular compromise, focal necrosis, embryo resorption, inflammation, and oxidative stress. Surviving embryos develop increased body weight, altered blood pressure, and increased susceptibility to pressure overload-induced heart failure as adults. In combination, these data provide strong evidence that *in utero* exposure to fine particular air pollution may have significant effects on adult susceptibility to cardiovascular disease and heart failure. Although we have previously reported that combined *in utero* and early life exposure to DE can promote adult susceptibility to heart failure [28], these data reported here extend our earlier findings by demonstrating that *in utero* exposure to DE alone is sufficient to affect adult susceptibility to heart failure, supporting a fetal origin for this susceptibility to adult disease, through a mechanism that involves placental injury.

Placental insufficiency results from placental injury and has been reported to promote fetal hypoglycemia, hypoinsulinemia, acidosis, hypoxia, and decreased branched chain amino acid transfer [36,37,38,39,40,41] that can result in 'reprogramming' events that alter metabolic, cellular, and physiological function

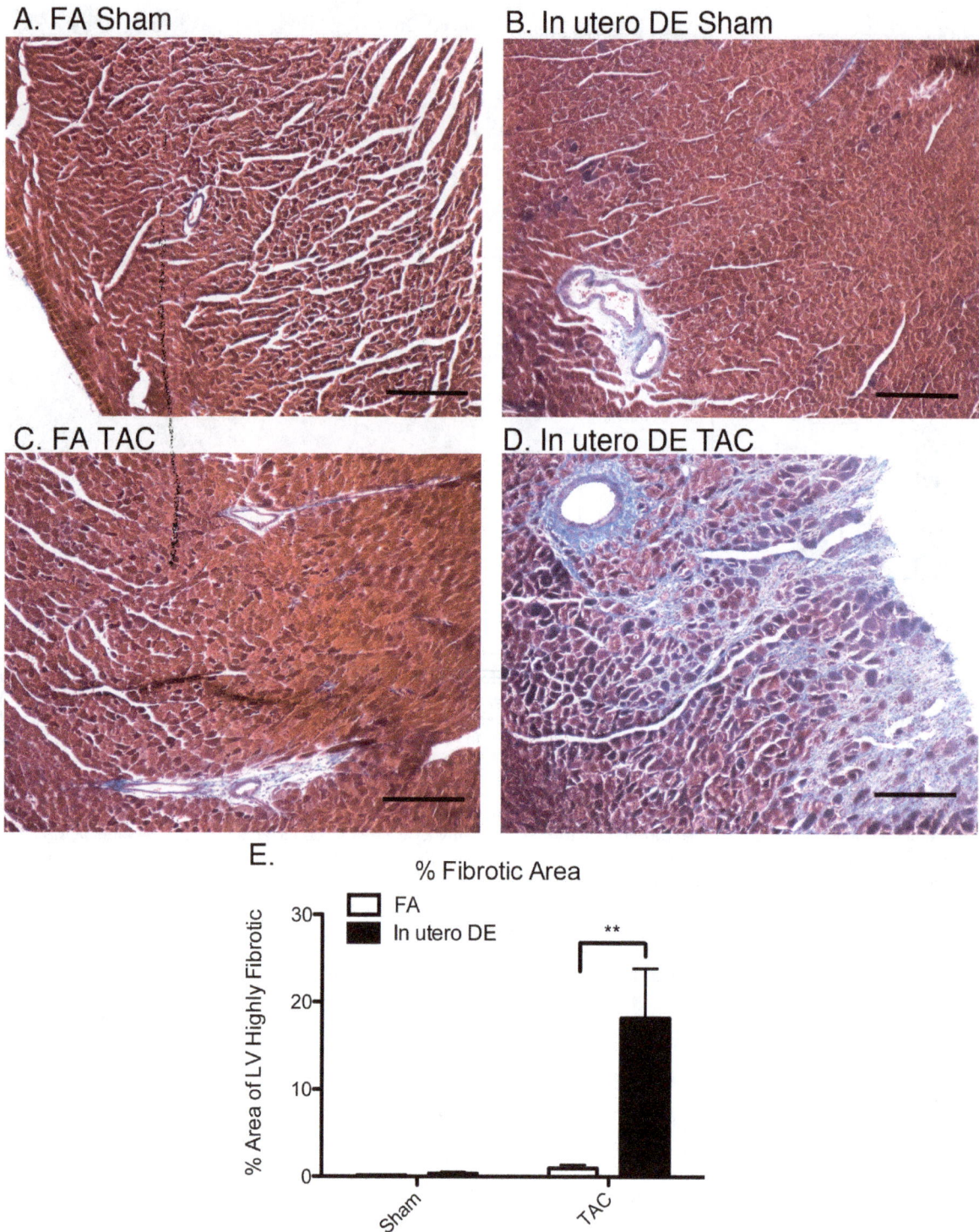

Figure 9. Assessment of myocardial fibrosis in the left ventricle wall of FA Sham (A), *in utero* **DE Sham (B), FA TAC (C), and** *in utero* **DE TAC (D) mice.** Blue staining indicates fibrotic regions. Percentage of LV area that is highly fibrotic was quantified (E). FA sham (n = 4), FA TAC (n = 6), DE sham (n = 5), DE TAC (n = 6). Scale bars = 200 μm.

throughout life [39,41,42,43,44,45,46]. Importantly, it has been suggested that many of these altered metabolic states are due to stable changes in gene expression resulting from epigenetic modifications such as DNA and histone methylation [39,43]. Fetal hypoxia, a potential complication of placental injury, has been reported to promote fetal inflammation and oxidative stress, resulting in elevated DNA methylation of the promoter region of PKCε, predisposing adult myocardium to ischemia-reperfusion injury [47,48]. Our findings are consistent with such a model, as the increased body weight to tibia length, decreased blood

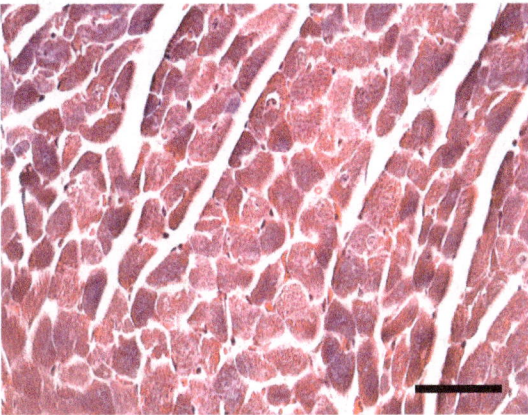

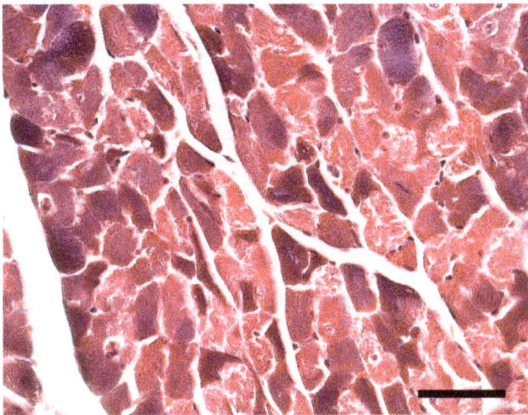

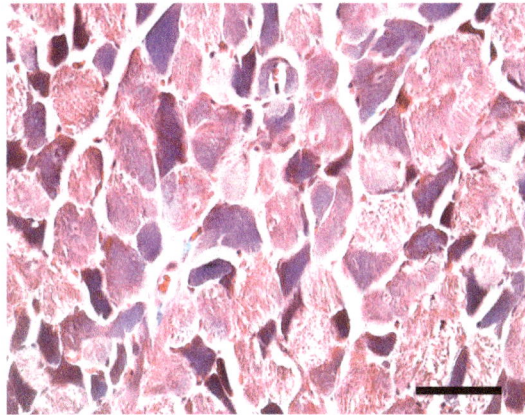

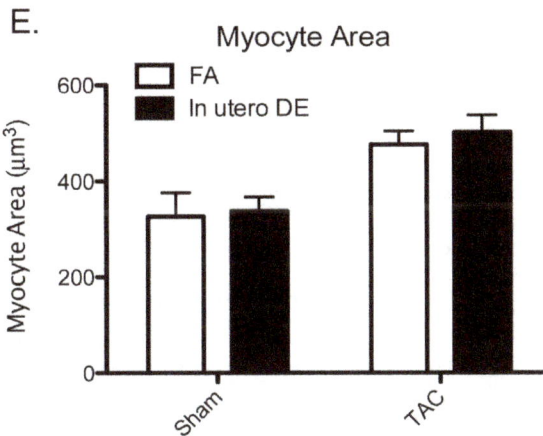

Figure 10. Assessment of individual cardiomyocyte hypertrophy in the left ventricle wall of FA Sham (A), *in utero* DE Sham (B), FA TAC (C), and *in utero* DE TAC (D) mice. Average myocyte area was quantified (E). FA sham (n = 4), FA TAC (n = 6), DE sham (n = 5), DE TAC (n = 6). Scale bars = 50 μm.

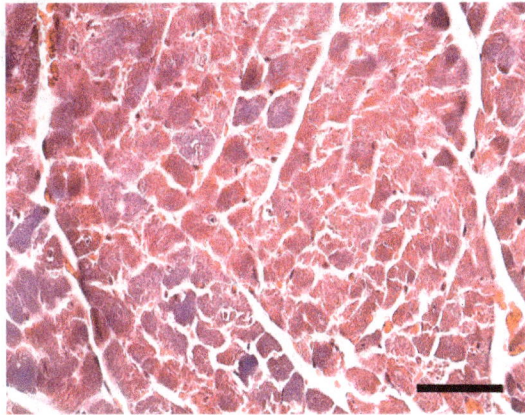

pressure, and increased susceptibility to heart failure in male mice exposed *in utero* is consistent with metabolic and physiologic reprogramming. Whether our effect is due to epigenetic reprogramming events that occur during *in utero* development is currently under investigation.

In utero exposure to other toxic agents have also been reported to promote adult cardiovascular disease. *In utero* exposure to AZT and 3TC reportedly leads to heart failure in adult female mice associated with increased cardiac mitochondrial DNA content and altered mitochondrial and myofibril ultrastructure [49,50]. *In utero* exposure to caffeine also reportedly predisposes to increased adult body fat composition and adult heart failure, possibly through inhibition of A1AR and HIF1α activity [51]. *In utero* exposure to cocaine has been shown to promote fetal cardiac myocyte apoptosis, adult susceptibility to ischemia-reperfusion and adult hypertension [52,53,54]. In our study, we have not yet assessed mitochondrial or myofibril ultrastructure, myocardial apoptosis or the potential role of A1AR or HIF1α. It is important to note, however, that antiretroviral drugs and caffeine exposure both result in baseline ventricular dysfunction, which we do not observe in our study. Cocaine exposure results in adult hypertension, while we observe reduced blood pressure, suggesting independent pathophysiological mechanisms. Reduced blood pressure has also been observed in adult rats with diet-induced metabolic syndrome after acute exposure to $PM_{2.5}$ and ozone, through a presumed autonomic mechanism [55]. Further investigation is required to determine whether there may be overlapping toxicological mechanisms in common between these reports and our study.

In our study, we tested the effect of DE exposure, at ≈ 300 μg/m^3 $PM_{2.5}$, 6 hours/day, 5 days/week. A 300 μg/m^3 $PM_{2.5}$ concentration for this exposure paradigm equates to a time weighted average of 53 μg/m^3/hr $PM_{2.5}$. This hourly exposure is significantly higher than the current U.S. EPA clean air standard of 12 μg/m^3 $PM_{2.5}$ per year. We chose to test this DE concentration as it is consistent with other investigations on DE and vascular function [56,57,58], and it is very relevant to population level exposures in highly polluted urban areas where the level of exposure can approach 500 μg/m^3 at any given time [59].

The relevance of our findings to human populations remains to be determined. A recent meta-analysis of PM exposure and birthweight observed $PM_{2.5}$ and PM_{10} exposure to be associated with reduced birthweight [17] at concentrations observed in developed countries. In addition, $PM_{2.5}$ exposure during pregnancy has been associated with decreased placental mitochondrial DNA content (a marker of mitochondrial oxidative stress) [60] and placental DNA hypomethylation [61]. These observations indicate that human exposure to $PM_{2.5}$, at concentrations relevant to those living in developed countries, directly impact the developing fetus, and likely through a placental insufficiency mechanism. These findings suggest that effects on adult human cardiovascular disease susceptibility are plausible, although further epidemiologic and clinical studies will be necessary to make this determination.

The mechanism by which inhaled PM can elicit placental inflammation and oxidative stress remains unclear. It has been reported that inhaled particles can elicit systemic inflammation, and in vitro models have shown that soluble factors released from diesel exhaust particulate exposed macrophages are able to rapidly incite vascular endothelial inflammation and oxidative stress [62,63]. In addition, it has been observed that human plasma collected from individuals exposed to DE for only short periods of time (1 hr) is proinflammatory to endothelial cells in vitro [64], further suggesting that soluble, proinflammatory mediators are circulating in the blood following the inhalation of DE. In our system, it remains to be determined whether there are maternal circulating proinflammatory mediators that are responsible for our observed effects. Although it is possible that circulating factors are causal to these effects, we cannot exclude the possibility that

maternal DE exposure promotes the activation of neuronal transient receptor potential (TRP) channels that line the airway and which have been observed to be activated by DE and promote systemic vascular and cardiac effects [65,66]. TRP channels are highly expressed within female reproductive organs and the placenta, and their activation has been suggested to play important roles in placental development and regulating the feto-maternal interface [67]. If DE exposure can result in systemic activation of TRP channels, it is possible that placental TRP channels are also activated and may mediate our observed effects. This possibility has yet to be investigated.

Taking into account the potential for circulating factors as well as neuronal TRP activation in mediating these systemic effects, we observed increased inflammatory cell infiltration into the maternal decidua layer of the placenta, suggesting inflammation to arise from signals originating from the mother. Interestingly, our observed increase in placental vascular oxidative stress did not overlap with areas of inflammatory cell infiltration and were largely limited to the fetal labyrinth layer. At present, a causal relationship between the observed increases in vascular oxidative stress and inflammatory cell infiltration has not been established and it remains possible that they are unrelated. Regardless of their relationship, they both provide evidence of significant placental injury in response to DE exposure.

Conclusions

In this report, we demonstrate that in utero exposure to diesel exhaust air pollution in mice promotes placental injury, creating long lasting effects on weight gain, blood pressure, and susceptibility to heart failure in surviving embryos. The results from this study raise the question of whether human susceptibility to adult cardiovascular disease may be exacerbated by in utero exposure to air pollution and suggest the need for future studies to address this question. Such studies will likely guide future regulatory policies that will address environmental exposures and public health.

Acknowledgments

We thank Mr. James Stewart, Drs. Timothy V. Larson, Julie R. Fox, and Timothy Gould for their expert technical assistance with diesel exhaust exposures. We thank Ms. Yu-Chi (Rachel) Chang and Ms. Jackie Coburn for their technical assistance and Dr. Piper Treuting for consultation.

Author Contributions

Conceived and designed the experiments: CSW MTC. Performed the experiments: CSW YL. Analyzed the data: CSW HDL MTC. Wrote the paper: CSW MTC. Overall conception and coordination of entire project: MTC.

References

1. Miller KA, Siscovick DS, Sheppard L, Shepherd K, Sullivan JH, et al. (2007) Long-term exposure to air pollution and incidence of cardiovascular events in women. N Engl J Med 356: 447–458.

2. Pope CA, Burnett RT, Thun MJ, Calle EE, Krewski D, et al. (2002) Lung cancer, cardiopulmonary mortality, and long-term exposure to fine particulate air pollution. JAMA 287: 1132–1141.

3. Pope CA, Burnett RT, Thurston GD, Thun MJ, Calle EE, et al. (2004) Cardiovascular mortality and long-term exposure to particulate air pollution: epidemiological evidence of general pathophysiological pathways of disease. Circulation 109: 71–77.

4. Lim SS, Vos T, Flaxman AD, Danaei G, Shibuya K, et al. (2013) A comparative risk assessment of burden of disease and injury attributable to 67 risk factors and risk factor clusters in 21 regions, 1990–2010: a systematic analysis for the Global Burden of Disease Study 2010. Lancet 380: 2224–2260.

5. Brook RD, Rajagopalan S, Pope CA, Brook JR, Bhatnagar A, et al. (2010) Particulate matter air pollution and cardiovascular disease: An update to the scientific statement from the American Heart Association. Circulation 121: 2331–2378.

6. Bai N, Kido T, Suzuki H, Yang G, Kavanagh TJ, et al. (2011) Changes in atherosclerotic plaques induced by inhalation of diesel exhaust. Atherosclerosis 216: 299–306.

7. Campen MJ, Lund AK, Knuckles TL, Conklin DJ, Bishop B, et al. (2010) Inhaled diesel emissions alter atherosclerotic plaque composition in ApoE(−/−) mice. Toxicology and Applied Pharmacology 242: 310–317.

8. Quan C, Sun Q, Lippmann M, Chen L-C (2010) Comparative effects of inhaled diesel exhaust and ambient fine particles on inflammation, atherosclerosis, and vascular dysfunction. Inhalation Toxicology 22: 738–753.

9. Sun Q, Wang A, Jin X, Natanzon A, Duquaine D, et al. (2005) Long-term air pollution exposure and acceleration of atherosclerosis and vascular inflammation in an animal model. JAMA 294: 3003–3010.

10. Kampfrath T, Maiseyeu A, Ying Z, Shah J, Deiuliis JA, et al. (2011) Chronic Fine Particulate Matter Exposure Induces Systemic Vascular Dysfunction via NADPH Oxidase and TLR4 Pathways. Circulation Research: 1–29.

11. Sun Q, Yue P, Ying Z, Cardounel AJ, Brook RD, et al. (2008) Air Pollution Exposure Potentiates Hypertension Through Reactive Oxygen Species-Mediated Activation of Rho/ROCK. Arteriosclerosis, Thrombosis, and Vascular Biology 28: 1760–1766.

12. Wold LE, Ying Z, Hutchinson KR, Velten M, Gorr MW, et al. (2012) Cardiovascular remodeling in response to long-term exposure to fine particulate matter air pollution. Circulation Heart failure 5: 452–461.

13. Ying Z, Yue P, Xu X, Zhong M, Sun Q, et al. (2009) Air pollution and cardiac remodeling: a role for RhoA/Rho-kinase. AJP: Heart and Circulatory Physiology. H1540–H1550.

14. Arck PC, Hecher K (2013) Fetomaternal immune cross-talk and its consequences for maternal and offspring's health. Nature medicine 19: 548–556.

15. Ward JM, Elmore SA, Foley JF (2012) Pathology methods for the evaluation of embryonic and perinatal developmental defects and lethality in genetically engineered mice. Veterinary pathology 49: 71–84.

16. Webster WS, Abela D (2007) The effect of hypoxia in development. Birth defects research Part C, Embryo today : reviews 81: 215–228.

17. Dadvand P, Parker J, Bell ML, Bonzini M, Brauer M, et al. (2013) Maternal exposure to particulate air pollution and term birth weight: a multi-country evaluation of effect and heterogeneity. Environmental Health Perspectives 121: 267–373.

18. Lewtas J (2007) Air pollution combustion emissions: characterization of causative agents and mechanisms associated with cancer, reproductive, and cardiovascular effects. Mutat Res 636: 95–133.

19. Auten RL, Gilmour MI, Krantz QT, Potts EN, Mason SN, et al. (2012) Maternal diesel inhalation increases airway hyperreactivity in ozone-exposed offspring. American Journal of Respiratory Cell and Molecular Biology 46: 454–460.

20. Bolton JL, Smith SH, Huff NC, Gilmour MI, Foster WM, et al. (2012) Prenatal air pollution exposure induces neuroinflammation and predisposes offspring to weight gain in adulthood in a sex-specific manner. The FASEB journal : official publication of the Federation of American Societies for Experimental Biology 26: 4743–4754.

21. Barker DJ, Osmond C, Golding J, Kuh D, Wadsworth ME (1989) Growth in utero, blood pressure in childhood and adult life, and mortality from cardiovascular disease. BMJ (Clinical research ed) 298: 564–567.

22. Eriksson J, Forsén T, Tuomilehto J, Osmond C, Barker D (2000) Fetal and childhood growth and hypertension in adult life. Hypertension 36: 790–794.

23. Feldt K, Räikkönen K, Eriksson JG, Andersson S, Osmond C, et al. (2007) Cardiovascular reactivity to psychological stressors in late adulthood is predicted by gestational age at birth. Journal of human hypertension 21: 401–410.

24. Godfrey KM, Barker DJ (2001) Fetal programming and adult health. Public health nutrition 4: 611–624.

25. Law CM, de Swiet M, Osmond C, Fayers PM, Barker DJ, et al. (1993) Initiation of hypertension in utero and its amplification throughout life. BMJ (Clinical research ed) 306: 24–27.

26. Martyn CN, Barker DJ, Jespersen S, Greenwald S, Osmond C, et al. (1995) Growth in utero, adult blood pressure, and arterial compliance. British heart journal 73: 116–121.

27. Vijayakumar M, Fall CH, Osmond C, Barker DJ (1995) Birth weight, weight at one year, and left ventricular mass in adult life. British heart journal 73: 363–367.

28. Weldy CS, Liu Y, Chang Y-C, Medvedev IO, Fox JR, et al. (2013) In utero and early life exposure to diesel exhaust air pollution increases adult susceptibility to heart failure in mice. Particle and fibre toxicology 10: 59.

29. Gould T, Larson T, Stewart J, Kaufman JD, Slater D, et al. (2008) A controlled inhalation diesel exhaust exposure facility with dynamic feedback control of PM concentration. Inhalation Toxicology 20: 49–52.

30. Weldy CS, Luttrell IP, White CC, Morgan-Stevenson V, Cox DP, et al. (2013) Glutathione (GSH) and the GSH synthesis gene Gclm modulate plasma redox and vascular responses to acute diesel exhaust inhalation in mice. Inhalation Toxicology 25: 444–454.

31. Yin F, Lawal A, Ricks J, Fox JR, Larson T, et al. (2013) Diesel exhaust induces systemic lipid peroxidation and development of dysfunctional pro-oxidant and pro-inflammatory high-density lipoprotein. Arteriosclerosis, Thrombosis, and Vascular Biology 33: 1153–1161.

32. Liu Y, Chien W-M, Medvedev IO, Weldy CS, Luchtel DL, et al. (2013) Inhalation of diesel exhaust does not exacerbate cardiac hypertrophy or heart failure in two mouse models of cardiac hypertrophy. Particle and fibre toxicology 10: 49.

33. Weldy CS, Luttrell IP, White CC, Morgan-Stevenson V, Bammler TK, et al. (2012) Glutathione (GSH) and the GSH synthesis gene Gclm modulate vascular reactivity in mice. Free Radical Biology and Medicine 53: 1264–1278.

34. Liu Y, Yu M, Wu L, Chin MT (2010) The bHLH transcription factor CHF1/Hey2 regulates susceptibility to apoptosis and heart failure after pressure overload. American journal of physiology Heart and circulatory physiology 298: H2082–2092.

35. Yu M, Liu Y, Xiang F, Li Y, Cullen D, et al. (2009) CHF1/Hey2 promotes physiological hypertrophy in response to pressure overload through selective repression and activation of specific transcriptional pathways. Omics : a journal of integrative biology 13: 501–511.

36. Bussey ME, Finley S, LaBarbera A, Ogata ES (1985) Hypoglycemia in the newborn growth-retarded rat: delayed phosphoenolpyruvate carboxykinase induction despite increased glucagon availability. Pediatric research 19: 363–367.

37. Economides DL, Nicolaides KH (1989) Blood glucose and oxygen tension levels in small-for-gestational-age fetuses. American journal of obstetrics and gynecology 160: 385–389.

38. Economides DL, Nicolaides KH, Gahl WA, Bernardini I, Bottoms S, et al. (1989) Cordocentesis in the diagnosis of intrauterine starvation. American journal of obstetrics and gynecology 161: 1004–1008.

39. MacLennan NK, James SJ, Melnyk S, Piroozi A, Jernigan S, et al. (2004) Uteroplacental insufficiency alters DNA methylation, one-carbon metabolism, and histone acetylation in IUGR rats. Physiological genomics 18: 43–50.

40. Ogata ES, Bussey ME, Finley S (1986) Altered gas exchange, limited glucose and branched chain amino acids, and hypoinsulinism retard fetal growth in the rat. Metabolism: clinical and experimental 35: 970–977.

41. Ogata ES, Bussey ME, LaBarbera A, Finley S (1985) Altered growth, hypoglycemia, hypoalaninemia, and ketonemia in the young rat: postnatal consequences of intrauterine growth retardation. Pediatric research 19: 32–37.

42. Bagley HN, Wang Y, Campbell MS, Yu X, Lane RH, et al. (2013) Maternal docosahexaenoic acid increases adiponectin and normalizes IUGR-induced changes in rat adipose deposition. Journal of obesity 2013: 312153.

43. Ke X, Lei Q, James SJ, Kelleher SL, Melnyk S, et al. (2006) Uteroplacental insufficiency affects epigenetic determinants of chromatin structure in brains of neonatal and juvenile IUGR rats. Physiological genomics 25: 16–28.

44. Lane RH, Kelley DE, Gruetzmacher EM, Devaskar SU (2001) Uteroplacental insufficiency alters hepatic fatty acid-metabolizing enzymes in juvenile and adult rats. American journal of physiology Regulatory, integrative and comparative physiology 280: R183–190.

45. Lane RH, Kelley DE, Ritov VH, Tsirka AE, Gruetzmacher EM (2001) Altered expression and function of mitochondrial beta-oxidation enzymes in juvenile intrauterine-growth-retarded rat skeletal muscle. Pediatric research 50: 83–90.

46. Tsirka AE, Gruetzmacher EM, Kelley DE, Ritov VH, Devaskar SU, et al. (2001) Myocardial gene expression of glucose transporter 1 and glucose transporter 4 in response to uteroplacental insufficiency in the rat. The Journal of endocrinology 169: 373–380.

47. Patterson AJ, Chen M, Xue Q, Xiao D, Zhang L (2010) Chronic prenatal hypoxia induces epigenetic programming of PKC{epsilon} gene repression in rat hearts. Circulation Research 107: 365–373.

48. Patterson AJ, Xiao D, Xiong F, Dixon B, Zhang L (2012) Hypoxia-derived oxidative stress mediates epigenetic repression of PKCε gene in foetal rat hearts. Cardiovascular Research 93: 302–310.

49. Torres SM, Divi RL, Walker DM, McCash CL, Carter MM, et al. (2010) In utero exposure of female CD-1 mice to AZT and/or 3TC: II. Persistence of functional alterations in cardiac tissue. Cardiovascular toxicology 10: 87–99.

50. Torres SM, March TH, Carter MM, McCash CL, Seilkop SK, et al. (2010) In utero exposure of female CD-1 Mice to AZT and/or 3TC: I. Persistence of microscopic lesions in cardiac tissue. Cardiovascular toxicology 10: 37–50.

51. Wendler CC, Busovsky-McNeal M, Ghatpande S, Kalinowski A, Russell KS, et al. (2009) Embryonic caffeine exposure induces adverse effects in adulthood. The FASEB journal : official publication of the Federation of American Societies for Experimental Biology 23: 1272–1278.

52. Bae S, Gilbert RD, Ducsay CA, Zhang L (2005) Prenatal cocaine exposure increases heart susceptibility to ischaemia-reperfusion injury in adult male but not female rats. The Journal of Physiology 565: 149–158.

53. Bae S, Zhang L (2005) Prenatal cocaine exposure increases apoptosis of neonatal rat heart and heart susceptibility to ischemia-reperfusion injury in 1-month-old rat. British Journal of Pharmacology 144: 900–907.

54. Xiao D, Yang S, Zhang L (2009) Prenatal cocaine exposure causes sex-dependent impairment in the myogenic reactivity of coronary arteries in adult offspring. Hypertension 54: 1123–1128.

55. Wagner JG, Allen K, Yang H-YY, Nan B, Morishita M, et al. (2013) Cardiovascular Depression in Rats Exposed to Inhaled Particulate Matter and Ozone: Effects of Diet-Induced Metabolic Syndrome. Environmental Health Perspectives 122: 27–33.

56. Cherng TW, Campen MJ, Knuckles TL, Gonzalez Bosc L, Kanagy NL (2009) Impairment of coronary endothelial cell ETB receptor function after short-term inhalation exposure to whole diesel emissions. AJP: Regulatory, Integrative and Comparative Physiology 297: R640–R647.

57. Cherng TW, Paffett ML, Jackson-Weaver O, Campen MJ, Walker BR, et al. (2011) Mechanisms of diesel-induced endothelial nitric oxide synthase dysfunction in coronary arterioles. Environmental Health Perspectives 119: 98–103.

58. Knuckles TL, Lund AK, Lucas SN, Campen MJ (2008) Diesel exhaust exposure enhances venoconstriction via uncoupling of eNOS. Toxicology and Applied Pharmacology 230: 346–351.

59. Wang J-F, Hu M-G, Xu C-D, Christakos G, Zhao Y (2013) Estimation of citywide air pollution in Beijing. PloS one 8: e53400.

60. Janssen BG, Munters E, Pieters N, Smeets K, Cox B, et al. (2012) Placental mitochondrial DNA content and particulate air pollution during in utero life. Environmental Health Perspectives 120: 1346–1352.

61. Janssen BG, Godderis L, Pieters N, Poels K, Kici Ski M, et al. (2013) Placental DNA hypomethylation in association with particulate air pollution in early life. Particle and fibre toxicology 10: 22.

62. Shaw CA, Robertson S, Miller MR, Duffin R, Tabor CM, et al. (2011) Diesel exhaust particulate–exposed macrophages cause marked endothelial cell activation. American Journal of Respiratory Cell and Molecular Biology 44: 840–851.

63. Weldy CS, Wilkerson H-W, Larson TV, Stewart JA, Kavanagh TJ (2011) Diesel particulate exposed macrophages alter endothelial cell expression of eNOS, iNOS, MCP1, and glutathione synthesis genes. Toxicology in vitro : an international journal published in association with BIBRA 25: 2064–2073.

64. Channell MM, Paffett ML, Devlin RB, Madden MC, Campen MJ (2012) Circulating factors induce coronary endothelial cell activation following exposure to inhaled diesel exhaust and nitrogen dioxide in humans: evidence from a novel translational in vitro model. Toxicological Sciences 127: 179–186.

65. Fariss MW, Gilmour MI, Reilly CA, Liedtke W, Ghio AJ (2013) Emerging mechanistic targets in lung injury induced by combustion-generated particles. Toxicological Sciences 132: 253–267.

66. Hazari MS, Haykal-Coates N, Winsett DW, Krantz QT, King C, et al. (2011) TRPA1 and sympathetic activation contribute to increased risk of triggered cardiac arrhythmias in hypertensive rats exposed to diesel exhaust. Environmental Health Perspectives 119: 951–957.

67. Dörr J, Fecher-Trost C (2011) TRP channels in female reproductive organs and placenta. Advances in experimental medicine and biology 704: 909–928.

20

Burnt Sugarcane Harvesting – Cardiovascular Effects on a Group of Healthy Workers, Brazil

Cristiane Maria Galvão Barbosa[1,2], Mário Terra-Filho[1], André Luis Pereira de Albuquerque[1], Dante Di Giorgi[3], Cesar Grupi[4], Carlos Eduardo Negrão[5], Maria Urbana Pinto Brandão Rondon[5], Daniel Godoy Martinez[5], Tânia Marcourakis[6], Fabiana Almeida dos Santos[6], Alfésio Luís Ferreira Braga[7,8], Dirce Maria Trevisan Zanetta[9], Ubiratan de Paula Santos[1]*

1 Pulmonary Division - Heart Institute(InCor), Hospital das Clínicas da Faculdade de Medicina da Universidade de São Paulo, São Paulo, São Paulo, Brazil, 2 FUNDACENTRO, São Paulo, São Paulo, Brazil, 3 Hypertension Unit, Heart Institute(InCor), Hospital das Clínicas da Faculdade de Medicina da Universidade de São Paulo, São Paulo, São Paulo, Brazil, 4 Electrocardiology Unit, Heart Institute(InCor), Hospital das Clínicas da Faculdade de Medicina da Universidade de São Paulo, São Paulo, São Paulo, Brazil, 5 Unit of Cardiovascular Rehabilitation and Exercise Physiology, Heart Institute (InCor), Hospital das Clínicas da Faculdade de Medicina da Universidade de São Paulo, São Paulo, São Paulo, Brazil, 6 Department of Clinical and Toxicological Analyses, University of São Paulo Pharmacological Sciences School, São Paulo, Brazil, 7 Environmental Epidemiology Study Group, Laboratory of Experimental Air Pollution, Department of Pathology, Faculdade de Medicina da Universidade de São Paulo, São Paulo, São Paulo, Brazil, 8 Environmental Exposure and Risk Assessment Group, Catholic University of Santos, Santos, São Paulo, Brazil, 9 Department of Epidemiology, University of São Paulo School Public Health, São Paulo, Brazil

Abstract

Background: Brazil is the world's largest producer of sugarcane. Harvest is predominantly manual, exposing workers to health risks: intense physical exertion, heat, pollutants from sugarcane burning.

Design: Panel study to evaluate the effects of burnt sugarcane harvesting on blood markers and on cardiovascular system.

Methods: Twenty-eight healthy male workers, living in the countryside of Brazil were submitted to blood markers, blood pressure, heart rate variability, cardiopulmonary exercise testing, sympathetic nerve activity evaluation and forearm blood flow measures (venous occlusion plethysmography) during burnt sugarcane harvesting and four months later while they performed other activities in sugar cane culture.

Results: Mean participant age was 31±6.3 years, and had worked for 9.8±8.4 years on sugarcane work. Work during the harvest period was associated with higher serum levels of Creatine Kinase – 136.5 U/L (IQR: 108.5–216.0) vs. 104.5 U/L (IQR: 77.5–170.5), (p=0.001); plasma Malondialdehyde–7.5±1.4 µM/dl vs. 6.9±1.0 µM/dl, (p=0.058); Glutathione Peroxidase – 55.1±11.8 Ug/Hb vs. 39.5±9.5 Ug/Hb, (p<0.001); Glutathione Transferase– 3.4±1.3 Ug/Hb vs. 3.0±1.3 Ug/Hb, (p=0.001); and 24-hour systolic blood pressure – 120.1±10.3 mmHg vs. 117.0±10.0 mmHg, (p=0.034). In cardiopulmonary exercise testing, rest-to-peak diastolic blood pressure increased by 11.12 mmHg and 5.13 mmHg in the harvest and non-harvest period, respectively. A 10 miliseconds reduction in rMSSD and a 10 burst/min increase in sympathetic nerve activity were associated to 2.2 and 1.8 mmHg rises in systolic arterial pressure, respectively.

Conclusion: Work in burnt sugarcane harvesting was associated with changes in blood markers and higher blood pressure, which may be related to autonomic imbalance.

Editor: John E. Mendelson, California Pacific Medical Center Research Institute, United States of America

Funding: The authors have no support or funding to report.

Competing Interests: The authors have declared that no competing interests exist.

* E-mail: pneubiratan@incor.usp.br

Introduction

Brazil is the world's largest producer of sugar and ethanol from sugarcane, with 570 million tons in 2007/2008 harvest [1]. Although industrial harvesting processes utilize technological methods, manual harvesting is still the predominant method of harvesting sugarcane, and it employs nearly 500,000 workers throughout the country. This is a seasonal activity. For seven months per year, to receive an average monthly wage of US$ 700.00 a sugar cane worker must cut approximately 10 tons of sugarcane daily, in journeys of eight hour and twenty minutes, six days a week, under high temperatures in the fields, due to the climate and the heat from burning sugarcane, and receiving inappropriate reposition of water and electrolytes. Moreover, they are exposed to pollutants released during the cutting of burnt sugarcane.

Exposure to air pollution is associated with increased cardio-respiratory morbimortality, and most studies on the subject are related to urban pollution (industrial/vehicular origin) [2,3,4,5]. The studies on outdoor air pollution caused by biomass burning have focused more on the respiratory effects [6,7] than on the cardiovascular effects [8,9]. Study carried out with sugar cane workers found in their urine levels of 1-hydroxipirene, an exposure markers for polycyclic aromatic hydrocarbons, 10 times higher during the harvest season [10].

There are currently no published studies on the cardiovascular risks associated with manual harvesting of sugarcane, which combines physical and thermal overload as well as exposure to pollutants under conditions that are present in countries such as Brazil, India, Philippines, Latin and Central America countries. It has been reported sugar cane workers' diseases and sudden deaths in the last decade [11,12].

Our objective was to evaluate the occurrence of cardiovascular effects and the possible mechanisms involved in these events associated with the harvesting of burnt sugarcane.

Methods

Study population and period

This is an observational study with repeated measures conducted on 28workers at a sugar and ethanol mill. All participants were Caucasian, male, healthy, between 18 and 50 years of age and had no clinical history or use of medications for cardiopulmonary disease.

The study participants were evaluated at two periods: at the end of burnt sugarcane harvest (October–November 2007) and at the end of period when burnt sugarcane was not being harvested (March–April 2008), when the cutters performed cleanup and planting of unburned sugarcane. As they did not earn by productivity in this period, activity was physically less intense. These time points are henceforth referred to as the harvest and non-harvest periods, respectively.

The Research Ethics Committee of the University of São Paulo Medical School approved the study and all participants signed consent forms.

Examination procedures

Because there are many risk factors as physical and thermal overload as well as exposure to air pollutants, we decide to use a wide range of effect indicators. These markers have been used in exercise, environmental, and occupational health studies. During both periods, workers were divided into groups of five or six. After working all week, they were brought from the countryside to the Heart Institute in São Paulo city, where they underwent several examinations over five consecutive days. Evaluations were conducted sequentially to avoid any changes between exams (Figure 1).

1. The participants answered a questionnaire that was developed for the study and involved the following: data regarding the work during both periods, work time, prior occupational exposures, smoking, and the presence of general and respiratory symptoms. Anthropometric measurements were taken at this moment.

2. Blood markers: fibrinogen, thrombin time (TT), prothrombin time (PT), platelet count, creatine kinase (CK), lactate dehydrogenase (LDH), lipid profile, serum calcium, serum sodium and C-reactive protein (CRP) by high sensitive immunology assay (Dade Behring Marburg GmbH, Germany), erythrocyte glutathione peroxidase (GPx), Glutathione-S-Transferase (GST) and glutathione reductase (GR) measured by spectrophotometry [13,14] and plasma malondialdehyde (MDA) measured by high-performance liquid chromatography [15].

3. Twenty-four-hour ambulatory blood pressure monitoring (ABPM-24 hour) was performed using a Spacelabs-90207 monitor (Spacelabs Medical Inc., USA). Measurements were taken every 10-minutes during the day (5a.m. to 10 p.m.) and every 20-minutes at night (10 p.m. to 5 a.m.), according to standards that have been defined and used with our services [16].

4. Twenty-four-hour electrocardiography monitoring (ECG-24 hour): was performed according to a method that was previously used in another study by our group [17]. Records were analyzed to obtain heart rate variability (HRV) indicators, including the standard deviation of normal RR intervals (SDNN), the standard deviation of sequential five-minute RR interval means (SDANN), and the root mean square of differences between NN adjacent intervals (rMSSD).

5. Cardiopulmonary exercise testing (CPET): a ramp symptom-limited CPET was performed on a cycle (Corival, The Netherlands), consisting in a 2-min period of rest, a 2-min period of warm-up (unloaded pedaling) followed by an incremental work-rate period (increase of 20 W/min) [18]. Oxygen saturation (SpO_2) by pulse oximetry (NONIN-ONYX, model 9500, Plymouth, MN, USA) and electrocardiography (Welch Allyn CardioPerfect, Inc, NY) were monitored continuously. The following variables were recorded breath-by-breath (CardiO₂ System, MGC): oxygen consumption (VO_2), minute-ventilation (V_E), carbon dioxide output (VCO_2), tidal volume (V_T), respiratory rate, respiratory exchange rate (RER) and heart rate (HR).

6. Muscle sympathetic nerve activity (MSNA) was directly measured from the peroneal nerve using the technique of microneurography, as described elsewhere [19]. Multiunit postganglionic muscle sympathetic nerve recordings were made using a tungsten microelectrode (tip-diameter 5–15 μm). Muscle sympathetic bursts were identified by visual inspection and were expressed as burst per 100 heartbeats.

7. Forearm blood flow (FBF) was measured by venous occlusion plethysmography (Hokanson, Bellevue, WA, USA), according to methods that have been described elsewhere [19]. FBF (mL/min/100 mL of tissue) was determined based on a minimum of four separate readings. Forearm vascular conductance (units) was calculated by dividing the forearm blood flow by the mean blood pressure (oscillometrically measured).

Work and environment assessment

During both periods, we recorded the concentration of Particulate Matter (PM) with a diameter of 2.5 ($PM_{2.5}$) using a DustTrak Aerosol Monitor, model-8520 (TSI-Inc., MN, USA) that was calibrated and adjusted prior to the measurements with a flow rate of 1.7 l/min [20]. On the same days and locations, the temperature (°C) and relative air humidity (%) were measured using a digital thermohygrometer (Dataloger-TFA, 3030.15, Germany). To record environmental variables, an activity cycle was sought to best represent the typical activities of the workers (Figure 2).

During the harvest period, activities began approximately 4–6 hours after the end of burning. PM and climate variables were measured in the sugarcane field at three 6-hour periods for three consecutive days during the harvest period and during the cutting and weeding of non-burnt sugarcane in the non-harvest period, when there is exposure to resuspended soil particulate matter.

To estimate the level of heat exposure to which the workers were subjected, we obtained the Wet Bulb Globe Temperature

Evaluation Timeline

2007 harvest ▬▬▬ ▬▬ ▬▬ ▬▬ ▬▬ ▬▬ ▬▬ ▬▬ ▬▬ ▬▬ ➤

| Day | 1 | 2 | 3 | 4 | 5 |

| ▪8 a.m.: Heart Institute ▪8-10 a.m.: Clinical examination •9-11 a.m.: 24-h ABPM and 24-h ECG | ▪ Remove 24-h ABPM and 24-h Holter | •Blood markers •MSNA and FBF (G1*) | •MSNA and FBF (G2) •Cardiopulmonary exercise testing (G1*) | •Cardiopulmonary exercise testing (G2*) |

| Day | 1 | 2 | 3 | 4 | 5 |

2008 non harvest ▬▬▬ ▬▬ ▬▬ ▬▬ ▬▬ ▬▬ ▬▬ ▬▬ ▬▬ ➤

*G1 and G2. Subgroups of 3 subjects; MSNA: Muscle sympathetic nerve activity; FBF: Forearm blood flow

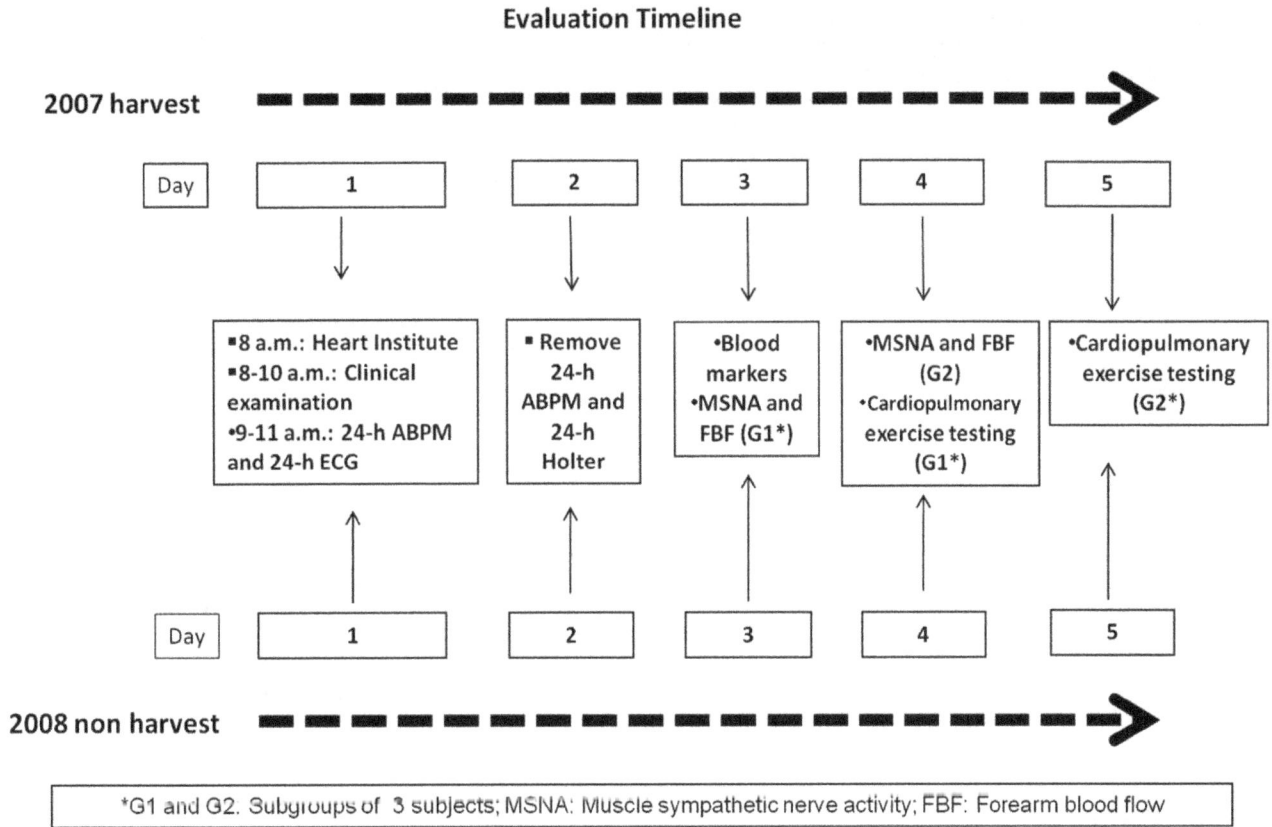

Figure 1. Flow Chart - Evaluation sequence: five groups of six participants.

(WBGT) Index, a heat exposure indicator [21]. We used a Thermal Stress Meter, model-500 by Quest. The equipment was placed near the worker and readings were made during one workday, according to ACGIH recommendations [22].

Statistical analysis

Categorical variables are presented as absolute numbers and percentages, whereas continuous variables are given as the mean±standard deviation (SD) or as the median and the interquartile range (IQR). Descriptive analysis were performed

Figure 2. Worker cutting burnt sugarcane.

for the study variables and the results obtained during the harvest and non-harvest periods were compared by statistical tests for repeated measurements (paired t-test, Wilcoxon rank-test or McNemar's test, as appropriate). Differences of $PM_{2.5}$ and climate variables measurements, between harvest and non-harvest periods, were tested using Mann-Whitney U test.

When the p-value of the differences in measurements for both periods evaluated was≤0.10, linear regression analysis was carried out using a generalized estimating equation (GEE) with robust standard error estimators to evaluate the effects of harvesting work. Adjustments were made for age, body mass index (BMI) and smoking (nonsmoking as a reference), assuming equal correlation of the measurements for each subject (exchangeable correlation). Socioeconomic status was quite similar among the participants and was not included in the analysis. The effects of HRV and MSNA on blood pressure (BP) were evaluated controlling for harvest, age, BMI and smoking (***BP~HRV or MSNA+harvesting+age+smoking+BMI***). The GEE function was obtained from StatLib (http://lib.stat.cmu.edu/) and the analysis was performed on S-Plus® 8.0 for Windows Statistics software, Insightful Corp., Seattle, WA.

Results

Particulate matter, climate and heat exposure

Table 1 shows data for $PM_{2.5}$ concentrations, temperature and relative air humidity (RAH) at the sugarcane fields during the two periods. The $PM_{2.5}$ concentration was higher during the harvest period. There was no significant difference in temperature, and RAH was significantly higher during the non-harvest period.

Table 1. Descriptive analyses of fine particles (PM$_{2.5}$), ambient temperature and relative humidity during harvest and non-harvest periods at the sugarcane field.

Variables	Harvest	Non-harvest	p-value[a]
	Median (IQR[b])	Median (IQR)	
PM$_{2.5}$ (μg/m^3)	87.0(70.0–100,0)	50.0(40.5–61.5)	<0.001
Ambient Temperature (°C)	28.1(25.6–33.0)	28.2(25.6–31.8)	0.500
Relative Humidity (%)	49.0(40.0–59.0)	65.0(61.0–72.8)	<0.001

[a]: Mann-Whitney test;
[b]: interquartile range.

WBGT varied between 18.1 and 28.4°C, with the highest value observed between 11a.m.and 12p.m.

Individual evaluations

The mean worker age was 31±6.3 years (range: 21–45 years), and the average length of time spent employed in sugarcane harvesting was 9.8±8.4 years. Nineteen workers (68%) were never-smokers and nine (32%) were light smokers (7±4.2 pack-years). All workers had less than eight years of schooling, earned between 500 and 800 US dollars per month, according to the amount of cut cane. The housing conditions were similar, living on the outskirts of small towns in the region. None had the automotive vehicle itself.

As the similar cultural habits and socioeconomic conditions, these workers have the same food habits. They carry to work on the field the meal prepared the day before and stored in bowls partially saving thermal food temperature. However, the fact that there is loss of heat by the time of the meal gave this group the nickname of *"cold-meal workers"*. Usually they take rice and beans, potatoes and some animal protein in the form of egg, meat, or chicken

History of previous chronic disease was absent in all participants and none presented infectious or traumatic events when evaluated.

Table 2 presents the general data for the participants in both periods. The workers showed significantly lower body weight during the harvest period as well as smaller abdominal circumference and BMI. The serum HDL presented higher levels during harvest when compared to non-harvest period. On the other hand, diastolic and, in a lesser extent, the systolic blood pressure were higher during harvest period.

Sixteen workers (51.7%) reported nasal itching and rhinorrhea during the harvest period, versus four (14.3%) in the non-harvest period (p<0.01); ten workers (35.7%) mentioned cramping during the harvest period, compared with three (10.7%) in the non-harvest period (p = 0.02); seven workers (25.0%) reported a dry cough during the harvest period, versus only one (3.6%) in the non-harvest period (p = 0.02).

During the harvest period, the workers reported a daily water intake of 5–10 liters. The average weight of the sugarcane that was cut daily, as reported by the workers, was 11 tons (range: 7–14), differently from non-harvest period, when workers receive a pre-defined payment by month and, therefore, the job is less intensive.

Table 3 and Figure 3 shows the blood marker results for the workers during both periods. The CK and LDH levels were higher, while CRP, calcium, TT and PT were significantly lower during the harvest period. Sixteen (57%) workers showed sodium levels below 140 meq/L during the harvest period, versus six

(21.5%) in the non-harvest period. Fibrinogen levels were higher in the harvest period, but the differences were not significant.

The enzymes GR, GST and GPX activity, as well as MDA levels, were higher in the harvest period and the difference was significant for the GST and GPX (Table 4).

Blood pressure and heart rate variability

The 24 hours Systolic and Mean BP were significantly higher during the harvest period (Table 5). HRV indicators tended to be higher in the harvest period, although the difference in multiple regression analysis was significant only for SDANN, the estimate of long-term components of HRV. The frequency domain variables did not present statistically significant variations.

Sympathetic nerve and vascular measures

Among the evaluated individuals, 25 adequately completed these tests during both stages. The values for MSNA and FBF at rest and during handgrip exercise tended to be higher during the harvest period, although the differences were not significant (Table 6).

Cardiopulmonary exercise testing

Among the evaluated individuals, 24 adequately completed this test during both stages. The mean duration of exercise was 12 minutes. The peak O$_2$ uptake (peak VO$_2$), the O$_2$ pulse and the peak systolic/diastolic BP were significantly higher during the harvest period (Table 7). Figure 4 shows that the systolic (SBP) and diastolic (DBP) blood pressures were higher in the harvest period during all test stages. DBP showed an 11.1 mmHg increase at the end of exercise during the harvest period (p<0.001), versus 5.1 mmHg in the non-harvest period (p = 0.064).

Multiple analyses showed that the BP values recorded by ABPM during the harvest period were significantly associated with reductions in HRV and increases in MSNA (Table 8). A 10 ms reduction in SDNN and in rMSSD and a 10 burst/min increase in sympathetic nerve activity were associated with 0.7, 2.2, and 1.8 mmHg rises in 24-hours SBP, respectively.

Discussion

This study revealed that sugarcane workers are subjected to several risks, such as physical overload in hot conditions and exposure to pollutants. Combined, these conditions, which were observed during the harvest period, are related to changes in cardiovascular and blood markers.

In agreement with other authors [11], the present findings indicated the adverse conditions to which sugarcane harvesters are subjected. Studies of the movements of workers during sugarcane harvesting have estimated that cutters bend their backs approximately 4,000 times and make close to 3,800 machete strikes during an 8-hour work day [23].

There are currently few published studies evaluating sugarcane workers [10,20]. The existing studies have evaluated pollutants in the area where the sugarcane is burned and in nearby cities [24]. It is also likely that the PM associated with sugarcane burning presents distinct characteristics during the different periods, as can be observed by the blackish color of the workers' clothes in the harvest period. During burning, (Figure 2) we observed suspended matter present in the sugar cane and on the surface of the top layer of soil, resulting from the burning of waste straw that is released by machete strikes, whereas during the non-burning period the suspended PM contains mainly particles suspended from soil.

The WBTG value was high, reaching 28.43°C, similar to the findings of another study on sugarcane harvesting [23] which that

Table 2. General Characteristics of participants during harvest and non-harvest periods in 28 participants.

Variables	Harvest	Non-harvest	p-value
BMI[a][kg/m^2 (mean $\pm$ SD[b])]	22.6±2.7	23.4±2.9	<0.001[c]
Weight [kg (median; IQR[d])]	64.5 (61.0–69.5)	67.0 (2.0–73.5)	<0.001[e]
Abdcirc[f][cm (median; IQR)]	80.0 (75.0–84.0)	83.0 (78.0–87.5)	0.002[e]
Total Cholesterol [mg/dL(mean± SD)]	159.9±38.1	151.2±33.5	0.041[c]
HDL[g] Cholesterol [mg/dL(mean± SD)]	50.0±10.7	42.5±8.2	<0.00[c]
LDL[h] Cholesterol [mg/dL(mean± SD)]	90.0±26.1	88.8±27.0	0.689[c]
Triglycerides [mg/dL(mean± SD)]	102.6±64.2	106.7±91.4	0.663[c]
Heart rate [BPM[i](mean ± SD]	60.8±8.8	58.7±9.1	0.237[c]
SBP[j][mmHg (mean ± SD)]	125.36±14,81	118.07±17.8	0.046[c]
DBP[k][mmHg (mean ± SD)]	78.00±12.96	70.57±13.88	0.013[c]

[a]: body mass index;
[b]: standard deviation;
[c]: paired t test;
[d]: interquartile range;
[e]: Wilcoxon rank test;
[f]: abdominal circumference;
[g]: High Density Lipoprotein;
[h]: Low Density Lipoprotein;
[i]: beats per minute;
[j]: systolic blood pressure;
[k]: diastolic blood pressure.

registered 27.9°C. These values surpass the recommended limits for continuous labor [21], suggesting that heat stress can occur over the working months, which demonstrates the need for breaks during the workday. Several studies have related the effect of variations of temperature in the cardiovascular system [24], a condition experienced by workers on a daily basis. In this study, the WBTG varied from 18°C to 28°C, what could explain the reports of hospitalization and deaths that have been described in the sugar cane plantation [20].

Higher levels of CK and LDH are compatible with labor under conditions of physical overload and hydro-electrolyte imbalance, as evidenced in several studies [25,26]. LDH and CK are biomarkers of muscular damage and may increase during situations of intense exercise, in which cell membranes become more permeable and enzymes are released into the interstitial matrix and reabsorbed via the lymphatic system to enter the bloodstream [26].The observation that CK remained high, even in blood samples measured 60 hours after exercise suggests a state of persistent hyperCKemia, much like that which occurs in

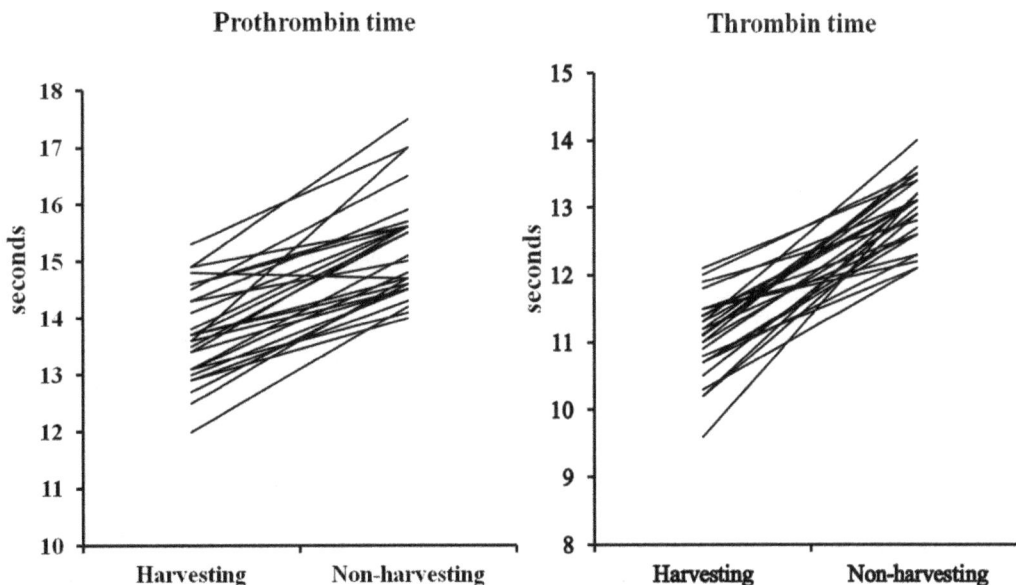

Figure 3. Prothrombin and Thrombin time in harvest and non-harvest periods (n:28).

Table 3. Descriptive analyses of blood markers during harvest and non-harvest periods and the estimated effects of harvest using regressions analysis for repeated measures in 28 participants.

Variables	Periods		Univariate Analysis	Multiple Analysis[a]	
	Harvest	Non-harvest	p-value	RC (95%CI)[b]	p-value
Sodium (mEq/L)	139.6±2.3[c]	141.6±1.6[c]	0.046[d]	−0.86 (−1.84; 0.12)	0.085
Calcium (mg/dl)	9.1±0.4[c]	9.3±0.3[c]	0.013[d]	−0.21(−0.36; −0.07)	0.004
CK[e](U/L)	136.5 (108.5–216.0)[f]	104.5 (77.5–170.5)[f]	<0.001[g]	39.06 (19.24; 58.87)	<0.001
LDH[h](U/L)	156.6±22.0[c]	148.1±23.9[c]	0.028[d]	9.44 (2.10; 16,78)	0.012
CRP[i](mg/L)	2.5 (1.6–5.3)[f]	4.4 (1.7–15.2)[f]	<0.001[g]	−4.46 (−6,89; −2.02)	<0.001
PT[j](s)	13.7±0.8[c]	15.3±0.9[c]	<0.001[d]	−1.50 (−1.77; −1.23)	<0.001
TT[k](s)	11.1±0.6[c]	13.0±0.5[c]	<0.001[d]	−1.86 (−2.10; −1.62)	<0.001

[a]: multiple analysis adjusted for age, body mass index and smoking;
[b]: RegressionCoefficientand95%Confidence interval;
[c]: mean and standard deviation;
[d]: paired t test;
[e]: creatine kinase;
[f]: median and interquartile range;
[g]: Wilcoxon runk test;
[h]: lactate dehydrogenase;
[i]: C-Reactive Protein;
[j]: prothrombin time;
[k]: thrombin time.

athletes [26]. Performing exhaustive exercise in an excessively warm environment increases the risk of muscle lesion [26].

Although serum sodium is a less sensitive marker, the lower sodium and calcium levels observed during the harvest period can arise due to possible hydro-electrolyte disturbances associated with intense sweating and the intake of large quantities of water, which could explain the higher frequency of reported cramping in this period [25,27]. Another study [28] reported the occurrence of hyponatremia among military personnel during intensive training, due to excessive water intake.

The changes observed in TT and PT may be related to exposure to pollutants, as suggested in other studies [29] and likely indicate increased blood viscosity. A study [30] involving 38 individuals showed that metals present in the water-soluble fraction of air pollution particles have an important role in decreasing the whole-blood coagulation time. These findings [30]

and those of a recently published study [31] analyzing the composition and effects of PM arising from vehicular emissions and sugarcane burning found a higher concentration of metals (Ni, Fe, Zn, Mn) in the latter, supporting the findings of short PT and TT observed in our study.

The lower CRP values during the harvest period do not concur with the findings of several studies on pollution [32,33]. However, other studies did not show an association between exposure to pollutants and CRP [34,35]. Our findings may be due to the positive effect of increased physical activity during the harvest period. Studies performed on athletes have shown a reduction in CRP after intensive physical training [36,37]. The same was observed in the lipid profile. The only statistically significant variation was observed for HDL cholesterol, which increased during the harvest period, probably due to the more intense physical activity performed during this period.

Table 4. Glutathione transferase (GST), glutathione peroxidase (GPX), glutathione reductase (GR), and malondialdehyde (MDA) during harvest and non-harvest periods in 28 participants.

Variables	Periods		Paired t test	Multiple analysis[a]	
	Harvest (mean±SD[b])	Non-harvest (mean±SD)	p-value	RC (95%CI)[c]	p-value
GST[d](Ug/Hb)	3.38±1.27	3.01±1.31	0.019	0.39(0.16; 0.63)	0.01
GPX[e] (Ug/Hb)	55.06±11.84	39.48±9.45	0.001	15.31(9.81; 20.82)	<0.001
GR[f] (Ug/Hb)	3.05±0.97	2.91±1.13	0.401	0.23(−0.08; 0.54)	0.143
MDA[g] (μM/dL)	7.50±1.42	6.89±1.01	0.088	6.4(−0.26; 13.06)	0.057

[a]: multiple analysis adjusted for age, body mass index and smoking;
[b]: standard deviation;
[c]: regression coefficient and 95% confidence interval;
[d]: Glutathione - S- Transferase;
[e]: Glutathione Peroxidase;
[f]: Glutathione Reductase;
[g]: Malondialdehyde.

Table 5. Ambulatory Blood Pressure Monitoring (ABPM) and Heart Rate Variability (HRV) during harvest and non-harvest periods in 28 participants.

Variables	Periods (mean±SD[a])		Paired t test	Multiple analysis[b]	
	Harvest	**Non-harvest**	**p-value**	**RC (95%CI)[c]**	**p-value**
24 h ABPM[d]					
SBP[e] (mmHg)	120.1±10.3	117.0±10.0	0.110	3.69 (0.27; 7.11)	0.034
MBP[f] (mmHg)	86.8±9.4	84.4±8.6	0.103	2.95 (0.15; 5.75)	0.039
24 h ECG-HRV[g]					
SDNN[h] (ms[i])	187.1±37.7	178.6±40.5	0.030	5.81 (−2.77; 14.39)	0.184
SDANN[j] (ms)	161.6±30.4	149.0±34.0	0.006	11.22 (2.40; 20.03)	0.013
RMSSD[k] (ms)	47.5±19.1	47.6±19.4	0.957	−2.52 (−7.19;2.15)	0,290

[a]: standard deviation;
[b]: multiple analysis adjusted for age, body mass index and smoking;
[c]: regression coefficient and 95% confidence interval;
[d]: Twenty-four-hour Ambulatory blood pressure monitoring;
[e]: Systolic blood pressure;
[f]: Mean blood pressure;
[g]: Twenty-four-hour electrocardiogram-heart rate variability;
[h]: standard deviation of normal RR intervals;
[i]: milliseconds;
[j]: Standard Deviation of Sequential Five-Minute R-R Interval Means;
[k]: root mean square of successive differences in adjacent NN intervals.

The increased activities of antioxidant enzymes during the harvest period may be associated with a greater stimulus by oxidant radicals produced by exercise [38,39] and inhalation of pollutants [40,41]. The increased levels of MDA, a lipid peroxidation marker [39], during harvest (Table 4), although not statistically significant (p = 0.057), may suggest cell damage even with increases of GST and GPX enzymes, indicating an upper regulation of antioxidant defense mechanism [42]. This suggests that both processes may be concomitant. Poorly planned physical activity that is excessive in intensity and pace and is associated with exhaustion – similar to what occurs during the harvest period – may also induce the development of oxidative stress [38,39].

Table 6. Descriptive analyses of muscle sympathetic nerve activity (MSNA) and forearm blood flow (FBF) at rest and during handgrip exercise in 25 participants.

Variable	Harvest mean±SD[a]	Non-harvest mean±SD	p-value[b]
MSNA (bursts/100HB[c])[d]			
Baseline	38.9±11.0	36.1±8.6	0.056
1′ exercise	38.2±10.9	36.0±8.9	0.339
2′ exercise	40.9±9.3	39.0±8.9	0.440
3′ exercise	45.5±10.1	43.5±10.4	0.460
FBF (ml/min/100 ml of tissue)			
Baseline	2.9±1.4	2.8±1.2	0.657
1′ exercise	3.3±1.5	3.1±1.4	0.559
2′ exercise	3.3±1.8	3.0±1.3	0.356
3′ exercise	3.3±1.9	3.2±1.4	0.602

[a]: standard deviation;
[b]: p-value: Paired T-test;
[c]: heart beats;
[d]: during 30% of maximum voluntary contraction.

The reduction of SDANN observed in this study, which has matched the ultralow frequency variability, suggests negative influence on cardiac autonomic balance, a known cardiovascular risk factor [43], it has been associated with air pollution exposures [44,45,46].

Higher O_2 pulse and peak VO_2 levels observed in CPET during the harvest period may be associated with an improvement in the physical conditioning of workers during that period [18]. Peak VO_2 is an index used to measure performance in athletes from whom high VO_2max values are expected. One study on sugarcane harvesters in Colombia showed a VO_2max of 42 ml/kg/min, which is close to the value found in our study [47]. The values found in this study were similar to those found in sedentary individuals [48] but lower than those in endurance athletes and higher than those in strength athletes [49]. This results are compatible with the activities of sugar cane workers whose labor requires a combination of physical strength and endurance to cut around tens tons per day of sugar cane and carry them from the field to the trucks.

Blood pressure levels raised more during the harvest period, both at rest and during exercise, which may be associated with several mechanisms. Experimental studies in animals [50] and humans [35] as well as epidemiological studies [17,51,52] have shown an association between exposure to vehicular and biomass air pollutants and both acute and chronic elevated BP [53]. The main hypothesis about the effect of air pollution involves oxidative stress and pulmonary inflammation [40] inducing systemic inflammation that could cause vascular dysfunction, endothelial dysfunction, vasoconstriction and cardiac remodeling [50,54]. An imbalance of the autonomic nervous system (ANS) can occur due to the stimulation of intrapulmonary receptors of the ANS. This imbalance can induce changes in the cardiovascular system, resulting in adrenergic vasoconstriction, increased cardiac output and activation of the renin-angiotensin system, leading to endothelial dysfunction and vasoconstriction [54].

By revealing an increase in ABPM-measured blood pressure, an abnormal DBP response during exercise, and the presence of lipid

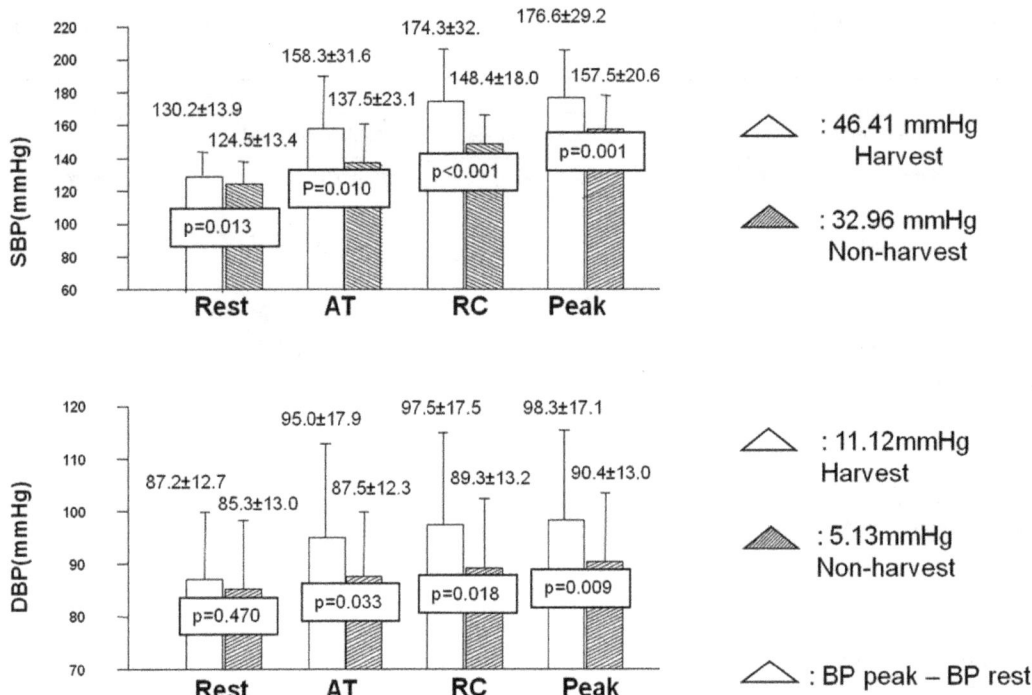

Figure 4. Systolic (SBP) and diastolic (DBP) blood pressure variation during cardiopulmonary exercise testing in each period, mean±SD, (n:24). AT: Anaerobic Threshold; RC: respiratory compensation; Δ = differences between peak and rest in mmHg.

peroxidation days after the end of exposure, our data suggest a subacute effect associated with systemic stress and inflammation [54]. These effects could account for the ANS imbalance evidenced by the relationship between reduced HRV and increased MSNA and higher BP (Table8), as suggested in a recent study [35].

Although the effects of pollution on the sodium balance in the kidneys are unknown, a possible hydro-electrolyte imbalance resulting from strenuous labor under thermal overload conditions may be an additional factor acting on the renin-angiotensin system and thus contributing to higher BP [55] possibly leading to synergism between PM exposure and high temperatures, as suggested in some studies [56,57]. Another possible explanation

could be a chronic rise in vasopressin levels resulting from the rhythm of water intake after intense sweating – in other words, thirst, which happens daily during work. This increase may be associated with higher BP, as suggested in other studies [58].

The persistently inadequate response of DBP during exercise, even during the non-harvest period, suggests the presence of subacute and chronic changes in the cardiovascular system.

This investigation was delineated to evaluate the possible causes associated to sudden deaths that are observed among sugar cane workers. Besides strenuous outdoor work under unsatisfactory feeding and hydration conditions, they are exposed to particulate matter that could increase adverse effects. Different from other study [59] that showed a small systemic proinflammatory response

Table 7. Univariate analysis and the estimated effects of harvest using regression models for repeated measures on cardiopulmonary exercise testing in 24 participants.

Variables	Periods		Paired t test	Multipleanalysis[a]	
	Harvest[b]	Non-harvest[b]	p-value	RC (95%CI)[c]	p-value
Peak VO$_2$ (ml/Kg/min)	40.4±6.9	36.5±6,3	0.005	2.86 (0.68; 5.05)	0.010
Peak O$_2$ (ml/beat)	16.3±3.0	14.9±3.2	0.009	1.51 (0.64; 2.38)	0.001
SBP[d]rest(mmHg)	130.2±13.9	124.5±13.5	0.013	6,94 (3.02; 10.86)	0.001
SBPpeak(mmHg)	176.6±29.9	157.6±20.6	0.001	21.59 (11.58; 31.60)	<0.001
DBP[e]rest(mmHg)	87.2±12.7	85.3±13.0	0.470	3.69 (−1.12;8.50)	0.133
DBP peak (mmHg)	98.3±17.1	90.4±13.0	0.009	10.15 (4.98; 15.32)	<0.001

[a]: multiple analysis adjusted for age, body mass index and smoking;
[b]: mean± standard deviation;
[c]: regression coefficient and 95% confidence interval;
[d]: systolic blood pressure;
[e]: diastolic blood pressure.

Table 8. Effects on blood pressure associated with heart rate variability (HRV) and muscle sympathetic nerve activity (10×).

Variable	24 h SBP[a]		24 h DBP[b]		24 h MBP[c]	
	Regression Coefficients (95%CI[d])	p value	Regression Coefficients (95%CI)	p value	Regression Coefficients (95%CI)	p value
HRV[e] (n:28)						
SDNN[f] (ms)	−0.7 (−1.5; 0.03)	0.058	−0.62 (−1.2; −0.1)	0.029	−0.7 (−1.3; −0.1)	0.02
RMSSD[g] (ms)	−2.2 (−4.1; −0.4)	0.019	−1.6 (−3.0; −2.5)	0.021	−2.4 (−3.9; −0.8)	0.009
HF[h] (ms)	−5.6 (−9.4; −1.8)	0.004	−3.7 (−6.8; −0.6)	0.021	−4.8 (−8.1; −1.4)	0.005
LF[i] (ms)	−3.7 (−7.7; 0.2)	0.065	−3.0 (−5.9; −0.06)	0.046	−3.5 (−6.8; −0.1)	0.044
LF/HF	28.3 (17.3; 39.3)	<0.001	−7.4 (−26.4; 11.5)	0,519	26.4 (18.8; 33.9)	<0.001
MSNA[j]/MVC[k] (n = 25)	1.8 (0.1; 3.5)	0.035	1.4 (0.05; 2.7)	0.049	1.6 (0.8; 2.4)	0.041

Multiple analysis adjusted for age, BMI and smoking;
[a]: systolic blood pressure;
[b]: diastolic blood pressure;
[c]: men blood pressure;
[d]: confidence interva;
[e]: heart rate variability;
[f]: standard deviation of normal RR intervals;
[g]: root mean square of differences between NN adjacent intervals;
[h]: high frequence;
[i]: low frequence;
[j]: Muscle sympathetic- nerve activity-bursts/100 Heart beats;
[k]: maximum voluntary contraction at 30%

induced only by intermittent moderate-intense exercise, we observed effects on blood pressure and coagulation markers inversely related with regular physical exercise [60]. Increases of blood pressure (mean, systolic, and diastolic, the last during exercise test) and shortening of thrombin and prothrombin times during harvest period could be attributed to pollution exposure.

In summary, it was not possible to evaluate the isolated impact of each risk factor (air pollution, heat and exertion) on the alterations found (changes in coagulation, oxidative stress and blood pressure). However, these alterations suggest an important impact of the work environment on health. As such, these changes can account for the reports of morbimortality among these workers, which affects the most susceptible and/or those who put forth the most effort to achieve a higher income.

The great number of agricultural workers – as well as other workers, such as miners – facing similar conditions around the world makes this problem a public health issue that deserves more attention from different public administration sectors.

Limitations

There are limitations in our study. One result from our inability to better describe individual worker exposure during a longer environmental evaluation period. However, the great homogeneity of the work involved in cutting sugarcane and the minor changes in meteorological variables suggest that the variation in $PM_{2.5}$ is not relevant, considering that our analyses compared the effects of harvest and non-harvest periods. The existence of a control group could have helped the comparison of certain outcomes, but the challenges in creating a group to compare several conditions (temperature, strenuous exercise, pollution and socioeconomic condition) could not be overcome. The fact that we used repeated measurements to compare the effects of different conditions on the same subjects minimizes this issue.

Conclusions

Sugarcane work during the harvesting period exposes workers to higher levels of particulate matter, thermal overload, and intense physical exertion, inducing muscle lesion, changes in blood coagulation and in heart rate variability, systemic oxidative stress, and high blood pressure. The autonomic imbalance seems to be one of the mechanisms involved in blood pressure changes.

Author Contributions

Conceived and designed the experiments: CMGB UPS. Performed the experiments: ALPA DDG CG CEN MUPBR DGM TM FAS. Analyzed the data: ALFB DMTZ. Wrote the paper: CMGB UPS ALPA DDG CG CEN MUPBR DGM TM ALFB DMTZ MT-F. Critically revised the manuscript: MT-F.

References

1. FAO Food and Agriculture Organization of the United Nations. Available: http://faostat.fao.org/site/339/default.aspx. Accessed on 2012 Apr 30.
2. World Health Organization (2006) WHO Air quality guidelines for particulate matter, ozone, nitrogen dioxide and sulfur dioxide. Global update 2005. Summary of risk assessment. Geneva.
3. Krewski D, Jerrett M, Burnett RT, Ma R, Hughes E, et al. (2009) Extended follow-up and spatial analysis of the American Cancer Society study linking particulate air pollution and mortality. Res Rep Health Eff Inst: 5–114; discussion 115–136.
4. Brunekreef B, Beelen R, Hoek G, Schouten L, Bausch-Goldbohm S, et al. (2009) Effects of long-term exposure to traffic-related air pollution on respiratory and cardiovascular mortality in the Netherlands: the NLCS-AIR study. Res Rep Health Eff Inst: 5–71; discussion 73–89.
5. Feng J, Yang W (2012) Effects of particulate air pollution on cardiovascular health: a population health risk assessment. PLoS One 7: e33385.
6. Naeher LP, Brauer M, Lipsett M, Zelikoff JT, Simpson CD, et al. (2007) Woodsmoke health effects: a review. Inhal Toxicol 19: 67–106.
7. Cançado JED, Saldiva PH, Pereira LAA, Lara LBSS, Artaxo P, et al. (2006) The impact of sugar cane-burning emissions on the respiratory system of children and the elderly. Environ Health Perspect 114: 725–729.

8. Arbex MA, Saldiva PH, Pereira LAA, Braga ALF (2010) Impact of outdoor biomass air pollution on hypertension hospital admissions. J Epidemiolo Community Health 64: 573–579.

9. Henderson SB, Brauer M, Macnab YC, Kennedy SM (2011) Three measures of forest fire smoke exposure and their associations with respiratory and cardiovascular health outcomes in a population-based cohort. Environ Health Perspect 119: 1266–1271.

10. Bosso RMV, Amorin LMF, Andrade SJ, Rossini A, Marchi MMR, et al. (2006) Effects of genetic polymorphisms CYP1A1, GSTM1 and GSTP1 on urinary 1-hidroxypyrene levels in sugar cane workers. Science of the Total Enviromental: 382–390.

11. Alves F (2006) Por que morrem os cortadores de cana. Saúde e Sociedade 15: 90–98.

12. Ribeiro H (2008) Sugar cane burning in Brazil: respiratory health effects. Rev Saúde Pública 42: 1–6.

13. Habig WH, Pabst MJ, Jakoby WB (1974) Glutathione S-transferases the first enzymatic step in mercapturic acid formation. J Biol Chem 22: 7130–7139.

14. Flohé L, Günzler WA (1984) Assays of glutathione peroxidase. Methods Enymol 105: 114–121.

15. Sim AS, Salonikas C, Naidoo D, Wilcken DE (2003) Improved method for plasma malondialdehyde measurement by high-performance liquid chromatography using methyl malondialdehyde as an internal standard. J Chromatogr B 785: 337–344.

16. Brazilian Society of Cardiology, Brazilian Society of Hypertension, and Brazilian society of Nephrology (2005) IV Guideline for ambulatory blood pressure monitoring. IV ABPM/II HBPM. Arquivos Brasileiros de Cardiologia 85: 1–18.

17. Santos UP, Braga ALF, Giorgi DM, Pereira LAA, Gruppi CJ, et al. (2005) Effects of air pollution on blood pressure and heart rate variability: a panel study of the vehicular traffic controllers in the city of São Paulo, Brazil. Eur Heart J 26: 193–200.

18. ATS ACPP (2003) Statement on cardiolopulmonary exercise testing. Am J Respir Crit Care Med 167: 211–277.

19. Negrão CE, Rondon MUPB, Tinucci T, Alves MJN, Roneda F, et al. (2001) Abnormal neurovascular control during exercise is linked to heart failure severity. Am J Physiol 280: H286–292.

20. Goto DM, Obuti CA, Barbosa CMG, Saldiva PHN, Zanetta DMT, et al. (2011) Effects of biomass burning on nasal mucuciliary clearance and mucus properties after sugarcane harvesting. Environmental Research 111: 664–669.

21. ACGIH@ (2010) American Conference of Governamental Industrial Hygienist. TLVs@ and BEIs@ - Threshold Limit values and Biological Exposure Indices.

22. ACGIH (2010) American Conference of Governamental Industrial Hygienist. TLVs@ and BEIs@ - Threshold Limit values and Biological Exposure Indices.

23. Laat EF, Vilela RAG, Silva AJN, Luz VG (2008) Impact over the working conditions:physical wear of sugar-cane cutters. In: Plataforma Banco Nacional de Desenvolvimento Econômico e Social (BNDES). Instituto Brasileiro de Análises Sociais e Econômicas (IBASE), eds Impactos da indústria canavieira no Brasil. Rio de Janeiro: IBASE. pp. 36–46.

24. Cançado JED, Braga ALF, Pereira LAA, Arbex MA, Saldiva PH, et al. (2006) Clinical repercussions of exposure to atmospheric pollution. J Bras Pneumol 32: s5–11.

25. Shieh SD, Lin YF, Lu KC, Li BL, Chu P, et al. (1992) Role of creatine phosphokinase in predicting acute renal failure in hypocalcemic exertional heat stroke. Am J Nephrol 12: 252–258.

26. Brancaccio P, Maffuli N, Limogelli FM (2007) Creatinine kinase monitoring in sport medicine. British Medical Bulletin 81: 287–230.

27. Hew-Butler T (2010) Arginine vasopressin, fluid balance and exercise: is exercise-associated hyponatraemia a disorder of arginine vasopressin secretion? Sports Med 40: 459–479.

28. Garigan TP, Ristedt DE (1999) Death from hyponatremia as result of acute water intoxication in a Amy Basic Trainee. Mil Med 164: 234–238.

29. Baccarelli A, Zanobetti A, Martinelli I, Grillo P, Hou L, et al. (2007) Effects of exposure to air pollution on blood coagulation. J of Thrombosis and hemostasis 5 (2): 252–260.

30. Sangani RG, Soukup JM, Ghio AJ (2010) Metals in air pollution particles decrease whole-blood coagulation time. Inhal Toxicol 22: 621–626.

31. Mazzoli-Rocha F, Magalhães CB, Malm O, Saldiva PH, Zin WA, et al. (2008) Comparative respiratory toxicity of particles produced by traffic and sugar cane burning. Environ Research 108: 35–41.

32. Peters A, Frobtich M, Doning A, Immervoll T, Wichmann HE, et al. (2001) Particulate air pollution is associated with an acute phase response in men. Results from the MONICA-Augsburg Study. Eur Heart Journal 22: 1198–1204.

33. Hoffmann B, Moebus S, Dragano N, Stang A, Mohlen-Kamp S, et al. (2009) Chronic residential exposure to particulate matter air pollution and systemic inflammatory markers. Environ Health Perspect 117: 1302–1308.

34. Forbes LJL, Patel MD, Rudnicka AR, Alicja R, Cook DJ, et al. (2009) Chronic exposure to out door air pollution and marker of systemic inflammation. Epidemiology 20: 245–253.

35. Brook RD, Urch B, Dovonch JT, Bard RL, Speck M, et al. (2009) Insights into mechanisms and mediators of the effects of air pollution exposure on blood pressure and vascular function in healthy humans. Hypertension 54: 659–667.

36. Anderson J, Jansson JH, Hellsten G, Nilsson TK, Hallmans G, et al. (2009) Effects of heavy endurance physical exercision inflammatory markers in non athletes. Atherosclerosis 10: 1–5.

37. Lakka TA, Lakka HM, Rankinen T, Leon AS, Rao DC, et al. (2005) Effect of exercise training on plasma levels of C-reactive-protein in healthy adults: The HERITAGE Family Study. Eur Heart J 19: 2018–2025.

38. Tanskanen M, Atalay M, Uusitalo A (2010) Altered oxidative stress in overtrained athletes. J of Sports Sciences 28: 309–317.

39. Marzatico F, Pansarasa O, Bertorelli L, Sonezini L (1997) Blood free radical antioxidant enzymes and lipidic peroxides following long distance and lactacidemic performances on highly trained aerobic and sprint athletes. J Sports Med Phys Fitness 37: 235–239.

40. Romieu I, Castro-Giner F, Kunzli N, Sunyer J (2008) Air pollution, oxidative stress and dietary supplementation: a review. Eur Respir J 31: 179–197.

41. Li N, Hao M, Phalen RF, Hinds WC, Nel AE (2003) Particulate air pollutants and asthma. A paradigm for the role of oxidative stress in PM-induced adverse health effects. Clinical Imunology 109: 250–265.

42. Gomes EC, Silva AN, de Oliveira MR (2012) Oxidants, antioxidants, and the beneficial roles of exercise-induced production of reactive species. Oxid Med Cell Longev 2012: 756132.

43. Task Force of Eur Soc Cardiology, North Am Society, Eletrophysiology (1996) Heart rate variability. Satandarts of measurement physiological interpretation and clinical use. Circulation 93: 1043–1065.

44. Pope CA 3rd, Verrier RL, Lovet EG, Larson AC, Raizenne ME, et al. (1999) Heart variability associated with particulate air pollution. Am Heart J 138: 804–807.

45. Gold DR, Litonjua A, Schwartz J, Lovett E, Larson A, et al. (2000) Ambient pollution and heart rate variability. Circulation 101: 1267–1273.

46. Liao D, Duan Y, Whitsel EA, Zheng ZJ, Heiss G, et al. (2004) Association of higher levels of ambient criteria pollutants with impaired cardiac autonomic control: a population-based study. Am J Epidemiol 159: 768–777.

47. Spurr GB, Maksud MG, Barac-Nieto M (1977) Energy expenditure, productivity and physical work capacity of sugarcane loaders. The Am J of Clinical Nutrition 30: 1740–1746.

48. Herdy AH, Uhlendorf D (2011) [Reference values for cardiopulmonary exercise testing for sedentary and active men and women.]. Arq Bras Cardiol 96: 54–59.

49. Anastasakis A, Kotsiopoulou C, Rigopoulos A, Theopistou A, Protonotarios N, et al. (2005) Similarities in the profile of cardiopulmonary exercise testing between patients with hypertrophic cardiomyopathy and strength athletes. Heart 91: 1477–1478.

50. Yin Z, Yue P, Xu X, Zhong M, Sun Q, et al. (2009) Air pollution and cardiac remodeling: a role for Rho\Rho-Kinase. Am J Physiol Heart Circ Physiol 296: H1540–H1550.

51. Delfino RJ, Tjoa T, Gillen DL, Staimer N, Polidori A, et al. (2010) Traffic-related air pollution and blood pressure in elderly subjects with coronary artery disease. Epidemiology 21: 396–404.

52. Baumgartner J, Schauer JJ, Ezzati M, Lu L, Cheng C, et al. (2011) Indoor Air Pollution and Blood Pressure in Adult Women Living in Rural China. Environ Health Perspect.

53. Fuks K, Moebus S, Hertel S, Viehmann A, Nonnemacher M, et al. (2011) Long-Term Urban Particulate Air Pollution, Traffic Noise and Arterial Blood Pressure. Environ Health Perspect.

54. Brook RD, Rajagopalan S, Pope-CA r, Brook JR, Bhatnagar A, et al. (2010) Particulate matter and air pollution and cardiovascular disease: An update to the scientific statement from the American Heart Association. Circulation 121: 2331–2378.

55. Alpérovitch A, Lacombe JM, Hanon O, Dartigues JF, Ritchie K, et al. (2009) Relationship between blood pressure and outdoor temperature in a larger sample of eldery individuals: the three city study. Arch Intern Med 169: 78–80.

56. Gliner JA, Raven PB, Horvath SM, Drinkwater BL, Sutton JC (1975) Man's physiologic response to long-term work during thermal and pollutant stress. J Appl Physiol 39: 628–632.

57. Qian Z, He Q, Lin HM, Kong L, Bentley CM, et al. (2008) High temperatures enhanced acute mortality effects of ambient particle pollution in the "oven" city of Wuhan, China. Environ Health Perspect 116: 1172–1178.

58. Bankir L, Perucca J, Weinberg MH (2007) Ethnic differences in urine concentration: possible relationship to blood pressure. Clin J Am Soc Nephrol 2: 304–312.

59. Donde A, Wong H, Frelinger J, Power K, Balmes JR, et al. (2012) Effects of exercise on systemic inflammatory, coagulatory, and cardiac autonomic parameters in an inhalational exposure study. J Occup Environ Med 54: 466–470.

60. Haskell WL, Lee IM, Pate RR, Powell KE, Blair SN, et al. (2007) Physical activity and public health: updated recommendation for adults from the American College of Sports Medicine and the American Heart Association. Circulation 116: 1081–1093.

21

Heat-Related Mortality in India: Excess All-Cause Mortality Associated with the 2010 Ahmedabad Heat Wave

Gulrez Shah Azhar[1,2]*, Dileep Mavalankar[1,2], Amruta Nori-Sarma[1,3], Ajit Rajiva[1], Priya Dutta[1], Anjali Jaiswal[4], Perry Sheffield[5], Kim Knowlton[3,4], Jeremy J. Hess [6,7], on behalf of the Ahmedabad HeatClimate Study Group¶

1 Indian Institute of Public Health, Ahmedabad, Gujarat, India, **2** Public Health Foundation of India, New Delhi, India, **3** Columbia Mailman School of Public Health, New York, New York, United States of America, **4** Natural Resources Defense Council, New York, New York, United States of America, **5** Icahn School of Medicine at Mount Sinai, New York, New York, United States of America, **6** Department of Emergency Medicine, Emory University School of Medicine, Atlanta, Georgia, United States of America, **7** Department of Environmental Health, Emory University School of Public Health, Atlanta, Georgia, United States of America

Abstract

AbstractIntroduction: In the recent past, spells of extreme heat associated with appreciable mortality have been documented in developed countries, including North America and Europe. However, far fewer research reports are available from developing countries or specific cities in South Asia. In May 2010, Ahmedabad, India, faced a heat wave where the temperatures reached a high of 46.8°C with an apparent increase in mortality. The purpose of this study is to characterize the heat wave impact and assess the associated excess mortality.

Methods: We conducted an analysis of all-cause mortality associated with a May 2010 heat wave in Ahmedabad, Gujarat, India, to determine whether extreme heat leads to excess mortality. Counts of all-cause deaths from May 1–31, 2010 were compared with the mean of counts from temporally matched periods in May 2009 and 2011 to calculate excess mortality. Other analyses included a 7-day moving average, mortality rate ratio analysis, and relationship between daily maximum temperature and daily all-cause death counts over the entire year of 2010, using month-wise correlations.

Results: The May 2010 heat wave was associated with significant excess all-cause mortality. 4,462 all-cause deaths occurred, comprising an excess of 1,344 all-cause deaths, an estimated 43.1% increase when compared to the reference period (3,118 deaths). In monthly pair-wise comparisons for 2010, we found high correlations between mortality and daily maximum temperature during the locally hottest "summer" months of April (r = 0.69, p<0.001), May (r = 0.77, p<0.001), and June (r = 0.39, p<0.05). During a period of more intense heat (May 19–25, 2010), mortality rate ratios were 1.76 [95% CI 1.67–1.83, p<0.001] and 2.12 [95% CI 2.03–2.21] applying reference periods (May 12–18, 2010) from various years.

Conclusion: The May 2010 heat wave in Ahmedabad, Gujarat, India had a substantial effect on all-cause excess mortality, even in this city where hot temperatures prevail through much of April-June.

Editor: Suminori Akiba, Kagoshima University Graduate School of Medical and Dental Sciences, Japan

Funding: This work is funded by a grant from the Climate Knowledge Development Network. The funders had no role in study design, data collection and analysis, decision to publish, or preparation of the manuscript.

Competing Interests: The authors have declared that no competing interests exist.

* E-mail: gsazhar@iiphg.org

¶ Membership of the Ahmedabad Heat and Climate Study Group is provided in the Acknowledgments

Introduction

Weather extremes can have significant public health impacts [1,2,3]. The global frequency of extreme weather events, including extreme precipitation, drought and resulting crop failure, tropical cyclones, and flooding has been increasing in recent years, consistent with anthropogenic climate change [4,5]. These trends, which exhibit significant regional variability, are expected to continue in future as climate change becomes more pronounced [6]. A national assessment conducted by the Indian government on climate change projects increasing temperatures for India

through the 21st century, including increasing extreme heat events [7].

Temperature extremes are a major underlying weather-related cause of mortality in much of the world and the leading cause of directly-mediated weather-related mortality [8,9,10,11,12,13]. Heat related morbidity and mortality can be due to either direct or indirect effects [14]. Direct effects include a spectrum of heat illness ranging from heat exhaustion to heat stroke; the indirect effects occur when heat exposure stresses underlying physiological systems and results in other specific manifestations such as renal insufficiency, acute cerebrovascular disease, and exacerbations of pulmonary disease [15].

A heat wave is a prolonged period of unusually and excessively hot weather, which may also be accompanied by high humidity. Definitions vary, in part because a heat wave is measured relative to the usual weather in the area and relative to normal temperatures for the season, and in part because there is no single best indicator from a public health perspective [12,13][16,17]. The US National Oceanic & Atmospheric Administration (NOAA) defines a heat wave as a period of abnormally and uncomfortably hot and unusually humid weather lasting two or more days, and advisories are issued when these conditions are forecast.[18]. On a population basis, the impacts of high temperatures on mortality vary by location [19]. Heat wave conditions are known to amplify the effect of temperature on mortality [20] and several other effect modifiers have been identified [10,21,22]. According to the Indian Meteorology Department (IMD), a heat wave in India is declared when either there is an excess of 5°C over a normal daily historical maximum temperature (30 year average) of less than 40°C; or an excess of 4°C over a normal historical maximum temperature of more than 40°C. If the actual maximum temperature is above 45°C, is a heat wave is declared irrespective of the normal historical maximum temperature [18],[23].

Cities and urban areas experience higher levels of heat exposure than surrounding rural areas, due to the urban heat island effect whereby temperatures in urban areas are on average 3.5–12°C higher than those found outside city limits [24]. Similarly, urban microclimates have a role in creating higher urban temperatures in some parts of cities [25]. Urbanization can exacerbate heat exposures for residents of urban core areas, especially for developing countries where in-migration of rural poor and unplanned development of urban service systems may not be able to keep pace with demand. However, this on-going development also provides opportunities for municipalities to implement specific and targeted actions to mitigate the impacts of rising temperatures.

Among historic heat waves, some have been associated with large numbers of all-cause and cause-specific deaths, notably in France, Europe more broadly, Chicago and California [26,27,28,29]. During the 2003 heat wave in France, excess mortality in 13 French cities varied from 4% to 142%, with a majority of cities experiencing excess mortality in the 20–50% range [26]. During the 2006 heat wave in California, there were an estimated 16,166 excess emergency department visits and 1,182 excess hospitalizations state-wide.[28]. In 2010, Moscow and Western Russia experienced a heat wave with ambient temperatures exceeding 39°C, and a high number of deaths, though specific estimates of excess morbidity and mortality have not been done [30].

India has had several historic heat waves. Most notably, in May 1998, India experienced a severe heat wave over a 2-week period, which was considered to be the worst in the previous 50 years [31]. The following year, a similar record-breaking event occurred in 1999 in north-west and central India. According to Kalsi et al (2001), during the summer of 1999, India experienced unprecedented heat in April, with maximum temperatures of 40°C or above for more than 14 days [32]. Another heat wave in 2003 caused more than 3,000 deaths in Andhra Pradesh [33].

In light of mounting epidemiologic evidence and the perception of a worsening threat to public health, several developed and even some developing countries have instituted prevention strategies and preparedness plans to minimize the human costs of increasing heat [34]. These include best practices for reducing municipal heat vulnerability and workplace heat-health promotion strategies, in cities from New York City to Abu Dhabi [35,36,37].

While the health impacts of heat waves, particularly impacts on mortality, have been explored for many regions of the world, there is relatively little information on specific impacts and characteristics of the relationship between excess heat exposure and health impacts in South Asia. We pursued the present analysis in an effort to begin filling this gap. Perhaps because extreme heat has not been recognized as a significant public health risk in India, or perhaps because the risk is now perceived to be increasing with climate change, heat health promotion strategies in India have only recently been made a matter of policy at the city government level. This study was done as part of a collaboration working to research and develop strategies for climate change adaptation in India. One of the first steps towards adaptation was to conduct the current analysis and risk assessment.

When considered with projected population densities, significant increases in premature heat-related mortality pose a threat to public health in India [38]. However, a discussion of the health effects of extreme heat is currently absent from India's national action plan for climate change [39].

In May 2010, Ahmedabad, India, a rapidly-growing city in the western state of Gujarat, experienced a heat wave. According to the India Meteorological Department criteria, the days of April 17–18 and May 13– 15, 17, and 20–25 in 2010, qualified as a "heat wave" with daily maximum temperatures varying between 44.5 – 46.8°C. There was an apparent increase in May 2010 all-cause mortality, though the true increase in mortality counts may have been under-reported [40,41]. The objective of this ecological study is to characterize the heat wave's impacts on all-cause mortality and assess excess mortality associated with extreme heat exposure.

Methods

To explore the relationship between the 2010 heat wave and mortality in Ahmedabad, we conducted an ecological analysis to evaluate potential relationships between daily all-cause mortality and maximum daily temperatures. We chose the May 2010 study period because of Ahmedabad's unprecedented heat wave, with maximum daily temperatures peaking at 46.8°C, a record high. The study was reviewed and approved by the Emory University Institutional Review Board and the Ethics committee at the Indian Institute of Public Health Gandhinagar.

Ahmedabad is the largest city in the state of Gujarat and the sixth largest city in the country, with an urbanized population of 5.571 million (2011) and with a metropolitan regional population of 6.35 million [42]. Extended population includes persons living outside of the Ahmedabad city limits but still within the municipal district of Ahmedabad. People living outside of the city limits are likely to have similar heat-related mortality risks as those in the urban core as these areas are also urbanized but not part of the municipality. The city's weather is usually dry and is hottest starting in March, with a seasonal monsoon rain period in July. The post-monsoon season is hot and wet and lasts until the end of October.

The Ahmedabad Municipal Corporation (AMC) is responsible for registering births and deaths in the city. We acquired anonymized and de-identified daily mortality data (day-wise death counts) from the AMC Office of the Registrar of Births and Deaths for the years 2009 to 2011. These deaths are inclusive of only those that occurred within city limits and not those in the extended city. The records for the daily mortality dataset were given to the authors under the necessary provision that the data is not distributed to the public and only the results obtained from the dataset may be used for publication purposes. Temperature data was obtained from the Indian Meteorology Department's Meteorological Aerodrome Report (METAR - a syntax used by the World Meteorological Organization or WMO), station at Ahmedabad airport (Station ID - VAAH, WMO ID 42647), located on the outskirts of the city, 8 Kilometres from the railway station.

Estimating Excess All-Cause Mortality during the 2010 Heat Wave

Adapting the methodology of Anderson et al. [43], we applied 7-day moving averages for each day in May 2010 and compared them to averages from 2009 and 2011. For each day in May 2010, we compared the counts of all-cause deaths against a reference period comprised of the mean of all-cause death counts from corresponding days in May 2009 and May 2011. These years were chosen to control for population changes at the ecological level, since the population would be most similar in size and other demographic characteristics in the years immediately preceding and following the heat wave year. The daily number of excess all-cause deaths during May 2010 was calculated as the difference between the total monthly deaths in May 2010 minus the total reference period death counts for May from 2009 and 2011, again using an averaging method and a 7-day moving average method as described above. Percentage increases in May 2010 monthly excess mortality were also estimated, relative to the 2009 and 2011 reference period.

To estimate increases in mortality rate ratios (RR), applying the methodology of Lan et al. [44] and Rothman et al. [45], we considered the week of May 19–25, 2010, which included an acute 4-day extreme heat period from May 20–23, 2010 (as per the IMD's heat wave definition of daily maximum temperatures above 45 degrees Celsius) as the extreme heat wave period (H). The immediately preceding period, May 12–18, 2010 was considered as the reference period R1 and an alternative reference period R2 of May 19–25 from 2009 and 2011. RR were calculated for both reference periods R1 and R2. We assumed that the population size changed little over this period, cancelling the person-time units from the numerator and denominator. Thus, the simplified rate ratio "RR" was calculated using the formula $RR = H/R$. The 95% confidence intervals (CI) for RR were calculated as exp (ln (RR) $\pm$ 1.96$(H^{-1}+R^{-1})^{\wedge}0.5$). To minimize the heat's potential effects on all-cause mortality in the reference period, we also used an alternative reference period to calculate RRs, applying averaged all-cause mortality from May 19–25, 2009 and 2011 to re-estimate RRs.

Monthly correlations in 2010 for monthly maximum temperature and total monthly counts of premature all-cause mortality were also estimated.

Statistical analysis software SPSS 20 and Microsoft Excel (2013) was used for all analyses including generation of descriptive statistics, moving averages, and mortality rate ratios. Statistical significance of differences in mean values (temperatures and mortality) were calculated using two-tailed paired students T-tests.

Results

The average daily mortality during the extreme heat wave period (as shown in Table 1) was estimated at 143.9 ±48.13, significantly higher than the average of 100.6 ±10.34 for the reference period (p<0.001 for the difference). This yields an estimated excess mortality in May 2010 of 1,344 deaths from May 1–31, 2010 an increase of 43.1% above the reference period.

Other analytical approaches yielded similar results: comparison of daily counts for the extreme heat wave period compared with daily averages from the reference period yielded an excess of 1,353.5 deaths (a 42.88% increase), while a comparison using the 7-day moving average method yielded 1,334.9 deaths (a 42.81% increase). All results are consistent and suggest slightly over 43% excess deaths during May 2010. Figure 1 illustrates the daily mortality counts in May 2010 heat wave, versus corresponding days in 2009 and 2011.

Using mortality RR calculations, the RR is 1.76 [95% CI 1.67–1.83] (p<0.05) compared to a reference period (R1) in the preceding week (May 12–18, 2010). Applying the alternative reference period of May 19–25 from 2009 and 2011 (R2), the RR was significantly higher at 2.12 [95% CI 2.03–2.21] (p<0.001).

Table 2 shows correlations to quantify the strength of observed relationships between temperature and mortality. We found moderate to high correlations between monthly total all-cause deaths and monthly maximum temperature in the summer months while the same were negative for winter months. For May 2010, when the heat wave was recorded in Ahmedabad, the correlation was the highest (r = 0.775; p<0.001). The gender distribution as shown in Table 3 highlights significantly more female deaths in the summer months and the heat wave period.

For 2010, yearly mean maximum temperature was 33.9±5.0°C. This is significantly lower than the yearly average maximum temperature for 2009 and 2011, which was 35±4.8°C (p<0.001). For May 2010, the mean maximum temperature was 42.8±1.3°C. This is significantly higher than the May average maximum temperature for 2009 and 2011 which was 41.0±0.7°C (p<0.001) (Table 1). Similarly, the 2010 yearly mean and median daily mortality were 108±23 and 106 [IQR = 24] respectively and the May mean and median mortality were 143±48 and 125 respectively.

As illustrated in Figure 1, in May 2010 the number of daily deaths (though higher than average) increased greatly on May 18, 2010, reached the highest level on May 21, 2010, and returned to a more typical range on May 28, 2010. This pattern corresponds to increases in daily maximum temperature that occur during the May 2010 heat wave period, with an apparent increased association when the temperature crosses a threshold (43 °C for maximum or 36°C for mean temperature). There may be some mortality displacement on May 31, 2010 at which time daily mean mortality fell below the average, perhaps in short-term response to the large numbers of heat-vulnerable who had perished in prior days. Figure 2 shows the time course of both temperatures and death counts in both the 2010 study period and the reference period.

Discussion

Although similar heat-mortality studies have been conducted in several European, American and Chinese cities, this is the first time that this relationship has been documented for any city in India. The findings are consistent with prior work that shows heat wave periods associated with overall excess all-cause mortality. Using a smoothed 7-day average calculation as well as a rough average estimate to calculate baseline measures, we estimate that the mortalities occurring during the May 2010 heat wave represent a 43% increase in mortality when compared with the same time period of other years in Ahmedabad. This amounts to an additional 1,350 deaths in the city during the heat wave period.

There was a similarly elevated mortality rate ratio of 1.76 [1.67–1.83 95% CI] during the shorter extreme heat wave period H

Table 1. Daily temperatures and all-cause deaths for the month of May in Ahmedabad, Gujarat, India.

	Standard analysis			7-day moving average analysis	
	2010	Average for 2009 and 2011	Excess in 2010 (%)	Average for 2009 and 2011	Excess in 2010 (%)
Total Deaths	4462	3118	1344 (43.10)	3120.07	1334.93 (42.81)
Average daily Mortalities (deaths / day)	143.94±48.13	100.58±10.34	43.36 (43.11)	100.65±2.66	43.29 (43.01)
Maximum Temperature (°C)	42.81±1.25	41.05±1.27	1.76 (4.28)		
Minimum Temperature (°C)	28.45±1.50	28.25±1.41	0.19 (0.68)		
Mean Temperature (°C)	35.65±1.33	34.39±1.21	1.26 (3.66)		

from May 19–25, 2010. This estimate is considerably higher than one for China's 2010 heat wave [46] of 1.41 [95% CI 1.22–1.63].

One issue with our approach to the RR calculation is that it may underestimate the effects of extreme heat exposure as temperatures were already elevated in the May 12–18 reference period. To minimize the potential impact of the abnormally warm May 2010 temperatures on estimates of risk for all-cause mortality in the reference period, we applied an alternative non-2010 reference period to calculate RRs, and found a significant increase in the RR of 2.12 (95% CI 2.03–2.21; p<0.001).

Using historical exposure-response functions for temperature-heat relationships and various approaches to incorporating future adaptation activities, several projections of climate change impacts on heat mortality in various settings around the world have projected significantly increased risk in coming decades. These studies show an almost two-fold increase in the projected mortalities[47]. Our findings suggest that there is likely to be concern for similarly increased risk in the Ahmedabad region.

We observed (Figure 2) that the increase in temperatures and mortalities were concurrent, *i.e.*, there was no apparent lag time from the increase in temperatures to the increase in mortalities for the heat wave period in May 2010. Future time series analyses can

investigate lag structures more fully, perhaps comparing mortality temperature lag times between tropical and temperate regions.

The analysis has several limitations. First, while the outcome data are generally reliable, there are several potential issues with mortality data in India. Reporting conventions are not uniform across the country. Even in one location reporting conventions are not always adhered to uniformly across time. In some settings some causes of deaths, though not the deaths themselves, are not accurately reported if they are considered sensitive. This has the potential to bias our results, likely toward the null, as it could lead to an understatement of the mortality risk during a heat wave period if not all of the deaths during that period were accounted for in the data set at a differential rate compared with the baseline. Assuming that reporting conventions remained stable during the study period, however, these issues should not affect our results, but they may be important if our results are compared with those from other areas in India.

A related concern is that not all deaths are reported. In the urban region of Ahmedabad, deaths taking place at home are likely to have been missed, and deaths among the homeless are unlikely to have been registered, potentially resulting in an underestimation of the effect if a greater proportion of deaths

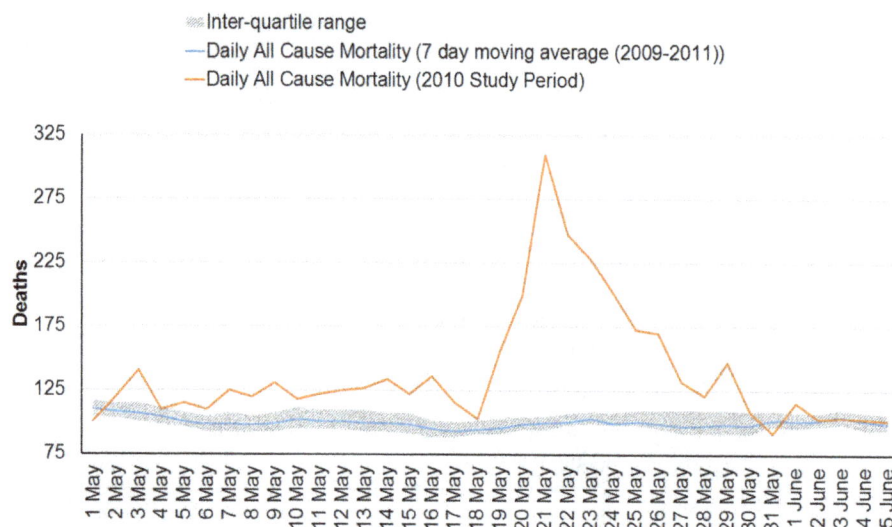

Figure 1. Daily mortality counts in May 2010 heat wave, versus corresponding days in 2009 and 2011. Daily mortality counts during the Ahmedabad, India May 2010 heat wave study period (in red), compared with average May all-cause mortality rates. Shown for comparison as a reference period are the mean (blue line) and interquartile range (hatched area) of a 7-day moving average of mortality in May of the preceding (2009) and following (2011) years.

Table 2. Month-wise correlations between monthly maximum temperature (°C) and total monthly all-cause mortality counts, 2010.

	January	February	March	April	May	June	July	August	September	October	November	December
Pearson Correlation Coefficient	−0.028#	−0.160#	0.360*	0.701**	0.775**	0.393*	0.503**	0.414*	0.083#	0.477**	−0.371*	0.290#

Note: *p<0.05, **p<0.01, #non-significant.

occurred at home or among the homeless during the heat wave period. This reporting bias may also have served as a confounder if there was increased likelihood of heat exposure and death and decreased likelihood of death reporting for a particular population, *e.g.*, daily laborers, during the exposure period. Based on the data available, we were not able to control for these potential effects.

Second, the outcome is all-cause mortality, and as such our estimate may be an overestimate of the true effect of the heat wave on mortality as we were not able to perform a cause-specific analysis and isolate heat-related deaths. As we compared all-cause mortality in 2010 with a running average derived from seasonally similar time periods, however, this bias is likely to be relatively small as using the Integrated Disease Surveillance Project (IDSP) data (the disease surveillance system in India) during April to June 2010 we ruled out any major disease outbreaks for that period. Moreover, risks for cause-specific mortality would have been interesting to compare if the causes of death were available. Evaluations of whether a lagged effect of extreme heat on mortality exists in Ahmedabad could be a part of future studies that involve time-series analysis.

Third, because we are not able to stratify the deaths by demographics (apart from gender, where differences were insignificant) or socioeconomic status, our findings provide little specific insight into factors that may affect risk. Similarly, data disaggregated on socio-economic and socio-demographic variables would have provided useful insights on vulnerability. More detailed investigation and documentation of individual cases with cause of death and linkage of location and demographic and other factors will be needed to conduct a more nuanced evaluation of specific risk factors.

Fourth, a further complication exists in the placement of the temperature gauge that has been used historically to collect data in Ahmedabad. The daily maximum temperatures recorded within the city where residents are exposed to extreme heat may be higher than daily reports, as the current local source of daily temperature data is from an IMD monitoring station on the outskirts of the city. The degree of Ahmedabad's urban heat island effect is starting to be evaluated and may offer opportunities to fine-tune local temperature reporting. Future work that considers possible alternatives to the definition of a heat wave, with more health-relevant measures included, are areas of on-going research for the Ahmedabad Heat and Climate Study Group.

Fifth, for those persons who were indoors, these measures of ambient temperature are proxies of indoor conditions. Indoor residential temperatures would depend on a number of built environment factors including building materials, ventilation, nearby vegetation, and additional heat sources and sinks. A survey of relative measures of indoor versus outdoor temperatures across Ahmedabad was beyond the scope of this study. It is unknown whether ambient daily temperature measures are an under- or over-estimate of indoor room temperature for those individuals who died indoors, as neither specific information regarding the location of the decedents nor indoor exposure information is available

Finally, correlation in an ecological analysis alone does not indicate causation. There might be some role of confounding variables, such as air pollution, and / or other sources of bias. Air pollution effects were not the focus of this descriptive study of the May 2010 heat wave's effect on excess mortality in Ahmedabad. There could be possible interactive effects of heat and particulate or ozone air pollution, which would be a promising area for future study. The authors do not have access to the archived 2010 daily ozone or particulate air pollution data for the study period, though this data might help establish typical concentrations for

Table 3. Gender distribution of decedents during the Heat Wave and the reference periods.

		Men	Women	Total
	Total (2009–2011)	68977	47021	115998
	Average deaths/day (2009 & 2011)	62.73	42.14	104.87
2010	Average deaths per day	63.52	44.55	108.07
	Excess deaths	287.50	881.00	1168.50
	Average excess deaths per day	0.79	2.41	3.20
May 2010	Deaths	2462	2000	4462
	Excess deaths	639	705	1344
	Average Excess deaths per day	20.61	22.74	43.35
	Average deaths excluding May 2010	62.04	42.69	104.74
Heat Wave Period (19–25th May2010)	Deaths	791	724	1515
	Excess deaths	373.50	427	800.50
	Average excess deaths per day	53.36	61	114.36
	Ratio (19–25th May 2010)	1.89	2.44	2.12
P value Compared with 2009 & 2011	p for 19–25th May 2010	0.002521923	0.000647	0.000887
	p for May 2010	6.35662E-05	3.11E-05	2.4E-05
	p for entire year	0.332967427	0.000688	0.018211

comparison with the heat wave period. For particulate concentrations, Guttikunda and Jawahar (2012)[48] show monthly average PM10 concentrations from Ahmedabad for 2009–2010. May 2010 showed the highest monthly values for PM10 concentrations, but not to an extent that is likely to account for the dramatic increase in mortality that Ahmedabad experienced in May 2010. Moreover, prior studies including Anderson and Bell (2009)[49] and Hajat et al. (2006)[20] support the case that heat has a substantial, independent effect on mortality, which is independent of air pollution. Other potential confounders that were considered include infectious disease outbreaks (none were reported for the state of Gujarat in the weeks 2010 around the

May heat wave), holidays, and other administrative issues that might have affected reporting.

Despite these potential concerns, the results of this study pose interesting questions and make findings regarding health protection against extreme heat in the region. This study is innovative and provides valuable data analysis since it is the first to examine the effects of extreme heat in a developing country setting. Such studies and findings are absent from the literature, hampering efforts to protect public health and to project the potential health impacts of climate change in developing countries.

One key question relates to the optimal definitions of extreme heat from the public health perspective and for early warning systems. This study evaluated mortality rate ratios during time

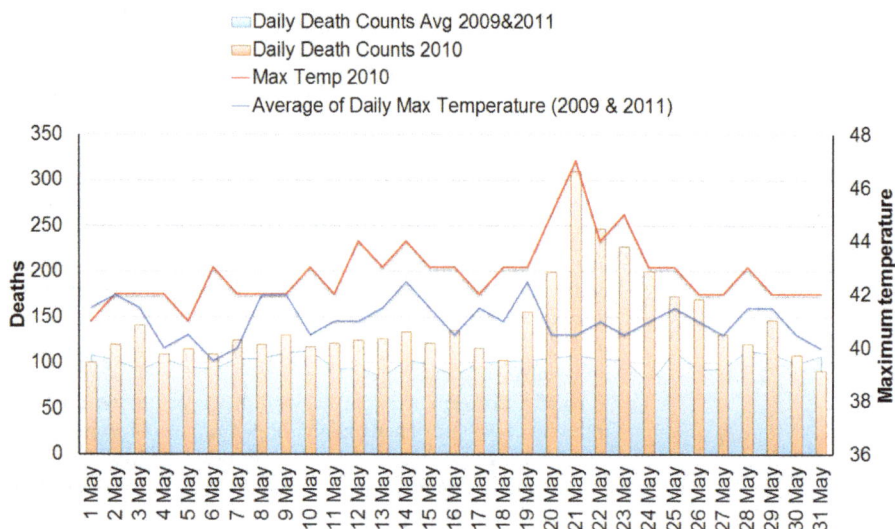

Figure 2. Temperatures and daily all-cause mortality, 2010 study period vs. 2009–2011 reference period. Daily maximum temperature versus daily all-cause mortality for the study period (1–31 May 2010), and the reference period (mean of corresponding values from days in May 2009 and 2011.

periods that fell both within and outside the currently operational definitions of a "heat wave" and "extreme heat days" according to the Indian Meteorology Department, as described in the Introduction. Our findings suggest that the IMD definitions may underestimate the impacts of extreme heat on health because under the current systems IMD threshold does not formally account for public health effects of extreme heat. For public health purposes the IMD definitions may not be as useful as a definition with lower thresholds that are observed to correlate with public health outcomes more directly. Short and medium-range temperature forecasts, along with appropriate thresholds based on observed population health effects, could be used to generate early warnings of extreme heat that might significantly reduce the number of heat-related deaths.

To be most effective, as examined as part of this study, early warning systems also require measures to build preparedness and response capacity for medical and public health professionals and improved coordination and sensitization of staff working across city government and the medical facilities. Public awareness about the harmful impact of exposure to heat on human health is also critical to saving lives. Public awareness messages must be simple and easy to understand. They may include suggestions to check weather forecasts, stay in the shade, maintain hydration, wear appropriate clothing, avoid physical activity in the hottest part of the day, and check on neighbors and vulnerable members of the community.

In addition to early warning systems and preparedness plans, partnerships and information-sharing between various government agencies and key stakeholders needs to be developed and nurtured to facilitate the interdisciplinary work and effective implementation. Interventions also must be coupled with evaluation plans to determine their efficacy in the local context. Information and data-sharing channels have to be kept open and working to facilitate operations and evaluation activities. Appropriate health system improvements have to be made to ensure adequate response to health emergencies. International partnerships can also provide a strong base for knowledge-sharing and technical support, as evidenced by this collaborative study.

Conclusions

In May 2010, the city of Ahmedabad in the state of Gujarat, India, experienced a heat wave with record-breaking maximum temperatures. During this heat wave an estimated excess 1,344 deaths occurred relative to a combined May 2009 and May 2011 reference period. This finding is consistent across several methods used for gauging the extra deaths. The May 2010 heat wave period represent a 43% increase in mortality when compared with the same time period of 2009 and 2011. The RR of an "extreme heat period" from 19–25 May was 1.76 (95% CI 1.67–1.83, p< 0.001). Correlation coefficients between monthly maximum temperatures and monthly mortality show a significant relationship for the months of April (r = 0.701, p<0.001), May (r = 0.775, p<0.001). This paper aims to draw attention to extreme heat as a relatively under-appreciated public health risk in India, and provide some insight into the temperatures at which this risk appears to be marked on a population level. Given the trends associated with climate change, dangerous periods of extreme heat are likely to occur more frequently, suggesting the need for measures to reduce population vulnerability currently and in future. Heat wave-related mortality merits further analysis in order to reduce harmful health effects among India's most vulnerable and to help India adapt to the effects of climate change by increasing resilience to extreme heat.

Acknowledgments

We would like to acknowledge that this paper was developed in conjunction with the on-going research collaboration formalized under a memorandum of understanding among the Ahmedabad Municipal Corporation, the Gujarat Government, the Public Health Foundation of India, Indian Institute of Public Health-Gandhinagar, and the Natural Resources Defense Council (NRDC).

The Ahmedabad Heat and Climate Study Group, consists of (in alphabetical order). Gulrez Shah Azhar (IIPH G), Bhaskar Deol (NRDC), Priya Shekhar Dutta (IIPH-G), Jeremy Hess (Emory University), Anjali Jaiswal (NRDC), Radhika Khosla (NRDC), Kim Knowlton (NRDC and Mailman SPH, Columbia University), Dileep Mavalankar (IIPH-G), Ajit Rajiva (IIPH-G), Amruta Sarma (Fulbright Student Research Scholar), and Perry Sheffield (Icahn SOM at Mount Sinai).

The authors are grateful to the officials of the Ahmedabad Municipal Corporation and the India Meteorology Department who provided them with data. The authors would like to thank Peter Webster, Violeta Toma, and Georgia Institute of Technology for their temperature forecasts for the Ahmedabad heat-health early warning system.

Author Contributions

Conceived and designed the experiments: GSA DM AS AJ KK JH PS. Performed the experiments: GSA AR AS PD DM. Analyzed the data: GSA AR PD AS DM. Contributed reagents/materials/analysis tools: AS AR PD GSA KK JH PS AJ. Wrote the paper: GSA KK JH PS DM AJ AR AS. Obtained the data and permissions to use the data: DM GSA AJ AS.

References

1. Meehl GA, Karl T, Easterling DR, Changnon SA, Pielke RA, et al. (2000) An introduction to trends in extreme weather and climate events: observations, socioeconomic impacts, terrestrial ecological impacts, and model projections. Bulletin-American Meteorological Society 81: 413–416.

2. IPCC Working Group II (2011) Special Report on Managing the Risks of Extreme Events and Disasters to Advance Climate Change Adaptation (SREX). Geneva: Intergovernmental Panel on Climate Change.

3. Bush KF, Luber G, Kotha SR, Dhaliwal R, Kapil V, et al. (2011) Impacts of climate change on public health in India: future research directions. Environmental Health Perspectives 119: 765.

4. Hansen J, Sato M, Ruedy R (2012) Perception of climate change. Proceedings of the National Academy of Sciences 109: E2415–E2423.

5. Min S-K, Zhang X, Zwiers FW, Hegerl GC (2011) Human contribution to more-intense precipitation extremes. Nature 470: 378–381.

6. IPCC editor (2012) Managing the Risks of Extreme Events and Disasters to Advance Climate Change Adaptation. A Special Report of Working Groups I and II of the Intergovernmental Panel on Climate Change. Cambridge, UK: Cambridge University Press.

7. Ministry of Environment & Forests, Government of India (2010) Climate Change and India: A 4×4 Assessment. A Sectoral and Regional Analysis for 2030s INCCA. New Delhi.

8. Centers for Disease Control and Prevention (2004) Extreme heat: a prevention guide to promote your personal health and safety. Atlanta, GA: Centers for Disease Control and Prevention.

9. De U, Dube R, Rao GP (2005) Extreme weather events over India in the last 100 years. Journal of the Indian Geophysical Union 9: 173–187.

10. Basu R (2009) High ambient temperature and mortality: a review of epidemiologic studies from 2001 to 2008. Environ Health 8: 40.

11. Bell ML, O'Neill MS, Ranjit N, Borja-Aburto VH, Cifuentes LA, et al. (2008) Vulnerability to heat-related mortality in Latin America: a case-crossover study in Sao Paulo, Brazil, Santiago, Chile and Mexico City, Mexico. Int J Epidemiol 37: 796–804.

12. Fowler DR, Mitchell CS, Brown A, Pollock T, Bratka LA, et al. (2013) Heat-Related Deaths After an Extreme Heat Event-Four States, 2012, and United States, 1999-2009. MMWR-Morb Mortal Wkly Rep 62: 433–436.

13. Matthies F BG, Marin NC, Hales S (2008) Heat-Health Action Plans: Guidance. Geneva: World Health Organization.

14. Kovats RS, Hajat S (2008) Heat stress and public health: a critical review. Annual Review of Public Health 29: 41–55.

15. Ellis FP (1977) Heat illness. II. Pathogenesis. Transactions of the Royal Society of Tropical Medicine & Hygiene 70: 412–418.

16. Meehl GA, Tebaldi C (2004) More intense, more frequent, and longer lasting heat waves in the 21st century. Science 305: 994–997.

17. Robinson PJ (2001) On the definition of a heat wave. Journal of Applied Meteorology 40: 762–775.
18. National Oceanic and Atmospheric Administration (2013) National Weather Service Glossary. Silver Springs, Maryland.
19. McMichael AJ, Wilkinson P, Kovats RS, Pattenden S, Hajat S, et al. (2008) International study of temperature, heat and urban mortality: the 'ISO-THURM' project. International Journal of Epidemiology 37: 1121–1131.
20. Hajat S, Armstrong B, Baccini M, Biggeri A, Bisanti L, et al. (2006) Impact of high temperatures on mortality: is there an added heat wave effect? Epidemiology 17: 632–638.
21. Anderson GB, Bell ML (2011) Heat waves in the United States: mortality risk during heat waves and effect modification by heat wave characteristics in 43 US communities. Environmental Health Perspectives 119: 210.
22. Medina-Ramón M, Schwartz J (2007) Temperature, temperature extremes, and mortality: a study of acclimatisation and effect modification in 50 US cities. Occupational and environmental medicine 64: 827–833.
23. Kamaljit Ray AT, Apte NY, Chicholikar JR (2012) Climate of Ahmedabad. Ahmedabad: Meteriological Center Ahmedabad.
24. Wong K, Paddon A, Jimenez A (2013) Review of World Urban Heat Islands: Many Linked to Increased Mortality. J Energy Resour Technol 135(2) 135.
25. Khosla R (2010) The relevance of rooftops: Analyzing the microscale surface energy balance in the Chicago region. Chicago: University of Chicago. 168 p.
26. Vandentorren S, Suzan F, Medina S, Pascal M, Maulpoix A, et al. (2004) Mortality in 13 French cities during the August 2003 heat wave. American Journal of Public Health 94: 1518.
27. Baccini M, Biggeri A, Accetta G, Kosatsky T, Katsouyanni K, et al. (2008) Heat effects on mortality in 15 European cities. Epidemiology 19: 711.
28. Knowlton K, Rotkin-Ellman M, King G, Margolis HG, Smith D, et al. (2009) The 2006 California heat wave: impacts on hospitalizations and emergency department visits. Environmental Health Perspectives 117: 61.
29. Anderson GB, Bell ML (2011) Heat waves in the United States: mortality risk during heat waves and effect modification by heat wave characteristics in 43 U.S. communities. Environ Health Perspect 119: 210–218.
30. Dole R, Hoerling M, Perlwitz J, Eischeid J, Pegion P, et al. (2011) Was there a basis for anticipating the 2010 Russian heat wave? Geophysical Research Letters 38.
31. Sidhu RK (1998) Severe heat wave over the Indian subcontinent in 1998, in perspective of global climate. Current science 75.
32. Kalsi S, Pareek R (2001) Hottest April of the 20th century over north-west and central India. Current science 80: 867–873.
33. Parry ML (2007) 8.2.1.1 - Heatwaves. Cambridge University Press. 0521880106 0521880106.
34. Kovats RS, Hajat S (2008) Heat stress and public health: a critical review. Annu Rev Public Health 29: 41–55.
35. Frumkin H, Hess J, Luber G, Malilay J, McGeehin M (2008) Climate change: the public health response. American Journal of Public Health 98: 435.
36. Joubert D, Thomsen J, Harrison O (2011) Safety in the Heat: A Comprehensive Program for Prevention of Heat Illness Among Workers in Abu Dhabi, United Arab Emirates. American Journal of Public Health 101: 395.
37. Hajat S, O'Connor M, Kosatsky T (2010) Health effects of hot weather: from awareness of risk factors to effective health protection. Lancet 375: 856–863.
38. Takahashi K, Honda Y, Emori S (2007) Assessing mortality risk from heat stress due to global warming. Journal of Risk Research 10: 339–354.
39. Ghosh P (2009) National Action Plan on Climate Change. Prime Minister's Council on Climate Change.
40. Adhyaru-Majithia P (2010) 8 die as worst heat wave in 94 years scorches Ahmedabad. DNA. Ahmedabad.
41. Dave J (2010) Dying heat wave kills 8 more in Ahmedabad. DNA. Ahmedabad.
42. Government of India (2011) Census of India. New Delhi.
43. Anderson GB, Bell ML (2012) Lights Out: Impact of the August 2003 Power Outage on Mortality in New York, NY. Epidemiology 23: 189.
44. Lan L, Cui G, Yang C, Wang J, Sui C, et al. (2012) Increased Mortality During the 2010 Heat Wave in Harbin, China. EcoHealth: 1–5.
45. Rothman KJ, Greenland S, Lash TL (2008) Modern Epidemiology. Philadelphia: Lippincott-Raven.
46. Lan L, Cui G, Yang C, Wang J, Sui C, et al. (2012) Increased Mortality During the 2010 Heat Wave in Harbin, China. EcoHealth 9: 310–314.
47. Huang C, Barnett AG, Wang X, Vaneckova P, FitzGerald G, et al. (2011) Projecting future heat-related mortality under climate change scenarios: a systematic review. Environmental Health Perspectives 119: 1681.
48. Guttikunda SK, Jawahar P (2012) Application of SIM-air modeling tools to assess air quality in Indian cities. Atmospheric Environment 62: 551–561.
49. Anderson BG, Bell ML (2009) Weather-related mortality: how heat, cold, and heat waves affect mortality in the United States. Epidemiology (Cambridge, Mass) 20: 205–213.

22

Spatial Cluster Detection of Air Pollution Exposure Inequities across the United States

Bin Zou[1]*, Fen Peng[1], Neng Wan[2], Keita Mamady[3], Gaines J. Wilson[4]

1 School of Geosciences and Info-Physics, Central South University, Changsha, Hunan, China, **2** Department of Geography, University of Utah, Salt Lake City, Utah, United States of America, **3** Department of Epidemiology and Health Statistics, School of Public Health, Central South University, Changsha, Hunan, China, **4** Department of Biological Sciences, Huston-Tillotson University, Austin, Texas, United States of America

Abstract

Air quality is known to be a key factor in affecting the wellbeing and quality of life of the general populous and there is a large body of knowledge indicating that certain underrepresented groups may be overexposed to air pollution. Therefore, a more precise understanding of air pollution exposure as a driving cause of health disparities between and among ethnic and racial groups is necessary. Utilizing 52,613 urban census tracts across the United States, this study investigates age, racial, educational attainment and income differences in exposure to benzene pollution in 1999 as a case. The study examines spatial clustering patterns of these inequities using logistic regression modeling and spatial autocorrelation methods such as the Global Moran's I index and the Anselin Local Moran's I index. Results show that the age groups of 0 to 14 and those over 60 years old, individuals with less than 12 years of education, racial minorities including Blacks, American Indians, Asians, some other races, and those with low income were exposed to higher levels of benzene pollution in some census tracts. Clustering analyses stratified by age, education, and race revealed a clear case of disparities in spatial distribution of exposure to benzene pollution across the entire United States. For example, people aged less than 4 years from the western south and the Pacific coastal areas exhibit statistically significant clusters. The findings confirmed that there are geographical-location based disproportionate pattern of exposures to benzene air pollution by various socio-demographic factors across the United States and this type of disproportionate exposure pattern can be effectively detected by a spatial autocorrelation based cluster analysis method. It is suggested that there is a clear and present need for programs and services that will reduce inequities and ultimately improve environmental conditions for all underrepresented groups in the United States.

Editor: Jaymie Meliker, Stony Brook University, Graduate Program in Public Health, United States of America

Funding: The research reported in this paper was funded by the National Natural Science Foundation of China (Project No. 41201384, http://www.nsfc.gov.cn/Portal0/default152.htm), the Hunan Provincial Natural Science Foundation of China (Project No. 12JJ3034, http://www.hnst.gov.cn/zzjg/nsjg/hnszrkxjjwyhbgs/), the State Key Laboratory of Resources and Environmental Information System (http://www.lreis.ac.cn/sc/index.aspx). Bin Zou would also like to thank the grant from the Key Laboratory of Geo-informatics of State Bureau of Surveying and Mapping (Project No. 201328, http://www.casm.ac.cn/), as well as the NieYing Talent Program of Central South University (www.csu.edu.cn). The funders had no role in study design, data collection and analysis, decision to publish, or preparation of the manuscript.

Competing Interests: The authors have declared that no competing interests exist.

* E-mail: 210010@csu.edu.cn

Introduction

Environmental injustice may be defined as a type of injustice when a particular social group is disproportionately burdened with environmental hazards [1]. The underlying contributors to environmental injustices can be political, economic, historical, and social [2].

Air pollution, the most common type of pollutant in environmental injustice studies, can be traced back to the industrialization-urbanization nexus beginning in the 19th century. Evidence indicates that air pollution exposure is more serious than previously thought, in terms of adverse health impacts such as reduced life expectancy, increased daily mortality and hospital admissions, birth outcomes, and asthma [3]. These effects have been shown to exist in both economically developing and developed countries [4]. Systematic efforts to control air pollution and to protect public health commenced mostly in the second half of the 20th century and have intensified since the 1960s [5].

Exposure to air pollution, however, may vary spatially within a city [6] and these variations may follow social gradients that influence susceptibility to environmental exposures [7]. Residents of poorer neighborhoods may live closer to point sources of industrial pollution or roadways with higher traffic density [8]. International research has shown that air pollution exposure varies by socio-economic status, with lower socio-economic groups being disproportionately exposed to air pollution and to environmental mechanisms that lead to inequities in health [9]. For example, there is consistent evidence in California that patterns of disproportionate exposure to air pollution among minority and lower-income communities exists [10]. These communities also face other challenges associated with low socioeconomic status, including psychosocial stressors, which make it more difficult to cope with these exposures [9].

Meanwhile, although current research has confirmed the relations between social-demographic characteristics (e.g., education, age, race etc.) and disease [11], they are still inadequate in explaining the underlying reasons for disease disparities. Thus,

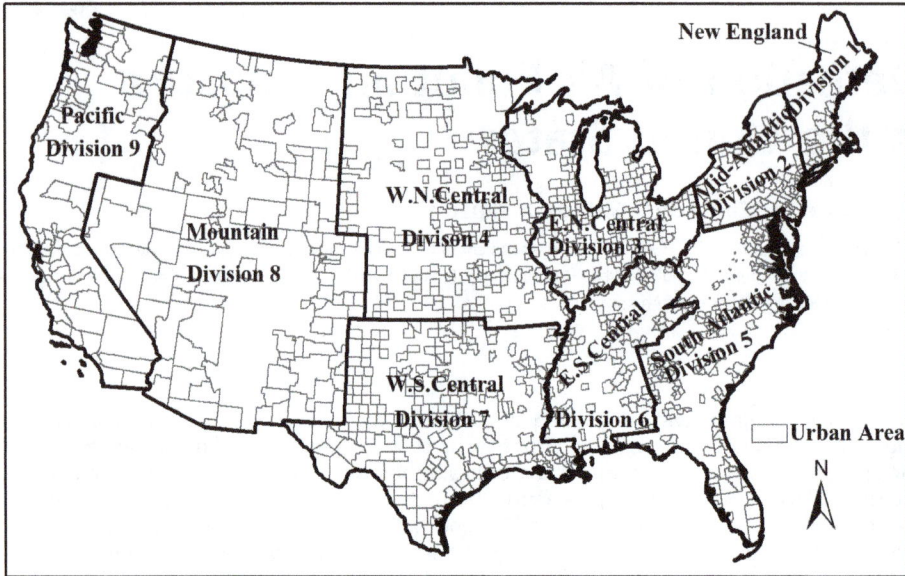

Figure 1. United States census divisions of urban-designated counties containing 52, 613 tracts used in this study. The study area focuses on all urban census tracts within the United States, which is further classified into nine divisions. It consists of an aggregate number of 52,613 census tracts within 48 contiguous states and Washington DC.

further understanding of the role of socio-demographic status as a component of susceptibility to the adverse health effects of air pollution is necessary in the process of setting ambient air quality standards and implementing programs and policy that lead to adherence to these standards.

Today, air pollution is still a major environmental health issue in the United States, directly affecting people's wellbeing and quality of life with adverse health impacts such as excess respiratory, cardiovascular morbidity and higher mortality [12]. International survey data showed a 7–10% premature birth rate in industrialized countries, and specifically 9–12% in United States in

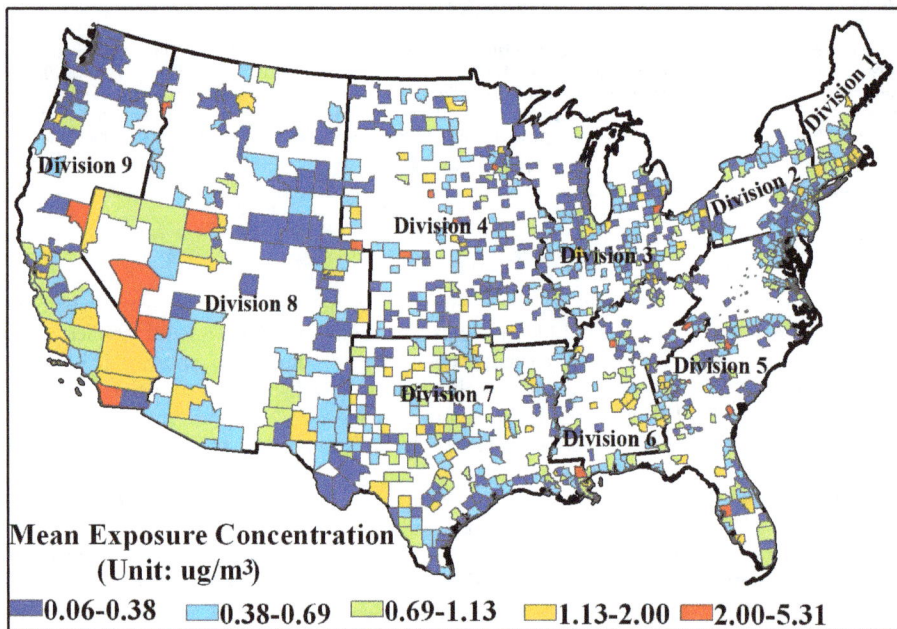

Figure 2. Annual exposure concentrations of total benzene at the census tract level in the United States. Annual exposure concentrations of census tracts have been utilized to calculate county level mean exposure concentration values, which was used as a 'relative exposure level' metric to evaluate benzene pollution exposure inequities. Division 1 is New England; Division 2 is Mid-Atlantic; Division 3 is East North Central; Division 4 is West North Central; Division 5 is South Atlantic; Division 6 is East South Central; Division 7 is West South Central; Division 8 is Mountain; Division 9 is Pacific.

Table 1. Standards of categorization and reference categories for socio-demographic measurements.

Characteristics	Level 1	Level 2	Level 3	Level 4	Level 5
Age	0–14	14–60*	>60	—	—
Race	White*	black	American Indian	Asian	Other races
Education attainment (years)	0–4	5–8	9–12	>12*	—
Income (US$)	<19000	> = 19000**	—	—	—

*Reference category for comparison based on existing studies in the environmental justice literature.
**The classification standard for income is detailed in the text.

recent years, with the trend for both showing an increase [13]. In this way, a broader understanding of the causes of population health disparities by race/ethnicity, socioeconomic status, and geographic location is necessary for achieving better solutions to population health problems caused by the complex cocktail of air pollution found in the United States. This study aims to investigate

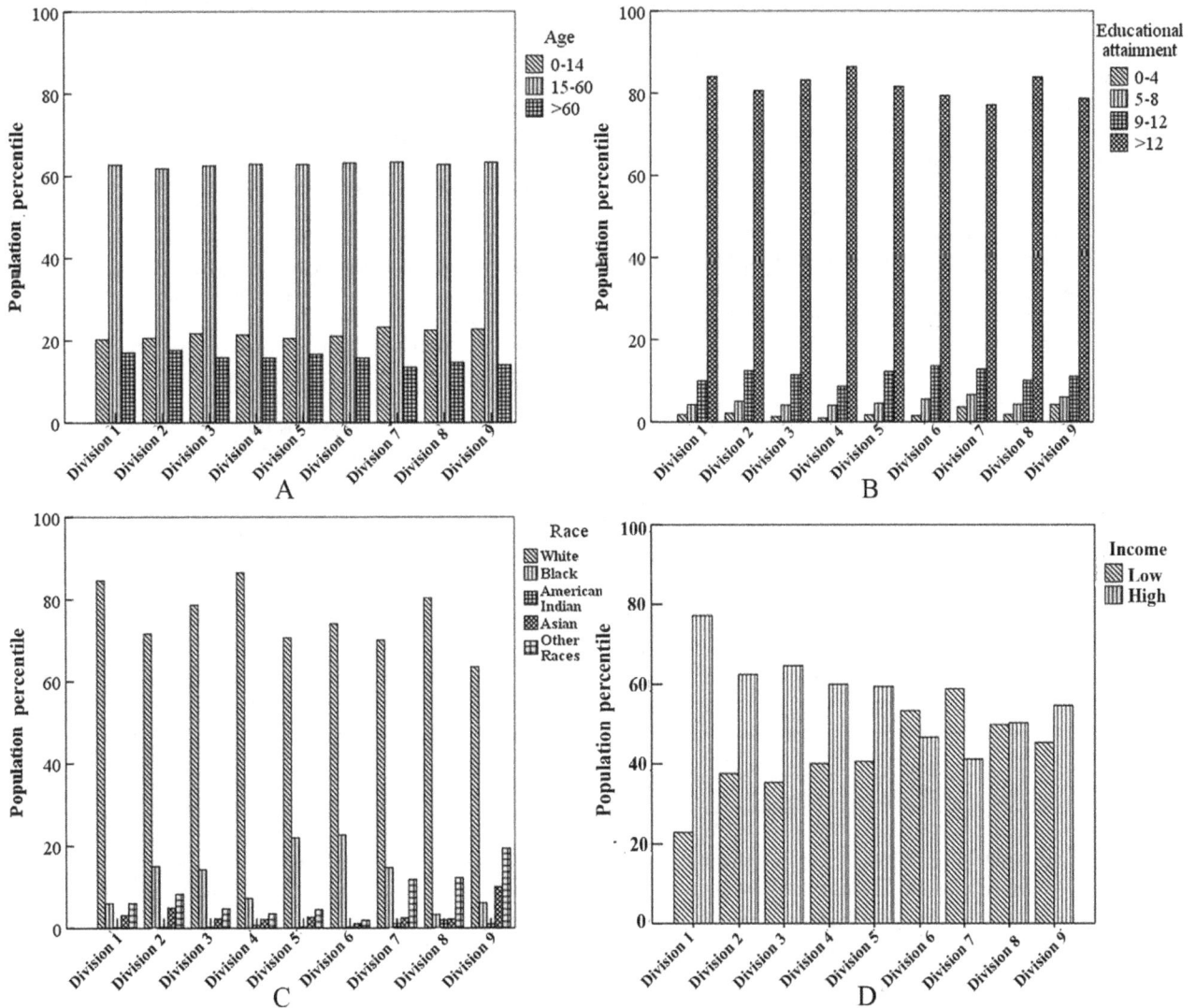

Figure 3. Population percentiles based on socio-demographic characteristics in the nine divisions. (A): Population percentile based on age characteristic in the nine divisions. Age group 15 to 60 have the highest population percentage. (B): Population percentile pertaining to educational attainment characteristic in the nine divisions. Educational attainment more than 12 years have the highest population percentage. (C): Population percentile of race characteristic in the nine divisions, the white have the highest population percentage. (D): Population percentile of income characteristic in the nine divisions. Division 1 is New England; Division 2 is Mid-Atlantic; Division 3 is East North Central; Division 4 is West North Central; Division 5 is South Atlantic; Division 6 is East South Central; Division 7 is West South Central; Division 8 is Mountain; Division 9 is Pacific.

census tract level exposure to air pollution by these factors and to examine the spatial clustering patterns of the disparities at county level.

Data and Methods

Study Area

This study focuses on all urban census tracts within the United States, which is further classified into four census regions (e.g. Northeast, Midwest, West, South), and nine divisions [14]. This regional and divisional classification, as defined by the United States Census Bureau, is based upon factors such as employment, crime, health, consumer expenditures, and housing. The demographic differences between these divisions are suitable to be utilized for analyzing the air pollution exposure inequities across the entire country. We chose to use census tracts because this was the smallest level of aggregation at which air quality information for benzene was available and it was generally utilized as the standard spatial scale for environmental justice studies due to its relatively homogeneous characteristics relative to socio-demographic status and living conditions [15], [16]. The study area consists of an aggregate number of 64,890 census tracts, 3,109 counties within 48 contiguous states and Washington DC. The number of counties included in our study is 29 for New England (Division 1), 81 for Mid-Atlantic (Division 2), 174 for East North Central (Division 3), 187 for West North Central (Division 4), 214 for South Atlantic (Division 5), 97 for East South Central (Division 6), 194 for West South Central (Division 7), 101 for Mountain (Division 8), and 69 for Pacific (Division 9) (Fig. 1). After filtering out rural census tracks, we were left with 52,613 urban census tracts that account for 80.5% of the total 64,890 census tracts in the United States.

Data Sources and Analysis

The Environmental Hazard Data were ascertained from the US Environmental Protection Agency's (EPA) NATA (National Air Toxics Assessments) website [17]. The NATA data is the EPA's ongoing comprehensive evaluation of air toxics in the U.S. EPA developed the NATA as a state-of-the-science screening tool for state, local, and tribal agencies to prioritize pollutants, emission sources, and locations of interest and for researchers to gain a better understanding of environmental risks. These datasets are particularly suitable for environmental justice research, not only because they allow researchers to estimate the potential health risks associated with specific environmental hazards and analytical spatial units, but also because the data modeling takes into account a number of factors such as wind speed, wind direction, air turbulence, smokestack height and the rate of chemical decay and deposition [18]. Another important advantage of the NATA data is their spatial compatibility with socio-demographic census data: the modeled risk estimates are available for census units (e.g., tracts), which also include demographic characteristics of residential population.

The annual benzene pollution concentration for census tracts was used to represent air pollution. Benzene is a ubiquitous chemical in the environment that causes acute leukemia and probably other hematological cancers [19]. Meanwhile, recent studies reported an association between higher benzene exposure concentrations with lower social economy status and social class [20], [21]. While other air pollutants (e.g. sulfur dioxide) have experienced a downward trend in use over the past few decades, benzene is still one of the key toxic air pollutants produced by today's petrochemical industry and can be found in gasoline petroleum tanks throughout urban areas. Benzene exposure data from 1999 NATA have been utilized for air pollution exposure equity analysis [22], [23]. We calculated county level mean exposure concentration values based on exposure concentrations of census tracts (Fig. 2). Because recent studies have focused on the effects of continuous exposure to low concentrations of benzene [24], [25], [26], we used a 'relative exposure level' metric to evaluate benzene pollution exposure inequities in this study [27]. In this way, population in census tracts with exposure concentrations higher than a county level mean exposure concentration value are recognized as 'high' exposure concentration, whereas as low exposure concentration is assigned to census tracts below average.

The population data at census tract and county levels in this study were retrieved from the US Census 2000 Summary File 1 [28], while the geographic boundaries of spatial scale were acquired from the Census 2000 Topologically Integrated Geographic Encoding and Referencing (TIGER)/Line dataset [29]. Following previous studies [30], [31], [32], [33], we selected age, race, educational attainment, and income as the socio-demographic indicators in this study. These characteristics were

Table 2. Global Moran's I statistic values by age, race, and educational attainment in the United States and the nine divisions.

	Age		Race				Educational attainment			Income
	<14	>60	Black	American Indian	Asian	Other races	<4	5–8	9–12	Low income
United States	0.046*	0.021*	0.033*	0.031*	0.003	0.028*	0.036*	0.021*	0.011	0.022*
Division 1	−0.071	−0.062	−0.105	−0.060	−0.061	−0.091	−0.065	0.083	−0.057	−0.007
Division 2	−0.118	0.084*	0.070*	−0.073	0.120*	−0.003	−0.019	0.213*	0.035	−0.069
Division 3	0.012	0.042	0.041*	−0.004	0.016	−0.002	0.026	−0.022	−0.018	−0.063
Division 4	−0.019	−0.010	0.032	0.025	0.054*	0.010	−0.005	0.050	0.007	0.051
Division 5	0.010	−0.022	−0.006	0.035*	0.044*	0.031	0.025	−0.022	−0.034	0.0004
Division 6	−0.007	0.159*	0.024	−0.175*	−0.005	−0.048	−0.016	0.090	0.088	0.033
Division 7	0.090*	0.050*	0.008	0.021	0.020	0.104*	0.026	0.114*	0.058*	−0.108
Division 8	−0.020	−0.019	0.009	0.053	0.004	−0.060	0.034	0.027	−0.000	−0.114
Division 9	0.110*	−0.027	0.051	0.226*	−0.015	0.088	−0.016	0.078	0.195*	−0.073

* $p < 0.05$.

Table 3. Frequency of ORs greater than 1 by age characteristic at the county level in the United States and the nine divisions.

	Age	Count (percentage)	Minimum (95%CI)	Maximum (95%CI)	Mean
United States	0–14	306(26.7%)	1.017(1.006,1.029)	5.911(4.935,7.081)	1.189
	>60	537(46.9%)	1.010(1.001,1.019)	5.400(4.856,6.005)	1.285
Division 1	0–14	10(31.0%)	1.018(1.005,1.032)	1.322(1.286,1.359)	1.131
	>60	11(37.9%)	1.027(1.004,1.050)	1.451(1.415,1.488)	1.184
Division 2	0–14	10(12.3%)	1.028(1.008,1.049)	1.180(1.125,1.237)	1.098
	>60	44(54.3%)	1.010(1.001,1.019)	1.407(1.383,1.431)	1.183
Division 3	0–14	52(29.9%)	1.020(1.005,1.034)	1.472(1.404,1.543)	1.106
	>60	84(48.3%)	1.030(1.025,1.035)	1.730(1.505,1.990)	1.220
Division 4	0–14	22(11.8%)	1.043(1.002,1.085)	1.302(1.154,1.470)	1.135
	>60	82(43.9%)	1.055(1.009,1.104)	2.243(2.142,2.348)	1.301
Division 5	0–14	66(30.8%)	1.017(1.001,1.032)	5.026(4.502,5.610)	1.209
	>60	96(44.9%)	1.029(1.016,1.042)	5.400(4.856,6.005)	1.333
Division 6	0–14	27(27.8%)	1.028(1.013,1.042)	1.334(1.276,1.395)	1.118
	>60	57(58.8%)	1.039(1.007,1.073)	1.937(1.866,2.010)	1.293
Division 7	0–14	79(40.7%)	1.034(1.002,1.066)	5.911(4.935,7.081)	1.296
	>60	92(47.4%)	1.026(1.011,1.041)	3.406(2.947,3.937)	1.324
Division 8	0–14	11(10.9%)	1.017(1.006,1.029)	1.373(1.264,1.493)	1.141
	>60	47(46.5%)	1.047(1.027,1.067)	2.750(2.685,2.817)	1.358
Division 9	0–14	30(43.5%)	1.025(1.017,1.033)	2.611(2.462,2.770)	1.175
	>60	24(34.8%)	1.017(1.006,1.028)	1.903(1.779,2.035)	1.187

CI: confident interval;
Percentage was derived by the number of geographic units for each age level divided by the total number of counties at each geographic division.

categorized into different levels based on the reference categories of existing studies [34] (Table 1). We reclassified the census tract level individual incomes into high or low levels (groups) by using the computed national wide median income values as standards. Population in census tracts with income values higher than the nation-wide median income value were categorized in the 'high' income group, whereas the 'low income' group was assigned to census tracts below that national average. Figure 3 shows the population percentiles based on socio-demographic characteristics in the nine divisions. It can be seen that the socio-demographic characteristics including age, race, education attainment, and income fluctuate significantly across the nine divisions. This again emphasizes the necessity of conducting demography-based analysis of air pollution exposure inequities.

Spatial Cluster Analysis

Spatial autocorrelation is an optimal method for systematically ascertaining spatial patterns of air pollution exposure inequities [35]. For the purpose of detecting spatial clusters of environmental inequity across the United States, the spatial cluster analytical strategy used in this study is designed to include three sub-processes, including global autocorrelation analysis, logistic regression modeling, and local hot spot detection. Since we are interested in spatial patterns based on a large data set in the study area, it is reasonable that spatial dependence exists at the global scale because of the continuous characteristic of terrain in developed or open areas. Global autocorrelation analysis is therefore adopted to preliminarily explore the spatial autocorrelations of benzene pollution concentration as well as socio-demographic indicators. Odds ratios (ORs) were calculated for each county across the entire study area to further diagnose

whether the environmental inequities were caused by the interactions among these different global scale spatial autocorrelations. Logistic regression modeling was used to calculate the ORs. Finally, local hotpot detection was employed to pinpoint the statistically significant hot spots or cluster areas based on the ORs of counties. The methodological principles and implementation details of these sub-processes are described as follows:

1. Global autocorrelation analysis. At present, there are many ways to test the global autocorrelations of events. The most popular one among them is Moran's I statistic, which has been used to test the null hypothesis that the spatial autocorrelation of a variable is zero [36], [37]. If the null hypothesis is rejected, the variable would be considered spatially autocorrelated. Moran's I statistic of spatial autocorrelation is presented by Cliff and Ord 1981 as formulas (1–2) [38]:

$$I = \sum_i^n \left(x_i - \overline{X}\right) \sum_{j=1}^n w_{ij}\left(x_j - \overline{X}\right) \bigg/ S_i^2 \sum_i^n \sum_{j=1}^n w_{ij} \quad (1)$$

$$S_i^2 = \sum_i^n \left(x_i - \overline{X}\right)^2 \bigg/ n \quad (2)$$

where the global Moran's I index indicates the extent of global spatial autocorrelation of a variable, with the value ranging from -1.0 to $+1.0$, n denotes the number of all spatial units, x_i and x_j are the attribute values of a variable at spatial unit i and j, respectively, $\overline{X}$ is the mean of attribute values of x, S_i is the deviation of an attribute value at spatial unit i from its mean $\left(x_i - \overline{X}\right)$, w denotes

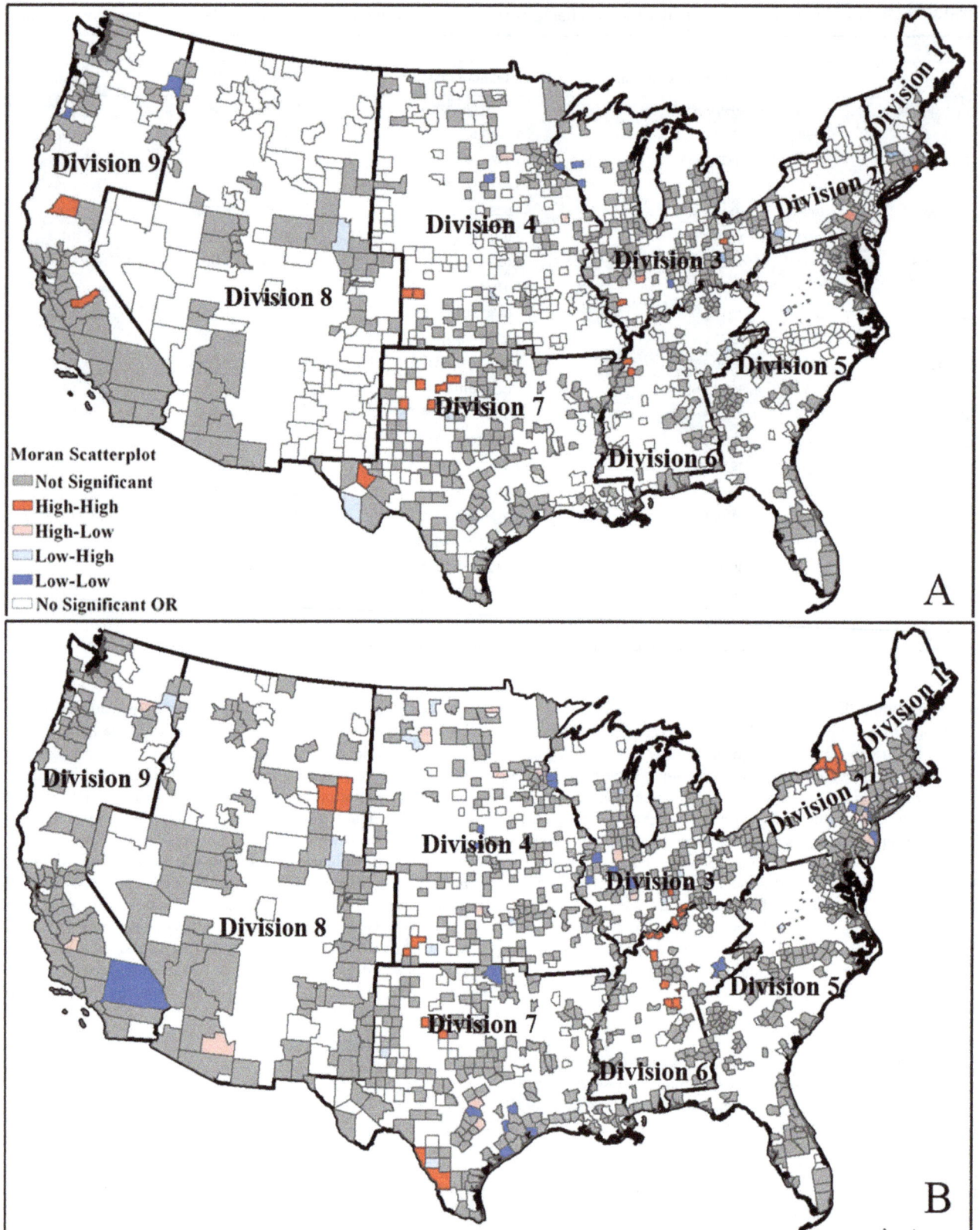

Figure 4. Clusters for benzene pollution exposure by age at county level. A local autocorrelation method is used to identify statistically significant hot spots, or cluster areas. High-High areas indicate high values near high values; Low-Low areas indicate low values near low values; High-Low areas indicate high values near low values; Low-High areas indicate low values near high values. (A) age (<14); (B) age (>60). Division 1 is New England; Division 2 is Mid-Atlantic; Division 3 is East North Central; Division 4 is West North Central; Division 5 is South Atlantic; Division 6 is East South Central; Division 7 is West South Central; Division 8 is Mountain; Division 9 is Pacific.

Table 4. Frequency of *OR*s greater than 1 by race at the county level in the United States and the nine divisions.

	Race	Count (percentage)	Minimum (95%CI)	Maximum (95%CI)	Mean
United States	Black	795(69.4%)	1.041(1.021,1.061)	56.589(48.884,65.510)	3.714
	American Indian	485(42.3%)	1.040(1.006,1.074)	14.726(10.112,21.445)	1.965
	Asian	679(59.2%)	1.044(1.016,1.072)	246.341(89.734,676.264)	3.157
	Other races	815(71.1%)	1.017(1.009,1.025)	11.887(8.812,16.034)	2.058
Division 1	Black	25(86.2%)	1.117(1.101,1.133)	9.809(9.470,10.160)	2.866
	American Indian	19(65.5%)	1.105(1.020,1.197)	3.161(2.945,3.394)	1.788
	Asian	24(82.8%)	1.220(1.185,1.257)	3.444(2.964,4.002)	2.038
	Other races	25(86.2%)	1.169(1.064,1.285)	6.110(5.941,6.284)	2.511
Division 2	Black	61(75.3%)	1.101(1.063,1.140)	17.066(16.205,17.973)	3.437
	American Indian	40(49.4%)	1.140(1.014,1.281)	2.788(2.497,3.114)	1.827
	Asian	63(77.8%)	1.044(1.016,1.072)	6.845(6.200,7.558)	2.009
	Other races	64(79.0%)	1.017(1.009,1.025)	7.097(6.915,7.283)	2.502
Division 3	Black	135(77.6%)	1.071(1.051,1.092)	37.255(26.109,53.160)	4.675
	American Indian	90(51.7%)	1.040(1.006,1.074)	3.824(1.769,8.269)	1.785
	Asian	94(54.0%)	1.049(1.014,1.086)	12.118(5.956,24.652)	2.891
	Other races	141(81.0%)	1.053(1.002,1.106)	6.173(4.613,8.261)	2.064
Division 4	Black	128(68.4%)	1.139(1.075,1.207)	50.856(7.124,363.056)	4.898
	American Indian	90(48.1%)	1.140(1.037,1.254)	14.726(10.112,21.445)	2.460
	Asian	110(58.8%)	1.075(1.023,1.132)	42.609(13.700,132.520)	4.063
	Other races	136(72.7%)	1.101(1.020,1.189)	11.886(9.798,14.419)	2.355
Division 5	Black	142(66.4%)	1.059(1.048,1.069)	56.500(40.804,65.510)	2.762
	American Indian	76(35.5%)	1.120(1.014,1.238)	14.161(8.893,22.550)	1.990
	Asian	130(60.7%)	1.050(1.035,1.066)	9.057(4.871,16.841)	2.061
	Other races	132(61.7%)	1.086(1.051,1.123)	7.285(6.750,7.862)	1.901
Division 6	Black	68(70.1%)	1.058(1.029,1.088)	16.902(13.287,21.502	4.093
	American Indian	30(30.9%)	1.117(1.005,1.241)	13.214(6.956,25.100)	2.254
	Asian	68(70.1%)	1.201(1.067,1.353)	246.341(89.734,676.264)	6.483
	Other races	67(69.1%)	1.157(1.021,1.312)	11.887(8.812,16.034)	2.023
Division 7	Black	130(67.0%)	1.060(1.020,1.102)	37.270(5.194,267.413)	3.631
	American Indian	60(30.9%)	1.096(1.012,1.188)	8.159(1.089,61.142)	1.817
	Asian	86(44.3%)	1.105(1.010,1.210)	20.688(2.850,150.144)	3.389
	Other races	126(64.9%)	1.041(1.010,1.073)	5.782(3.813,8.767)	1.939
Division 8	Black	55(54.5%)	1.127(1.098,1.157)	10.894(5.371,22.096)	2.691
	American Indian	41(40.6%)	1.200(1.101,1.308)	4.209(2.995,5.913)	1.892
	Asian	57 (56.4%)	1.066(1.023,1.111)	24.251(7.696,76.415)	2.845
	Other races	67(66.3%)	1.060(1.022,1.101)	4.781(4.079,5.605)	1.768
Division 9	Black	51(73.9%)	1.041(1.021,1.061)	13.205(8.974,19.430)	2.399
	American Indian	39(56.5%)	1.062(1.033,1.092)	2.192(2.131,2.255)	1.494
	Asian	47(68.1%)	1.064(1.043,1.086)	8.475(7.168,10.021)	1.848
	Other races	57(82.6%)	1.039(1.020,1.059)	3.091(3.049,3.133)	1.649

CI: confident interval;
Percentage was derived by the number of geographic units for each race level divided by the total number of counties at each geographic division.

the space matrix, and w_{ij} represents the spatial weight between spatial unit i and j.

In this study, we use the census tract as the base spatial unit. Moran's I index means the extent of global spatial autocorrelations of benzene pollution concentration as well as socio-demographic indicators (i.e. age, race, educational attainment, and income). The variable x in formulas (1) and (2) is therefore the attribute value of either 'benzene pollution concentration' or 'a socio-demographic indicator' such as 'age'. w_{ij} is determined based on the adjacency standard. Agency standard is when a shared side occurs between two adjacent census tracts i and j, then $w_{ij} = 1$, otherwise $w_{ij} = 0$. In order to verify the necessity of detecting local spatial clusters of potential environmental inequities, the global autocorrelation analyses in this study were implemented for entire

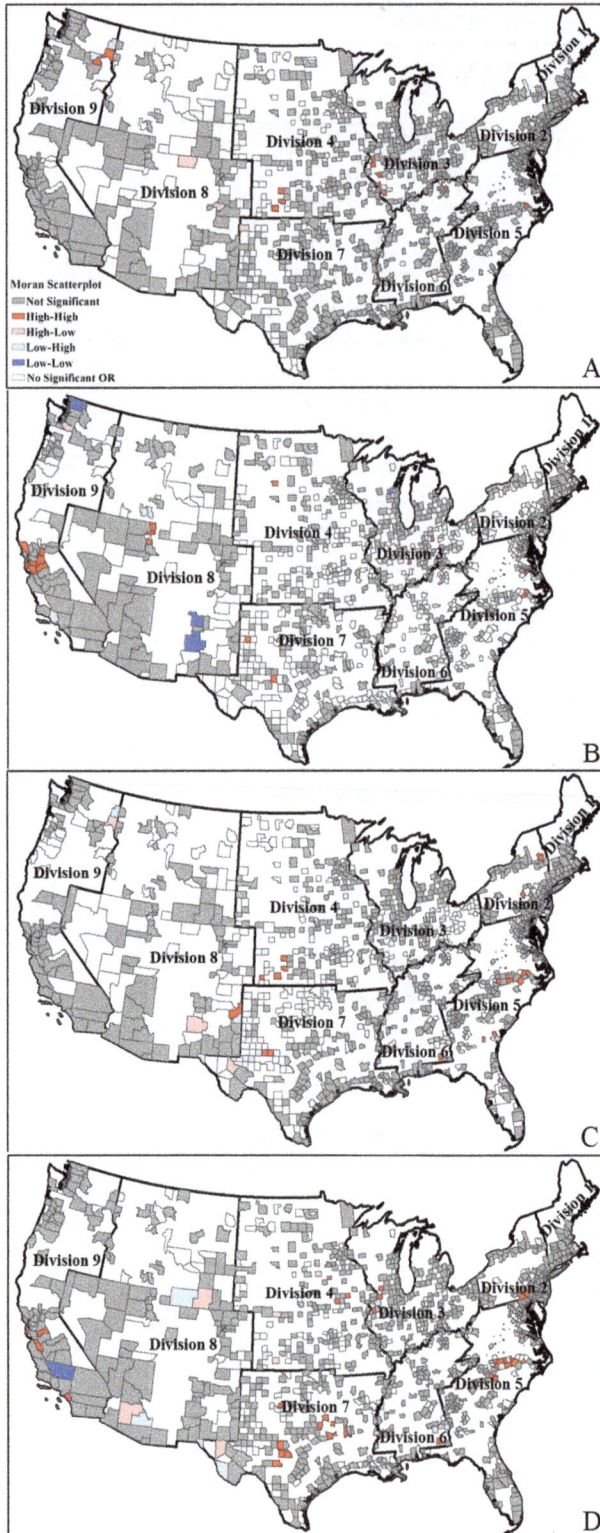

Figure 5. Clusters for benzene pollution exposure by race at county level. (A): Black; (B): American Indian; (C): Asian; (D): Other races. Division 1 is New England; Division 2 is Mid-Atlantic; Division 3 is East North Central; Division 4 is West North Central; Division 5 is South Atlantic; Division 6 is East South Central; Division 7 is West South Central; Division 8 is Mountain; Division 9 is Pacific.

United States and each Division separately. The analyses were conducted using the 'Spatial Statistic Tools' in ArcGIS 10.0.

2. Logistic regression modeling. Logistic regression is a mathematical modeling technique that describes the relationship between several independent variables and a dichotomous dependent variable [39]. Most environmental justice studies use logistic regression to derive ORs based on the following formulas (3–8):

$$odds = P/(1-P) \tag{3}$$

$$\log(odds) = \log it(P) = \ln[P/(1-P)] \tag{4}$$

$$\ln[P/(1-P)] = a + bX \tag{5}$$

$$P/(1-P) = e^{a+bX} \tag{6}$$

$$P = e^{a+bX}/(1 + e^{a+bX}) \tag{7}$$

$$OR = [P_1/(1-P_1)]/[P_2/(1-P_2)] \tag{8}$$

where 'odds' is the probability of the dichotomous dependent variable equals an event (i.e. the case or control group being exposed to air pollution) (i.e., 'p') divided by the probability of the event not to occur (i.e., '$p/1-p$'). OR denotes odd ratio, indicating the relative value by which the 'odds' of the outcome increases (i.e., OR greater than 1.0) or decreases (i.e., OR less than 1.0). 'e' is the exponential constant, equal to 2.71828. 'P_1' denotes the probability of the case group being exposed to air pollution. 'P_2' denotes the probability of the control or reference group being exposed to air pollution. 'X' represents the explanatory variables which are either interval-level or 'dummy', a, b represents partial regression coefficients of the independent variable 'X'.

The logistic regression modeling in this study was implemented in SPSS version 17. In this process, the census tract level benzene pollution concentration was dichotomized as the dependent variable and coded as either '1' (i.e. above) or '0' (i.e. below) based on the mean concentrations at the county level. Consequently, age, race, educational attainment, and income were selected as independent variables and recoded (e.g., the reference category was coded as '0'). Meanwhile, the population amount of above/below pollution concentrations in each category by different socio-demographic indicators were input correspondingly as weight cases while the 'indicator option' in SPSS was set first as the reference category. In addition, we assess whether there is any significant relationship between the dependent variable Y (i.e. benzene pollution concentration) and independent variables X (socio-demographic indicators). More specifically, if any of the null hypotheses that $b = 0$ is valid, then X is statistically insignificant in the logistic regression model. However, it was difficult for us to eliminate the potential bias of the logistic regression modeling for each type of demographic variable (e.g., age) by inputting the remaining variables (e.g., race, education attainment, income) as confounding factors, because the attribute values for those variables were aggregated values rather than individual level ones.

3. Local hot spot detection. When underlying global autocorrelation is detected, the question about how to identify

Table 5. Frequency of ORs greater than 1 by educational attainment at the county level in the United States and the nine divisions.

	Educational attainment	Count (percentage)	Minimum (95%CI)	Maximum (95%CI)	Mean
United States	0–4	660(57.6%)	1.075(1.021,1.132)	26.923(6.594,109.924)	2.427
	5–8	586(51.1%)	1.045(1.001,1.092)	21.318(13.042,34.847)	1.757
	9–12	625(54.5%)	1.023(1.002,1.045)	9.982(8.204,12.146)	1.515
Division 1	0–4	19(65.5%)	1.272(1.099,1.472)	4.294(3.690,4.997)	2.355
	5–8	22(75.9%)	1.118(1.025,1.220)	2.606(2.422,2.804)	1.757
	9–12	21(72.4%)	1.051(1.014,1.089)	2.052(2.002,2.103)	1.510
Division 2	0–4	61(75.3%)	1.081(1.003,1.166)	4.946(4.790,5.108)	1.906
	5–8	51(63.0%)	1.059(1.014,1.105)	2.958(2.890,3.028)	1.566
	9–12	51(63.0%)	1.064(1.019,1.110)	2.303(2.257,2.350)	1.479
Division 3	0–4	109(62.6%)	1.167(1.052,1.296)	7.555(5.171,11.039)	2.310
	5–8	96(55.2%)	1.092(1.003,1.190)	3.030(2.706,3.394)	1.566
	9–12	108(62.1%)	1.041(1.001,1.083)	3.182(3.087,3.280)	1.450
Division 4	0–4	99(52.9%)	1.152(1.008,1.317)	17.858(12.621,25.268)	3.309
	5–8	80(42.8%)	1.106(1.021,1.198)	3.308(2.929,3.736)	1.616
	9–12	91(48.7%)	1.046(1.023,1.069)	2.685(2.518,2.864)	1.469
Division 5	0–4	117(54.7%)	1.144(1.042,1.255)	11.495(9.706,13.615)	2.120
	5–8	112(52.3%)	1.075(1.008,1.146)	21.318(13.042,34.847)	1.905
	9–12	120(56.1%)	1.023(1.002,1.045)	9.982(8.204,12.146)	1.590
Division 6	0–4	43(44.3%)	1.093(1.029,1.161)	3.184(2.804,3.616)	1.765
	5–8	34(35.1%)	1.112(1.007,1.227)	3.076(2.846,3.323)	1.590
	9–12	44(45.4%)	1.064(1.004,1.128)	2.083(1.961,2.213)	1.381
Division 7	0–4	110(56.7%)	1.143(1.004,1.301)	7.281(3.523,15.051)	2.295
	5–8	96(49.5%)	1.076(1.001,1.156)	5.169(3.627,7.367)	1.931
	9–12	102(52.6%)	1.042(1.010,1.074)	5.187(3.816,7.050)	1.586
Division 8	0–4	56(55.4%)	1.246(1.132,1.371)	26.923(6.594,109.924)	3.333
	5–8	48(47.5%)	1.291(1.184,1.408)	3.568(3.492,3.645)	2.038
	9–12	49(48.5%)	1.084(1.030,1.140)	2.350(2.049,2.697)	1.604
Division 9	0–4	46(66.7%)	1.075(1.021,1.132)	5.470(4.389,6.816)	2.141
	5–8	47(68.1%)	1.045(1.001,1.092)	3.220(3.179,3.262)	1.723
	9–12	39(56.5%)	1.068(1.025,1.112)	2.504(2.479,2.529)	1.472

CI: confident interval;
Percentage was derived by the number of geographic units for each education level divided by the total number of counties at each geographic division.

more local patterns emerges. This leads to the challenge of finding an appropriate test for local spatial autocorrelations in the presence of global spatial autocorrelation. Local Moran's I based cluster mapping has been suggested as an effective method in detecting the hot spots or cluster areas of environmental exposure inequity based on spatial autocorrelation theory [40]. Formulas (9–11) present the basic principle of local Moran's I statistic.

$$I_i = \left[(x_i - \overline{X})/S_i^2\right] \sum_{j=1, j \neq i}^{n} w_{ij}(x_j - \overline{X}) \qquad (9)$$

$$\overline{X} = \sum_{i=1}^{n} x_i/n \qquad (10)$$

$$S_i^2 = \frac{\sum_{j=1, j \neq i}^{n} w_{ij}}{(n-1)} - \overline{X}^2 \qquad (11)$$

where the designations for the letters such as n, x_i, x_j, $\overline{X}$ are similar to those in formulas 3–8, S_i is the deviation between an attribute value at spatial unit i and its mean $\overline{X}$, I_i is the Moran's I index which indicates the extent to which neighboring spatial units congregate with each other in terms of attributes. If the attribute values in the dataset tend to cluster spatially (i.e., high values near high values; low values near low values), the Moran's I index will be positive. When high values repel other high values, or tend to be near low values, the index value will be negative. If the values in the dataset tend to scatter spatially, the index will be near zero. The range of the index value falls between −1.0 and +1.0 [40].

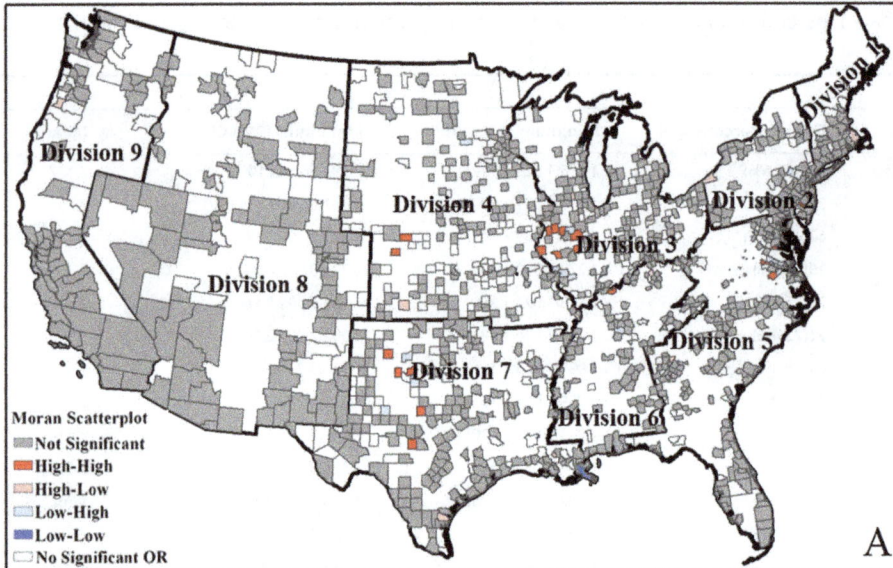

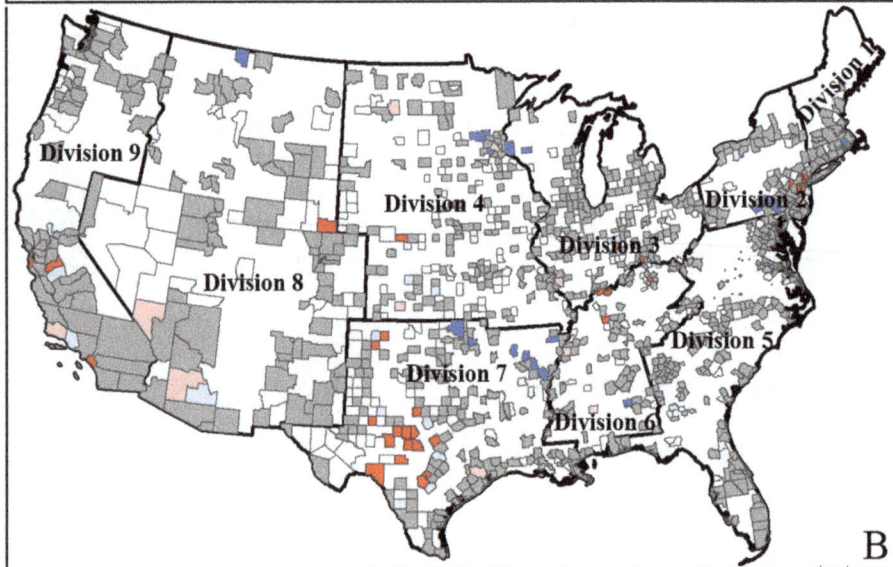

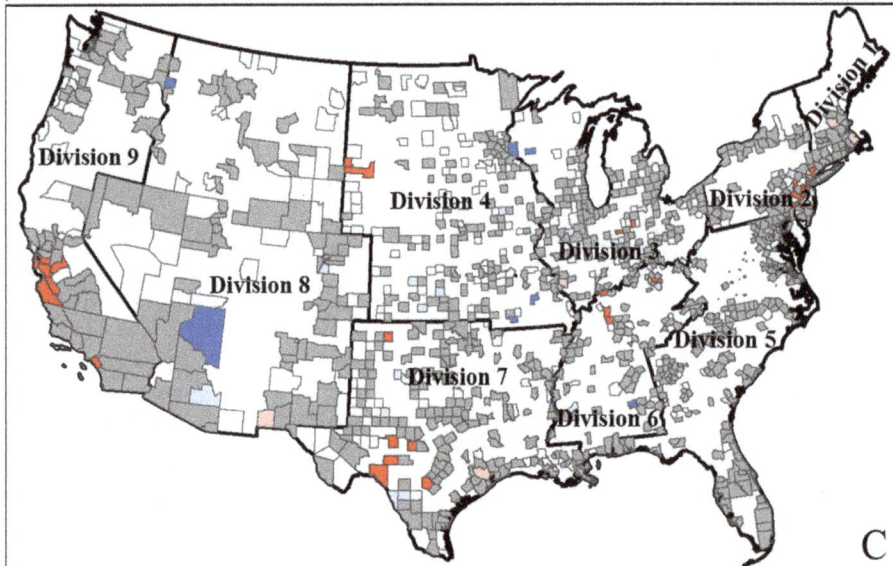

Figure 6. Clusters for benzene pollution exposure by education at county level. (A): Educational attainment (0–4); (B): Educational attainment (5–8); (C): Educational attainment (9–12). Division 1 is New England; Division 2 is Mid-Atlantic; Division 3 is East North Central; Division 4 is West North Central; Division 5 is South Atlantic; Division 6 is East South Central; Division 7 is West South Central; Division 8 is Mountain; Division 9 is Pacific.

We utilized cluster and outlier analysis (Anselin Local Moran's *I*) functions in 'Spatial Statistic Tools' within ArcGIS 10.0 to identify the local hot spots or cluster areas of benzene exposure inequity in this study. In this process, x is the *OR* value of each county. w_{ij} is determined based on the adjacency standard, where $w_{ij} = 1$ when there is a shared side between adjacent two counties, and 0 otherwise. The significance of the hot spots or cluster areas is determined by the Z-score and *P* value. That is, a high positive Z-score for benzene exposure inequities of a county with *P* value at 0.05 level indicates the surrounding features have the either high or low *OR* values (i.e., High-high, or Low-low). Inversely, a low negative Z-score for benzene exposure inequities of a county with *P* value at 0.05 level indicates a significant spatial outlier (i.e., High-low, or Low-high).

Results

Global Autocorrelation Analysis of Air Pollution Exposure

Table 2 delineates the values derived from the Global autocorrelation calculation for the nine divisions and the entire United States. With Global Moran's *I* index, people aged less than 4 years from Divisions 7 (0.090) and 9 (0.110) exhibit statistically significant clusters and have larger positive index values than the entire United States (0.046). Furthermore, even though Global Moran's *I* index values for Asians with educational attainment of 9-12 years appeared not to be statistically significant for the United States, Asians for Divisions 2, 4 and 5 (0.120, 0.054, 0.044), and education level of 9-12 years for Divisions 7 and 9 (0.058, 0.195) show significant cluster patterns.

Spatial Clustering of Air Pollution Exposure Inequity by Age

Table 3 delineates frequency of *OR*s greater than 1 by age characteristic at the county level in the United States and by the nine divisions. From Table 3, it can be seen that people belonging to age groups 0 to 14 and 60+ years old were exposed to higher levels of benzene pollution in some counties across the United States. For the age group of 60 years and older, Division 6 had the highest proportion (58.8%), followed by Division 2 (54.3%) and Division 7 (47.4%). The smallest proportion for that age group was found in Division 9 (34.8%). For the age group of 0–14, Division 9 displayed the greatest exposure (43.5%) followed by Divisions 7 and 1 (40.7%; 31.0% respectively). Division 8 has the lowest exposure in that age group (10.9%). We also observed that the proportion of counties exposed to higher levels of benzene pollution by division is mostly less than 50% for the United States and the nine divisions, except for the age group of 60 years and older in Divisions 2 and 6.

Figure 4 delineates the county level spatial clusters of benzene pollution exposure inequity based on data from Table 3. From Figure 4, the high-risk areas for the age group of 0–14 are located in Divisions 1, 3, 4, 6, 7, and 9, which includes the number of spatial cluster county units of 1, 2, 2, 2, 6, 2 respectively (Fig. 4A). Figure 4B shows high-risk areas for people age 60 years and over. It can be seen that these clusters were mainly located in Divisions 2–4 and 6–8, which includes the number of spatial cluster county units, which was 4, 7, 2, 7, 4, 2 respectively.

Spatial Clustering of Air Pollution Exposure Inequity by Race

Table 4 delineates frequency of *OR*s greater than 1 by race characteristic at the county level in the United States and the nine divisions. It can be seen that racial minorities such as Blacks, American Indians, and Asians were exposed to higher levels of benzene pollution in some counties. For Blacks, Division 1 had the highest proportion (86.2%), followed by Division 3 (77.6%) and Division 2 (75.3%). The smallest proportion for Blacks was found in Division 8 (54.5%). For American Indians, Division 1 had the highest proportion (65.5%), followed by Division 9 (56.5%) and Division 3 (51.7%). The smallest proportion for American Indians was found in Division 6 and Division 7 (30.9%). For Asians, Division 1 had the highest proportion (82.8%), followed by Division 2 (77.8%) and Division 6 (70.1%). The smallest proportion for Asians was found in Division 7 (44.3%). For other races, Division 1 showed the highest level of exposure with (86.2%) followed by Divisions 9 and 3 (82.6%; 81.0%). The lowest exposure in this racial group was in Division 5 (61.7%). It could also be observed that the proportion of counties exposed to higher levels of benzene pollution by divisions is mostly more than 50% for the United States and the nine divisions, except for the American Indians in Divisions 2, 4, 5, 6, 7, and 8 and Asians in Division 7.

Figure 5 shows the county level clusters of benzene pollution exposure inequity based on the results from Table 5. High-risk areas for Blacks were found in Divisions 3, 4, 5, 6, 9, which included the number of spatial cluster county units of 5, 3, 1, 3, 2 respectively (Fig. 5A). Figure 5B shows the high-risk clusters for American Indians. These cluster areas are mainly located in Divisions 3, 4, 5, 7, 8, 9, which included the number of spatial cluster county units is 2, 1, 6, 2, 3, 7 respectively. High-risk spatial cluster areas for Asians are located in Divisions 2, 4, 5, 6, 7, 8, with spatial cluster county units of 2, 4, 8, 1, 2, and 1 (Fig. 5C). High-risk spatial cluster areas of other races are located in Divisions 2, 3, 4, 5, 6, 7, 9, which included the spatial cluster county units of 3, 3, 3, 10, 1, 11, and 4 respectively(Fig. 5D).

Spatial Cluster of Air Pollution Exposure Inequity by Education

Table 5 delineates frequency of *OR*s greater than 1 by education characteristic at the county level in the United States and the nine divisions. Results indicate that individuals with less than 12 years education were exposed to higher levels of benzene pollution in some counties of the United States. For those with less than 4 years education, Division 2 had the highest proportion (75.3%), followed by Division 9 (66.7%) and Division 1 (65.5%). The smallest proportion for this same education group was found in Division 6 (44.3%). For the education level of 5 to 8 years, Division 1 had the highest proportion (75.9%), followed by Division 9 (68.1%) and Division 2 (63.0%). The smallest proportion for this education group was found in Division 6 (35.1%). For the education level of 9 to 12 years, Division 1 bore the greatest exposure with (72.4%) followed by Divisions 2 (63.0%) and 3 (62.1%). The lowest exposure for this age group was in Division 6 (45.4%). We also observed that the proportion of the total number of counties exposed to high levels of benzene pollution by divisions was more than 50% for the United States and the nine divisions, except for the education levels of 5 to 8 years in Divisions 4, 6, 7, 8 and the education levels ranging from 9 to 12 years in Divisions 4, 6, 8.

Figure 6 shows the county levels inequality of benzene pollution exposure based on information in Table 5. High-risk areas for education level less than 4 years were located in Divisions 2, 3, 4, 5, 6, 7, which included the number of spatial cluster county units of 1, 7, 2, 6, 1, 5, respectively (Fig. 6A). Figure 6B shows that high-risk areas for people of the 5–8 years of education level were mainly located in Divisions 2, 3, 4, 6, 7, 8, 9, which are associated with spatial cluster county units of 8, 1, 1, 5, 15, 1, 3 respectively. High-risk areas for education level between 9 and 12 years were located in Divisions 2, 3, 4, 6, 7, 9, which included the number of spatial cluster county units of 10, 2, 2, 5, 6, 7 respectively (Fig. 6C).

Spatial Cluster of Air Pollution Exposure Inequity by Income

Table 6 delineates frequencies of ORs greater than 1 by income characteristics at the county level in the United States and by the nine US Census Bureau divisions. From Table 6, it can be seen that people belonging to low income groups were exposed to higher levels of benzene pollution in some counties across the United States. For the low-income group, Division 1 had the highest proportion of residents with high exposure (65.5%), followed by Division 2 (60.5%) and Division 3(42.5%). The smallest proportion for the low-income group was found in Division 7 (10.3%). We also observed that the proportion of counties exposed to higher levels of benzene pollution by Division is mostly less than 50% for the United States and the nine divisions, except for the low-income group in Divisions 1 and 2.

Figure 7 details the county level spatial clusters of benzene pollution exposure inequity based on data from Table 6. As shown in Figure 7, the high-risk areas for the low-income groups are located in Divisions 3, 4, 5 and 9, which include the number of spatial cluster county units of 2, 1, 1, 2 respectively.

Discussion

This study is among the first spatial assessments of the inequities of air pollution exposure across the entire continental United States at the census tract scale. The results demonstrated that disparities in benzene air pollution exposure could help explain health disparities by age, race, educational attainment, and income. Although there has been a national decrease in health

disparities between 1990 and 1998 [41], some divisions have reported an increase in disparities during the same period [42]. Marshall [34] found environmental inequities of air pollution exposure in California's South Coast Air Basin, which persisted even after accounting for covariates such as population density, travel distance, mean differences between whites and nonwhites were 16–40% among the five pollutants.

A unique insight of this current study is that it highlighted spatial clusters of air pollution exposure inequity by race. Previous studies have shown that hazardous waste and industrial facilities were commonly located in or close to communities with populations that are of disproportionately higher proportions of minority or low-income individuals [43]. Our study extends the findings of previous studies by incorporating the spatial perspective of these inequities.

Minority neighborhoods tend to have higher rates of mortality, morbidity, and are more likely to be influenced by health risk factors than white neighborhoods, even after accounting for economic and other characteristics [44]. According to Gee and Takeuchi [45], differential residential locations come with differential levels of exposure to health risks. In particular, neighborhood stressors and pollution sources are related to adverse health conditions, which are counterbalanced by neighborhood resources. When community stressors and pollution sources outweigh neighborhood resources, levels of community stress manifest or increase. Community stress is a state of ecological vulnerability that may translate into individual stressors, which in turn may lead to individual stress. Individual stress may then make individuals more vulnerable to illness when they are exposed to environmental hazards. Furthermore, compromises in individual and community health may further weaken community resources, leading to a vicious cycle [46].

In addition, a key finding in our study is the significant inequities of air pollution exposure by educational attainment and income in the United States. For educational attainment based inequities, the results followed those of a previous study of 20 US cities which revealed strong (although not statistically significant) associations between PM_{10} and mortality for less educated subjects [47] as well as a study from Shanghai, China that showed an association between lower education and greater impact of air pollution-attributed mortality [48]. As to income attainment-based inequities, although the income in most census tracts across United States in this study exceeds the national poverty guideline for the

Table 6. Frequency of ORs greater than 1 by income at the county level in the United States and the nine divisions.

	Income	Count (percentage)	Minimum (95%CI)	Maximum (95%CI)	Mean
United States	Low income	346(30.2%)	1.029(1.006, 1.053)	9.809(9.558,10.067)	3.340
Division 1	Low income	19(65.5%)	1.149(1.136,1.163)	9.809(9.558,10.067)	4.116
Division 2	Low income	49(60.5%)	1.040(1.018,1.062)	9.653(9.380,9.935)	3.263
Division 3	Low income	74(42.5%)	1.064(1.041,1.088)	9.497(9.300,9.699)	3.673
Division 4	Low income	47(25.1%)	1.057(1.041,1.073)	9.095(8.680,9.530)	3.437
Division 5	Low income	58(27.1%)	1.056(1.046,1.065)	9.269(9.026,9.520)	9.269
Division 6	Low income	31(32.0%)	1.029(1.006,1.053)	9.436(9.316,9.558)	3.039
Division 7	Low income	20(10.3%)	1.107(1.088,1.127)	8.754(8.393,9.130)	2.642
Division 8	Low income	20(19.8%)	1.087(1.074,1.101)	9.175(9.126,9.224)	3.936
Division 9	Low income	28(40.6%)	1.178(1.150,1.208)	8.609(8.458,8.764)	3.079

CI: confident interval;
Percentage was derived by the number of geographic units for low income level divided by the total number of counties at each geographic division.

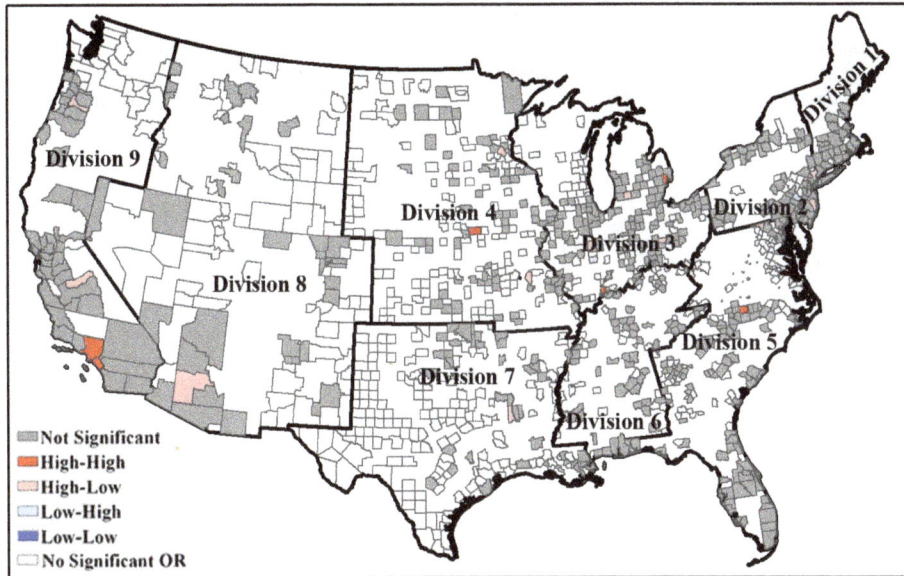

Figure 7. Clusters for benzene pollution exposure by income at county level. Fig. 7 shows High-High areas are high values cluster areas, in which people with low income exposed to higher level of benzene pollution than those with high income. Division 1 is New England; Division 2 is Mid-Atlantic; Division 3 is East North Central; Division 4 is West North Central; Division 5 is South Atlantic; Division 6 is East South Central; Division 7 is West South Central; Division 8 is Mountain; Division 9 is Pacific.

same period, significant and large *OR*s were observed for counties with relatively low income. This would indicate that people belonging to low income groups were more likely to be exposed to higher levels of benzene pollution in the United States relative to their higher income counterparts.

It should be noted that as this study is fairly unique in the methodology employed (e.g. spatial autocorrelation) for investigating environmental and socio-demographic inequities (geographic unit, methods of statistical analysis, exposure assessment procedures and definition of deprivation), our results are difficult to compare to other studies in relativistic terms. As more studies using this type of methodology are performed, a more comprehensive comparison will be possible. However, the results provided in this study would be highly applicable in other areas of research such as causal analysis of disease clusters, environmental policy targeting, and human rights policy making over large geographical areas.

Similar to previous analyses, the results of this study must be interpreted with caution. For example, since this study only examined a single type of air pollutant (i.e. benzene), our findings may not be generalizable to the cumulative effect of all other types of air pollutants. Further, our racial disparity analysis was only restricted to the classification of Blacks, American Indians, Asians and "Other races". Thus, we do not know if the interactive relationships uncovered here would hold true for Pacific Islanders who were probably combined with Asians or whether the results would change, which might make it be reasonable to identify Pacific Islanders as 'Other race' in the categorization. This study may also mask rural/urban characteristics when analyzing racial inequities in air pollution exposure. Similar to other ecological studies, this paper used aggregate data (e.g. census tract level) and could not incorporate individual-level information such as individual migration, time length of residence, and location exposure differences between work, recreation and living. Finally, as this study does not test any causal hypotheses, we could not

explain how or why race, age, educational attainment and income interact to produce air pollution inequity.

Another limitation of our data source is that, in Canada and the United States, census tracts are often referred to as a representation of the neighborhood [49]. However, it has been demonstrated that these census units do not represent underlying social boundaries and may depict the artifacts of administrative rules of a putative system [50]. Hence, it is sometimes difficult to tease out if the results of the analysis are representative of the reality or if they are the results of using a certain type of geographical unit [51].

To remedy the limitations of current studies, this paper identifies a set of overarching recommendations. Based on our results, scientists and community leaders should work in partnership to prioritize research needs, gather data, assess other air pollutants beyond benzene, and test interventions that will influence public policy in order to protect the health of all, including those living in communities of color and places that are economically deprived. Policy-makers can also enhance existing services that assist vulnerable groups and/or susceptible individuals to help close the disparity of exposure.

Conclusions

In summary, this study revealed that there are disproportionate exposures to benzene air pollution by a range of factors including age, race, education attainment and income in the United States. Spatial autocorrelation was also shown to be a valuable tool in this study to analyze how socio-demographic variables can influence the spatial patterns of air pollution exposure. However, further work is needed to inform policy-makers so that they can respond to the challenges and expectations that will improve environmental conditions for all underrepresented groups in the United States and beyond.

Acknowledgments

We thank Dr. Jaymie Meliker as well as the two anonymous referees for helpful comments on this manuscript. Sincere thanks would also go to Prof. Limin Jiao in Wuhan University for his great suggestions on 'spatial cluster analysis' method.

Author Contributions

Conceived and designed the experiments: BZ. Performed the experiments: FP BZ. Analyzed the data: BZ FP NW. Contributed reagents/materials/analysis tools: KM. Wrote the paper: BZ FP NW KM GJW.

References

1. Pellow DN (2000) Environmental inequality formation: toward a theory of environmental justice. Ame Behav Sci 43: 581–601.

2. Cole L (1992) Empowerment as the key to environmental protection: the need for environmental poverty law. Ecol Law Quart 19: 634–683.

3. The Lancet (2012) Global burden of disease study 2010. Available: http://www.thelancet.com/themed/global-burden-of-disease. Accessed 26 April 2013.

4. Schwartz J, Zanobetti A (2000) Using meta-smoothing to estimate dose-response trends across multiple studies, with application to air pollution and daily death. Epidemiology 11: 666–672.

5. Reitze AW (1999) The Legislative History of U.S. Air Pollution Control. Houston Law Rev 36: 679–741.

6. Briggs D, de Hoogh C, Gulliver J, Wills J, Elliott P, et al. (2000) A regression-based method for mapping traffic-related air pollution: application and testing in four contrasting urban environments. Sci Total Environ 253:151–167.

7. Jerrett M, Burnett R, Willis A, Krewski D, Goldbery MS, et al. (2003) Spatial analysis of the air pollution mortality association in the context of ecologic confounders. J Toxicol Environ Health 66:1735–1777.

8. Jerrett M, Burnett RT, Ma R, Pope III CA, Krewski D, et al. (2005) Spatial analysis of air pollution and mortality in Los Angeles. Epidemiology 16(6):727–736.

9. O'Neill MS, Jerrett M, Kawachi I, Levy JI, Cohen AJ, et al. (2003) Health, Wealth, and Air Pollution: Advancing Theory and Methods. Environ Health Perspect 111: 1861–1870.

10. Pulido L (1996) A critical review of the methodology of environmental racism research. Antipode 28:142–159.

11. Tian J, Wilson JG, Zhan FB (2010) Female breast cancer mortality clusters within racial groups in the United States. Health & Place 16: 209–218.

12. Health. 11 U.S. Cities With the Worst Air Pollution. Available: http://www.health.com/health/gallery/0204-90855,00.html. Accessed 29 May 2013.

13. Ponce NA, Hoggatt KJ, Wilhelm M, Ritz B (2005) Preterm birth: the interaction of traffic-related air pollution with economic hardship in Los Angeles neighborhoods. Am J Epidemiol 162:140–148.

14. Wikipedia, the free encyclopedia. List of regions of the United States. Available: http://en.wikipedia.org/wiki/United_States_Census_Bureau. Accessed 23 January 2013.

15. Bowen WM, Salling MJ, Haynes KE, Cyran EJ (1995) Toward environmental justice: spatial equity in Ohio and Cleveland. Ann Assoc Am Geogr 85(4), 641–663.

16. Buzzelli M, Jerrett M, Burnett R, Finklestein N (2003) Spatiotemporal perspectives on air pollution and environmental justice in Hamilton, Canada, 1985—1996. Ann Assoc Am Geogr 93(3), 557–573.

17. U.S. Environmental Protection Agency. 1999 National Air Toxics Assessments. Available: http://www.epa.gov/ttn/atw/nata1999/tables.html. Acessed 22 May 2012.

18. Chakraborty J, Maantay JA (2011) Proximity analysis for exposure assessment in environmental health justice research. In: Maantay JA, McLafferty S, editors. Geospatial analysis of environmental health. Netherlands: Springer. pp.111–138.

19. Smith MT (2010) Advances in understanding benzene health effects and susceptibility. Public Health 31: 133–148.

20. Fernández-Somoano A, Tardon A (2013) Socioeconomic status and exposure to outdoor NO2 and benzene in the Asturias INMA birth cohort, Spain. J Epidemiol Community Health doi:10.1136/jech-2013-202722

21. Morrens B, Bruckers L, Hond ED, Nelen V, Schoeters G, et al. (2013) Social distribution of internal exposure to environmental pollution in Flemish adolescents. Int J Hyg Environ Health 215(4): 474–81.

22. Chakraborty J (2012) Cancer risk from exposure to hazardous airpollutants: spatial and social inequities in Tampa Bay, Florida. Int J Environ Res 22(2):165–183.

23. Pastor M, Morello-Frosch R, Sadd JL (2005) The air is always cleaner on the other side: race, space, and ambient air toxics exposures in California. J Urban Aff 27(2):127–148.

24. Duarte-Davidson R, Courage C, Rushton L, Levy L (2001) Benzene in the environment: an assessment of the potential risks to the health of the population. Occup Environ Med 58(1): 2–13.

25. Bollati V, Baccarelli A, Hou L (2007) Change in DNA methylation patterns in subjects exposed to low-dose benzene. Cancer Res 67: 876–880.

26. Marchetti F, Eskenazi B, Weldon RH, Li G, Zhang L, et al. (2012) Occupational exposure to benzene and chromosomal structural aberrations in the sperm of Chinese men. Environ Health Perspect 120(2): 229–234.

27. Zou B (2010) How should environmental exposure risk be assessed? A comparison of four methods for exposure assessment of air pollutions. Environ Monitor and Assess 166: 159–167.

28. U.S. Bureau of the Census. (2001a) Census 2000 Summary File1 (SF1) Texas. Washington, DC: US Bureau of the Census.

29. U.S. Bureau of the Census. (2001b) Census 2000 TIGER/Line files Texas. Washington, DC: US Bureau of the Census.

30. Zanobetti A, Schwartz J (2000) Race, gender and social status as modifiers of the effects of PM10 on mortality. J Occup Environ Med 42: 469–474.

31. Gwynn RC, Thurston GD (2001) The burden of air pollution: impacts among racial minorities. Environ Health Perspect 109: 501–506.

32. Pope CA, Burnett RT, Thun MJ, Calle EE, Krewski D, et al. (2002) Lung cancer, cardiopulmonary mortality and long-term exposure to fine particulate air pollution. JAMA 287: 1132–1141.

33. Ou CQ, Hedley AJ, Chung RY, Thach TQ, Chau YK, et al. (2008) Socioeconomic disparities in air pollution– associated mortality. Environ Res 107: 237–244.

34. Marshall JD (2008) Environmental inequality: Air pollution exposures in California's South Coast Air Basin. Atmos Environ 42: 5499–5503.

35. Fuller MM, Enquist BJ (2012) Accounting for spatial autocorrelation in null models of tree species association. Ecography 35(6):510–518.

36. Toan DTT, Hu W, Thai PQ, Hoat LN, Wright P, et al. (2013) Hot spot detection and spatio-temporal dispersion of dengue fever in Hanoi, Vietnam. Glob Health Action 6:18632.doi: 10.3402/gha.v6i0.18632.

37. Rogerson PA, Kedron P (2012) Optimal weights for focused tests of clustering using the Local Moran statistic. Geographical Analysis 44(2): 121–133.

38. Cliff A, Ord JK (1987) Spatial process: models and application. London: Pion.

39. Kleinbaum DG, Klein M (2010) Introduction to logistic regression. In: Logistic regression. Atlanta: Springer. pp. 1–38.

40. ESRI. ArcGIS desktop help EB/OL. Available: http://webhelp.esri.com/arcgisdesktop/9.3/index.cfm? TopicName = How Cluster and Outlier Analysis: Anselin Local Moran's I (Spatial Statistics) works. Acessed 28 January 2009.

41. Keppel KG, Pearcy JN, Wagener D (2002) Trends in racial and ethnic-specific rates for the United States indicators: United States, 1990–1998. In: Healthy People Statistical Notes No. 23. Hyattsville, MD: National Center for Health Statistics.

42. Margellos H, Silva A, Whitman S (2004) Comparison of health status indicators in Chicago: are black-white disparities worsening? Am J Public Health 94:116–121.

43. Ringquist EJ (2005) Assessing evidence of environmental inequities: A meta-analysis. J. Policy Anal. Manage 24(2):223–247.

44. Cubbin C, Hadden WC, Winkleby MA (2001) Neighborhood context and cardiovascular disease risk factors: the contribution of material deprivation. Ethn Dis 11:687–700.

45. Gee GC, Takeuchi DT (2004) Traffic stress, vehicular burden and well-being: a multilevel analysis. Soc Sci Med 59(2):405–14.

46. Gee GC, Payne-Sturges DC (2004) Environmental Health Disparities: A Framework Integrating Psychosocial and Environmental Concepts. Environ Health Perspect 112:1645–1653.

47. Zeka A, Zanobetti A, Schwartz J (2006) Individual-level modifiers of the effects of particulate matter on daily mortality. Am J Epidemiol 163:849–859.

48. Kan H, London SJ, Chen G, Zhang Y, Song G, et al. (2008) Season, sex, age, and education as modifiers of the effects of outdoor air pollution on daily mortality in Shanghai, China: The Public Health and Air Pollution in Asia (PAPA) Study. Environ Health Perspect 116:1183–8.

49. Lebel A, Pampalon R, Villeneuve PY (2007) A multi-perspective approach for defining neighbourhood units in the context of a study on health inequalities in the Quebec city region. Int J Health Geogr 6:27. doi:10.1186/1476-072X-6-27. Online 5 July 2007.

50. Martin D (2004) Neighborhoods and area statistics in the post 2001 census era. Area 36(2): 136–145.

51. Mennis J (2003) Generating surface models of population using dasymetric mapping. Prof Geogr 55(1): 31–42.

23

Spatial/Temporal Variations and Source Apportionment of VOCs Monitored at Community Scale in an Urban Area

Chang Ho Yu[1], Xianlei Zhu[2], Zhi-hua Fan[1]*

1 Division of Exposure Science, Environmental and Occupational Health Sciences Institute, Rutgers University, Piscataway, New Jersey, United States of America, **2** College of Geosciences, China University of Petroleum, Beijing, People's Republic of China

Abstract

This study aimed to characterize spatial/temporal variations of ambient volatile organic compounds (VOCs) using a community-scale monitoring approach and identify the main sources of concern in Paterson, NJ, an urban area with mixed sources of VOCs. VOC samples were simultaneously collected from three local source-dominated (i.e., commercial, industrial, and mobile) sites in Paterson and one background site in Chester, NJ (located ~58 km southwest of Paterson). Samples were collected using the EPA TO-15 method from midnight to midnight, one in every sixth day over one year. Among the 60 analyzed VOCs, ten VOCs (acetylene, benzene, dichloromethane, ethylbenzene, methyl ethyl ketone, styrene, toluene, m,p-xylene, o-xylene, and p-dichlorobenzene) were selected to examine their spatial/temporal variations. All of the 10 VOCs in Paterson were significantly higher than the background site ($p < 0.01$). Ethylbenzene, m,p-xylene, o-xylene, and p-dichlorobenzene measured at the commercial site were significantly higher than the industrial/mobile sites ($p < 0.01$). Seven VOCs (acetylene, benzene, dichloromethane, methyl ethyl ketone, styrene, toluene, and p-dichlorobenzene) were significantly different by season ($p < 0.05$), that is, higher in cold seasons than in warm seasons. In addition, dichloromethane, methyl ethyl ketone, and toluene were significantly higher on weekdays than weekend days ($p < 0.05$). These results are consistent with literature data, indicating the impact of anthropogenic VOC sources on air pollution in Paterson. Positive Matrix Factorization (PMF) analysis was applied for 24-hour integrated VOC measurements in Paterson over one year and identified six contributing factors, including motor vehicle exhausts (20%), solvents uses (19%), industrial emissions (16%), mobile+stationery sources (12%), small shop emissions (11%), and others (22%). Additional locational analysis confirmed the identified sources were well matched with point sources located upwind in Paterson. The study demonstrated the community-scale monitoring approach can capture spatial variation of VOCs in an urban community with mixed VOC sources. It also provided robust data to identify major sources of concern in the community.

Editor: Yinping Zhang, Tsinghua University, China

Funding: This study was supported by the U.S. Environmental Protection Agency (subcontract through New Jersey Department of Environmental Protection Grant #SR05-035). Dr. Fan is supported in part by the National Institute of Environmental Health Sciences (NIEHS) sponsored Rutgers University Center for Environmental Exposures and Disease, Grant #NIEHS P30ES005022. The funders had no role in study design, data collection and analysis, decision to publish, or preparation of the manuscript.

Competing Interests: The authors have declared that no competing interests exist.

* E-mail: zfan@eohsi.rutgers.edu

Introduction

Volatile organic compounds (VOCs) are a group of air pollutants emitted from multiple types of anthropogenic sources, such as refineries, chemical factories, gas stations, dry cleaners, paint shops and diesel/gasoline-powered vehicles as well as biogenic sources. Previous studies have suggested associations between some VOCs in ambient air and adverse health outcomes, such as asthma [1,2,3]. As reported by many studies, "hot spots" of VOCs may exist due to presence of various local emission sources in urban communities [4,5,6,7,8,9,10,11]. However, VOCs data measured at community levels are limited. Thus, to better understand community exposures to ambient VOCs and associated health effects, monitoring of VOCs at community scale and characterization of their spatial/temporal variations are needed.

Many urban areas have mixed emission sources of VOCs, including mobile, commercial and industrial sources. However, gross industrial VOC emissions rather than speciated VOC emissions are usually reported to local air pollution control agencies [12]. Moreover, emission data are often obtained from estimation rather than true measurements, and many are not even available for small facilities. Therefore, the lack of detailed emission data prevents the evaluation of the impact of any emission sources on local VOC air pollution, and thus limits the development of effective controlling strategies. Furthermore, previous VOC source apportionment studies were extensively conducted using the measurements collected in the summer (e.g., Photochemical Assessment Monitoring Stations (PAMS)) [13,14,15]. Therefore, the results obtained from those studies primarily represented the sources of VOCs in the summer, not for other seasons. Given such, measurement of VOCs at community scale throughout one year and apportionment of their sources are needed.

This study aimed to characterize spatial and temporal variations of air toxics at community-scale in an urban area, i.e. Paterson, NJ, with mixed sources of VOCs. The emission sources included industrial, commercial, mobile and residential sources [12,16,17].

Among the monitored 60 VOCs, ten VOCs (acetylene, benzene, dichloromethane, ethylbenzene, methyl ethyl ketone, styrene, toluene, m,p-xylene, o-xylene and p-dichlorobenzene) that were detected over 75%, had toxicities and/or known sources in the study area were specifically selected for examining spatial and temporal variations. Also, the contributions from different VOC sources to local air pollution were estimated using Positive Matrix Factorization (PMF) analysis. Our study demonstrated that the community-scale monitoring approach could effectively capture local-dominated VOC sources in urban communities with mixed emission sources. In addition, to our best knowledge, this is the first attempt to conduct VOC source apportionment using measurements collected over a course of one year. Therefore, the major sources identified in our study reflected seasonal changes in the study area, and our approach provided more accurate estimate of the contribution of local VOC emission sources to community air pollution when compared to those obtained from the summer measurements only. Therefore, our study approach is more helpful for the development of effective strategies to control and reduce community air pollution.

Methods

Study Area

Paterson is located in Passaic county of NJ, with high population density ($6,826/km^2$ with a total population of 146,199 in US Census of 2010) and socio-economically disadvantaged populations [18]. It is composed of sections that are dominated by industrial (e.g., textiles, dyes, chemicals, metal fabrication/refinishing/recovery; plastics, printing, electronics, paper/food products, etc.), commercial (e.g., dry cleaners, fast food restaurants, photo labs, commercial heating/boilers, nail salons, print shops, etc.) and mobile sources (e.g., US I-80, Route 19 and County Route 649, 639 and 648).

Monitoring Sites

Three monitoring sites, i.e., commercial, industrial and mobile sampling sites, were selected for sampling based on Geographic Information System (GIS) layers of population density, road type, source proximity, traffic count and land use type, as well as accessibility, security of the sampling systems and availability of electricity. The site map can be found in our previous publication [17]. Briefly, the industrial site was located at a public school in northern Paterson, about 0.1–1.0 km south-southeast of a highly industrialized area known as Bunker Hill. This area hosts a variety of industrial facilities, emitting toluene, methyl ethyl ketone (MEK), methyl isobutyl ketone (MIBK), xylenes, ethylbenzene, and general VOCs. The monitoring site for the mobile source-dominated area was located at a public school in southwestern Paterson. Several major roadways are located within 0.8 km of the school, including the US Interstate Route 80 & 19, a major NJ Transit Bus Depot and an active rail yard/line. The commercial monitoring site was located at a health department building near the shopping district in downtown Paterson. There are many typical urban commercial sources, such as dry cleaners, fast food restaurants, photo finishing, commercial heating/boilers, nail salons, print shops, etc. Monitoring devices were placed on the rooftop (approximately 10~13 m above the ground) of the school/building given space and security restrictions. The background site was located in an open field in Chester, NJ, about 58 km west/southwest of Paterson. This site is designated as the background/rural site for the Urban Air Toxic Monitoring Program (UATMP) by the New Jersey Department of Environmental Protection (NJDEP) and has been in operation since 2001.

This study was jointly conducted by Rutgers University and the NJDEP. The NJDEP obtained the approval from the Board of Education of Paterson Public School District to place air toxics monitoring equipment on the roof of two school buildings, to capture mobile and industrial source-oriented emissions. The NJDEP also obtained approval from the Paterson Public Health Department to place the air monitors at the local building in downtown of Paterson. The NJDEP allowed the state-designated background site in Chester for air sampling. Due to confidentiality concerns, specific location information (e.g., GPS coordinates) is not provided in this manuscript.

Sample Collection and Measurement

One year field sampling was conducted from November 18, 2005 to December 19, 2006. The sampling frequency was one in six days, and the sampling duration was 24 hours, from midnight to midnight. The study was designed to represent community's exposure to air toxics in an urban community; thus, one full day monitoring (i.e., 24-hour sampling) was employed. The study employed the same UATMP sampling frequency and schedule, which aims to capture trends of air toxic pollutants, so that we could compare the data collected from this study to other UATMP urban sites (e.g., Camden, Elizabeth and New Brunswick) in NJ. Ambient VOCs were collected using a stainless steel canister with an air sampler (ATEC model 2200, Malibu, CA), following the EPA TO-15 method [19]. After sample collection, the canister was sent to Environmental Research Group (ERG, Morrisville, NC) for analysis. The delivered samples were analysed using gas chromatography-mass spectrometry (GC-MS) within 3 days. Besides the 10 target compounds, other 50 species were analysed by ERG. All sampling and analysis procedures, including canister cleaning, calibration of analytical system and quantification of target compounds, were exclusively conducted by ERG, an US EPA national contract laboratory. All quality assurance/quality control (QA/QC) procedures have been overseen and documented by the USEPA.

QA/QC

Twenty five duplicate samples (~10% out of the 209 regular samples) were collected side-by-side during the study period, and the measurement precision for each VOC species was evaluated by the absolute percent difference (%Diff) between the two co-located samples. The difference was calculated using the following equation (1):

$$\%Diff = \frac{abs(regular - duplicate)}{(regular + duplicate)/2} \times 100 \qquad (1)$$

Good precision was obtained for most VOCs, except acrolein. The %Diff was less than 20% for the majority of the target compounds. The precision of acrolein, however, was poor, with %Diff of 55%. It was suspected that the poor precision of acrolein may be partially contributed by artificial formation of acrolein in the canister during storage [20].

The method detection limit (MDL) was calculated as the product of the standard deviation (SD) of seven replicate analyses and the Student's t-test value for 99% [19]. The MDLs are reported in Table 1. Prior to field sampling, all canisters were cleaned at the analytical laboratory and delivered to sampling sites vacuumed. Therefore, field blank sample collection was not applicable for this type of canister method. Thus, field blanks were

Table 1. Ambient VOCs concentrations (μg/m³) measured at the three monitoring sites in Paterson and one background site in Chester.

VOC Compounds	N	Avg	SD	Min	Med	Max	MDL	>MDL(%)	ND(%)
1,1,1,-Trichloroethane[a]	209	0.15	0.09	0.05	0.11	0.98	0.02	99	0
1,2,4-Trimethylbenzene[a]	209	0.54	0.59	0.01	0.39	4.38	0.02	89	9
1,3,5-Trimethylbenzene[a]	209	0.17	0.17	0.01	0.15	1.23	0.02	85	12
1,3-Butadiene[a]	209	0.15	0.18	0.01	0.11	1.35	0.01	84	14
Acetonitrile[a]	209	0.53	0.83	0.08	0.20	6.54	0.17	52	44
Acetylene[a]	209	1.15	1.24	0.02	0.77	8.22	0.03	100	0
Acrolein[a]	209	0.79	0.74	0.11	0.62	3.86	0.25	76	20
Benzene[a]	209	1.13	0.89	0.22	0.90	6.52	0.02	100	0
Carbon Disulfide	209	0.72	1.46	0.01	0.09	16.0	0.03	73	27
Carbon Tetrachloride[a]	209	0.61	0.18	0.06	0.57	1.07	0.06	100	0
Chloroethane	209	0.04	0.06	0.01	0.03	0.58	0.02	59	34
Chloroform[a]	209	0.19	0.21	0.01	0.15	1.52	0.02	75	22
Chloromethane[a]	209	1.15	0.19	0.56	1.16	1.78	0.03	100	0
Dichlorodifluoromethane[a]	209	2.89	0.64	1.19	2.83	7.59	0.03	100	0
Dichloromethane[a]	209	0.91	1.17	0.03	0.56	7.18	0.06	98	1
Dichlorotetrafluoroethane	209	0.12	0.03	0.07	0.14	0.14	0.02	100	0
Ethylbenzene[a]	209	0.57	0.84	0.04	0.35	9.04	0.02	100	0
Methyl Ethyl Ketone[a]	209	1.88	2.07	0.07	1.18	14.0	0.13	91	8
Methyl Isobutyl Ketone[a]	209	0.37	0.48	0.01	0.25	3.57	0.03	78	21
Methyl tert-Butyl Ether[a]	209	0.45	1.09	<0.01	0.14	7.93	0.01	68	32
Propylene[a]	209	1.05	0.96	0.12	0.86	6.90	0.02	100	0
Styrene[a]	209	0.16	0.16	0.02	0.13	0.98	0.04	77	13
Tetrachloroethylene[a]	209	0.48	0.56	0.04	0.34	5.10	0.08	82	7
Toluene[a]	209	5.27	5.83	0.19	3.62	32.4	0.02	100	0
Trichloroethylene[a]	209	0.10	0.14	0.03	0.03	1.46	0.05	36	54
Trichlorofluoromethane[a]	209	2.02	1.19	0.68	1.69	11.2	0.04	100	0
Trichlorotrifluoroethane[a]	209	0.79	0.17	0.38	0.77	1.61	0.09	100	0
m,p-Xylene[a]	209	1.77	3.67	0.04	0.91	40.9	0.04	100	0
n-Octane[a]	209	0.20	0.20	0.01	0.14	1.50	0.03	85	13
o-Xylene[a]	209	0.56	0.72	0.01	0.39	6.65	0.02	97	1
p-Dichlorobenzene[a,b]	195	0.24	0.23	0.02	0.18	1.39	0.04	75	23

[a]These VOCs were selected for the PMF analysis based on S/N ratio >2 and detection >50% in pooled Paterson data.
[b]High concentrations (N = 14) monitored at the commercial site in the period of 9/26/2006~12/17/2006 were excluded.

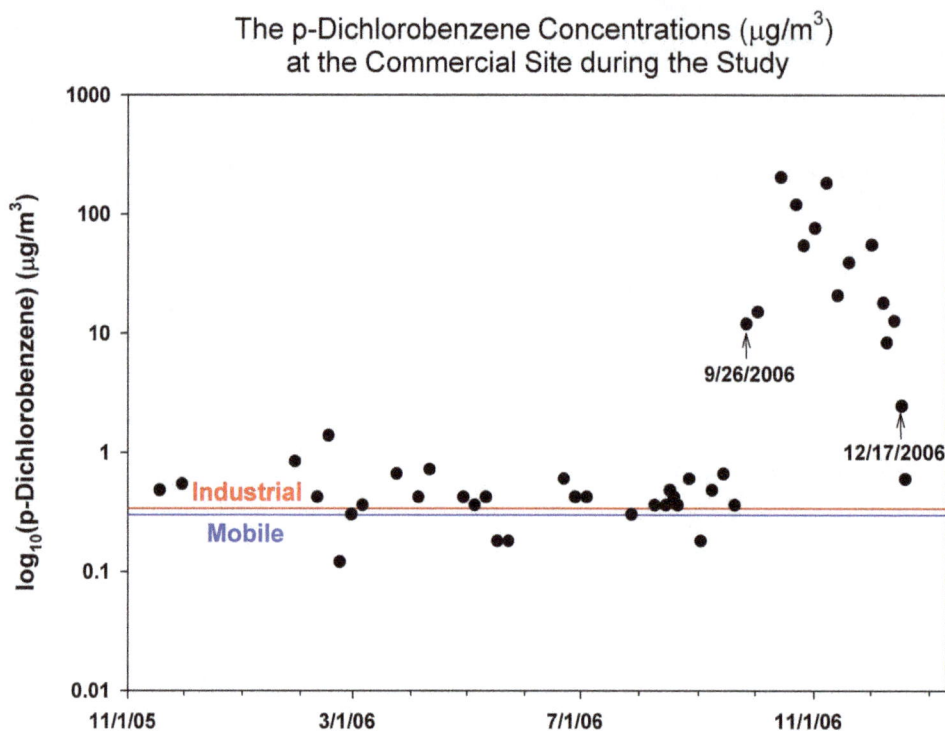

Figure 1. The monitored p-dichlorobenzene concentrations at the commercial site during the study period (11/18/2005~12/19/2006). The averaged p-dichlorobenzene concentrations at industrial (dashed line) and mobile (dotted line) sites for the monitoring period were added for references.

not collected and blank subtraction was not performed for the data reported in this study.

Data Analyses

Data selection and substitution. The number of samples (i.e., 209) presented in Table 1 included all of the samples collected from the three monitoring sites in Paterson and the one background site in Chester. Among the 60 VOCs analyzed by the EPA TO-15 method, ten VOCs (acetylene, benzene, dichloromethane, ethylbenzene, MEK, styrene, toluene, m,p-xylene, o-xylene and p-dichlorobenzene), which were detected over 75% during the course of monitoring period, toxic and/or having known sources in the study area, were specifically selected for analyzing their spatial/temporal variations. In addition to the 10 selected VOCs, twenty one VOCs that were detected over 50% or used for the PMF analysis were also reported in Table 1. For the non-detects (ND), we replaced them with a half of the MDL in data analysis. Since, for the 10 target species, more than 75% of the samples were detected above MDL, the substitution of ND with a half of MDL was not expected to significantly affect the spatial/temporal variations for these target species.

The data were not normally distributed; thus, non-parametric approaches were used for data analysis. Specific descriptions of each analysis are presented below.

Spatial and temporal variability. Descriptive statistics, including mean, standard deviation, minimum, median and maximum, were performed to characterize the distributions of the VOC concentrations. To examine site and seasonal differences, non-parametric Kruskal-Wallis test (two-sided) was conducted. If the difference was found to be significant ($p < 0.05$), pairwise multiple comparison tests were followed with the significance determined by Bonferroni's corrected alpha (i.e., $0.05/6 = 0.0083$).

For the difference between weekday and weekend, Wilcoxon Rank-sum test (two-sided) was conducted. To increase statistical power, temporal variability (i.e., seasonal differences and weekday and weekend differences) was conducted on pooled data from the three sites in Paterson. All statistical analyses were conducted by SAS (v. 9.2).

Positive Matrix Factorization (PMF) model. As the study measured VOC concentrations in Paterson over one year period, source-receptor relationships were explored using a mass balance approach to identify and to apportion the sources of ambient VOC concentrations in Paterson, with a consideration of seasonal variation in emission sources. A PMF model (v. 3.0) was used to provide source profiles and contributions to the measured data. Detailed explanations and equations used in the PMF analysis can be found elsewhere [21,22]. The predicted mass fractions and source factors obtained from the PMF analysis can be used to identify major sources that significantly contribute to the VOC air pollution in Paterson.

Prior to performing PMF analysis with pooled Paterson data, spatial correlations were examined first to check the pooled VOC data might be correlated each other spatially. Moran's I and Geary's C tests were conducted for the VOC data sets, and the results indicated that the spatial autocorrelations were not significant ($p > 0.05$). Therefore, daily arithmetic mean for each VOC species was calculated by averaging the measurements from the three monitoring sites in Paterson. This allowed to estimate VOC source contributions in Paterson over entire study period (i.e., ~1 year). The uncertainty of each species was calculated using MDLs and error fraction, assuming 20% for all species in this study [23]. Among the 60 species measured in the study, twenty eight compounds that were detected more than 50% and had a signal-to-noise (S/N) ratio greater than 2 [22,24] were

Table 2. Descriptive statistics and spatial differences for the 10 VOCs (µg/m³) monitored at the four local-source dominated sites in the study.

	N	Mean	SD	Min	Med	Max	p-value[a]	Multiple Comparison[b]
Acetylene								
Background	69	0.52	0.35	0.02	0.43	1.76	<.0001	A
Commercial	45	1.76	1.72	0.34	1.16	8.22		B
Industrial	45	1.20	1.10	0.20	0.97	6.89		B
Mobile	50	1.41	1.27	0.27	1.11	6.72		B
Benzene								
Background	69	0.52	0.22	0.22	0.48	1.19	<.0001	A
Commercial	45	1.81	1.15	0.54	1.44	6.52		B
Industrial	45	1.11	0.68	0.35	0.90	4.16		C
Mobile	50	1.36	0.82	0.35	1.21	4.99		B C
DCM								
Background	69	0.33	0.25	0.03	0.28	1.60	<.0001	A
Commercial	45	1.15	1.26	0.17	0.80	7.18		B
Industrial	45	1.31	1.46	0.17	0.70	6.34		B
Mobile	50	1.14	1.28	0.24	0.71	6.83		B
EB								
Background	69	0.14	0.08	0.04	0.13	0.43	<.0001	A
Commercial	45	1.28	1.50	0.17	0.83	9.04		B
Industrial	45	0.53	0.38	0.09	0.48	2.22		C
Mobile	50	0.54	0.34	0.13	0.48	2.17		C
MEK								
Background	69	1.08	1.22	0.07	0.91	8.09	<.0001	A
Commercial	45	2.76	2.11	0.07	2.27	10.0		B
Industrial	45	2.77	3.06	0.07	1.77	14.0		B
Mobile	50	1.41	1.09	0.07	1.03	4.72		A

	N	Mean	SD	Min	Med	Max	p-value[a]	Multiple Comparison[b]
Styrene								
Background	69	0.07	0.06	0.02	0.04	0.34	<.0001	A
Commercial	45	0.28	0.24	0.02	0.21	0.98		B
Industrial	45	0.16	0.11	0.02	0.13	0.60		C
Mobile	50	0.17	0.11	0.02	0.13	0.60		B C
Toluene								
Background	69	0.71	0.54	0.19	0.53	3.17	<.0001	A
Commercial	45	7.98	6.06	1.06	6.37	32.4		B
Industrial	45	6.46	4.94	0.45	5.47	25.6		B
Mobile	50	8.06	6.60	0.87	6.49	31.4		B
m,p-Xylene								
Background	69	0.30	0.21	0.04	0.26	1.13	<.0001	A
Commercial	45	4.62	6.99	0.43	2.39	40.9		B
Industrial	45	1.51	1.28	0.17	1.35	7.17		C
Mobile	50	1.45	1.09	0.26	1.24	6.82		C
o-Xylene								
Background	69	0.13	0.08	0.01	0.13	0.43	<.0001	A
Commercial	45	1.20	1.19	0.17	0.87	6.65		B
Industrial	45	0.55	0.42	0.09	0.48	2.48		C
Mobile	50	0.57	0.40	0.13	0.48	2.56		C
p-DCB								
Background	69	0.04	0.04	0.02	0.02	0.24	<.0001	A
Commercial	31	0.47	0.24	0.12	0.42	1.39		B
Industrial	45	0.34	0.20	0.02	0.30	1.02		C
Mobile	50	0.30	0.17	0.06	0.27	0.78		C

[a]Differences within the four sampling sites were conducted using the Kruskal-Wallis test and pairwise multiple comparison tests (Wilcoxon rank-sum test) were followed, if the difference was significant ($p < 0.05$).
[b]Different letters mean significant differences ($p < 0.0083$) among the four monitoring sites.
Abbreviation in the table: DCM: Dichloromethane, EB: Ethylbenzene, MEK: Methyl Ethyl Ketone, p-DCB: p-Dichlorobenzene.

(a) Methyl Ethyl Ketone

(b) Toluene

(c) p-Dichlorobenzene

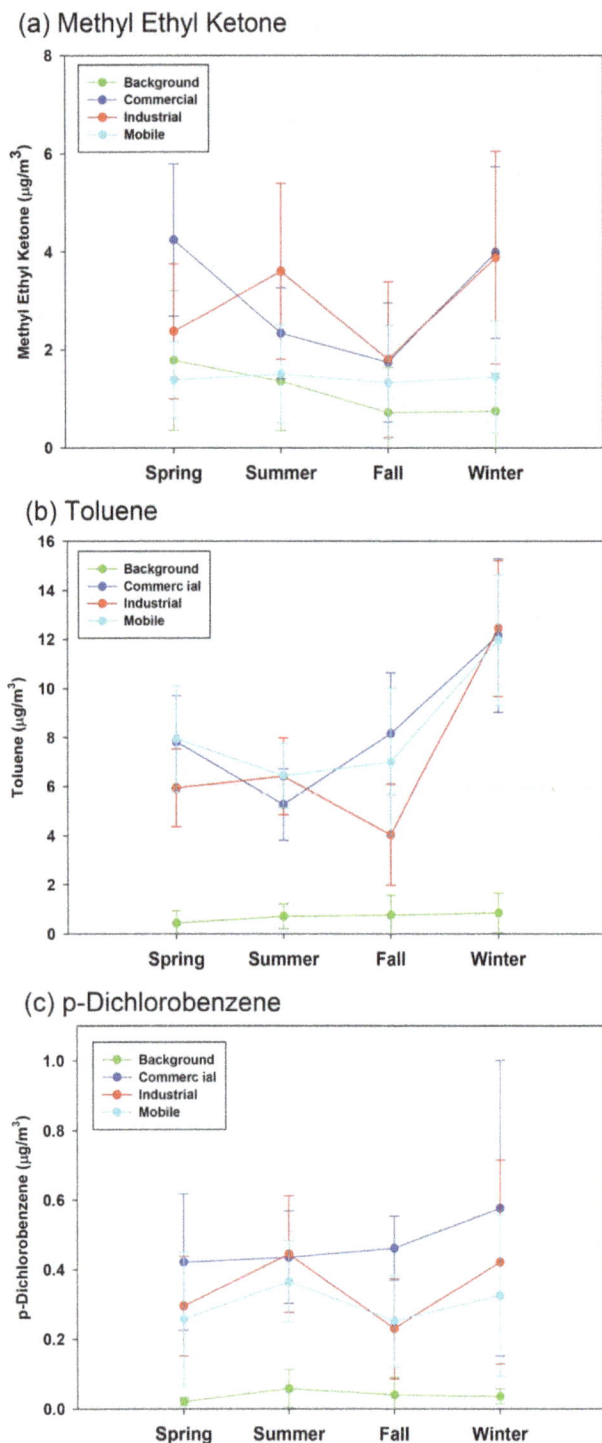

Figure 2. Seasonal variation (mean±SE [standard error]) of (a) MEK, (b) toluene, and (c) p-dichlorobenzene at each sampling site.

selected for PMF analysis (Table 1). Any non-detects in the input data were substituted with a half of the MDL. There was no missing sampling date for the PMF input database. In addition, the total VOC concentration, summing the 28 selected VOCs, were calculated and included as an independent variable in the PMF model to provide direct mass apportionments [13,14].

The PMF analysis suggested six factor profiles as appropriate source types in the source-receptor relationships. Different number of source profiles (e.g., five and seven factors) was additionally conducted during the analysis to identify proper number of sources in the given data sets [25]. In the seven-source model, the final model produced a negative constant, indicating too many sources were used. In the five-source model, the mobile and industrial emissions were not further separated. Therefore, the six-source model provided the most physically reasonable source profiles. As a part of finalizing best-fit source profiles, we utilized emission inventory (EI) data in Paterson as well as source identification results published previously for particulate matter less than 10 μm in diameter (PM_{10}) [17] and polycyclic aromatic hydrocarbons (PAHs) [16] that were concurrently measured in the same study. The rotational ambiguity was further investigated by inspecting pairs of the final factors in different FPEAK value range to avoid subjective bias in some extent. Sensitivity analyses, such as running a PMF with the 5-factor and 7-factor models as well as 5% extra modeling uncertainty, were conducted to verify whether the selected 6-factors were robust in the final form. The bootstrap running for the selected 6-factor model was repeated in 100 times to check if the factors were stable and consistent. Also, the positive FPEAK values (0.1~0.5) were used to sharpen the ambiguous source profiles in the base run model. After these additional tests, the final 6 factors remained constant. In this way, subjective bias was reduced significantly [26,27].

To help identifying the likely locations of the PMF-identified sources, a conditional probability function (CPF) was calculated. This approach was previously conducted by Kim et al. [28]. Briefly, daily fractional mass contribution from each source was used to minimize the effect of atmospheric dilution, rather than the absolute source contribution from all sources. The same daily fractional contribution was assigned to each hour in a given day to match the hourly wind data. Specifically, the CPF was defined as the following equation (2):

$$CPF = \frac{m_{\Delta\theta}}{n_{\Delta\theta}} \qquad (2)$$

where $m_{\Delta\theta}$ is the number of occurrence from wind sector $\Delta\theta$, and $n_{\Delta\theta}$ is the total number of data from the same wind direction. In this study, the highest 10 percentile of the daily fractional contribution from each source was chosen. Corresponding hourly wind data, except calm winds (<1 m/sec), were counted by the sector of 10 degrees. A CPF value close to 1.0 for a given sector indicates a high probability of a source located in that direction.

Results and Discussion

Spatial and Temporal Variability

Spatial variability. For p-dichlorobenzene, there was a striking difference among commercial (AVG±SD [Min–Max]: 18.7±45.2 [0.12–205] μg/m³), industrial (0.34±0.20 [0.04–1.02] μg/m³) and mobile (0.30±0.17 [0.06–0.78] μg/m³) sites. The mean concentration at the commercial site was two orders of magnitude higher than the other two sites. The high concentrations at the commercial site were driven by the high concentrations (N = 14; 59.0±66.1 [2.47–205] μg/m³) observed between September 26 and December 17, 2006 (Figure 1). Because of the measured high p-dichlorobenzene concentrations at the commercial site, additional VOC monitoring was conducted from April 2010 to May 2011 at this site. The study monitored VOC concentrations every six days over a course of one year at the commercial site, and five spatial saturation sampling (SSS)

The ratios of weekday/weekend for 10 VOCs in the study

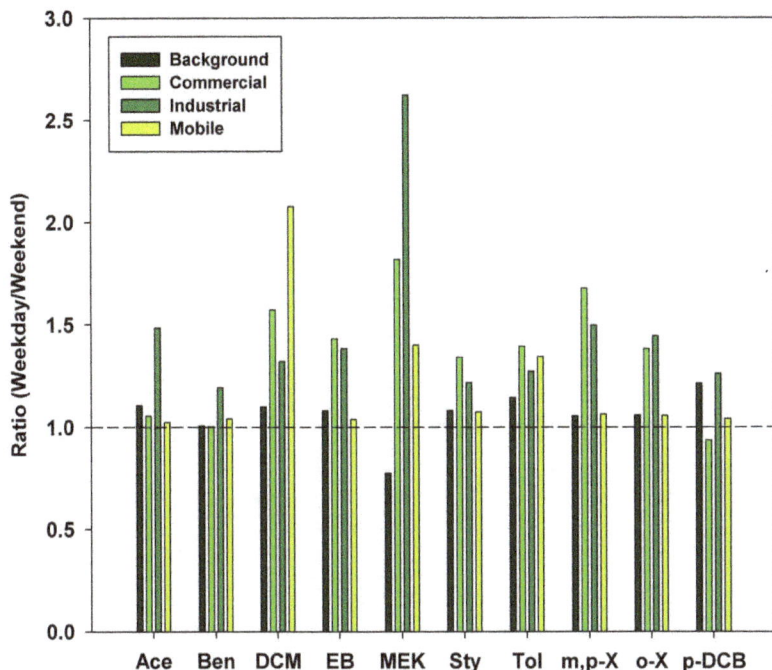

Figure 3. The ratios of weekday/weekend mean concentrations by each sampling site. The dotted line indicates the equivalent concentration for weekday and weekend measurements. Abbreviation in the figure: Ace: Acetylene, Ben: Benzene, DCM: Dichloromethane, EB: Ethylbenzene, MEK: Methyl Ethyl Ketone, Sty: Styrene, Tol: Toluene, m,p-X: m,p-Xylene, o-X. o-Xylene, p-DCB: p-Dichlorobenzene.

campaigns around the commercial site were carried out over the one year monitoring period. For the SSS sampling, organic vapor monitor (OVM, 3M, St. Paul MN) passive badge were deployed for three days in 23 locations within the city in a grid-like fashion around the commercial monitoring site. There was no spike p-dichlorobenzene concentrations measured in any sampling campaign (N = 37; 0.29±0.23 [0.03–1.28] μg/m^3), and the concentrations obtained from the SSS campaign were similar to other urban areas in NJ (i.e., Camden, Elizabeth, and New Brunswick). The study concluded that the high p-dichlorobenzene concentrations were one time event, and indoor sources, such as room deodorizer or moth repellent from the building, might result in those high measurements. Thus, the measurements from this particular period were excluded for the spatial/temporal variations analysis and PMF analysis.

The concentrations of the 10 selected VOC species at each sampling site and the comparison results among the 4 sampling sites are presented in Table 2. All of the 10 VOCs showed significant differences among the four sampling sites (p<0.05), particularly, the concentrations measured in Paterson were much higher than the background site in Chester. The VOC concentrations in Paterson were similar to those in other urban communities, i.e., Camden, Elizabeth and New Brunswick, across NJ [12]. Multiple comparison tests (i.e., Kruskal-Wallis test) confirmed that the spatial variability was resulted from significant differences between the background site and the three monitoring sites in Paterson. The difference between two geographical locations (i.e., higher concentrations in Paterson than in Chester) indicated the impact of local sources of VOC in Paterson, consistent with the VOC source information in Paterson documented by the NJDEP. As described in the Introduction, the NJDEP has identified many industrial sources of these species

in Paterson, such as emissions of toluene (10.5 tons/year), MEK (0.1 tons/year), xylenes (8.4 tons/year), ethylbenzene (1.6 tons/year), styrene (0.5 tons/year) and benzene (0.2 tons/year) from industrial facilities. In contrast, there are no identified industrial sources of air pollution near the Chester site [12].

Among the three sites in Paterson, ambient ethylbenzene, m,p-xylene, o-xylene and p-dichlorobenzene concentrations measured at the commercial site were significantly higher than at the industrial and mobiles sites (p<0.01). Specifically, benzene and styrene concentrations were significantly higher at the commercial site than at the industrial site (p<0.01) and marginally higher than at the mobile site (p = 0.04 and 0.01, respectively). These results indicated additional sources of these species at the commercial site. According to the NJDEP's EI database, industrial facilities in Paterson reported significant emissions of xylenes, ethylbenzene and styrene to ambient air. Acetylene, benzene, ethylbenzene, toluene, m,p-xylene and o-xylene can be emitted from gasoline-powered vehicles [29]. The commercial site was located in downtown, Paterson, close to busy local roads with high volume of traffic. No significant spatial differences were found for acetylene, dichloromethane, MEK and toluene in Paterson.

Seasonal variability. The seasonal differences were examined on the selected 10 VOCs. We found significant seasonal differences (p<0.05) for acetylene, benzene, dichloromethane, MEK, styrene, toluene and p-dichlorobenzene, and MEK, toluene and p-dichlorobenzene were selected for illustration (Figure 2). However, specific seasonal patterns were different by species. The winter concentrations of acetylene, benzene and toluene were higher than in other seasons; meanwhile, the summer p-dichlorobenzene concentrations were higher than in other seasons. Benzene is emitted from numerous industrial operations and mobile sources [30] and heating [31], which may explain the

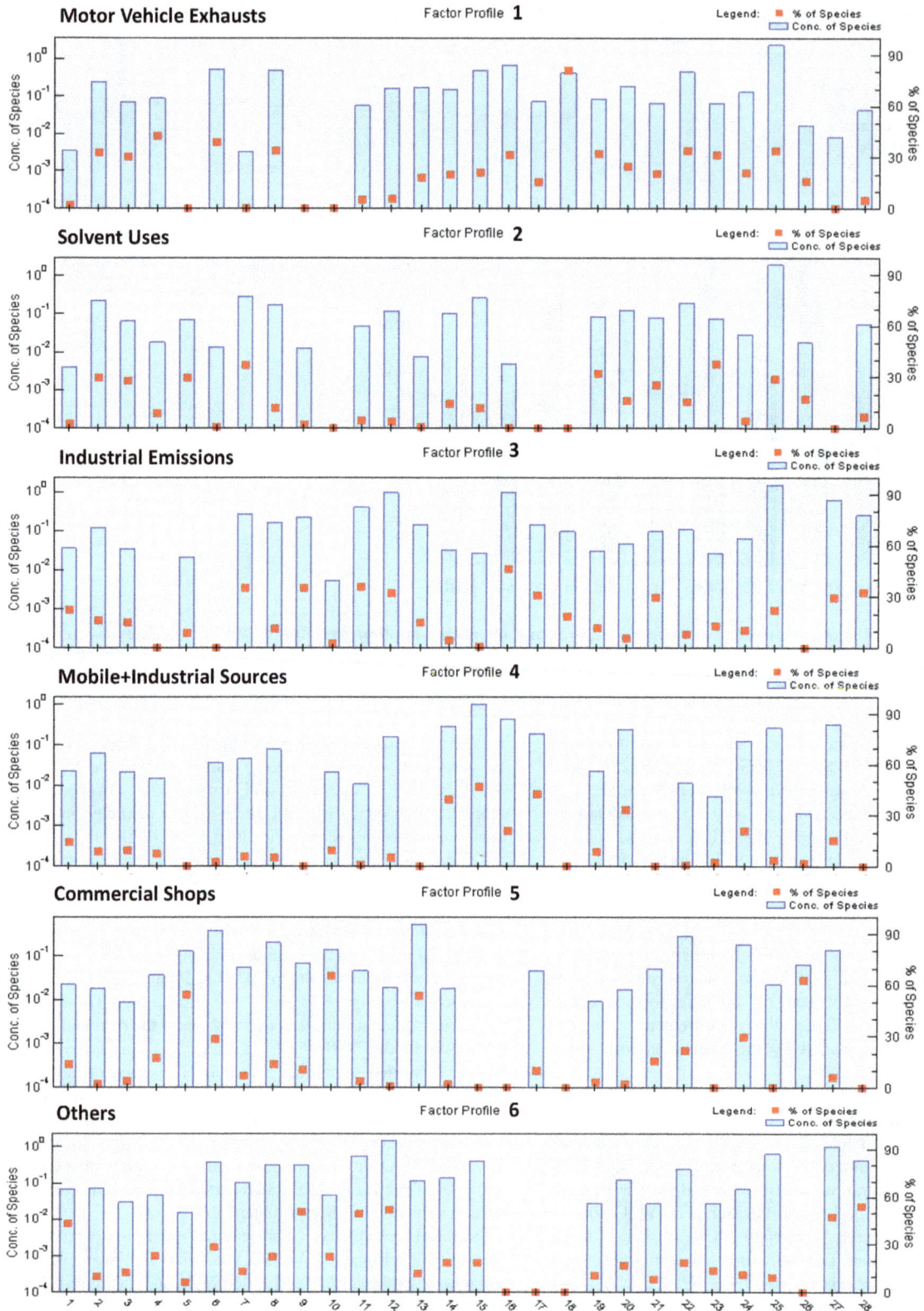

Figure 4. Factor profiles for ambient VOC data collected in 2005~2006 in Paterson, NJ. Abbreviation in the figure: 1: 1,1,1-Trichloroethane, 2: 1,2,4-Trimethylbenzene, 3: 1,3,5-Trimethylbenzene, 4: 1,3-Butadiene, 5: Acetonitrile, 6: Acetylene, 7: Acrolein, 8: Benzene, 9: Carbon Tetrachloride, 10: Chloroform, 11: Chloromethane, 12: Dichlorodifluoromethane, 13: Dichloromethane, 14: Ethylbenzene, 15: m,p-Xylene, 16: Methyl Ethyl Ketone, 17: Methyl Isobutyl Ketone, 18: Methyl tert-Butyl Ether, 19: n-Octane, 20: o-Xylene, 21: p-Dichlorobenzene, 22: Propylene, 23: Styrene, 24: Tetrachloroethylene, 25: Toluene, 26: Trichloroethylene, 27: Trichlorofluoromethane, 28: Trichlorotrifluoromethane.

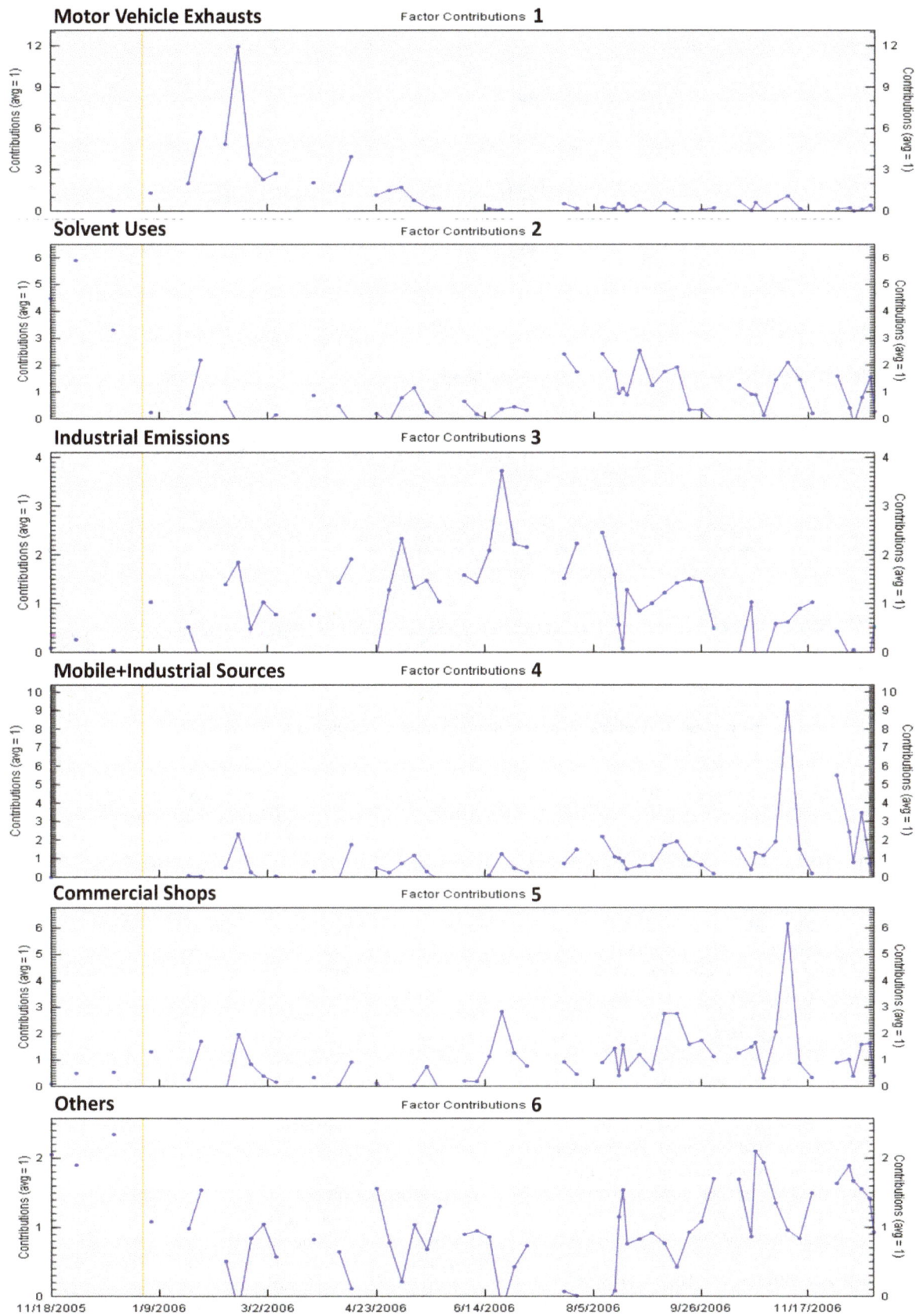

Figure 5. Factor contributions for the duration of the study (11/18/2005~12/19/2006) in Paterson, NJ.

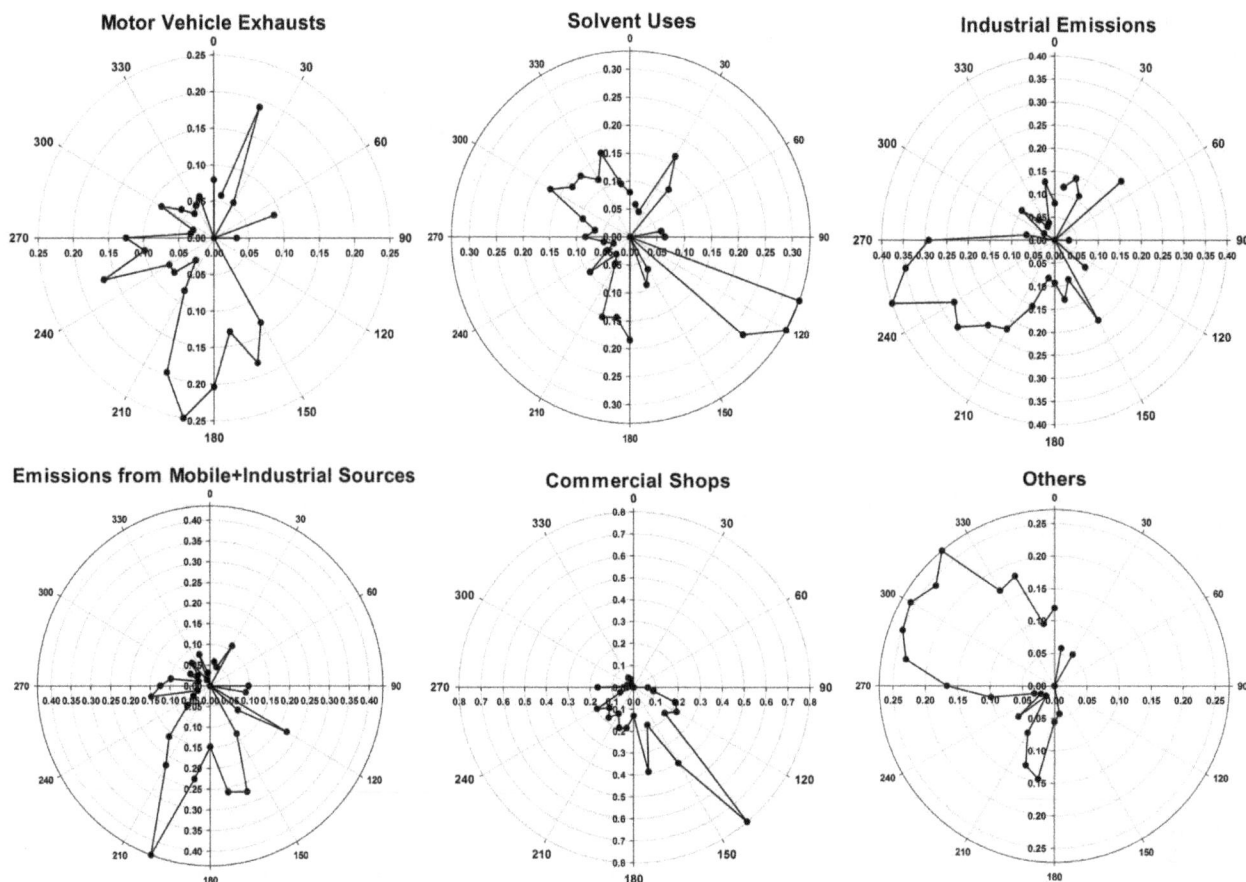

Figure 6. Hourly CPF plots for the highest 10% of the mass contribution from VOC sources in Paterson, NJ.

seasonal variation, i.e., higher emissions of benzene from combustion sources under low temperature in cold seasons. The higher winter concentrations were most likely due to winter heating, lower photochemical degradation and meteorological conditions (i.e., inversions and low mixing heights) in cold seasons [32,33,34]. The spring concentrations of dichloromethane, ethylbenzene, styrene, m,p-xylene and o-xylene were lower than in other seasons. On the other hand, MEK concentrations were the lowest in the fall and higher in warm seasons than in cold seasons. The winter MEK concentrations were also found to be high in the industrial and commercial areas. Main sources of MEK in ambient air are industrial sources (MEK is a common solvent); thus higher concentrations of MEK in warm seasons are expected due to evaporation. The high concentration in the winter is probably due to low photochemical decay rate and low mixing height by inversion, as stated above. In addition, the "large" increase of MEK (AVG±SD, 1.78 ± 2.03 µg/m^3) in the spring at the background site was driven by one high value (8.09 µg/m^3). If this suspected outlier (Grubbs' test for outliers, $p < 0.01$) was removed, the background MEK concentrations in the spring (1.30 ± 0.95 µg/m^3) were similar to those at the mobile site (1.38 ± 0.60 µg/m^3).

Weekday vs. weekend difference. The weekday vs. weekend difference was examined on the selected 10 VOCs. We found significant higher concentrations of dichloromethane, MEK and toluene on weekdays than those on weekends ($p < 0.05$). These VOCs are commonly used in industrial products. For example, large quantity of MEK and toluene was emitted to atmosphere

from industrial facilities located in Paterson area according to the NJDEP's EI database. In addition, dichloromethane is widely used as an industrial solvent/degreaser, paint stripper, aerosols and pesticides [30]. The ratios of weekday/weekend averages of the 10 VOC species at each sampling site are plotted in Figure 3. The bar charts above the dotted line (ratio of 1.0) mean that the average weekday concentration was higher than the weekend concentration. Most VOCs were measured higher on weekdays than on weekends at the three monitoring sites in Paterson, indicating greater industrial and commercial activities on weekdays. Elevated VOCs on weekdays suggested the impact of emissions from industrial facilities and commercial districts on VOC air pollution in Paterson. The analysis of weekday vs. weekend ratio confirmed the findings from the spatial analysis, i.e., significant impact was found from the emissions generated by traffic, commercial activities and the operation of industrial facilities located in Paterson.

Source Apportionment

We identified 6 VOC source profiles from our community-scale ambient VOC monitoring data. The resolved factor profiles and the source contributions are presented in Figures 4 and 5, respectively. In Figure 4, each source profile was displayed by a log-scaled mass concentration (µg/m^3) on the primary y-axis and %species on the secondary y-axis, respectively. The left axis represents a mass concentration apportioned by each species, and the right axis indicates the contribution of each species to the source profile. In Figure 5, the source contributions indicate

VOC Source Contributions in Paterson, NJ

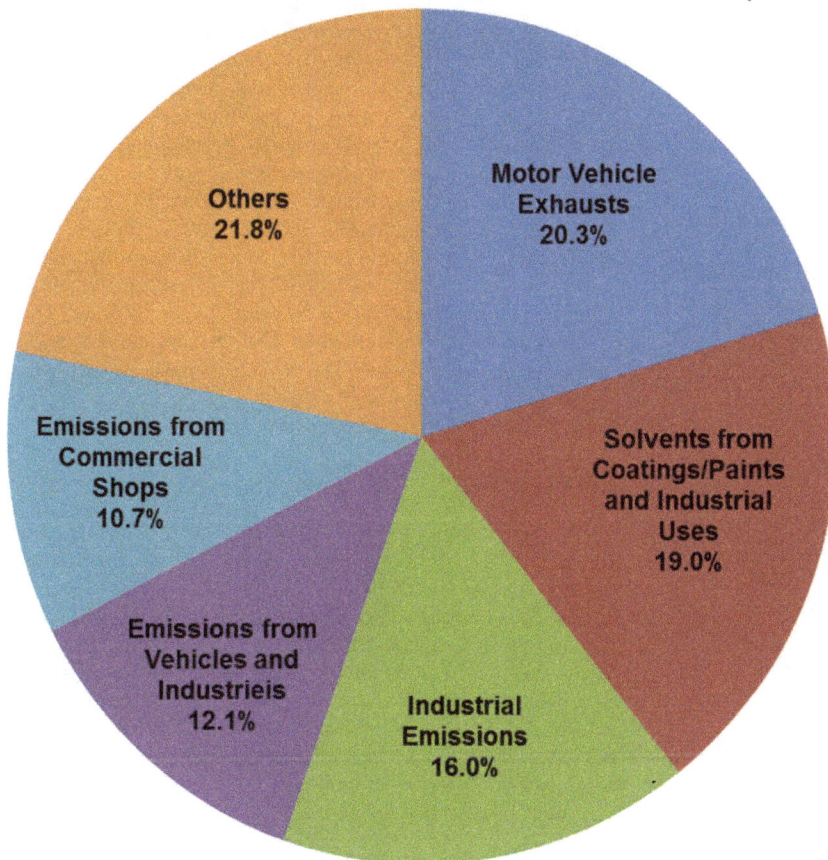

Figure 7. The VOC source contributions to ambient air in Paterson, NJ.

temporal changes in contribution influenced by meteorological factors and emission intensities. Hourly CPF for the highest 10% of the VOC mass contribution is plotted with wind direction data in Figure 6. Higher calculated CPF represents stronger impact from the point source in a given time and direction, suggesting potential locations for the PMF-identified sources. The overall contribution of each factor to total VOC mass is provided in Figure 7. The quality of the PMF solutions was evaluated by comparing the reconstructed VOC mass contributions (the sum of the contributions from the PMF resolved factors) with the measured VOC mass concentrations. The results showed a good agreement (slope = 0.94 and $r^2 = 0.95$) between the two VOC concentrations, indicating that the resolved factors well reproduced measured values and accounted for most of the variations in the measured VOC mass concentrations.

Based on the source profiles, the temporal patterns, i.e., higher VOCs concentrations in the winter and/or higher on weekdays, and the source information from the NJDEP's EI database, we identified possible sources for each factor profile. The most significant factor in Paterson was motor vehicle exhausts (20%), driven by MTBE, acetylene and benzene in the winter. Similar source profiles for ambient VOCs were also reported in previous studies [13,14]. The dominant source direction was south, suggested by the hourly CPF plot in Figure 6. Heavy traffics were concentrated on the highways located in the south of Paterson. The second most significant factor was solvent emissions from coatings/paints and industrial uses (19%). This source profile was

characterized by higher contributions of toluene, styrene and n-octane in the fall and on weekdays. These results are consistent with the source emission data. As well documented, toluene is widely used for solvents in coatings/paints and industrial processes [30]. Annual emissions of toluene (10.5 tons) and styrene (0.45 ton) were reported to the NJDEP for the year of 2006 in Paterson. The hourly CPF plot also indicated the source direction primarily from the northwest and southeast. The heavily industrialized area known as "Bunker Hill" was located in the north of Paterson city. In addition, many other industrial facilities were clustered in the south or southeast of Paterson. The third significant factor in Paterson was industrial emissions (16%), specifically dominated by MEK and MIBK in the source profile. According to the NJDEP's EI database, annual MEK and MIBK emissions were 9.1 tons and 3.2 tons in the air from the industrial facilities located in Paterson. The hourly CPF plot indicated that those MEK and MIBK emissions were dominant by the industrial facilities located in the southwest and northeast of Paterson. A hot stamping/metal foil manufacturing facility was located in the south of Paterson city, emitting 5.8 tons of MEK and 3.2 tons of MIBK per year. A coating and laminating facility was located in the north of Paterson, emitting MEK with a rate of 3.3 tons/year. The fourth contributing factor was the combined emissions from industrial and vehicle sources (12%), represented by stronger contributions from ethylbenzene, m,p-xylene and o-xylene in the profile. These results are consistent with the source emission data. According to the NJDEP's EI database in 2006, 8.4 tons xylenes and 1.6 tons ethylbenzene were emitted to the air annually in Paterson. Motor vehicle emissions also contain significant amount of these species [30]. The hourly CPF plot showed that the sources were significantly originated from the south of Paterson, where heavy trafficked highways and industrial facilities were located. A chemical plant was located in the southeast of Paterson, emitting significant amounts of xylenes and ethylbenzene (annually 8.2 tons and 1.6 tons, respectively) to ambient air in Paterson. The fifth contributing factor was emissions from commercial activities in downtown of Paterson. Included were dry cleaners, nail salons and printing press (11%). This source profile was apparently contributed by trichloroethylene, tetrachloroethylene and dichloromethane in heating seasons and on weekdays. Trichloroethylene and tetrachloroethylene are solvents widely used for degreasing in dry cleaning [30]. The hourly CPF plot supports the VOC emissions related to commercial activities in downtown of Paterson substantially located in the south of Paterson. The sixth contributing factor was noticeable by higher contributions from carbon tetrachloride and chloromethane seen in Figure 4 as well as relatively constant contribution during entire study period observed in Figure 5. Carbon tetrachloride (0.62 ± 0.17 $\mu g/m^3$) and chloromethane (1.15 ± 0.20 $\mu g/m^3$) in Paterson were similar to those at the background site in Chester (0.58 ± 0.18 and 1.14 ± 0.18 $\mu g/m^3$, respectively). They are very stable in the atmosphere and there are no significant local emission sources [30]. In particular, carbon tetrachloride has been phased out in consumer products since 1970. Thus, there are no significant anthropogenic sources for these two species in Paterson. We consider this factor profile indicated aged (background) VOCs in the air and collectively named as "Others" in the pie chart (22%).

Assuming each source contributes equally to the mixed source profiles (i.e., factor profiles 2 and 4), we can classify those contributing sources into three broad categories: industrial (32%), mobile (26%) and commercial (20%) in ambient air of Paterson. The remaining 22% represents aged VOCs in the atmosphere or un-identified potential VOC sources (e.g., evaporative emissions and liquid gasoline) in urban air. The results from the source identification showed that the impact from the land use type, i.e., commercial, industrial and mobile sources, on Paterson VOC air pollution is similar. Thus, we would not expect to see significant spatial differences for many VOCs in Paterson, as observed in the study (see the section of Spatial Variability). This study was the first effort to conduct source apportionment for the one year measurements of relatively stable VOCs in ambient air. The study demonstrated that the 24-hour averaged VOC data were successfully used for the PMF source apportionment. This is because the sources identified in the study were consistent with those from previous source apportionment studies in Northeast areas [15,35,36]. They were also consistent with the results derived from the PM$_{10}$ [17] and PAHs [16] measured concurrently. Lin et al. [16] indicated that diesel emissions, combustion of oil, coal and fossil fuels were dominant PAH sources in Paterson. Yu et al. [17] reported seven contributing sources to ambient PM$_{10}$ in downtown of Paterson: sulfate aerosols (26%), vehicle emissions (16%), residual oil combustions (16%), industrial emissions (12%), airborne soils (12%), road dusts (13%) and road salts (5%).

Limitations and Recommendation

There are some limitations of the study. First, the PMF model could not identify all of the contributing sources to local VOCs because some signature VOCs (e.g., ethane, pentane, propane, isoprene, etc.) of potential sources, such as evaporated gasoline vapor, liquid gasoline, natural gas and biogenic emissions, were not measured in the study. Second, VOC sampling in Paterson were conducted on the roof-top of the 2–4 story buildings, which may tend to underestimate the ground-level VOC emissions including traffic sources. However, the VOC concentrations measured on the rooftop (i.e., 10~13 meters above the ground) were not significantly lower than the measurements on the ground-level (approximately 2%). Thus, the underestimated VOC concentrations do not significantly affect the conclusions drawn in the study. A risk assessment study is recommended for further evaluation of potential health risks based on the measurements of the community-scale monitoring study.

Conclusions

This study monitored 60 VOCs in Paterson, NJ, an urban community with mixed sources of air pollution, and characterized spatial/temporal variations of 10 VOCs that were detected over 75% during the course of monitoring period, toxic and/or having known sources in the study area. The study demonstrated that monitoring VOCs at community scale was an effective approach to capture spatial/temporal variations of VOCs in an urban area with mixed VOC sources. The comparisons between sites and between weekday and weekend indicated the significant impact from anthropogenic VOC emissions on ambient air pollution in Paterson. These observations are consistent with the NJDEP EI database. The PMF source apportionment results confirmed the contributions from various sources of VOCs in Paterson, including industrial (32%), mobile (26%) and commercial (20%) emission sources. The estimated source contributions from our study are more accurate than those obtained from the summer measurements only. The findings from this study demonstrated the importance of the community-oriented air toxic monitoring approach because it was able to 1) capture spatial variation in VOCs, 2) identify sources of concern (contribution from different emission sources), 3) better assess community exposure to ambient VOC air pollution, and 4) develop effective controlling strategies for urban communities with mixed air pollution sources.

Acknowledgments

The authors wish to thank Dr. Linda Bonanno and other scientists from the NJDEP for their support for the project. The authors also thank Drs. Qingyu Meng and Kathy Black, and Ms. Martha Hernandez for their help with field sampling.

Author Contributions

Conceived and designed the experiments: CHY ZF. Performed the experiments: XZ. Analyzed the data: CHY. Wrote the paper: CHY. Revised the manuscript: ZF XZ.

References

1. Delfino RJ, Gong H Jr, Linn WS, Pellizzari ED, Hu Y (2003) Asthma symptoms in Hispanic children and daily ambient exposures to toxic and criteria air pollutants. Environmental Health Perspectives 111: 647.

2. Rumchev K, Spickett J, Bulsara M, Phillips M, Stick S (2004) Association of domestic exposure to volatile organic compounds with asthma in young children. Thorax 59: 746–751.

3. Wichmann FA, Müller A, Busi LE, Cianni N, Massolo L, et al. (2009) Increased asthma and respiratory symptoms in children exposed to petrochemical pollution. Journal of Allergy and Clinical Immunology 123: 632–638.

4. Jia C, Batterman S, Godwin C (2008) VOCs in industrial, urban and suburban neighborhoods–Part 2: Factors affecting indoor and outdoor concentrations. Atmospheric Environment 42: 2101–2116.

5. Miller L, Lemke LD, Xu X, Molaroni SM, You H, et al. (2010) Intra-urban correlation and spatial variability of air toxics across an international airshed in Detroit, Michigan (USA) and Windsor, Ontario (Canada). Atmospheric Environment 44: 1162–1174.

6. Mohamed MF, Kang D, Aneja VP (2002) Volatile organic compounds in some urban locations in United States. Chemosphere 47: 863–882.

7. Pankow JF, Luo W, Bender DA, Isabelle LM, Hollingsworth JS, et al. (2003) Concentrations and co-occurrence correlations of 88 volatile organic compounds (VOCs) in the ambient air of 13 semi-rural to urban locations in the United States. Atmospheric Environment 37: 5023–5046.

8. Smith LA, Stock TH, Chung KC, Mukerjee S, Liao XL, et al. (2007) Spatial analysis of volatile organic compounds from a community-based air toxics monitoring network in Deer Park, Texas, USA. Environmental Monitoring and Assessment 128: 369–379.

9. Touma JS, Cox WM, Tikvart JA (2006) Spatial and temporal variability of ambient air toxics data. Journal of the Air & Waste Management Association 56: 1716–1725.

10. Wu X, Fan ZT, Zhu X, Jung KH, Ohman-Strickland P, et al. (2012) Exposures to volatile organic compounds (VOCs) and associated health risks of socio-economically disadvantaged population in a "Hot Spot" in Camden, New Jersey. Atmospheric Environment 57: 72–79.

11. Zhu X, Fan Z, Wu X, Meng Q, Wang S, et al. (2008) Spatial variation of volatile organic compounds in a "Hot Spot" for air pollution. Atmospheric Environment 42: 7329–7338.

12. New Jersey Department of Environmental Protection (2010) Urban Community Air Toxics Monitoring Project, Paterson City, NJ (UCAMPP). Trenton, NJ: NJDEP.

13. Brown SG, Frankel A, Hafner HR (2007) Source apportionment of VOCs in the Los Angeles area using positive matrix factorization. Atmospheric Environment 41: 227–237.

14. Buzcu B, Fraser MP (2006) Source identification and apportionment of volatile organic compounds in Houston, TX. Atmospheric Environment 40: 2385–2400.

15. Choi Y-J, Ehrman SH (2004) Investigation of sources of volatile organic carbon in the Baltimore area using highly time-resolved measurements. Atmospheric Environment 38: 775–791.

16. Lin L, Fan Z-HT, Zhu X, Huang L, Korn L, et al. (2011) Characterization of Atmospheric Polycyclic Aromatic Hydrocarbons in a Mixed-use Urban Community in Paterson, NJ: Concentrations and Sources. In Press, Journal of the Air & Waste Management Association 61: 631–639.

17. Yu CH, Fan ZH, Meng Q, Zhu X, Korn L, et al. (2011) Spatial/Temporal Variations of Elemental Carbon, Organic Carbon, and Trace Elements in PM$_{10}$ and the Impact of Land-Use Patterns on Community Air Pollution in Paterson, NJ. Journal of the Air & Waste Management Association 61: 673–688.

18. US Census Bureau (2012) Census data for Paterson city. Available: http://factfinder.census.gov/servlet/GCTTable?_bm = y&-geo_id = 04000US34&-_box_head_nbr = GCT-PH1-R&-ds_name = DEC_2000_SF1_U&-_lang = en&-format = ST-7S&-_sse = on. Accessed 22 June 2012.

19. U.S. Environmental Protection Agency (1999) Determination of volatile organic compounds (VOCs) in air collected in specially-prepared canisters and analyzed by gas chromatography/mass spectrometry (GC/MS). Washington DC, USA: US EPA.

20. Swift J, Howell M, Tedder D, Merrill R (2006) Collection and Analysis of Acrolein using Compendium Method TO-15. Presented at National Environmental Monitoring Conference (NEMC), Washington, DC, USA.

21. Paatero P (1997) Least squares formulation of robust non-negative factor analysis. Chemometrics and Intelligent Laboratory Systems 37: 23–35.

22. U.S. Environmental Protection Agency (2008) EPA Positive Matrix Factorization (PMF) 3.0 Fundamentals and User Guide. Washington DC, USA: US EPA.

23. Polissar A, Paatero P, Hopke P, Malm W, Sisler J (1998) Atmospheric aerosol over Alaska 2. Elemental composition and sources. Journal of Geophysical Research 103: 19045–19057.

24. Paatero P, Hopke P (2003) Discarding or downweighting high-noise variables in factor analytic models. Analytica Chimica Acta 490: 277–289.

25. Kim E, Hopke PK, Pinto JP, Wilson WE (2005) Spatial variability of fine particle mass, components, and source contributions during the regional air pollution study in St. Louis. Environ Sci Technol 39: 4172–4179.

26. Kim E, Hopke P (2007) Comparison between sample-species specific uncertainties and estimated uncertainties for the source apportionment of the speciation trends network data. Atmospheric Environment 41: 567–575.

27. Kim E, Hopke P, Paatero P, Edgerton E (2003) Incorporation of parametric factors into multilinear receptor model studies of Atlanta aerosol. Atmospheric Environment 37: 5009–5021.

28. Kim E, Hopke PK, Edgerton ES (2003) Source identification of Atlanta aerosol by positive matrix factorization. Journal of the Air & Waste Management Association 53: 731–739.

29. U.S. Environmental Protection Agency (2011), SPECIATE data browser. Available: http://cfpub.epa.gov/si/speciate/. Accessed 22 June 2012.

30. U.S. Environmental Protection Agency (2009) National-scale air toxics assessment (NATA). Available: http://www.epa.gov/nata2002/. Acccessed 22 June 2012.

31. Gustafson P, Barregard L, Strandberg B, Sällsten G (2007) The impact of domestic wood burning on personal, indoor and outdoor levels of 1, 3-butadiene, benzene, formaldehyde and acetaldehyde. J Environ Monit 9: 23–32.

32. Cheng L, Fu L, Angle R, Sandhu H (1997) Seasonal variations of volatile organic compounds in Edmonton, Alberta. Atmospheric Environment 31: 239–246.

33. Na K, Kim YP (2001) Seasonal characteristics of ambient volatile organic compounds in Seoul, Korea. Atmospheric Environment 35: 2603–2614.

34. Parrish D (2006) Critical evaluation of US on-road vehicle emission inventories. Atmospheric Environment 40: 2288–2300.

35. Fujita E, Lu Z (1998) Analysis of data from the 1995 NARSTO Northeast study, vol. III: Chemical mass balance receptor modeling. Reno, NV: Desert Research Institute.

36. Watson JG, Chow JC, Fujita EM (2001) Review of volatile organic compound source apportionment by chemical mass balance. Atmospheric Environment 35: 1567–1584.

24

Asthma Prevalence Associated with Geographical Latitude and Regional Insolation in the United States of America and Australia

Goran Krstić*

Fraser Health, Environmental Health Services, New Westminster, Canada

Abstract

Background: It has been proposed that vitamin D deficiency may be responsible for an increase in the prevalence of allergic diseases and asthma worldwide. Human ability to generate physiologically required quantities of vitamin D through sun exposure is decreasing with increasing geographical latitude.

Objectives: Considering that vitamin D deficiency is usually due to lack of outdoor sun exposure, this study is designed to test the hypothesis that a higher prevalence of asthma should be expected at high relative to low geographical latitudes.

Methods: Linear regression analyses are performed on asthma prevalence in the U.S. adult population vs. geographical latitude, insolation, air temperature, and air pollution ($PM_{2.5}$) for 97 major metropolitan/micropolitan statistical areas of the continental United States of America and on general population asthma prevalence vs. geographical latitude in eight metropolitan areas of Australia.

Results: A $10°$ change in geographical latitude from southern to northern regions of the Eastern Seaboard is associated with a 2% increase in adult asthma prevalence ($p < 0.001$). Total insolation in winter months is almost as strong as latitude in its ability to explain the observed spatial variation in the prevalence of asthma ($r^2 = 0.43$; $p < 0.001$). Similar results are obtained using the Australian data ($r^2 = 0.73$; $p < 0.01$), suggesting a consistent association between the latitude/insolation and asthma prevalence worldwide.

Conclusions: The results of this study suggest that, as a known modulator of the immune response closely linked with the geographical latitude and erythemal UV irradiation, vitamin D may play an important role in the development/exacerbation of asthma.

Editor: Raymond J. Pickles, University of North Carolina at Chapel Hill, United States of America

Funding: The author has no support or funding to report.

Competing Interests: The author has declared that no competing interests exist.

* E-mail: Goran.Krstic@fraserhealth.ca

Introduction

Both positive [1,2] and negative associations [3,4] between the prevalence of asthma/allergies and geographical latitude have been reported in the published literature. This is an ecological study designed to test the proposed hypothesis that the prevalence of asthma increases with increasing latitude due to a decreasing intensity of solar irradiation which effectively reduces the individual's cutaneous generation of vitamin D.

Maternal vitamin D deficiency/insufficiency is associated with an increased probability of developing asthma and allergy-related symptoms in early life of their offspring. Camargo, Devereux and colleagues concluded that specifically in the northeastern United States of America, an increased maternal intake of vitamin D from diet or supplements during pregnancy may decrease the risk of wheeze symptoms in early childhood [5,6]. Low levels of serum vitamin D in adults have been associated with impaired lung function, increased airway hyperresponsiveness, and reduced glucocorticoid response in asthma [7]. Litonjua and Weiss proposed that vitamin D deficiency may be responsible for an increase in the prevalence of allergic diseases and asthma worldwide, as more time is spent indoors with less exposure to sunlight, leading to a decreased cutaneous vitamin D production [8]. Exposure to solar ultraviolet radiation within a wavelength band of 290–315 nm (UV-B) and production of vitamin D in the skin is the primary source of vitamin D for many people [9], particularly for those who do not receive adequate vitamin D doses through diet.

Higher rates of emergency department visits for acute allergic reactions have been observed in northeastern when compared to southern regions of the United States [10], suggesting an association with latitudinal difference in insolation. Kimlin and colleagues observed that the available levels of erythemal or vitamin D producing UV-B irradiation decreases dramatically as the latitude increases [11], particularly during the four cooler months (i.e., November to February). The efficiency of vitamin D production in the exposed skin depends on the dose of solar UV-B radiation, which can be curtailed by clothing, excess body fat, sunscreen, and the skin pigment melanin [12]. Higher prevalence of vitamin D deficiency has been found among inner-city African American youth with asthma when compared to non-asthmatic control subjects [13].

An intense annual insolation in the lower latitudes of the southern U.S. regions provides greater potential for the cutaneous vitamin D production in exposed individuals. It is estimated that over 95% of the variability in average daily UV dosages can be explained by the latitude and altitude, where the effect of the latitude on the UV irradiation on the Earth's surface is much more significant than the altitude. The longitude is not statistically significant in terms of its ability to predict the intensity of the UV irradiance [14]. Considering that vitamin D is suspected to have a role in asthma development [8,15] and that vitamin D deficiency is usually due to lack of outdoor sun exposure [11,12], it is hypothesized in this study that a higher prevalence of asthma should be expected at high relative to low geographical latitudes.

Materials and Methods

Asthma – U.S. data

The data on asthma prevalence in the U.S. adult population (i.e., age 18 and over) is obtained from the U.S. Department of Health and Human Services (DHHS), Centers for Disease Control and Prevention (CDC) Behavioral Risk Factor Surveillance System [16]. The mean annual asthma prevalence in U.S. "*adults who have been told they currently have asthma*" for the period from 2006 to 2008 by Metropolitan/Micropolitan Statistical Area (MMSA) are included in this study. The obtained data relates to simple prevalence by MMSA and contains no patient-specific information. Therefore, there was no need to request an ethics committee approval or a written consent from the patients. All asthma prevalence data used in this study are freely available to download from online resources. The citation recommended by the CDC is used to reference the source of asthma prevalence information for this research paper.

A continuity of geographical latitude and longitude is considered as important for reducing the influence of statistical outliers and implementing a meaningful regression analysis. Hence, the studied U.S. geographical area includes all continental states with the exception of Alaska. Asthma prevalence data are matched with the data on Air Quality Trends by Core Based Statistical Areas (CBSA), obtained from the U.S. Environmental Protection Agency (EPA) [17]. The resulting 97 matched and collated U.S. wide MMSAs/CBSAs are used in the statistical analyses presented in this paper.

Asthma – Australian data

The Australian general population asthma prevalence data by Local Government Area (LGA) for the period from 2004 to 2005 are obtained from the publically available online resource of the Australian Public Health Information Development Unit [18]. The studied metropolitan areas of Australia include: Sydney – New South Wales (NSW), Melbourne – Victoria (Vic), Brisbane – Queensland (Qld), Adelaide – South Australia (SA), Perth – Western Australia (WA), Hobart – Tasmania (Tas), Darwin – Northern Territory (NT), and Canberra – Australian Capital Territory (ACT).

Geographical coordinates data

Geographical latitudes and longitudes for the main cities in the MMSAs are obtained using Microsoft Research (MSR) Maps [19]. The geographical information and maps made available online by the MSR are supplied through their partnership with the U.S.

Geological Survey (USGS). The geographical coordinates for Australian metropolitan areas are obtained from the Guide to Australia at Charles Sturt University online resource, which is based on the 1996 data from the Australian Bureau of Statistics [20].

Air pollution and meteorological data

The air pollution data for fine airborne particulate matter, aerodynamic diameter of less than 2.5 μm (i.e., $PM_{2.5}$), are obtained from the U.S. EPA Air Trends online information resource [17]. Mean annual $PM_{2.5}$ concentrations expressed in $μg/m^3$ for the period from 1999 to 2008 by matched CBSAs/MMSAs are included in this study.

The mean annual insolation on horizontal surface and air temperature, expressed in $kWh/m^2/day$ and °C, respectively, for the MMSAs/CBSAs in the period from 1983 to 2005 are obtained from the National Aeronautics and Space Administration (NASA) Atmospheric Science Data Center [21]. Central latitudes and longitudes of cities/statistical areas, as presented by the USGS, are used to obtain the data on annual mean insolation and air temperature at specific geographical coordinates and regional elevation.

Statistical analysis

Descriptive statistics and scatter plot analysis revealed no evidence of non-normality in the distribution of asthma prevalence, geographical coordinates, insolation, air temperature, and air pollution data, allowing the use of linear regression analysis models to evaluate the relationships between the studied variables.

Linear regression and correlation analyses are implemented to evaluate the strength, direction, and statistical significance of regression/correlation coefficients. The selected independent variables (i.e., regional horizontal surface insolation, latitude, air temperature, and air pollution) are compared in terms of their ability to explain the observed variation in the dependent/response variable, the prevalence of asthma in the U.S. adult population, by calculating the coefficients of determination (r^2). Linear regression equations are developed and tested for their ability to predict the prevalence of asthma in 97 major metropolitan/micropolitan areas of the continental U.S. and a subset of 39 areas from the Eastern Seaboard in response to meteorology and air pollution at different latitudes. The same approach is applied to test the correlation between the general population asthma prevalence and the geographical latitude in eight metropolitan areas of Australia.

Results

The prevalence of asthma in the U.S. adult population, as presented in **Table 1** and **Figure 1**, is associated with geographical latitude ($r^2 = 0.22$; $p<0.001$), annual mean insolation on horizontal surface ($r^2 = 0.15$; $p<0.001$), and annual regional air temperature ($r^2 = 0.17$; $p<0.001$). The association of asthma prevalence with the annual mean air pollution as $PM_{2.5}$ is very weak and not statistically significant ($r^2 = 0.002$; $p = 0.66$). Although the annual air temperature appears to be a marginally better predictor of asthma prevalence than the annual mean insolation in the studied population, both insolation ($r^2 = 0.48$) and air temperature ($r^2 = 0.84$) are correlated with the geographical latitude (**Table 1**).

The best predictor of asthma prevalence among the studied variables is the geographical latitude with an ability to explain up to 22% of the variation in the prevalence of asthma in the continental U.S. adult population. Based on the linear regression

Table 1. Linear regression estimates of asthma prevalence in US adult population associated with latitude, annual mean insolation, air temperature and air pollution ($PM_{2.5}$).

Linear Regression Coefficient/Variable[‡]	Model 1 - Asthma vs. Latitude	Model 2 - Asthma vs. Insolation	Model 3 - Asthma vs. Temperature	Model 4 - Asthma vs. $PM_{2.5}$	Model 5 - Insolation vs. Latitude	Model 6 - Temperature vs. Latitude
Y-axis Intercept	3.74±0.90***	12.22±0.96***	9.77±0.34***	8.13±0.52***	7.08±0.31***	45.74±1.50***
Latitude (°)	0.12±0.02***	—	—	—	−0.07±0.01***	−0.87±0.04***
Annual Mean Insolation (kWh/m^2/day)	—	−0.92±0.23***	—	—	—	—
Annual Mean Air Temperature (°C)	—	—	−0.11±0.03***	—	—	—
Air Pollution ($PM_{2.5}$) ($\mu g/m^3$)	—	—	—	0.02±0.04	—	—
Number of Areas	97	97	97	97	97	97
Coefficient of Determination (r^2)	**0.22**	**0.15**	**0.17**	**0.002**	**0.48**	**0.84**

***$p < 0.001$.
[‡]Plus-minus values are linear regression coefficients and standard errors (i.e., ±SE).

analysis, a 10° change in the geographical latitude from southern to northern U.S. regions is associated with a 1.2% increase in the prevalence of adult asthma. The highest prevalence of adult asthma is observed in Detroit, MI at 10.97% (latitude: 42.35° North) and the lowest in Miami, FL at 5.03% (latitude: 25.81° North).

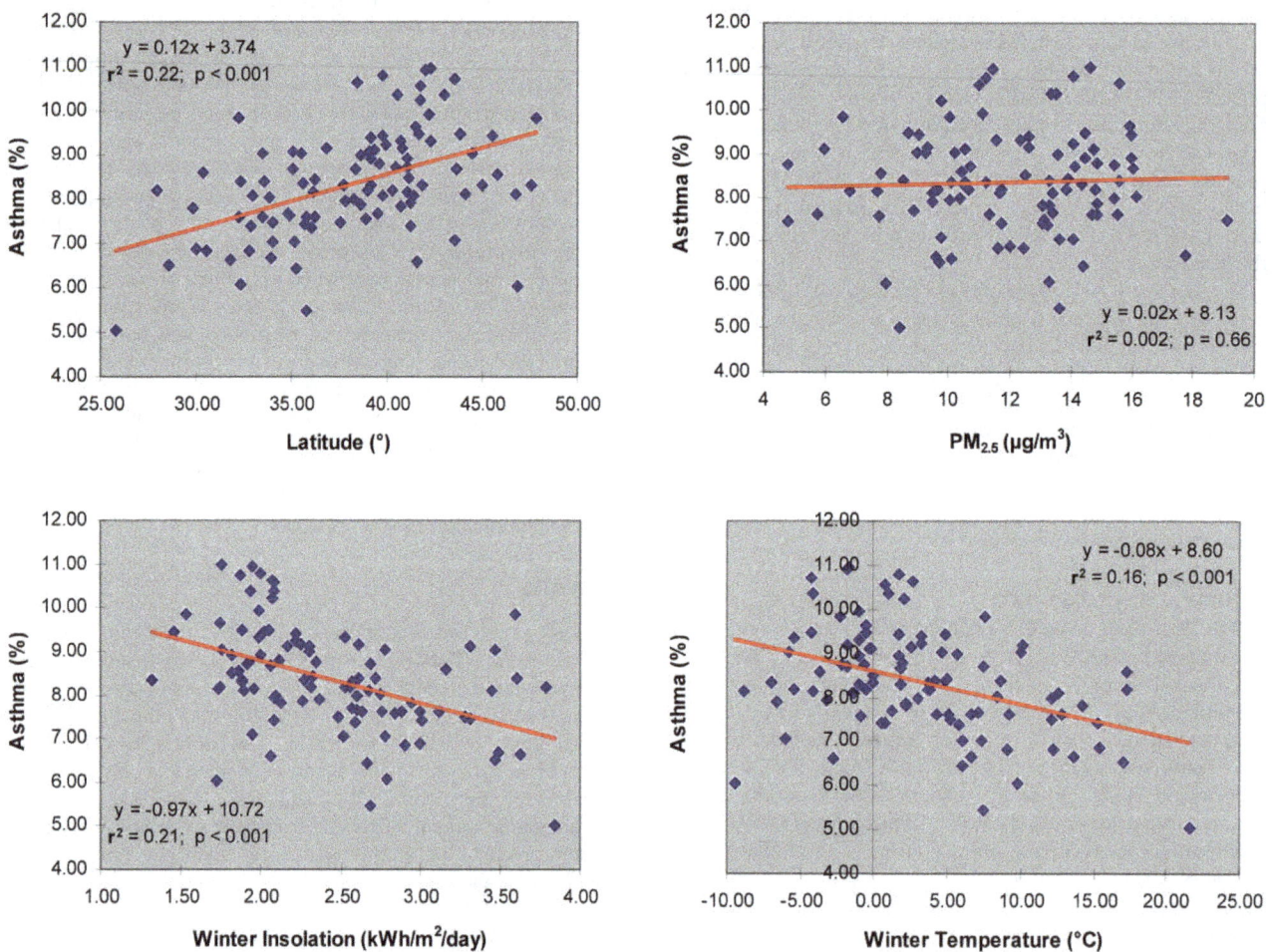

Figure 1. Asthma prevalence vs. latitude, air pollution ($PM_{2.5}$), winter insolation, and winter temperature in adult population of 97 major metropolitan/micropolitan areas of continental U.S.

Bothe the regional insolation on horizontal surface ($r^2 = 0.21$; $p<0.001$) and air temperature ($r^2 = 0.16$; $p<0.001$) in winter months (i.e., November to February) are statistically significant predictors of asthma prevalence (**Table 2, Figure 1**). However, regional insolation during winter appears to be almost as effective as latitude in terms of its ability to predict the prevalence of asthma in the U.S. adult population. When compared to the results of regression analyses using annual mean values, a stronger association between latitude and insolation ($r^2 = 0.80$; $p<0.001$) is observed for winter months, which is in agreement with the findings of Kimlin and colleagues [11].

The correlation matrix in **Table 3** shows that asthma prevalence is best explained by the variation in geographical latitude ($r = 0.47$) and winter insolation ($r = -0.46$), closely followed by annual air temperature ($r = -0.42$). Both the geographical longitude ($r = 0.15$) and air pollution ($r = 0.04$) showed weak and not statistically significant correlation with the prevalence of asthma in U.S. adult population. Latitude is best correlated with annual air temperature ($r = -0.92$), closely followed by winter insolation ($r = -0.90$).

The regression analysis of a subset of data covering 39 major metropolitan/micropolitan areas in the U.S. regions of the Eastern Seaboard (i.e., New England, Middle Atlantic, and South Atlantic) shows a very strong association between asthma prevalence and winter insolation ($r^2 = 0.43$; $p<0.001$). A $10°$ change in the geographical latitude from southern to northern regions of the Eastern Seaboard is associated with a 2% change in adult asthma prevalence (**Figure 2**). Considering that asthma prevalence estimates are more reliable for the U.S. states and regions with larger population size, having narrower 95% confidence intervals [16], and that the Eastern Seaboard is home to most highly populated U.S. cities, a regression analysis based on the data from this region is expected to be more reliable when compared to the data for all continental U.S. regions.

Multiple regression analyses are performed on asthma prevalence vs. insolation, air temperature, and air pollution in an attempt to identify independent effects of the studied predictors (**Table 4**). Air temperature and insolation are the only statistically significant predictors of asthma in the continental U.S., where the air temperature appears to be the best predictor based on total annual data (annual mean air temperature: $p<0.01$; annual mean insolation: $p = 0.17$) and the insolation when only winter data is applied (winter mean insolation: $p<0.05$; winter mean air temperature: $p = 0.31$). Air pollution is not statistically significant predictor of asthma in any of the models.

The observed correlation between asthma prevalence and absolute values of geographical latitude for eight Australian metropolitan regions is positive, statistically significant, and in agreement with the results obtained using the U.S. adult asthma prevalence data (**Figure 3**). The results show that up to 73% of the variation in asthma prevalence in the Australian general population could be explained by the variation in the geographical latitude ($r^2 = 0.73$; $p<0.01$). A $10°$ change in the geographical latitude from the North to the South is associated with approximately the same increase in asthma prevalence as presented in **Figure 2** for the U.S. Eastern Seaboard region from the South to the North (i.e., $\sim 2\%$). It is important to observe that latitudes in the southern hemisphere have negative values and that, although the relationship between the absolute latitude and asthma is positive, the correlation coefficient is mathematically negative.

Discussion

There are pros as well as cons associated with study designs in epidemiological research. An ecological study design, as presented in this and similar papers, has some known limitations in terms of its ability to provide reliable inferences about the population characteristics at individual level. As discussed by Robinson in 1950, ecological correlations cannot be validly used as substitutes for individual correlations [22]. Using the same principle one could argue that the opposite also applies, where one cannot make valid inferences about large populations and their possible interactions by focusing only on the individuals from those populations. Robinson indicated that the purpose of his paper is to prevent "*the future computation of meaningless correlations*". However, despite lacking the ability to deal with health risk factors at a small-scale individual level, well designed ecological studies provide meaningful correlations and inferences that are useful when dealing with national and international public health issues and health risk factors at a large-scale population level. In their revisiting of Robinson's paper, Subramanian et al. [23] concluded that "*… perils are posed by not only ecological fallacy but also individualistic fallacy*".

Although a similar study design was applied by Staples and colleagues [3], this is the first comprehensive study that includes the populations of the continental United States and Australia, covering over 100 metropolitan areas within a substantial latitudinal range from both the northern and the southern hemispheres. It is interesting that Staples and colleagues observed a negative correlation between the prevalence of asthma and latitude, which is in contrast to the results of the study on the basis of more recent data for both the U.S. and Australia as presented in

Table 2. Linear regression estimates of asthma prevalence in US adult population associated with insolation and air temperature in winter months (November to February).

Linear Regression Coefficient/ Variable‡	Model 7 - Asthma vs. Insolation	Model 8 - Asthma vs. Temperature	Model 9 - Insolation vs. Latitude	Model 10 - Temperature vs. Latitude
Y-axis Intercept	10.72±0.48***	8.60±0.13***	6.64±0.21***	48.68±2.61***
Latitude (°)	—	—	−0.11±0.01***	−1.19±0.07***
Winter Mean Insolation (kWh/m^2/day)	−0.97±0.19***	—	—	—
Winter Mean Air Temperature (°C)	—	−0.08±0.02***	—	—
Number of Areas	97	97	97	97
Coefficient of Determination (r^2)	**0.21**	**0.16**	**0.80**	**0.76**

***$p<0.001$.
‡Plus-minus values are linear regression coefficients and standard errors (i.e., ±SE).

Table 3. Correlation matrix for asthma prevalence in US adult population, latitude, longitude, air pollution ($PM_{2.5}$), annual insolation, winter insolation, annual air temperature, and winter air temperature.

	Latitude (°)	Longitude (°)	Air Pollution ($PM_{2.5}$) ($\mu g/m^3$)	Annual Insolation ($kWh/m^2/day$)	Winter Insolation ($kWh/m^2/day$)	Annual Temperature (°C)	Winter Temperature (°C)	Asthma (%)
Latitude (°)	1.00							
Longitude (°)	−0.04	1.00						
Air Pollution ($PM_{2.5}$) ($\mu g/m^3$)	−0.15	0.43	1.00					
Annual Insolation ($kWh/m^2/day$)	−0.69	−0.54	−0.29	1.00				
Winter Insolation ($kWh/m^2/day$)	−0.90	−0.24	−0.13	0.91	1.00			
Annual Temperature (°C)	−0.92	0.13	0.29	0.51	0.75	1.00		
Winter Temperature (°C)	−0.87	0.04	0.24	0.51	0.73	0.97	1.00	
Asthma (%)	0.47	0.15	0.04	−0.38	−0.46	−0.42	−0.40	1.00

this paper. It should be taken into consideration that Staples et al. (2003) study was based on the 1995 Australian National Health Survey of approximately 54,000 people from all states and territories and across all age groups. In addition, asthma prevalence data were based on state/territory while geographical latitudes represented smaller areas of corresponding capital

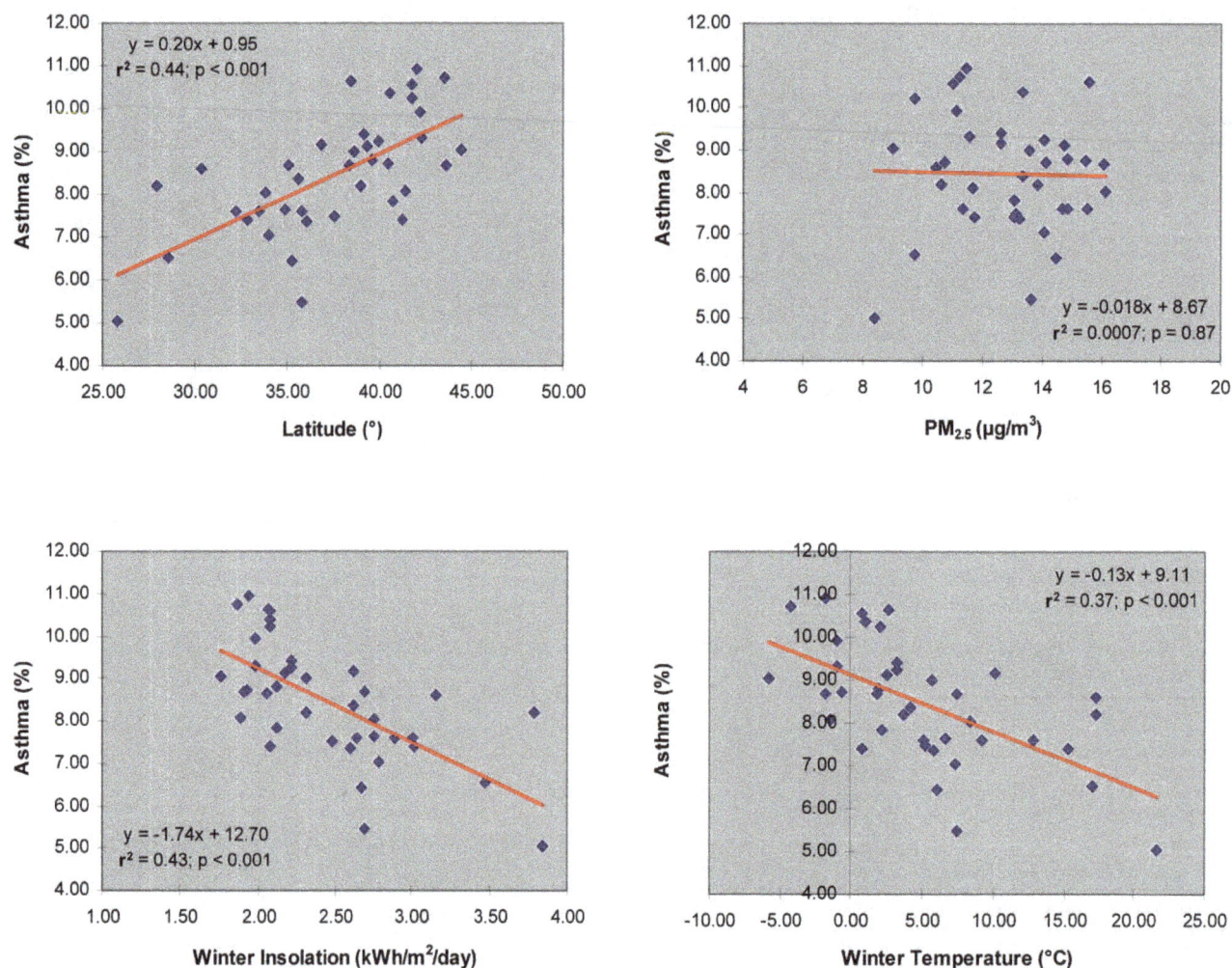

Figure 2. Asthma prevalence vs. latitude, air pollution ($PM_{2.5}$), winter insolation, and winter temperature in adult population of 39 major metropolitan/micropolitan areas of the eastern seaboard.

Table 4. Multiple regression estimates of asthma prevalence in US adult population associated with annual and winter mean insolation, annual and winter mean air temperature and air pollution ($PM_{2.5}$).

Multiple Regression Coefficient/Variable[‡]	Model 11 - Asthma vs. Annual Mean Insolation, Temperature, and $PM_{2.5}$	Collinearity Statistics		Model 12 - Asthma vs. Winter Mean Insolation, Temperature, and $PM_{2.5}$	Collinearity Statistics	
		Tolerance	VIF[#]		Tolerance	VIF[#]
Y-axis Intercept	10.86 ± 1.38***			10.01 ± 1.06***		
Annual Mean Insolation (kWh/m²/day)	-0.42 ± 0.30	0.53	1.89	—		
Annual Mean Air Temperature (°C)	-0.10 ± 0.03**	0.53	1.89	—		
Winter Mean Insolation (kWh/m²/day)	—			-0.72 ± 0.32*	0.36	2.75
Winter Mean Air Temperature (°C)	—			-0.03 ± 0.03	0.35	2.87
Air Pollution ($PM_{2.5}$) (µg/m³)	0.04 ± 0.05	0.65	1.53	0.02 ± 0.04	0.74	1.36
Number of Areas	97			97		
Coefficient of Determination (R^2)	**0.22**			**0.22**		
Adjusted R^2	**0.19**			**0.20**		

***$p < 0.001$;
**$p < 0.01$;
*$p < 0.05$.
[‡]Plus-minus values are multiple regression coefficients and standard errors (i.e., $\pm$SE).
[#]VIF – Variance Inflation Factor.

regions. The authors indicated that the latitude ranges for some states/territories spanned over 10° (up to 19° for Queensland). Yet a single latitude coordinate was applied for each of the states/territories, not necessarily representing the areas from which the cases of asthma were identified and clearly affecting the accuracy of the observed relationship between the asthma prevalence and geographical latitude in Australia.

Franco et al. [1] showed a statistically significant positive correlation between geographical latitude and active asthma prevalence in the eight International Study of Asthma and Allergies in Childhood (ISAAC) centres in North-East Brazil. The authors found no relation between the tropical weather and high prevalence of childhood asthma in the studied population.

Weiland et al. [2] studied the association between climate and atopic diseases using the data from 146 ISAAC centres worldwide

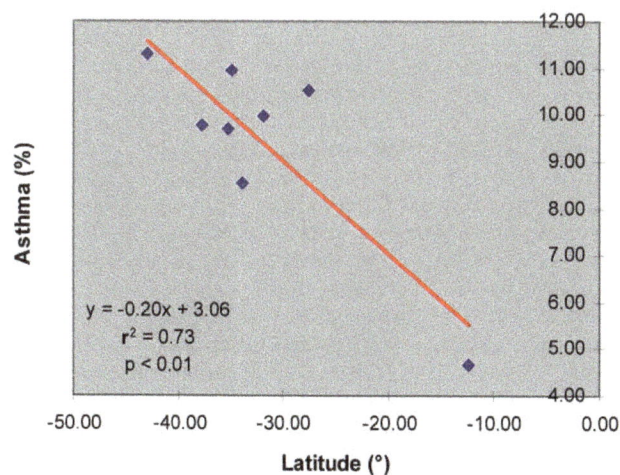

y = -0.20x + 3.06
r² = 0.73
p < 0.01

Figure 3. Asthma prevalence vs. latitude in the population of 8 major metropolitan areas of Australia.

and found that the prevalence of eczema symptoms was positively associated with latitude. The authors observed a similar association in both 6–7 and 13–14 years age-groups in Europe and also among 6–7 years old children worldwide, suggesting that "... *latitude and temperature affect the prevalence of eczema only indirectly, due to changes in behaviour and differences in sun exposure*" (emphasis added). They concluded that "*climate may affect the prevalence of asthma and atopic eczema in children*".

In an ecological analysis of geo-climatic variations in the prevalence of current asthma, allergic rhinitis and chronic cough, and phlegm in Italy, Zanolin et al. [4] found a negative correlation between asthma-like symptoms and geographical latitude. Rather than to focus specifically on the diagnosed asthma cases country-wide, the study is based on a random sample of 18,873 subjects with a response rate of 72.7% from different climatic regions, effectively representing less than 0.03% of the Italian total general population. The authors suggested that "*variations in the prevalence of respiratory symptoms according to geo-climatic factors could provide important clues to the knowledge of the aetiology of asthma*".

A report from the US CDC on The State of Childhood Asthma, United States, 1980–2005 [24] indicates that current asthma prevalence rates among children 0–17 years of age, by state, annual average for the period 2001–2005 are generally higher in the northeast region of the United States. These findings are in agreement with the results and observations for U.S. adult population, by metropolitan/micropolitan statistical area, presented in this paper.

Ethnic groups with darker skin have higher prevalence of asthma when compared to those with lighter skin pigmentation [13,25,26,27]. Darker skin pigmentation is a form of evolutionary adaptation, providing protection against potentially harmful effects of excessive UV irradiation doses, developed in populations inhabiting tropical regions [28]. Therefore, when compared to natives of higher latitudes with lighter skin pigmentation, natives of tropical regions migrating from lower to higher latitudes may become disadvantaged in terms of their ability to synthesize physiologically required quantities of vitamin D under low levels of annual/winter solar irradiation [29,30]. This may result in a potentially severe deficiency if sufficient doses of vitamin D are not obtained through diet and/or supplementation.

Different immunologic, genetic and environmental mechanisms are considered in the etiology of asthma and it is proposed that geographical variation in asthma prevalence could be due to gene-by-environment interactions [31]. Although cold air has been associated with worsened respiratory symptoms and exacerbation of asthma, ambient air temperature is expected to be a symptom trigger rather than a causal factor initiating respiratory diseases [32]. Asthma attacks can be triggered by cold air, irritating fumes, or fine airborne particulate matter [33].

Vitamin D has been recognized as an important immuno-modulating factor with dendritic cells as its primary targets [34]. Calcitriol (1,25-dihydroxycholecalciferol), the main vitamin D metabolite, inhibits dendritic cell maturation and T-helper1 (Th1) cell differentiation, which has been described as a key mechanism of allergy development [35]. The inhibition of Th1 cell differentiation leads to a predominance of Th2 cells, which has been implicated in asthma pathogenesis [36,37]. Wjst and Dold (1999) proposed that nutritional intake of vitamin D for rickets prophylaxis could be responsible for the increase in the prevalence of asthma in developed countries over the last three decades, suggesting that if protective antigen-reactive Th1 memory cells fail to develop the subsequent predominance of Th2 cells can trigger allergic reactions [38].

However, the mechanism of action of vitamin D and its active metabolites is quite complex and not yet fully understood. One should take into consideration that, in addition to the inhibition of dendritic cell maturation, activated vitamin D modulates the immune response through inhibition or enhancement on multiple levels of cell function, such as production of both pro- and anti-inflammatory cytokines, and inhibition of B cell differentiation, proliferation, and antibody secretion [34]. Activated vitamin D enhances the development of interleukin-IL-10- and reduces the number of IL-6- and IL-17-secreting cells [39]. High levels of pro-inflammatory IL-6 and IL-17 cytokines [40,41,42] and low levels of an anti-inflammatory cytokine IL-10 have been observed in asthmatic patients [43,44], suggesting that a functional vitamin D insufficiency/deficiency may be responsible for an increased probability of developing asthma. Allergies and asthma could be the result of unbalanced metabolic transformation or inadequate vitamin D receptor (VDR) binding, leading to plasma accumulation of active/inactive vitamin D metabolites, predominance of pro-inflammatory cytokines and an impaired immuno-modulation.

The observed correlation between geographical longitude and annual insolation ($r = -0.54$) presented in **Table 3** is in agreement with the estimates of monthly mean erythemal UV irradiation values for the U.S. and Canada [45], showing an increase of solar irradiation in the direction from northeastern to southwestern regions of North America. Considering that total insolation is almost perfectly correlated with the UV component of the total solar irradiation on the Earth's surface ($r > 0.96$, [46,47]), the correlation between latitude and total insolation presented in this paper could be used to predict the effect of erythemal UV irradiation on the skin production of vitamin D in the studied population.

The results of linear regression analyses performed in this study provide evidence that the geographical latitude can be used to predict the prevalence of asthma in the U.S. adult population. It is interesting that in winter months the association of asthma prevalence with air temperature is weaker than the association with insolation. If the factor associated with geographical latitude

and asthma prevalence is air temperature and not insolation, one would expect cold weather in winter months to yield a stronger correlation of asthma prevalence with air temperature than insolation. A contrary is observed in this study, where winter insolation is almost as good as geographical latitude in terms of its ability to predict the prevalence of asthma. In addition, the results of multiple regression analyses presented in **Table 4** confirm that, based on the data for winter insolation, winter air temperature, and air pollution, insolation is the best predictor of asthma.

Winter insolation is consistently a stronger predictor of asthma prevalence than winter air temperature in both the continental U.S. population and a subset of data for the Eastern Seaboard states. It appears that populations living in northern regions may have wider swings in vitamin D levels due to a significant decrease in vitamin D production during the winter months. When compared to those living in tropical areas, populations in the northern regions have much lower production of vitamin D in summer months which may lead to a weak physiological baseline at the beginning of winter. This condition could then progress into a vitamin D insufficiency or a severe deficiency in winter months, when regional insolation is at its annual minimum, potentially affecting the immune system and increasing the probability of respiratory infection and developing/exacerbating asthma.

The results of statistical analyses presented in this paper indicate that the probability for the observed association between latitude/insolation and asthma prevalence to be simply due to a chance is very small (i.e., $p < 0.01$). In addition, consistent results are obtained using the data from the U.S. and Australia further reducing the probability that this may be just a property of the data distribution and not a biological or a physiological response.

Camargo et al. (2011) studied the relationship between cord-blood levels of an active metabolite of vitamin D, 25-hydroxyvitamin D (25[OH]D), and the risk of respiratory infection, wheezing, and asthma [48]. The authors concluded that cord-blood levels of 25[OH]D are inversely associated with the risk of respiratory infection and childhood wheezing but no association with incident asthma is observed. These results suggest that active metabolites of vitamin D could have an effect on reducing the frequency of respiratory infections, which may lead to a reduction in exacerbation of symptoms in those suffering from asthma.

Allan et al. (2010) conducted a case-control study of vitamin D status and asthma in 160 adults aged between 15 and 80 years, 80 with physician-confirmed mild/moderate asthma and 80 age and gender-matched controls who had a smoking history of <10 pack-years [49]. The majority of controls (i.e., 70%) were recruited from local surgery units and the remainder through the advertising in local press. The study showed no significant difference in the serum 25-hydroxyvitamin D3 concentrations between cases and controls, and no association between 25-hydroxyvitamin D3 levels and asthma severity or lung function. The authors concluded that this study does not find evidence to support the use of vitamin D as an adjunct to conventional therapy in asthma in adults. However, this study is based on a rather small group of individuals who were confined to a small geographical area with only a snapshot in time for a vitamin D status. Although high doses of vitamin D may not be useful in the treatment of adult asthma as an existing condition, the study by Allan et al. does not address possible effects of a long-term vitamin D deficiency/insufficiency, an impaired immuno-modulation, and how these parameters may affect the prevalence of asthma in large populations and over large geographical areas with significantly different levels of annual UV-B insolation.

Air temperature and confounding factors that could be associated with the geographical latitude, such as socio-economic status, regional diet, demographic structure and ethnic origin, or some unknown effects of insolation and UV radiation cannot be eliminated based on the results of this study. However, in conjunction with other published studies suggesting a possible link between vitamin D and asthma, a significant decrease of vitamin D producing erythemal UV irradiation with an increase in latitude provides a plausible explanation for the observed geographical distribution of asthma prevalence in both the northern and the southern hemispheres. As a modulator of the immune response, vitamin D could have an influence on the frequency and severity of respiratory infections which may lead to exacerbation of symptoms from preexisting asthma. A U.S.-wide comprehensive study on the relationship between plasma vitamin D status and the prevalence of asthma in different age groups (e.g., children age 0–17 years), ethnic groups and geographical regions would provide a better understanding of the role that vitamin D may play in the frequency of respiratory infections, and the development/exacerbation of allergies and asthma. Although such a study may confirm or rule out vitamin D, in absence of other plausible leads in the published literature, it may be difficult to design a specific study to determine if there are other possible effects of insolation and UV radiation that could be associated with asthma.

Conclusions

At 21% for the continental U.S. and up to 43% for the Eastern Seaboard regions, total insolation in winter months is almost as strong as latitude in its ability to explain the observed spatial variation in the prevalence of asthma in the U.S. adult population.

Similar results are obtained using the Australian data, suggesting a consistent association between the latitude/insolation and asthma prevalence worldwide.

Taking into consideration confounding factors and possible limitations of an ecological study design, the results presented in this paper suggest that, as a known modulator of the immune response closely linked with geographical latitude and erythemal UV irradiation, vitamin D may play an important role in the development/exacerbation of asthma. Vitamin D is essential for the functional immune system and should be maintained at adequate levels not only in those suffering from asthma but also in the general population. Hence, community-level educational programs on asthma may benefit from including recommendations on adequate diet and vitamin D supplementation to prevent severe deficiencies, particularly among ethnic groups with darker skin pigmentation inhabiting high geographical latitudes.

Acknowledgments

I would like to thank the *Anonymous* reviewers for their constructive comments and suggested changes, and Dušica Krstić for the encouragement and enthusiastic support. The author had full access to all the data in the study and takes responsibility for the integrity of the data and the accuracy of the data analysis.

Author Contributions

Analyzed the data: GK. Wrote the paper: GK.

References

1. Franco JM, Gurgel R, Sole D, Lúcia França V, Brabin B, et al. (2009) Socio-environmental conditions and geographical variability of asthma prevalence in Northeast Brazil. Allergol Immunopathol (Madr) 37(3): 116–21.
2. Weiland SK, Husing A, Strachan DP, Rzehak P, Pearce N, et al. (2004) Climate and the prevalence of symptoms of asthma, allergic rhinitis, and atopic eczema in children. Occup Environ Med 61: 609–615.
3. Staples JA, Ponsonby AL, Lim LLY, McMichael AJ (2003) Ecologic Analysis of Some Immune-Related Disorders, Including Type 1 Diabetes, in Australia: Latitude, Regional Ultraviolet Radiation, and Disease Prevalence. Environ Health Perspect 111: 518–523.
4. Zanolin ME, Pattaro C, Corsico A, Bugiani M, Carrozzi L, et al. (2004) The role of climate on the geographic variability of asthma, allergic rhinitis and respiratory symptoms: results from the Italian study of asthma in young adults. Allergy 59(3): 306–14.
5. Camargo CA Jr., Rifas-Shiman SL, Litonjua AA, Rich-Edwards JW, Weiss ST, et al. (2007) Maternal intake of vitamin D during pregnancy and risk of recurrent wheeze in children at 3 y of age. Am J Clin Nutr 85: 788–95.
6. Devereux G, Litonjua AA, Turner SW, Craig LCA, McNeill G, et al. (2007) Maternal vitamin D intake during pregnancy and early childhood wheezing. Am J Clin Nutr 85: 853–9.
7. Sutherland ER, Goleva E, Jackson LP, Stevens AD, Leung DY (2010) Vitamin D levels, lung function, and steroid response in adult asthma. Am J Respir Crit Care Med 181(7): 699–704.
8. Litonjua AA, Weiss ST (2007) Is vitamin D deficiency to blame for the asthma epidemic? J Allergy Clin Immunol 120(5): 1031–5.
9. Grant WB (2007) Roles of solar UV radiation and vitamin D in human health and how to obtain vitamin D. Expert Rev Dermatol 2(5): 563–577.
10. Rudders SA, Espinola JA, Camargo CA, Jr. (2010) North-south differences in US emergency department visits for acute allergic reactions. Ann Allergy Asthma Immunol 104(5): 413–6.
11. Kimlin MG, Olds WJ, Moore MR (2007) Location and vitamin D synthesis: is the hypothesis validated by geophysical data? J Photochem Photobiol B 86(3): 234–9.
12. Mead MN (2008) Benefits of Sunlight: A Bright Spot for Human Health. Environ Health Perspect 116(4): A161–A167.
13. Freishtat RJ, Iqbal SF, Pillai DK, Klein CJ, Ryan LM, et al. (2010) High prevalence of vitamin D deficiency among inner-city African American youth with asthma in Washington, DC. J Pediatr 156(6): 948–52.
14. Wang X, Gao W, Davis J, Becky O, George J, et al. (2007) Dependence of erythemally weighted UV radiation on geographical parameters in the United States. Proc SPIE 6679: 667903.
15. Schauber J, Gallo RL (2008) Vitamin D deficiency and asthma: Not a strong link – yet. J Allergy Clin Immunol 121(3): 782–784.

16. U.S. Department of Health and Human Services (DHHS) - Centers for Disease Control (CDC) and Prevention (2008) Behavioral Risk Factor Surveillance System Survey: Prevalence and Trends Data by selected Metropolitan/Micropolitan Statistical Areas (MMSA). Atlanta, Georgia (Accessed in May 2010 at http://apps.nccd.cdc.gov/brfss-smart/SelMMSAPrevData.asp).
17. U.S. Environmental Protection Agency (2009) Air Trends: Air Quality Statistics by City and County (Accessed in May 2010 at http://epa.gov/air/airtrends/factbook.html).
18. Public Health Information Development Unit (2010) A Social Health Atlas of Australia – Local Government Area (LGA) Data (Accessed in June 2010 at http://www.publichealth.gov.au/data/a-social-health-atlas-of-australia_-2010.html).
19. Microsoft Research (MSR) Maps (Sponsored by U.S. Geological Survey (USGS). Microsoft Bay Area Research Center (Accessed in May 2010 at http://msrmaps.com/advfind.aspx).
20. Charles Sturt University, Guide to Australia – Latitude and Longitude Search, 1996 Census Data from the Australian Bureau of Statistics (Accessed in June 2010 at http://www.csu.edu.au/australia/latlong/index.html).
21. National Aeronautics and Space Administration (NASA), Atmospheric Science Data Center, Surface Meteorology and Solar Energy: Interannual Variability (Accessed in May 2010 at http://eosweb.larc.nasa.gov/sse/).
22. Robinson WS (1950) Ecological correlations and the behavior of individuals. Am Sociol Rev 15(3): 351–357.
23. Subramanian SV, Jones K, Kaddour A, Krieger N (2009) Revisiting Robinson: The perils of individualistic and ecologic fallacy. Int J Epidemiol 38: 342–360.
24. Akinbami LJ (2006) The State of Childhood Asthma, United States, 1980–2005. Advance data from vital and health statistics No. 381. U.S. Department of Health and Human Services, Centers for Disease Control and Prevention, National Center for Health Statistics (http://www.cdc.gov/nchs/data/ad/ad381.pdf).
25. Ginde AA, Espinola JA, Camargo CA Jr. (2008) Improved overall trends but persistent racial disparities in emergency department visits for acute asthma, 1993–2005. J Allergy Clin Immunol 122(2): 313–8.
26. Meng YY, Babey SH, Hastert TA, Brown ER (2007) California's racial and ethnic minorities more adversely affected by asthma. Policy Brief UCLA Cent Health Policy Res PB2007-3: 1–7.
27. Grant EN, Lyttle CS, Weiss KB (2000) The relation of socioeconomic factors and racial/ethnic differences in US asthma mortality. Am J Public Health 90(12): 1923–5.
28. Jablonski NG, Chaplin G (2000) The evolution of human skin coloration. J Hum Evol 39: 57–106.
29. Hintzpeter B, Scheidt-Nave C, Müller MJ, Schenk L, Mensink GB (2008) Higher prevalence of vitamin D deficiency is associated with immigrant background among children and adolescents in Germany. J Nutr 138: 1482–1490.

30. Genuis SJ, Schwalfenberg GK, Hiltz MN, Vaselenak SA (2009) Vitamin D status of clinical practice populations at higher latitudes: Analysis and applications. Int J Environ Res Public Health 6: 151–173.

31. Subbarao P, Mandhane PJ, Sears MR (2009) Asthma: epidemiology, etiology and risk factors. CMAJ 181(9): E181–E190.

32. Koskela HO (2007) Cold-air provoked respiratory symptoms: the mechanisms and management. Int J Circumpolar Health 66(2): 91–100.

33. George RB, Owens MW (1991) Bronchial asthma. Dis Mon 37(3): 137–96.

34. Cutolo M (2009) Vitamin D and autoimmune rheumatic diseases. Rheumatology 48: 210–212.

35. Wjst M (2006) The vitamin D slant on allergy. Pediatr Allergy Immunol 17(7): 477–83.

36. Bharadwaj AS, Bewtra AK, Agrawal DK (2007) Dendritic cells in allergic airway inflammation. Can J Physiol Pharmacol 85(7): 686–99.

37. Aiba S (2007) Dendritic cells: importance in allergy. Allergol Int 56(3): 201–8.

38. Wjst M, Dold S (1999) Genes, factor X, and allergens: what causes allergic diseases? Allergy 54: 757–759.

39. Correale J, Ysrraelit MC, Gaitán MI (2009) Immunomodulatory effects of vitamin D in multiple sclerosis. Brain 132: 1146–1160.

40. Wong CK, Ho CY, Ko FWS, Chan CHS, Ho ASS, et al. (2001) Proinflammatory cytokines (IL-17, IL-6, IL-18 and IL-12) and Th cytokines (IFN-gamma, IL-4, IL-10 and IL-13) in patients with allergic asthma. Clin Exp Immunol 125(2): 177–83.

41. Molet S, Hamid Q, Davoine F, Nutku E, Taha R, et al. (2001) IL-17 is increased in asthmatic airways and induces human bronchial fibroblasts to produce cytokines. J Allergy Clin Immunol 108(3): 430–8.

42. Neveu WA, Allard JL, Raymond DM, Bourassa LM, Burns SM, et al. (2010) Elevation of IL-6 in the allergic asthmatic airway is independent of inflammation but associates with loss of central airway function. Respir Res 11: 28.

43. Matsumoto K, Inoue H, Fukuyama S, Tsuda M, Ikegami T, et al. (2004) Decrease of interleukin-10-producing T cells in the peripheral blood of severe unstable atopic asthmatics. Int Arch Allergy Immunol 134(4): 295–302.

44. Tomita K, Lim S, Hanazawa T, Usmani O, Stirling R, et al. (2002) Attenuated production of intracellular IL-10 and IL-12 in monocytes from patients with severe asthma. Clin Immunol 102(3): 258–66.

45. Fioletov VE, Kimlin MG, Krotkov N, McArthur LJB, Kerr JB, et al. (2004) UV index climatology over the United States and Canada from ground-based and satellite estimates. J Geophys Res 109: D22308.

46. Trabea AA, Salem I (2001) Empirical Relationship for Ultraviolet Solar Radiation Over Egypt. Egypt J Sol 24(1): 123–132.

47. Murillo W, Cañadab J, Pedrós G (2003) Correlation between global ultraviolet (290–385 nm) and global irradiation in Valencia and Cordoba (Spain). Renew Energ 28(3): 409–418.

48. Camargo CA Jr., Ingham T, Wickens K, Thadhani R, Silvers KM, et al. (2011) Cord-blood 25-Hydroxyvitamin D levels and risk of respiratory infection, wheezing, and asthma. Pediatrics 127(1): e180–e187.

49. Allan K, Devereux G, McNeill G, Wilson A, Avenell A, et al. (2010) A case-control study of vitamin D status and asthma in adults. Proceedings of the Nutrition Society 69(OCE6): E475.

25

The Association between Cold Spells and Pediatric Outpatient Visits for Asthma in Shanghai, China

Yuming Guo[1⊚], Fan Jiang[2,3⊚], Li Peng[4], Jun Zhang[3], Fuhai Geng[4], Jianming Xu[4], Canming Zhen[4], Xiaoming Shen[2,3]*, Shilu Tong[1]*

1 School of Public Health and Institute of Health and Biomedical Innovation, Queensland University of Technology, Brisbane, Australia, 2 Department of Developmental and Behavioral Pediatrics, Shanghai Institute of Pediatric Translational Medicine, Shanghai Children's Medical Centre, Shanghai Jiaotong University, School of Medicine, Shanghai, China, 3 Ministry of Education-Shanghai Key Laboratory of Children's Environmental Health, Xinhua Hospital, Shanghai Jiaotong University, School of Medicine, Shanghai, China, 4 Shanghai Meteorological Bureau, Shanghai, China

Abstract

Background: Asthma is a serious global health problem. However, few studies have investigated the relationship between cold spells and pediatric outpatient visits for asthma.

Objective: To examine the association between cold spells and pediatric outpatient visits for asthma in Shanghai, China.

Methods: We collected daily data on pediatric outpatient visits for asthma, mean temperature, relative humidity, and ozone from Shanghai between 1 January 2007 and 31 December 2009. We defined cold spells as four or more consecutive days with temperature below the 5th percentile of temperature during 2007–2009. We used a Poisson regression model to examine the impact of temperature on pediatric outpatient visits for asthma in cold seasons during 2007 and 2009. We examined the effect of cold spells on asthma compared with non-cold spell days.

Results: There was a significant relationship between cold temperatures and pediatric outpatient visits for asthma. The cold effects on children's asthma were observed at different lags. The lower the temperatures, the higher the risk for asthma attacks among children.

Conclusion: Cold temperatures, particularly cold spells, significantly increase the risk of pediatric outpatient visits for asthma. The findings suggest that asthma children need to be better protected from cold effects in winter.

Editor: Sanja Stanojevic, Hospital for Sick Children, Canada

Funding: This work is supported by Shanghai Science and Technology Committee (10231203903), MOE-Shanghai Key Laboratory of Children's Environmental Health (10DZ2272200). YG is supported by the QUT Postgraduate Research Award (QUTPRA); ST is supported by a NHMRC Research Fellowship (#553043). The funders had no role in study design, data collection and analysis, decision to publish, or preparation of the manuscript.

Competing Interests: The authors have declared that no competing interests exist.

* E-mail: s.tong@qut.edu.au (ST); xmshen@shsmu.edu.cn (XS)

⊚ These authors contributed equally to this work.

Introduction

Asthma is one of the most important chronic diseases worldwide. It is estimated that there are about 300 million people with asthma currently in the world [1]. Asthma accounts for about one percent of all disability-adjusted life years lost worldwide, which reflects the severity and high prevalence of this disease [1]. The prevalence of asthma has been increasing in both children and adults around the world in recent decades [2,3]. However, there is no treatment to cure it. Therefore, it is urgently required to identify the causes and/or risk factors for the onset of asthma so that effective control and prevention strategies can be developed.

There is sufficient evidence suggesting that air pollution (e.g., ozone) decreases lung function, triggers exacerbations of asthma [4], and increases hospital admissions for asthma [5–7], especially for children [8–10]. Cold temperature is also one of major environmental factors that exacerbate chronic inflammatory airway diseases (for example, chronic obstructive pulmonary

disease and asthma) [11]. Studies have shown that weather conditions play an important role in asthma attacks [12]. For example, there is a seasonal pattern in asthma admissions, with more during the wet season than the dry season in Mexico city [13]. The emergency room visitors for asthma in Oulu, Finland are higher in winter than summer [14]. For the short-term (day-to-day) effect of temperature, cold temperatures are related to acute exacerbations of asthma symptoms, while hot temperatures are associated with increased asthma prevalence which might be related to higher levels of allergen exposure [15,16]. Some weather conditions like extremely hot or cold temperatures, changes in barometric pressure or humidity and wind can trigger asthma [17–19]. However, there is little data available on the effects of extreme cold temperatures on childhood asthma. This study investigated the association between cold spells and pediatric outpatient visits for asthma in Shanghai, China.

Materials and Methods

Study population

Shanghai is located on the east tip of Yangtze River Delta and along China's eastern coastline, at latitude 31° 14' N and longitude 121° 29' E. Shanghai covers an area of 6,341 square kilometers. Shanghai is the largest city by population in China, with a total population of over 23 million in 2010 including 1.99 million children (0–14 years) [20]. Weather in Shanghai is generally mild and moist, with four distinct seasons: a warm spring, a hot rainy summer, a cool autumn and a cold winter. The hottest time in Shanghai is usually between July and August while the coldest time is from the late January to early February.

Data on pediatric outpatient visitors for asthma

We collected the retrospective data on daily counts of pediatric outpatient visits for asthma between 1 January 2007 and 31 December 2009 from the Shanghai Children's Medical Center (SCMC) affiliated to Shanghai Jiao Tong University School of Medicine. The SCMC is one of the largest pediatric research institutes in China. The primary diagnoses of daily outpatients were coded according to the International Classification of Disease, 9th revision for asthma (ICD9: code 493).

Data on air pollution and weather conditions

Data on daily mean temperature, relative humidity and ozone (O_3), were obtained from the Shanghai Meteorological Bureau. Daily mean temperature, relative humidity and O_3 were calculated using the records from monitors throughout the urban area of Shanghai.

Data analysis

1. Cold spell definition. There is no standard definition of a cold spell worldwide [21]. Many methods were used to define cold spells [21–23]. For example, Huynen et al. defined a cold spell by using a period of at least 9 days with a minimum temperature of $-5°C$ or lower [23]. We analysed several possible definitions of a cold spell, such as two or more consecutive days with temperature below the 1st, 2.5th or 5th percentile of the temperature distribution. Finally, we defined a cold spell as four or more consecutive days with mean temperature below the 5th percentile of the distribution during 2007–2009 as sensitivity analyses suggested it to be an appropriate cold spell definition for Shanghai (results not shown).

We used mean temperature (not minimum or maximum temperature) to define a cold spell, because it gave a best model fit as judged by quasi-Poisson Akaike Information Criterion (Q-AIC). The other reason is that mean temperature represents the exposure throughout the whole day, while minimum or maximum temperature only reflects the exposure for a short period. So mean temperature can be more easily interpreted for decision making purposes [24].

2. Temperature impacts on pediatric outpatient visits for asthma. Both cold and hot temperatures increase the risk of morbidity, and the temperature-morbidity relationships are generally U-, V-, and J- shaped [25,26]. Therefore, we only used data in cold season (from December to April) to explore the effect of temperature on pediatric outpatient visits for asthma. We used a time series model to explore the relationship between current day's temperature (continuous variable) and pediatric outpatient visits for asthma in cold seasons. We assumed that the daily number of outpatient visits was over-dispersed by quasi-Poisson function. As previous studies have shown that the temperature impact on morbidity was non-linear [25,26], we used a spline with 3 degrees of freedom for temperature. We controlled for relative humidity and O_3 using spline with 3 degrees of freedom. We controlled for day of the week as a category variable. We used a spline with 3 degrees of freedom for calendar day to control for season and a long-term trend. We plotted the relationship between temperature and pediatric outpatient visits for asthma.

3. The effect of cold spells on pediatric outpatient visits for asthma. We used a time series model to estimate relative risks (RRs) (with 95% confidence intervals (CIs)) of pediatric outpatient visits for asthma by comparing the number of the visits during the cold spells with those during non-cold spells in cold seasons. Mean temperature was used to derive a dummy variable to categorise cold spell or non-cold spell days. We controlled for a similar range of confounding factors as indicated above.

4. The effect of an intense cold spell on pediatric outpatient visits for asthma in 2008. Our preliminary analysis showed that there was an intense and long cold spell (20 days) in 2008. Thus, to make the best use of a good opportunity of this natural experiment, we examined the relation between exposure to this cold spell and pediatric outpatient visits for asthma through comparing its RRs with those during the same periods in 2007 and 2009, after adjustment for a similar range of confounding factors as indicated above. We did not select the reference period immediately after or before this cold spell, as studies have found that the cold spell-related morbidity (or mortality) is usually followed by temporary reduction in morbidity (or mortality) in subsequent weeks. The demographic characteristics were unlikely to change substantially in the neighbouring years. Therefore we selected the same period of the cold spell in the neighbouring winters as the reference. This method has been widely used to examine the effects of heat waves and cold spells on morbidity and mortality in previous studies [22,27].

5. Lag effects of cold spells. Many studies have shown that the effects of cold temperature on morbidity and mortality last more than weeks [25,26]. So it is necessary to examine the lag effects of cold spells on pediatric outpatient visits for asthma. We examined the lag effects (lag 0, lag 1–2, lag 3–6, lag 7–14, lag 15–30, and lag 0–30) of cold spells on pediatric outpatient visits for asthma in this study [28].

Sensitivity analyses were performed through changing degrees of freedom for calendar day, relative humidity and O_3. We also used different definitions for a cold spell. Values of $P<0.05$ (two-sided) were considered statistically significant. The R software (version 2.10.1, R Development Core Team 2009) was used to fit all models. The "mgcv" package was used to fit quasi-Poisson regression [29].

Results

There were 6 cold spells during 2007–2009 (Figure 1). Generally, in the extreme cold periods (not only cold spells), there were more paediatric outpatient visits for asthma than other periods (Figure 1).

Table 1 illustrates that during the 6 cold spells, the average temperature, relative humidity and O_3 were 1.4°C, 62.4%, and 13.3 ug/m^3, respectively, and there were 53 average (standard deviation: 19) pediatric outpatient visits for asthma. By contrast, in the non-cold spell days, the mean temperature, relative humidity and O_3 were 10.3°C, 70.5%, 20.6 ug/m^3, respectively, and there were 43 average (standard deviation: 22) pediatric outpatient visits for asthma. During the 20 days' cold spell in 2008, the number of pediatric outpatient visits for asthma was remarkably higher and mean temperature was significantly lower than that during the same periods in 2007 and 2009 (Table 1).

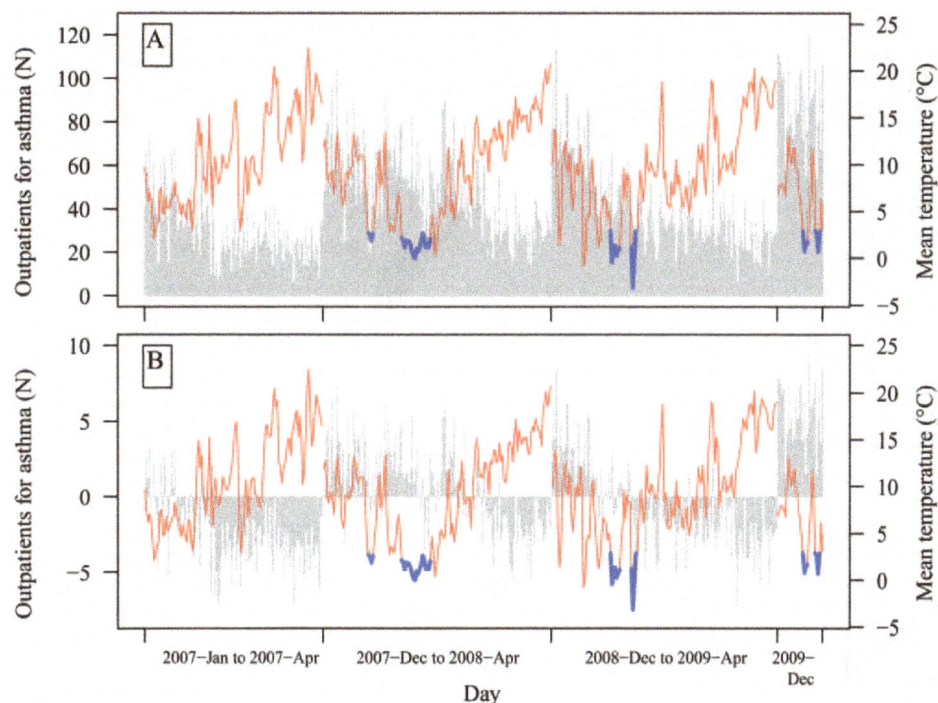

Figure 1. Time series of daily temperature and pediatric outpatient visits for asthma in cold seasons during 2007 and 2009 before (A) and after removing trends using a spline with 3 degrees of freedom for calendar time (B). The vertical bars are pediatric outpatient visits for asthma. The red lines are temperatures in non-cold spell days. The blue lines are temperatures in cold spell days.

Figure 2 shows the relationship between current day's temperature and pediatric outpatient visits for asthma in cold seasons during 2007 and 2009. There were apparent cold effects on pediatric outpatient visits for asthma. In general, the lower the mean temperature, the higher the risk of pediatric outpatient visits for asthma.

We examined the relative risk for children's asthma during cold spells compared with the non-cold spell days during 2007 and 2009 (Table 2). We found that broadly there was a significantly increased risk for children's asthma associated with cold spells across different lags, compared with the non-cold spell days. However, the impact of cold spells was generally attenuated and only significant at lag 7–14 days after adjustment for the effect of O_3.

We also examined the relative risk for children's asthma during the 20-day cold spell in 2008 compared with the same periods in 2007 and 2009 (Table 3). Results show that the 20-day cold spell

significantly increased the risk for children's asthma at different lags, even after adjustment for confounding factors. The severe cold spell also seemed to exhibit a quite long lagged effect on the onset of children's asthma.

Sensitivity analyses

Sensitivity analyses were conducted by changing the *df* of the splines for temperature, relative humidity and O_3 and the *df* of smoothing for time. The results remained broadly similar (results not shown).

When different definitions for a cold spell were used, we found there were similar estimated effects of cold spells on pediatric outpatient visits for asthma. However, the definition of four or more consecutive days with temperature below the 5th percentile of temperature gave a best model fit as judged by quasi-Poisson Akaike information criterion.

Table 1. Statistic summary for weather conditions, air pollution and pediatric outpatient visits for asthma during the cold spell and non-cold spell periods.

| Period | Mean (Minimum, Maximum) | | | |
	Mean temperature	Relative humidity	O_3	Asthma outpatients
All Cold spells	1.4 (−3.2, 2.9)	62.4 (29.5, 95)	13.3 (4.8, 33.9)	53 (20, 104)
Non-cold spells during cold season	10.3 (−0.8, 22.5)	70.5 (34.5, 96.8)	20.6 (0.6, 59.5)	43 (7, 121)
20 days cold spell in 2008	1.4 (0, 2.6)	75.8 (59, 95)	16.1 (4.8, 33.9)	53 (24, 82)
The same period in 2007	7.3 (3.1,14.9)	65.4 (48.0, 91.3)	12.1 (6.1, 20.5)	38 (18, 68)
The same period in 2009	6.8 (−3.2, 12.3)	73.3 (21.5, 91.8)	14.4 (3.3, 35.4)	38 (19, 62)

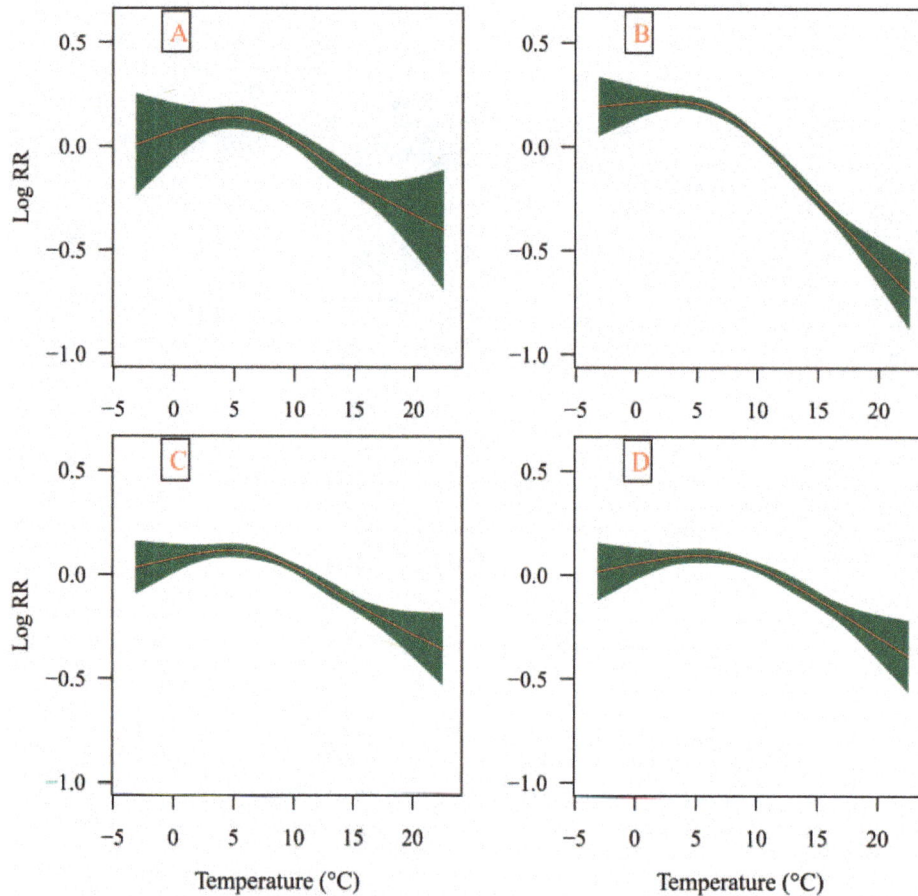

Figure 2. The relationship between temperature and pediatric outpatient visits for asthma in cold seasons during 2007 and 2009 after adjustment for trends and day of the week (A); trends, day of the week and relative humidity (B); trends, day of the week and O_3 (C); trends, day of the week, relative humidity and O_3 (D).

Table 2. The relative risks (RRs) for pediatric outpatient visits for asthma during cold spells compared with non-cold spell days in cold seasons, 2007–2009[a].

Lags	RR (95% CI)			
	Model 1[b]	Model 2[c]	Model 3[d]	Model 4[e]
Lag 0	1.25 (1.08, 1.44)**	1.25 (1.08, 1.45)**	1.06 (0.93, 1.20)	1.05 (0.92, 1.20)
Lag 1–2	1.29 (1.13, 1.47)**	1.30 (1.13, 1.48)**	1.11 (0.98, 1.25)	1.09 (0.96, 1.23)
Lag 3–6	1.29 (1.14, 1.46)**	1.29 (1.14, 1.47)**	1.09 (0.97, 1.22)	1.07 (0.95, 1.20)
Lag 7–14	1.26 (1.12, 1.42)**	1.26 (1.13, 1.42)**	1.12 (1.02, 1.25)*	1.12 (1.02,1.25)*
Lag 15–30	1.06 (0.95, 1.19)**	1.06 (0.95, 1.19)**	1.02 (0.92, 1.13)	1.03 (0.93, 1.14)
Lag 1–30	1.26 (1.14, 1.38)**	1.25 (1.14, 1.39)**	1.08 (0.98, 1.18)	1.06 (0.97, 1.17)

[a]All the models were adjust for day of the week, season and long-term trend;
[b]Model 1 did not controlled for relative humidity and O_3;
[c]Model 2 controlled for relative humidity;
[d]Model 3 controlled for O_3;
[e]Model 4 controlled for relative humidity and O_3;
*<0.05;
**P<0.01.

Table 3. The relative risk for pediatric outpatient visits for asthma during 20-day cold spell in 2008 compared with the same periods in 2007 and 2009[a].

Lag	RR (95% CI)			
	Model 1[b]	Model 2[c]	Model 3[d]	Model 4[e]
Lag 0	1.40 (1.21, 1.61)	1.37 (1.18, 1.58)	1.45 (1.26, 1.67)	1.41 (1.22, 1.63)
Lag 1–2	1.44 (1.25, 1.66)	1.42 (1.22, 1.64)	1.51 (1.30, 1.75)	1.46 (1.25, 1.72)
Lag 3–6	1.47 (1.29, 1.68)	1.46 (1.27, 1.67)	1.49 (1.30, 1.71)	1.47 (1.27, 1.72)
Lag 7–14	1.71 (1.46, 1.99)	1.66 (1.41, 1.96)	1.75 (1.49, 2.06)	1.70 (1.43, 2.02)
Lag 15–30	1.68 (1.50, 1.93)	1.71 (1.43, 2.03)	1.72 (1.48, 2.00)	1.65 (1.36, 2.00)
Lag 1–30	1.62 (1.46, 1.79)	1.63 (1.46, 1.82)	1.65 (1.48, 1.84)	1.64 (1.46, 1.84)

[a]All the models were adjusted for day of the week, season and long-term trend; and all RRs were statistically significant ($P<0.01$);
[b]Model 1 did not controlled for relative humidity and O_3;
[c]Model 2 controlled for relative humidity;
[d]Model 3 controlled for O_3;
[e]Model 4 controlled for relative humidity and O_3;

Discussion

A number of studies have assessed the climate impact on respiratory mortality [30], but relatively few studies have examined the relationship between extreme cold temperatures and respiratory morbidity [25,26]. In addition, no identified study has investigated the association between cold spells and pediatric outpatient visitors for asthma. According to our knowledge, this is the first study to investigate the association between cold spells and pediatric outpatient visits for asthma in Shanghai, China. In 2008, there was a severe cold spell in winter in Shanghai, which provided a good opportunity to examine the cold effect on asthma morbidity.

In this study, we found that cold temperature may significantly increase the risk of the outpatient for children's asthma. We assessed the relative risk for children's asthma during the 20-day cold spell in 2008 compared with the same periods in 2007 and 2009. Results show that the 20-day cold spell remarkably increased the risk of pediatric outpatient visitors for asthma. We also examined the relative risks of pediatric outpatient visitors for asthma during cold spells compared with non-cold spell days in cold seasons during 2007 and 2009. There was still a significant impact on the utilisation of health services for asthma but this impact was alleviated after controlling for the confounding effects of O_3. These findings suggest that extreme cold temperature may trigger asthma.

Some epidemiological studies have found that cold temperature is associated with respiratory morbidity and mortality [25,26]. Cold temperature is related to an increased risk of significant exacerbation of asthma [19]. Cold temperature–related exacerbation of those diseases is often followed by a subsequent increase in bacterial and viral infections of the airway, infiltration of inflammatory factors, and mucus secretion [11,31,32]. Cold temperature is associated with increased occurrence of respiratory tract infections, and a decrease in temperature often proceed the onset of the infections [33].

Study shows that cold temperature induces mucin hypersecretion from normal human bronchial epithelial cells *in vitro* through a transient receptor potential melastatin 8 [11]. A radionuclide perfusion study shows that rabbits exposed to cold temperature had lesser lung perfusion than controls. Cold temperature can induce contraction of the trachea and decrease pulmonary circulation and lung perfusion. This summation of acute cooling for tracheal smooth muscle and pulmonary circulation might be the reason why severe cold temperature induces contraction [34].

When examining the effects of cold spells on pediatric outpatients for asthma, we found that the cold effects were attenuated after controlling for O_3 effects. However, when we only assessed the risk for children's asthma during the 20-day cold spell in 2008 compared with the same periods in 2007 and 2009, the cold effects were still significant, even after controlling for O_3. The reason might be because there is an interactive effect of O_3 and temperature on respiratory health [35]. In cold seasons, the O_3 concentration was higher in non-cold spell days than cold spell days (Table 2). This situation will affect the assessment of the effects of cold spells on the risk for children's asthma. By contrast, the O_3 concentration in the 20-day cold spell was similar as the same periods in 2007 and 2009 (Table 2). The impact of temperature in those periods was unlikely to be modified by O_3.

This study found that cold spells had a long lagged effect on children asthma. Although no data is available on the relationship between cold spells and children asthma, several studies have reported that cold temperatures had a relatively long lagged effect on mortality and morbidity (including non-external, cardiovascular, and respiratory) [25,36,37]. These findings suggest that the lagged effects of cold spells are important and should not be overlooked. When healthcare providers and public health authorities develop response plans to protect children with asthma from cold temperatures, it is necessary to consider the delayed impact of cold spells on asthma exacerbations.

In this study, we used Poisson regression model to examine the effect of cold spells on pediatric outpatient visits for asthma. Alternatively, case-crossover design can also be used to examine the short-term effect of temperature (or air pollution) on mortality and morbidity. Previous studies have compared these two designs, and found that they are comparable and produce similar effect estimates [38,39]. To make sure our results are reliable, we used a time-stratified case-crossover analysis to examine the relationship between cold spells and pediatric outpatient visits for asthma. The results are similar as those from Poisson regression model.

We used multiple comparisons to examine the effects of cold spells on pediatric outpatient visits for asthma. All these analyses have shown that cold spells were associated with the increased pediatric outpatient visits for asthma. The multiple comparisons may produce spurious statistical associations because a large dataset was used. However, our sensitivity analysis using case-crossover study also gave the similar results (results not shown).

We are aware that well controlled patients do not necessarily attend a hospital clinic, and may see a general practitioner, change beheviour (i.e. won't go outside) or take medications. It means that the effects of cold temperatures on pediatric outpatient visits for asthma might be under-estimated. Additionally, even though a clear and consistent relation between cold spells and pediatric outpatient visitors for asthma was observed, we cannot conclude that the increase in child asthmatic attacks was directly caused by cold spells. There are some confounding factors that we are unable to control for. For example, is asthma exacerbation caused by the cold weather *per se* or triggered by viral infections during cold weather? Because data on viral infections were not available, we could not adjust for its influence. Therefore, the mechanisms underlying the relationship between cold spells and pediatric asthmatic attacks should be examined in future research.

Strengths and Limitations

According to the best of our knowledge, this is the first study to examine the impact of cold spells on children's asthma. We used the opportunity of a natural experiment that there was a long cold spell in Shanghai in 2008. We found that cold temperatures, particularly severe cold spells, significantly increased the risk of the onset of children's asthma. This evidence is useful for controlling and preventing children's asthma in cold seasons, especially in developing countries. Secondly, we examined the lagged effects of cold spells on children's asthma in detail and found that extreme cold temperatures had quite long delayed effects on the onset of children's asthma.

This study also has some limitations. We only used the data on pediatric outpatient visits for asthma from one children's hospital in one city. Even though we controlled for the effect of O_3, other air pollutants were not adjusted for in the model (the data on other air pollutants were incomplete for the study period). However, previous studies showed that temperature effects on health outcomes are generally robust and independent to air pollution [37,40,41]. This is an ecological study. Individual exposure data was not available, and thus measurement bias is inevitable to some extent. The cold spell definition used in this study is difficult to be generalised to other countries, especially for higher latitude regions, because people can adapt to their local climate.

Conclusion

Cold temperatures significantly increased the risk of pediatric outpatient visits for asthma. The findings suggest that extreme temperature events such as cold spells may trigger asthmatic attacks. Children with asthma need to be well protected from cold effects in winter.

Author Contributions

Conceived and designed the experiments: YG XS ST. Performed the experiments: FJ LP JZ FG JX CZ. Analyzed the data: YG. Contributed reagents/materials/analysis tools: YG FJ. Wrote the paper: YG FJ LP JZ FG JX CZ XS ST.

References

1. Masoli M, Fabian D, Holt S, Beasley R (2004) The global burden of asthma: executive summary of the GINA Dissemination Committee report. Allergy 59: 469–478.
2. Wong GWK, Ko FWS, Hui DSC, Fok TF, Carr D, et al. (2004) Factors associated with difference in prevalence of asthma in children from three cities in China: multicentre epidemiological survey. Bmj 329: 486.
3. Eder W, Ege MJ, von Mutius E (2006) The asthma epidemic. New England Journal of Medicine 355: 2226–2235.
4. Mortimer KM, Neas LM, Dockery DW, Redline S, Tager IB (2002) The effect of air pollution on inner-city children with asthma. Eur Respir J 19: 699–705.
5. Tatum AJ, Shapiro GG (2005) The effects of outdoor air pollution and tobacco smoke on asthma. Immunology and allergy clinics of North America 25: 15–30.
6. Friedman MS, Powell KE, Hutwagner L, Graham LRM, Teague WG (2001) Impact of changes in transportation and commuting behaviors during the 1996 Summer Olympic Games in Atlanta on air quality and childhood asthma. JAMA: the journal of the American Medical Association 285: 897.
7. Andersen ZJ, Bonnelykke K, Hvidberg M, Jensen SS, Ketzel M, et al. (2011) Long-term exposure to air pollution and asthma hospitalisations in older adults: a cohort study. Thorax.
8. Bates DV (1995) The effects of air pollution on children. Environmental Health Perspectives 103: 49.
9. Islam T, Gauderman WJ, Berhane K, McConnell R, Avol E, et al. (2007) Relationship between air pollution, lung function and asthma in adolescents. Thorax 62: 957–963.
10. Hwang BF, Lee YL, Lin YC, Jaakkola JJ, Guo YL (2005) Traffic related air pollution as a determinant of asthma among Taiwanese school children. Thorax 60: 467–473.
11. Li M, Li Q, Yang G, Kolosov VP, Perelman JM, et al. (2011) Cold temperature induces mucin hypersecretion from normal human bronchial epithelial cells in vitro through a transient receptor potential melastatin 8 (TRPM8)-mediated mechanism. J Allergy Clin Immunol 128: 626–634 e625.
12. Lee YL, Shaw CK, Su HJ, Lai JS, Ko YC, et al. (2003) Climate, traffic-related air pollutants and allergic rhinitis prevalence in middle-school children in Taiwan. Eur Respir J 21: 964–970.
13. Rosas I, McCartney H, Payne R, Calderón C, Lacey J, et al. (1998) Analysis of the relationships between environmental factors (aeroallergens, air pollution, and weather) and asthma emergency admissions to a hospital in Mexico City. Allergy 53: 394–401.
14. Rossi O, Kinnula V, Tienari J, Huhti E (1993) Association of severe asthma attacks with weather, pollen, and air pollutants. Thorax 48: 244.
15. Hales S, Lewis S, Slater T, Crane J, Pearce N (1998) Prevalence of adult asthma symptoms in relation to climate in New Zealand. Environmental health perspectives 106: 607–610.
16. Epton MJ, Martin IR, Graham P, Healy PE, Smith H, et al. (1997) Climate and aeroallergen levels in asthma: a 12 month prospective study. Thorax 52: 528–534.
17. Lin S, Luo M, Walker RJ, Liu X, Hwang SA, et al. (2009) Extreme high temperatures and hospital admissions for respiratory and cardiovascular diseases. Epidemiology 20: 738–746.
18. Harju T, Makinen T, Nayha S, Laatikainen T, Jousilahti P, et al. (2010) Cold-related respiratory symptoms in the general population. Clin Respir J 4: 176–185.
19. Abe T, Tokuda Y, Ohde S, Ishimatsu S, Nakamura T, et al. (2009) The relationship of short-term air pollution and weather to ED visits for asthma in Japan. Am J Emerg Med 27: 153–159.
20. Shanghai Statistic Bureau (2010) The sixth national population census of Shanghai.
21. Kysely J, Pokorna L, Kyncl J, Kriz B (2009) Excess cardiovascular mortality associated with cold spells in the Czech Republic. BMC public health 9: 19.
22. Ma W, Yang C, Chu C, Li T, Tan J, et al. (2012) The impact of the 2008 cold spell on mortality in Shanghai, China. International journal of biometeorology.
23. Huynen MM, Martens P, Schram D, Weijenberg MP, Kunst AE (2001) The impact of heat waves and cold spells on mortality rates in the Dutch population. Environmental health perspectives 109: 463–470.
24. Guo Y, Punnasiri K, Tong S, Aydin D, Feychting M, et al. (2012) Effects of temperature on mortality in Chiang Mai city, Thailand: a time series study. Environmental Health 11: 36.
25. Guo Y, Barnett AG, Pan X, Yu W, Tong S (2011) The impact of temperature on mortality in Tianjin, China: a case-crossover design with a distributed lag nonlinear model. Environ Health Perspect 119: 1719–1725.
26. Ye X, Wolff R, Yu W, Vaneckova P, Pan X, et al. (2012) Ambient temperature and morbidity: a review of epidemiological evidence. Environ Health Perspect 120: 19–28.
27. Knowlton K, Rotkin-Ellman M, King G, Margolis HG, Smith D, et al. (2009) The 2006 California heat wave: impacts on hospitalizations and emergency department visits. Environmental health perspectives 117: 61–67.
28. Huynen MM, Martens P, Schram D, Weijenberg MP, Kunst AE (2001) The impact of heat waves and cold spells on mortality rates in the Dutch population. Environ Health Perspect 109: 463–470.
29. Wood SN (2001) mgcv: GAMs and generalized ridge regression for R. R news 1: 20–25.
30. Guo Y, Barnett AG, Yu W, Pan X, Ye X, et al. (2011) A Large Change in Temperature between Neighbouring Days Increases the Risk of Mortality. PLoS One 6: e16511.
31. Donaldson GC, Seemungal T, Jeffries DJ, Wedzicha JA (1999) Effect of temperature on lung function and symptoms in chronic obstructive pulmonary disease. Eur Respir J 13: 844–849.

32. Larsson K, Tornling G, Gavhed D, Muller-Suur C, Palmberg L (1998) Inhalation of cold air increases the number of inflammatory cells in the lungs in healthy subjects. Eur Respir J 12: 825–830.

33. Mäkinen TM, Juvonen R, Jokelainen J, Harju TH, Peitso A, et al. (2009) Cold temperature and low humidity are associated with increased occurrence of respiratory tract infections. Respiratory medicine 103: 456–462.

34. Khadadah M, Mustafa S, Elgazzar A (2011) Effect of acute cold exposure on lung perfusion and tracheal smooth muscle contraction in rabbit. European journal of applied physiology: 1–5.

35. Ren C, Williams GM, Morawska L, Mengersen K, Tong S (2008) Ozone modifies associations between temperature and cardiovascular mortality: analysis of the NMMAPS data. Occupational and environmental medicine 65: 255.

36. Goodman PG, Dockery DW, Clancy L (2004) Cause-Specific Mortality and the Extended Effects of Particulate Pollution and Temperature Exposure. Environmental Health Perspectives 112: 179–185.

37. Anderson G, Bell M (2009) Weather-related mortality: how heat, cold, and heat waves affect mortality in the United States. Epidemiology 20: 205–213.

38. Tong S, Wang XY, Guo Y (2012) Assessing the Short-Term Effects of Heatwaves on Mortality and Morbidity in Brisbane, Australia: Comparison of Case-Crossover and Time Series Analyses. PloS one 7: e37500.

39. Guo Y, Barnett AG, Zhang Y, Tong S, Yu W, et al. (2010) The short-term effect of air pollution on cardiovascular mortality in Tianjin, China: Comparison of time series and case-crossover analyses. Science of the Total Environment 409: 300–306.

40. O'Neill MS, Zanobetti A, Schwartz J (2003) Modifiers of the temperature and mortality association in seven US cities. Am J Epidemiol 157: 1074–1082.

41. Pattenden S, Nikiforov B, Armstrong BG (2003) Mortality and temperature in Sofia and London. J Epidemiol Community Health 57: 628–633.

Permissions

The contributors of this book come from diverse backgrounds, making this book a truly international effort. This book will bring forth new frontiers with its revolutionizing research information and detailed analysis of the nascent developments around the world.

We would like to thank all the contributing authors for lending their expertise to make the book truly unique. They have played a crucial role in the development of this book. Without their invaluable contributions this book wouldn't have been possible. They have made vital efforts to compile up to date information on the varied aspects of this subject to make this book a valuable addition to the collection of many professionals and students.

This book was conceptualized with the vision of imparting up-to-date information and advanced data in this field. To ensure the same, a matchless editorial board was set up. Every individual on the board went through rigorous rounds of assessment to prove their worth. After which they invested a large part of their time researching and compiling the most relevant data for our readers.

The editorial board has been involved in producing this book since its inception. They have spent rigorous hours researching and exploring the diverse topics which have resulted in the successful publishing of this book. They have passed on their knowledge of decades through this book. To expedite this challenging task, the publisher supported the team at every step. A small team of assistant editors was also appointed to further simplify the editing procedure and attain best results for the readers.

Apart from the editorial board, the designing team has also invested a significant amount of their time in understanding the subject and creating the most relevant covers. They scrutinized every image to scout for the most suitable representation of the subject and create an appropriate cover for the book.

The publishing team has been an ardent support to the editorial, designing and production team. Their endless efforts to recruit the best for this project, has resulted in the accomplishment of this book. They are a veteran in the field of academics and their pool of knowledge is as vast as their experience in printing. Their expertise and guidance has proved useful at every step. Their uncompromising quality standards have made this book an exceptional effort. Their encouragement from time to time has been an inspiration for everyone.

The publisher and the editorial board hope that this book will prove to be a valuable piece of knowledge for researchers, students, practitioners and scholars across the globe.

List of Contributors

Yungling Leo Lee
Institute of Epidemiology and Preventive Medicine, College of Public Health, National Taiwan University, Taipei, Taiwan
Department of Public Health, College of Public Health, National Taiwan University, Taipei, Taiwan

Bing-Fang Hwang
Department of Occupational Safety and Health, College of Public Health, China Medical University, Taichung, Taiwan

Yu-An Chen
Institute of Epidemiology and Preventive Medicine, College of Public Health, National Taiwan University, Taipei, Taiwan

Jer-Min Chen and Yi-Fan Wu
Institute of Epidemiology and Preventive Medicine, College of Public Health, National Taiwan University, Taipei, Taiwan
Department of Family Medicine, Taipei City Hospital, Renai Branch, Taipei, Taiwan

Nicola Martinelli, Domenico Girelli, Davide Cigolini, Marco Sandri and Oliviero Olivieri
Section of Internal Medicine, Department of Medicine, University of Verona, Verona, Italy

Giorgio Ricci and Giampaolo Rocca
Emergency Department, Hospital of Verona, Verona, Italy

Jacqueline MacDonald Gibson, Elizabeth Harder and Nicholas DeFelice
Department of Environmental Sciences and Engineering, Gillings School of Global Public Health, University of North Carolina–Chapel Hill, Chapel Hill, North Carolina, United States of America

Jens Thomsen
Health Authority–Abu Dhabi, Abu Dhabi, United Arab Emirates

Frederic Launay
Environment Agency–Abu Dhabi, Abu Dhabi, United Arab Emirates

Itai Kloog, Antonella Zanobetti, Petros Koutrakis and Joel D. Schwartz
Exposure, Epidemiology and Risk Program, Department of Environmental Health, Harvard School of Public Health, Boston, Massachusetts, United States of America

Brent A. Coull
Department of Biostatistics, Harvard School of Public Health, Boston, Massachusetts, United States of America

Ghislaine Rosa, Fiona Majorin, Sophie Boisson and Miles Kirby
Department of Disease Control, London School of Hygiene and Tropical Medicine, London, United Kingdom

Christina Barstow
Department of Civil, Environmental and Architectural Engineering, University of Colorado at Boulder, Boulder, Colorado, United States of America

Michael Johnson
Berkeley Air Monitoring Group, Berkeley, California, United States of America

Fidele Ngabo
Ministry of Health, Government of Rwanda, Kigali, Rwanda

Evan Thomas
Department of Mechanical and Materials Engineering, Portland State University, Portland, Oregon, United States of America

Thomas Clasen
Department of Disease Control, London School of Hygiene and Tropical Medicine, London, United Kingdom
Department of Environmental Health, Emory University, Atlanta, Georgia, United States of America

Ya-Li Huang and Kai-Jen Chuang
Department of Public Health, School of Medicine, College of Medicine, Taipei Medical University, Taipei, Taiwan

School of Public Health, College of Public Health and Nutrition, Taipei Medical University, Taipei, Taiwan

Hua-Wei Chen
Department of Cosmetic Application and Management, St. Mary's Junior College of Medicine, Nursing and Management, Yilan, Taiwan

Bor-Cheng Han
School of Public Health, College of Public Health and Nutrition, Taipei Medical University, Taipei, Taiwan

Chien-Wei Liu
Department of Information Management, St. Mary's Junior College of Medicine, Nursing and Management, Yilan, Taiwan

Hsiao-Chi Chuang
Division of Pulmonary Medicine, Department of Internal Medicine, Shuang Ho Hospital, Taipei Medical University, Taipei, Taiwan
School of Respiratory Therapy, College of Medicine, Taipei Medical University, Taipei, Taiwan

Lian-Yu Lin
Department of Internal Medicine, Division of Cardiology, National Taiwan University Hospital, Taipei, Taiwan

Guglielmo Longo, Michelanna Trovato and Veronica Mazzei
Dipartimento di Scienze Biologiche, Geologiche e Ambientali, Universitá di Catania, Catania, Italy,

Margherita Ferrante and Gea Oliveri Conti
Dipartimento di Anatomia, Biologia e Genetica, Medicina Legale, Neuroscienze, Patologia Diagnostica, Igiene e Sanitá Pubblica "G. F. Ingrassia", Universitá di Catania, Catania, Italy

Jennifer A. Borcherding, Juan C. Caraballo, Alejandro A. Pezzulo, Joseph Zabner and Alejandro P. Comellas
Department of Internal Medicine, Carver College of Medicine, University of Iowa, Iowa City, Iowa, United States of America

Haihan Chen
Department of Chemical and Biochemical Engineering, Iowa City, Iowa, United States of America

Jonas Baltrusaitis
Central Microscopy Research Facility, University of Iowa, Iowa City, Iowa, United States of America

Vicki H. Grassian
Department of Chemistry, University of Iowa, Iowa City, Iowa, United States of America

Teresa Tamayo and Wolfgang Rathmann
Institute of Biometrics and Epidemiology, German Diabetes Center, Leibniz Center for Diabetes Research at Heinrich Heine University Düsseldorf, Düsseldorf, Germany

Ursula Krämer and Dorothea Sugiri
Institute for Environmental Medicine (IUF), Leibniz Center at Heinrich Heine University Düsseldorf, Düsseldorf, Germany

Matthias Grabert and Reinhard W. Holl
Institute for Epidemiology and Medical Biometry, University of Ulm, Ulm, Germany

Jennifer D. Roberts and Brandon Knight
Department of Preventive Medicine and Biometrics, F. Edward Hebert School of Medicine, Uniformed Services University, Bethesda, Maryland, United States of America

Jameson D. Voss
Department of Preventive Medicine and Biometrics, F. Edward Hebert School of Medicine, Uniformed Services University, Bethesda, Maryland, United States of America
Epidemiology Consult Service, United States Air Force School of Aerospace Medicine, Wright-Patterson Air Force Base, Ohio, United States of America

Robert Böhm
Center for Empirical Research in Economics and Behavioral Sciences, University of Erfurt, Erfurt, Germany

Bettina Rockenbach
Department of Economics, University of Cologne, Cologne, Germany

Xiao-Bo Yu and Gao Chen
Department of Neurosurgery, the Second Affiliated Hospital of Zhejiang University School of Medicine, Hangzhou, P.R. China

Jun-Wei Su
Key Laboratory of Infectious Diseases Ministry of Public Health of China, the First Affiliated Hospital of Zhejiang University School of Medicine, Hangzhou, P.R. China

Xiu-Yang Li
Department of Public Health, Zhejiang University, Hangzhou, P.R. China

Suji Lee and Bo Yeon Kwon
Department of Public Health, Graduate School, Korea University, Seoul, South Korea
Graduate School of Public Health, Graduate School, Korea University, Seoul, Korea

Eunil Lee
Department of Public Health, Graduate School, Korea University, Seoul, South Korea
Department of Preventive Medicine, College of medicine, Korea University, Seoul, Korea
Graduate School of Public Health, Graduate School, Korea University, Seoul, Korea

Man Sik Park and Hana Kim
Department of Statistics, Sungshin Women's University College of Natural sciences, Seoul, Korea

Dea Ho Jung and Kyung Hee Jo
Graduate School of Public Health, Graduate School, Korea University, Seoul, Korea

Myung Ho Jeong
Department of Cardiology, Chonnam National University Hospital, Gwangju, Korea

Seung-Woon Rha
Cardiovascular Center, Korea University Guro Hospital, Seoul, Korea

Deliang Tang, Joan Lee, Loren Muirhead, Lirong Qu, Jie Yu and Frederica Perera
Department of Environmental Health Sciences, Columbia Center for Children's Environmental Health, Mailman School of Public Health, Columbia University, New York, New York, United States of America

Ting Yu Li
Department of Pediatrics, Chongqing Medical University, Chongqing, China

Lara P. Clark and Julian D. Marshall
Department of Civil Engineering, University of Minnesota, Minneapolis, Minnesota, United States of America

Dylan B. Millet
Department of Civil Engineering, University of Minnesota, Minneapolis, Minnesota, United States of America
Department of Soil, Water and Climate, University of Minnesota, Minneapolis, Minnesota, United States of America

Peter Scarborough, Steven Allender and Mike Rayner
British Heart Foundation Health Promotion Research Group, Department of Public Health, University of Oxford, Headington, Oxford, United Kingdom

Michael Goldacre
Unit of Health Care Epidemiology, Department of Public Health, University of Oxford, Headington, Oxford, United Kingdom

Hyunok Choi and John D. Spengler
Department of Environmental Health, Harvard School of Public Health, Boston, Massachusetts, United States of America

Lu Wang
Department of Biostatistics, The University of Michigan, Ann Arbor, Michigan, United States of America

Xihong Lin
Department of Biostatistics, Harvard School of Public Health, Boston, Massachusetts, United States of America

Frederica P. Perera
Columbia Center for Children's Environmental Health, Mailman School of Public Health, New York, New York, United States of America

John P. McCracken and Joel Schwartz
Department of Environmental Health, Harvard School of Public Health, Boston, Massachusetts, United States of America

Anaite Diaz
Center for Health Studies, Universidad del Valle de Guatemala, Guatemala City, Guatemala

Nigel Bruce
Department of Public Health and Policy, University of Liverpool, Liverpool, United Kingdom

Kirk R. Smith
School of Public Health, University of California, Berkeley, California, United States of America

Chad S. Weldy and Michael T. Chin
Division of Cardiology, Department of Medicine, University of Washington School of Medicine, Seattle, Washington, United States of America
Department of Pathology, University of Washington School of Medicine, Seattle, Washington, United States of America

Yonggang Liu
Division of Cardiology, Department of Medicine, University of Washington School of Medicine, Seattle, Washington, United States of America

H. Denny Liggitt
Department of Comparative Medicine, University of Washington School of Medicine, Seattle, Washington, United States of America

Cristiane Maria Galvão Barbosa
Pulmonary Division - Heart Institute(InCor), Hospital das Clínicas da Faculdade de Medicina da Universidade de São Paulo, São Paulo, São Paulo, Brazil
FUNDACENTRO, São Paulo, São Paulo, Brazil

Mário Terra-Filho, André Luis Pereira de Albuquerque and Ubiratan de Paula Santos
Pulmonary Division - Heart Institute(InCor), Hospital das Clínicas da Faculdade de Medicina da Universidade de São Paulo, São Paulo, São Paulo, Brazil

Dante Di Giorgi
Hypertension Unit, Heart Institute(InCor), Hospital das Clínicas da Faculdade de Medicina da Universidade de São Paulo, São Paulo, São Paulo, Brazil

Cesar Grupi
Electrocardiology Unit, Heart Institute (InCor), Hospital das Clínicas da Faculdade de Medicina da Universidade de São Paulo, São Paulo, São Paulo, Brazil

Carlos Eduardo Negrão, Maria Urbana Pinto Brandão Rondon and Daniel Godoy Martinez
Unit of Cardiovascular Rehabilitation and Exercise Physiology, Heart Institute (InCor), Hospital das Clínicas da Faculdade de Medicina da Universidade de São Paulo, São Paulo, São Paulo, Brazil

Tânia Marcourakis and Fabiana Almeida dos Santos
Department of Clinical and Toxicological Analyses, University of São Paulo Pharmacological Sciences School, São Paulo, Brazil

Alfésio Luís Ferreira Braga
Environmental Epidemiology Study Group, Laboratory of Experimental Air Pollution, Department of Pathology, Faculdade de Medicina da Universidade de São Paulo, São Paulo, São Paulo, Brazil
Environmental Exposure and Risk Assessment Group, Catholic University of Santos, Santos, São Paulo, Brazil

Dirce Maria Trevisan Zanetta
Department of Epidemiology, University of São Paulo School Public Health, São Paulo, Brazil

Gulrez Shah Azhar and Dileep Mavalankar
Indian Institute of Public Health, Ahmedabad, Gujarat, India
Public Health Foundation of India, New Delhi, India

Amruta Nori-Sarma
Indian Institute of Public Health, Ahmedabad, Gujarat, India
Columbia Mailman School of Public Health, New York, New York, United States of America

Ajit Rajiva and Priya Dutta
Indian Institute of Public Health, Ahmedabad, Gujarat, India

Anjali Jaiswal
Natural Resources Defense Council, New York, New York, United States of America

Perry Sheffield
Icahn School of Medicine at Mount Sinai, New York, New York, United States of America

Kim Knowlton
Indian Institute of Public Health, Ahmedabad, Gujarat, India
Natural Resources Defense Council, New York, New York, United States of America

Jeremy J. Hess
Department of Emergency Medicine, Emory University School of Medicine, Atlanta, Georgia, United States of America
Department of Environmental Health, Emory University School of Public Health, Atlanta, Georgia, United States of America

Bin Zou and Fen Peng
School of Geosciences and Info-Physics, Central South University, Changsha, Hunan, China

Neng Wan
Department of Geography, University of Utah, Salt Lake City, Utah, United States of America

Keita Mamady
Department of Epidemiology and Health Statistics, School of Public Health, Central South University, Changsha, Hunan, China

Gaines J. Wilson
Department of Biological Sciences, Huston-Tillotson University, Austin, Texas, United States of America

Chang Ho Yu and Zhi-hua Fan
Division of Exposure Science, Environmental and Occupational Health Sciences Institute, Rutgers University, Piscataway, New Jersey, United States of America

Xianlei Zhu
College of Geosciences, China University of Petroleum, Beijing, People's Republic of China

Goran Krstić
Fraser Health, Environmental Health Services, New Westminster, Canada

Yuming Guo and Shilu Tong
School of Public Health and Institute of Health and Biomedical Innovation, Queensland University of Technology, Brisbane, Australia

Fan Jiang and Xiaoming Shen
Department of Developmental and Behavioral Pediatrics, Shanghai Institute of Pediatric Translational Medicine, Shanghai Children's Medical Centre, Shanghai Jiaotong University, School of Medicine, Shanghai, China
Ministry of Education-Shanghai Key Laboratory of Children's Environmental Health, Xinhua Hospital, Shanghai Jiaotong University, School of Medicine, Shanghai, China

Li Peng, Fuhai Geng, Jianming Xu and Canming Zhen
Shanghai Meteorological Bureau, Shanghai, China

Jun Zhang
Ministry of Education-Shanghai Key Laboratory of Children's Environmental Health, Xinhua Hospital, Shanghai Jiaotong University, School of Medicine, Shanghai, China

Index

www.ingramcontent.com/pod-product-compliance
Lightning Source LLC
Chambersburg PA
CBHW061247190326
41458CB00011B/3606